ALGAL METABOLITES

Biotechnological, Commercial, and Industrial Applications

Innovations in Biotechnology

ALGAL METABOLITES

Biotechnological, Commercial, and Industrial Applications

Edited by

Jeyabalan Sangeetha, PhD

Devarajan Thangadurai, PhD

Saher Islam, PhD

Ravichandra Hospet, PhD

AAP | APPLE ACADEMIC PRESS

First edition published 2023

Apple Academic Press Inc.
1265 Goldenrod Circle, NE,
Palm Bay, FL 32905 USA

760 Laurentian Drive, Unit 19,
Burlington, ON L7N 0A4, CANADA

CRC Press
6000 Broken Sound Parkway NW,
Suite 300, Boca Raton, FL 33487-2742 USA

4 Park Square, Milton Park,
Abingdon, Oxon, OX14 4RN UK

Library and Archives Canada Cataloguing in Publication

Title: Algal metabolites : biotechnological, commercial, and industrial applications / edited by Jeyabalan Sangeetha, PhD, Devarajan Thangadurai, PhD, Saher Islam, PhD, Ravichandra Hospet, PhD.
Names: Sangeetha, Jeyabalan, editor. | Thangadurai, Devarajan, 1976- editor. | Islam, Saher, editor. | Hospet, Ravichandra, editor.
Series: Innovations in biotechnology (Series)
Description: First edition. | Series statement: Innovations in biotechnology | Includes bibliographical references and index.
Identifiers: Canadiana (print) 20230158374 | Canadiana (ebook) 20230158412 | ISBN 9781774912737 (hardcover) | ISBN 9781774912744 (softcover) | ISBN 9781003346081 (ebook)
Subjects: LCSH: Marine algae—Industrial applications. | LCSH: Marine algae—Biotechnology. | LCSH: Marine algae—Therapeutic use. | LCSH: Metabolites.
Classification: LCC QK570.2 .A44 2023 | DDC 579.8/177—dc23

Library of Congress Cataloging-in-Publication Data

CIP data on file with US Library of Congress

ISBN: 978-1-77491-273-7 (hbk)
ISBN: 978-1-77491-274-4 (pbk)
ISBN: 978-1-00334-608-1 (ebk)

BOOK SERIES
Innovations in Biotechnology

Series Editor
Devarajan Thangadurai, PhD
Professor, Karnatak University, Dharwad, South India

Books in the Series

- **Fundamentals of Molecular Mycology**
  Authors: Devarajan Thangadurai, PhD, Jeyabalan Sangeetha, PhD, and Muniswamy David, PhD
- **Biotechnology of Microorganisms: Diversity, Improvement, and Application of Microbes for Food Processing, Healthcare, Environmental Safety, and Agriculture**
  Editors: Jeyabalan Sangeetha, PhD, Devarajan Thangadurai, PhD, Somboon Tanasupawat, PhD, and Pradnya Pralhad Kanekar, PhD
- **Phycobiotechnology: Biodiversity and Biotechnology of Algae and Algal Products for Food, Feed, and Fuel**
  Editors: Jeyabalan Sangeetha, PhD, Devarajan Thangadurai, PhD, Sanyasi Elumalai, PhD, and Shivasharana Chandrabanda Thimmappa, PhD
- **Biogenic Nanomaterials: Structural Properties and Functional Applications**
  Editors: Devarajan Thangadurai, PhD, Saher Islam, PhD, Jeyabalan Sangeetha, PhD, and Natália Martins, PhD
- **Seaweed Biotechnology: Biodiversity and Biotechnology of Seaweeds and Their Applications**
  Editors: Jeyabalan Sangeetha, PhD, and Devarajan Thangadurai, PhD
- **Microalgal Biotechnology: Bioprospecting Microalgae for Functional Metabolites towards Commercial and Sustainable Applications**
  Editors: Jeyabalan Sangeetha, PhD, Svetlana Codreanu, PhD, and Devarajan Thangadurai, PhD
- **Algal Metabolites: Biotechnological, Commercial, and Industrial Applications**
  Editors: Jeyabalan Sangeetha, PhD, Devarajan Thangadurai, PhD, Saher Islam, PhD, and Ravichandra Hospet, PhD

ABOUT THE EDITORS

Jeyabalan Sangeetha, PhD

Jeyabalan Sangeetha, PhD, is an Assistant Professor at Central University of Kerala, Kasaragod, South India. She has edited/co-edited several books in her research areas, which include environmental toxicology, environmental microbiology, environmental biotechnology, and environmental nanotechnology. She earned her BSc in Microbiology and PhD in Environmental Science from Bharathidasan University, Tiruchirappalli, Tamil Nadu, India. She holds an MSc in Environmental Science from Bharathiar University, Coimbatore, Tamil Nadu, India. She is the recipient of a Tamil Nadu Government Scholarship and a Rajiv Gandhi National Fellowship of the University Grants Commission (UGC), Government of India, for her doctoral studies. She served as Dr. D.S. Kothari Postdoctoral Fellow and UGC Postdoctoral Fellow at Karnatak University, Dharwad, South India, during 2012–2016 with funding from the University Grants Commission, Government of India, New Delhi.

Devarajan Thangadurai, PhD

Devarajan Thangadurai, PhD, is Professor at Karnatak University in South India. He has authored/edited over 30 books with national and international publishers and has visited 24 countries in Asia, Europe, Africa, and the Middle East for academic visits, scientific meetings, and international collaborations. He received his PhD in Botany from Sri Krishnadevaraya University in South India as a CSIR Senior Research Fellow with funding from the Ministry of Science and Technology, Government of India. He served as a Postdoctoral Fellow at the University of Madeira, Portugal; University of Delhi, India; and ICAR National Research Center for Banana, India. He is the recipient of a Best Young Scientist Award with a Gold Medal from Acharya Nagarjuna University, India, and the VGST-SMYSR Young Scientist Award of the Government of Karnataka, Republic of India.

Saher Islam, PhD

Saher Islam, PhD, is a Postdoctoral Researcher at the University of Texas Southwestern Medical Center, Dallas, USA. She worked at Cornell University, New York, and West Virginia State University, West Virginia, in the USA as Visiting Scholar. She has keen research interests in genetics, molecular biology, biotechnology, and bioinformatics and has ample hands-on experience in molecular marker analysis, whole-genome sequencing (WGS), and RNA sequencing. She has visited the USA, UK, Singapore, Germany, Italy, and Russia for academic and scientific training, courses, and meetings. She is the recipient of the 2016 Boehringer Ingelheim Fonds Travel Grant from the European Molecular Biology Laboratory, Germany. She is an author/co-author of 60 publications, including journal articles, book chapters, books, and conference presentations.

Ravichandra Hospet, PhD

Ravichandra Hospet, PhD, obtained his MSc and PhD degrees in Botany from Karnatak University, Dharwad, India. He has keen research interests in fermentation technology, agrobiotechnology, agronanotechnology, and biofumigation technology. His research expertise includes the improvement of agriculturally and industrially important microbial strains, the development of novel strategies for fruit wine production, and scaling up biofumigation technology for the protection of stored food commodities. He is also the recipient of the DBT-KUD IPLS Fellowship. He has authored/co-authored more than 30 publications in high-impact journals, including *Green Chemistry* and *Critical Reviews in Food Science and Nutrition.*

CONTENTS

CONTRIBUTORS

Paul Agastian
Research Department of Plant Biology and Biotechnology, Loyola College, Nungambakkam, Chennai, Tamil Nadu, India

Surendraraj Alagarsamy
Ecosystem-Based Management of Marine Resources Program, Environment and Life Sciences Research Center, Kuwait Institute for Scientific Research, P.O. Box–1638, Salmiya–22017, Kuwait

Naif Abdullah Al-Dhabi
Department of Biotechnology, Guru Nanak College (Autonomous), Velachery, Chennai–600042, Tamil Nadu, India

Pinheiro De Carvalho Miguel Ângelo Almeida
ISOPlexis, Center for Sustainable Agriculture and Food Technology, University of Madeira, Penteada Campus, Funchal–9050-290, Madeira, Portugal

Mariadhas Valan Arasu
Department of Botany and Microbiology, College of Science, King Saud University, P.O. Box 2455, Riyadh–11451, Saudi Arabia

Jayaseelan Arun
Center for Waste Management–'International Research Center,' Sathyabama Institute of Science and Technology, Jeppiaar Nagar (OMR), Chennai–600119, Tamil Nadu, India

Malik Badshah
Department of Microbiology, Faculty of Biological Sciences, Quaid-i-Azam University, Islamabad–45320, Pakistan

Namita Bedi
Amity Institute of Biotechnology, Amity University, Sector 125, Noida–201303, Uttar Pradesh, India

Neeta Bhagat
Amity Institute of Biotechnology, Amity University, Sector 125, Noida–201303, Uttar Pradesh, India

Barreto Maria Carmo
cE3c–Center for Ecology, Evolution, and Environmental Changes/Azorean Biodiversity Group and Faculty of Sciences and Technology, University of Azores, Ponta Delgada–9501-801, Portugal

Usman Ali Chaudhry
Infection Control and Disease Prevention Center, Ministry of Health, Tabuk, Kingdom of Saudi Arabia

Gourav Choudhir
Center for Rural Development and Technology, Indian Institute of Technology Delhi, Hauz Khas–110016, New Delhi, India

João Cotas
University of Coimbra, MARE-Marine, and Environmental Sciences Center, Department of Life Sciences, Coimbra–3001-456, Portugal

Meltem Conk Dalay
Department of Bioengineering, Faculty of Engineering, Ege University, Bornova, İzmir–35100, Turkey

Laurent Dufossé
University of Reunion, CHEMBIOPRO Chemistry and Biotechnology of Natural Products, ESIROI Food Department, Sainte-Clotilde F–97490, Reunion Island, Indian Ocean, France

Ayşegül Erdoğan
Ege University Application and Research Center for Testing and Analysis (EGE MATAL), Bornova, İzmir–35100, Turkey

Sara García-Poza
University of Coimbra, MARE-Marine, and Environmental Sciences Center, Department of Life Sciences, Coimbra–3001-456, Portugal

Ana Marta Mendes Gonçalves
University of Coimbra, MARE-Marine, and Environmental Sciences Center, Department of Life Sciences, Coimbra–3001-456, Portugal; Department of Biology and CESAM, University of Aveiro, Aveiro–3810-193, Portugal

Kannappan Panchamoorthy Gopinath
Department of Chemical Engineering, SSN College of Engineering, Kalavakkam–603110, Tamil Nadu, India

Sabeena Farvin Koduvayur Habeebullah
Ecosystem-Based Management of Marine Resources Program, Environment and Life Sciences Research Center, Kuwait Institute for Scientific Research, P.O. Box–1638, Salmiya–22017, Kuwait

David Elisha Henry
Department of Biotechnology, School of Applied Sciences, REVA University, Bengaluru–560064, Karnataka, India

Eduardo Jacob-Lopes
Department of Food Science and Technology, Federal University of Santa Maria (UFSM), Roraima Avenue, 1000, 97105-900, Santa Maria, RS, Brazil

Samiullah Khan
Department of Microbiology, Faculty of Biological Sciences, Quaid-i-Azam University, Islamabad–45320, Pakistan

Ameer Khusro
Department of Botany and Biotechnology, MSM College, Kayamkulam, Alappuzha, Kerala, India

Neelakanta Sarvashiva Kiran
Department of Biotechnology, School of Applied Sciences, REVA University, Bengaluru–560064, Karnataka, India

Ameet Kumar
Department of Microbiology, Faculty of Biological Sciences, Quaid-i-Azam University, Islamabad–45320, Pakistan

Sushil Kumar
Department of Botany, Shaheed Mangal Pandey Government Girls PG College, Madhavpuram, Meerut–250002, Uttar Pradesh, India

Adriana Leandro
University of Coimbra, MARE-Marine, and Environmental Sciences Center, Department of Life Sciences, Coimbra–3001-456, Portugal

R. Lekshmi
Department of Botany and Biotechnology, MSM College, Kayamkulam, Alappuzha, Kerala, India

A. Amala Lourthuraj
Department of Biotechnology, Guru Nanak College (Autonomous), Velachery, Chennai–600042, Tamil Nadu, India

João Carlos Marques
University of Coimbra, MARE-Marine, and Environmental Sciences Center, Department of Life Sciences, Coimbra–3001-456, Portugal

Prabhakar Mishra
Department of Biotechnology, School of Applied Sciences, REVA University, Bengaluru–560064, Karnataka, India

Pricila P. Nass
Department of Food Science and Technology, Federal University of Santa Maria (UFSM), Roraima Avenue, 1000, 97105-900, Santa Maria, RS, Brazil

Aneela Nawaz
Department of Microbiology, Faculty of Biological Sciences, Quaid-i-Azam University, Islamabad–45320, Pakistan

Marcele L. Nörnberg
Department of Food Science and Technology, Federal University of Santa Maria (UFSM), Roraima Avenue, 1000, 97105-900, Santa Maria, RS, Brazil

Nunes Nuno
ISOPlexis, Center for Sustainable Agriculture and Food Technology, University of Madeira, Penteada Campus, Funchal–9050-290, Madeira, Portugal; UBQ II, Biochemistry Unit, Lda. Rua Visconde de Anadia, Anadia Building 5° Floor CC, Funchal – 9050-020, Madeira, Portugal

Diana Pacheco
University of Coimbra, MARE-Marine, and Environmental Sciences Center, Department of Life Sciences, Coimbra–3001-456, Portugal

Leonel Pereira
University of Coimbra, MARE-Marine, and Environmental Sciences Center, Department of Life Sciences, Coimbra–3001-456, Portugal

R. Rajakrishnan
Department of Botany and Microbiology, College of Science, King Saud University, P.O. Box 2455, Riyadh–11451, Saudi Arabia

M. A. Rathi
Department of Biochemistry, Sree Narayana Guru College, Coimbatore–641105, Tamil Nadu, India

Aamer Ali Shah
Department of Microbiology, Faculty of Biological Sciences, Quaid-i-Azam University, Islamabad–45320, Pakistan

Ankit Singh
Center for Rural Development and Technology, Indian Institute of Technology Delhi, Hauz Khas–110016, New Delhi, India

Authilingam Swetha
Department of Chemical Engineering, SSN College of Engineering, Kalavakkam–603110, Tamil Nadu, India

Chidambaram Kulandaisamy Venil
Department of Biotechnology, Anna University, Regional Campus, Coimbatore–641046, Tamil Nadu, India

Sivakumar Shri Vigneshwar
Department of Chemical Engineering, SSN College of Engineering, Kalavakkam–603110, Tamil Nadu, India

Matheswaran Yaamini
Department of Biotechnology, Anna University, Regional Campus, Coimbatore–641046, Tamil Nadu, India

Leila Q. Zepka
Department of Food Science and Technology, Federal University of Santa Maria (UFSM), Roraima Avenue, 1000, 97105-900, Santa Maria, RS, Brazil

ABBREVIATIONS

AA	arachidonic acid
ABG	Azorean biodiversity group
ABTS	2,2-azino-bis-3-ethylbenzothiaz-oline-6-sulfonic acid
ACE	angiotensin-I converting enzyme
AD	Alzheimer's disease
AHLs	acyl homoserine lactones
ALA	α-linolenic acid
ALS	amyotrophic lateral sclerosis
AMD	age-related macular degeneration
AMPs	antimicrobial peptides
APCs	allophycocyanins
ASE	accelerated solvent extraction
BDDE	bis(2,3-dibromo-4,5-dihydroxy benzoyl) ether
BG	blue-green
BGA	blue-green algae
BHK	baby hamster kidney
BHT	butylated hydroxyl toluene
CAPS	cultivated algal production system
CAT	catalase
CE	catechol equivalents
CO_2	carbon dioxide
CVDs	cardiovascular diseases
DAD	diode array detector
DGCC	1,2-diacylglyceryl-3-O-carboxy-(hydroxymethyl)-choline
DGDG	digalactosyl diacylglycerol
DGDGs	digalactosyldiacylglycerides
DGTA	diacylglyceryl-3-O-2'-(hydroxymethyl)-(N,N,N-trimethyl)-β-alanine
DGTS	1,2-diacylglyceryl-3-O-4'-(N,N,N-trimethyl)-homoserine
DHA	docosahexaenoic acid
DOPA	3,4-dihydroxyphenethylamine
DOXP	1-deoxy-D-xylose-5-phosphate
DPA	docosapentaenoic acid
DPHC	diphlorethohydroxycarmalol

DPPH	2,2-diphenyl-1-picrylhydrazyl
EAA	essential amino acid
EAE	enzyme-assisted extraction
EEI	ecological evaluation index
EFAs	essential fatty acids
EFSA	European Food Safety Authority
EPA	eicosapentaenoic acid
ESBL	extended spectrum beta-lactamase
ETC	electron transport chain
EU	European Union
FAME	fatty acid methyl esters
FAO	Food and Agriculture Organization
FCSPs	fucose-containing sulphated polysaccharides
FCT	Foundation for Science and Technology
GABA	γ-aminobutyric acid
GAE	gallic acid equivalents
GC	gas chromatography
GC-MS	gas chromatography-mass spectrometry
GGDP	geranylgeranyl-diphosphate
GL	glycolipids
GLA	gamma-linolenic acid
GPX	glutathione peroxidase
GRAS	generally recognized as safe
GSH	glutathione
GTPase	guanosine triphosphatase
HAT	human African trypanosomiasis
HBV	hepatitis B virus
HCV	hepatitis C virus
HDL	high-density lipoprotein
HDL-C	high-density lipoprotein cholesterol
HIV	human immunodeficiency virus
HPLC	high-performance liquid chromatography
HPV	human papillomavirus
HSV	herpes simplex viruses
HTA	hexadecatrienoic acid
HTL	hydrothermal liquefaction
IA	image analysis
IAA	indole-3-acetic acid
IMTA	integrated multi-trophic aquaculture

IPP	isopentenyl pyrophosphate
IR	infrared
IRDiRC	International Rare Diseases Research Consortium
IRUD	initiative on rare and undiagnosed diseases
JEV	Japanese encephalitis virus
LA	linoleic acid
LEDs	light-emitting diodes
LS	liver-stage
MAA	mycosporine-like amino acid
MAE	microwave-assisted extraction
MCDA	multi-criteria decision analysis
MDR	multi-drug resistant
MERS-CoV	middle east respiratory syndrome coronavirus
MGDG	monogalactosyldiacylglycerol
MIR	mid-near-infrared spectroscopy
MRSA	methicillin-resistant *Staphylococcus aureus*
MS	mass spectrometry
MUFA	mono and polyunsaturated fatty acids
MVA	mevalonate
NIH	National Institute of Health
NIR	near-infrared spectroscopy
NMR	nuclear magnetic resonance
NMSC	non-melanoma skin cancer
NO	nitric oxide
OMIM	online mendelian inheritance in man
OMPs	orphan medicinal products
PA	palmitoleic acid
PA	phosphatidic acid
PBP	phycobiliproteins
PC	phycocyanin
PCR	polymerase chain reaction
PCs	phenolic compounds
PCs	phycocyanins
PE	phosphatidyl-ethanolamine
PEA	phenylethylamine
PEF	field-assisted extraction
PG	phosphatidylglycerol
PHB	polyhydroxybutyrate
PI	phosphatidylinositol

PL	phospholipids
PLE	pressurized liquid extraction
PR	pathogenesis-related
PrxR	peroxiredoxin
PS	phosphatidylserine
PUFA	polyunsaturated fatty acid
PVC	polyvinyl alcohol
RD	rare diseases
ROS	reactive oxygen species
R-PE	R-phycoerythrin
RSV	respiratory syncytial virus
SARS-CoV	severe acute respiratory syndrome–coronaviruses
SFA	saturated fatty acid
SHRs	spontaneously hypertensive rats
SQDG	sulfoquinovosyl diacylglycerol
TAG	triacylglycerols
TCC	total carotenoid content
TPC	total phenolic content
UAE	ultrasound-assisted extraction
UDNI	undiagnosed diseases network international
UDP	undiagnosed diseases program
UGC	University Grants Commission
UV	ultraviolet
VLD	very-low-density lipoprotein
VRE	vancomycin-resistant enterococcus
VZV	varicella zoster virus
WES	whole-exome sequencing
WGS	whole-genome sequencing
WHO	World Health Organization

PREFACE

The sustainable use of natural resources is increasing gradually with the increase in the world's human population. To meet the growing population's demand, algal resources are placed as a promising and alternative source for energy and health. Generally, algae produce energy in the form of food and novel metabolites through complex photosynthetic processes. The main advantage of algal cultivation is that it doesn't need a large land area. Algal systems exhibit high productivity with minimum nutrient availability. It is enriched with the most valuable nutrient content, such as proteins, lipids, fiber, and carbohydrates. Also, it plays a vital role as biocatalysts in the area of white biotechnology to ensure the sustainable and rapid production of algal metabolites and are well explored for their useful applications in the field of food, agriculture, medicine, and the environment.

Furthermore, algal cultivation in a controlled environment makes algal species more eco-friendly and economically viable. The recent research reports on bioactive compounds of algae such as vitamin E, vitamin B12, β-carotene, cyanovirin, phycocyanin (PC), zeaxanthin, and oleic acid exhibits antimicrobial, anticancer, anti-inflammatory, and antioxidant properties. Primary metabolism of proteins, vitamins, fatty acids, and pigments are the primary source for producing algal metabolites or synthesized through secondary metabolites such as flavonoids, steroids, alkaloids, and many other metabolites.

Algae are well known for their diverse ecosystem, which is the major reason for algae to synthesize highly variable biogenic active metabolites. The bioactive compounds are either stored in the algal biomass or excreted from the cell to the surrounding medium known as exometabolites. These exometabolites are used by the algal communities to withstand the competitive ecosystem as a defense mechanism. Algal metabolites are the potential gateway for researchers and industrialists to explore the possible utilization of these metabolites to develop non-toxic, very cheap, highly active medicinal compounds, food additives, and materials for cosmetics. Various cosmetics such as anti-aging creams, rejuvenating cream, sun protectants, and hair care creams are in the market derived from algal extracts. In addition, algal compounds are also used as a thickening and water-binding agents in cosmetics. Due to the diverse range of algal species, various biogenic active metabolic compounds can be well explored for potential applications.

This book contains 13 chapters written by outstanding subject experts. Part I with Chapters 1–4 aims to highlight the antibacterial potential of marine algae, commercial application of algae, halide metabolites and their therapeutic application, and metabolite profiling of algal biomass and its industrial applications. Part II with Chapters 5–8 provides an overview of potential resources for recovery of bioactive algal metabolites, bioaccessibility, the bioavailability of microalgal carotenoid, and its industrial application. Part III with Chapters 9–13 mainly focuses on current applications of macroalgae in daily life, the therapeutic potential of macroalgae in aquaculture, pharmaceutical, nutraceutical potential of seaweed, bioactive secondary metabolites from marine macroalgae, and its industrial importance.

All these chapters are well illustrated and contain accessible information for a wide audience, especially algal biologists, food biotechnologists, industrialists, and young researchers interested in algal metabolite studies. We are extremely thankful to all those eminent personalities who have imparted their knowledge and timely cooperation to complete this book.

—Editors

PART I

ALGAL METABOLITES: INTRODUCTION AND APPLICATION

CHAPTER 1

ANTIBACTERIAL POTENTIALS OF MARINE ALGAE

NEETA BHAGAT and NAMITA BEDI

Amity Institute of Biotechnology, Amity University, Sector 125, Noida–201303, Uttar Pradesh, India

ABSTRACT

Algae occupy many aquatic biomes, which bestows them with the potent capability to produce significant classes of bioactive compounds. Bioactive compounds from microalgae, serve as a promising source of natural products that have wide applications in pharmaceutical industries. Algae-derived bioactive compounds are used for drug screening, due to their tremendous structural diversity and biological availability. This chapter reviews various metabolites from algae-like lactones, polysaccharides, peptides, amino acids, alkaloids, pigments, terpenes, cladophores, and bromophenols for their antibacterial properties against pathogenic bacteria, including antibiotic-resistant strains. A comprehensive view of functional aspects of novel natural compounds from algae is presented here. Algae synthesized bioactive compounds can be contemplated for developing novel drugs, which can provide an alternative in overcoming antibiotic resistance. These diversified algal metabolites will be favorable lead compounds for researchers and pharmaceuticals to develop nontoxic, cost-effective, and promising antimicrobial agents.

1.1 INTRODUCTION

Marine algae are rich in a range of components, which have potential health benefits. Marine macro-algae are a primitive type of plant lacking true roots,

Algal Metabolites: Biotechnological, Commercial, and Industrial Applications. Jeyabalan Sangeetha, Devarajan Thangadurai, Saher Islam, & Ravichandra Hospet (Eds.)

stems, and leaves and have been classified based on types of pigments present into Phaeophyta (Brown), Rhodophyta (Red), and Chlorophyta (Green) types (Abu-Ghannam and Rajauria, 2013). These algae produce many biologically active metabolites such as polyunsaturated fatty acids (PUFAs), sterols, proteins, polysaccharides, peptides, terpenes, polyacetylenes, phlorotannins, sterols, fatty acids, indole alkaloids, aromatic organic acids, shikimic acid, polyketides, hydroquinone, alcohols, aldehydes, ketones, and halogenated furanones, alkanes, and alkenes (Chojnacka and Kim, 2015). The concentration and number of elements like protein, iodine, bromine, vitamins, and substances of stimulatory and antibiotics produced by marine algae are much higher than in terrestrial plants. Their ability to survive in complex habitats and extreme environmental conditions is due to the production of a variety of secondary metabolites which are usually not present in another organism.

Algal classes such as the Phaeophyceae (brown), Rhodophyceae (red), Chlorophyceae (green), Chrysophyceae (golden), and Bacillariophyceae (diatoms) are common sources of antibacterial compounds (Rizzo et al., 2016). *Chlorella vulgaris* was the first algae from which chlorellin was extracted which was found to have antibacterial activity against *Pseudomonas aeruginosa*, *Staphylococcus aureus*, *Streptococcus pyogenes*, *Bacillus subtilis*, and *Pseudomonas aeruginosa* (Pratt et al., 1944). As many as 27,000 metabolites have been isolated from marine organisms, with many of them defined as secondary metabolites (Harizani et al., 2016). The red algae (Rhodophyta), which comprises of 8,000 species, are the richest source of marine secondary metabolites. Several chemical functional groups in algae, such as phlorotannins, fatty acids, peptides, terpenes, polysaccharides, polyacetylenes, sterols, indole alkaloids, aromatic organic acids, shikimic acid, polyketides, hydroquinones, alcohols, aldehydes, ketones, and halogenated furanones have been reported as bacterial inhibitors. Various such metabolites from algal sources and their mechanism of antibacterial properties are discussed here and shown in Table 1.1.

1.2 ALGAL METABOLITES AS ANTIBACTERIAL

1.2.1 AMINO ACIDS AND PEPTIDES

Algae are a rich source of proteins as various types of essential amino acids (EAAs) are found in abundance in algal cells. This property makes them a source of single-cell protein and an excellent food supplement. *Palmaria palmata* is rich in high content of valine, methionine, and leucine and *Ulva*

TABLE 1.1 Source, Structure, and Name of Antibacterial Compound from Algal Source and as well its Mode of Action and Spectrum on Bacterial Species

Name of the Compound	Structure	Source	Mode of Action	Spectrum
Amino acids and peptides	Amino acid chain	*Tetraselmis suecica*	AMP being amphiphilic interact with both polar and nonpolar sites on bacterial cellular membrane to create pores which results in cellular leakage resulting in killing of cells	MRSA, *Bacillus cereus* and *Escherichia coli* (Guzmán et al., 2019)
Alkaloids	Oroidin	*Agelas sponges*	Disruption of the bacterial cell membrane targeting the response regulator protein, BfmR, and inhibition of enoyl reductases	Gram-positive bacteria, *S. aureus* and *E. faecalis* (Zidar et al., 2012)
Phenols	Bromophenols are organic compounds containing one or more bromine atoms covalently bonded to the phenol backbone	*Rhodomela confervoides* red marine algae *Kappaphycus* sp., *Osmundaria colensoi*	Bromophenols downregulate virulent proteins (PPAD, Rgp, and HgA) required for pathogenesis of the pathogenic bacteria	*Staphylococcus aureus*, *Pseudomonas fluorescens* and *Vibrio cholerae* (Keffer et al., 2013)
	Polyphenols(cladophores) 2,3,8,9-tetrahydroxybenzo[c] chromen-6-one	*Cladophora socialis*	Inhibits bacterial DNA gyrase	*Escherichia coli*, *Klebsiella pneumoniae*, *Pseudomonas aeruginosa*, *Acinetobacter baumannii*
Chrysophantins	Polyhydroxated polyhalogenated diarylbutene	*Chrysophaeum taylori*	Chrysophaentins inhibit enzymes, bind to the bacterial guanosine triphosphatase (GTPase), and inhibits FtsZ and Z ring formation in bacteria, thus interfering with cell division	Drug-resistant gram-positive bacteria like VRE, MRSA, and MDR *Staphylococcus aureus* (Keffer et al., 2013)

TABLE 1.1 *(Continued)*

Name of the Compound	Structure	Source	Mode of Action	Spectrum
Fatty acids	–	*Phaeodactylum tricornutum*	Free fatty acids found in algal cell inhibit ETC and oxidative phosphorylation, the nutrient uptake process and also interfere with enzyme activity and auto degradation of cellular products leading to bacterial lysis	Gram-positive MRSA, *Staphylococcus aureus*, *Staphylococcus epidermidis*, and *Bacillus weihenstephanensis*, and gram-negative bacteria like *Vibrio* sp. and *Listonella anguillarum* (Zheng et al., 2005; Smith et al., 2010)
Lectins	–	Red algae *Solieria filiformis*	Lectin binds to the mannan present in the cell surface of gram-negative bacteria to kill the bacteria	*Salmonella typhi*, *Klebsiella pneumoniae*, *Pseudomonas aeruginosa*, *Serratia marcescens*, *Enterobacter aerogenes*, and *Proteus* species (Liao et al., 2003; Holanda et al., 2005)
Lactones	Furanones which are cyclic organic compounds formed by esterification of hydrocarboxylic acids	*Delisea pulchra*, a red alga	Inhibit carABCDEFGH gene expression required for quorum sensing	*Pseudomonas aeruginosa* and *E. coli* (Liao et al., 2003)
Phlorotannins	Polymeric units of phloroglucinol, i.e., fucols, phlorethols, fucophloretols, fuhalols, and eckols	*Fucus vesiculosus* phlorotannins from brown algae *Ecklonia stolonifera*, *Ecklonia kurome* marine red algae *Ceratodictyon spongiosum*	Involves inhibition of oxidative phosphorylation, binding to enzymes, cell membrane and lysis of the cell	Gram-positive *Staphylococcus aureus* and *Streptococcus pneumoniae*, *B. cereus*, *Shigella flexneri*, *Streptococcus pyogenes*, *Klebsiella pneumoniae*,

TABLE 1.1 *(Continued)*

Name of the Compound	Structure	Source	Mode of Action	Spectrum
				Campylobacter spp., *C. jejuni*, *E. coli*, *S. enteritidis*, *S. typhimurium*, *V. parahaemolyticus* and *Vibrio* sp. (Nagayama et al., 2002)
Pigments	Fucoxanthin commercial extract	*Scenedesmus quadricauda*, *Chlamydomonas reinhardtii*, and *Chlorella vulgaris*	Inhibition of several pathogenic bacteria by increasing cell membrane permeability, leakage of cytoplasm and inhibition of nucleic acid.	Inhibition of *L. monocytogenes* (Deyab, 2013; Karpinski, 2019)
Poly-saccharides	Sulfated polysaccharides, xyloarabinogalactan bactericidal effect	Marine alga *Chaetomorpha*	Algal polysaccharides with glycoprotein receptors can bind to the molecules present on the bacterial cell wall, membrane, and nucleic acids. This disintegrates the cell wall leading to cell lysis.	Gram-positive *Bacillus subtilis* and *Micrococcus luteus* and drug-resistant *Staphylococcus aureus* (He et al., 2010; Pierre et al., 2011; Amorim et al., 2012)
	Fucoidans	*Kjellmaniella crassifolia*, *Polyopes affinis*, and *Cladosiphon okamuranus*	–	*Helicobactor pylori* (Yoo et al., 2020)
	Laminarin	*Laminaria hyperborea* and *Ascophyllum nodosum*	–	*Salmonella typhimurium*, *Escherichia coli*, *Listeria monocytogenes*, *Staphylococcus aureus*, *Micrococcus luteus*, *Klebsiella pneumoniae*,

TABLE 1.1 *(Continued)*

Name of the Compound	Structure	Source	Mode of Action	Spectrum
				Salmonella typhimurium, *Salmonella enterica*, and *Pseudomonas aeruginosa* (Yoo et al., 2020)
	Sulfated polysaccharide	*Sargassum swartzii* and *Chlamydomonas reinhardtii*	Inhibited biofilm formation	*Bacillus subtilis*, *Streptococcus* sp., *Escherichia coli*, and *Neisseria mucosa* (Vishwakarma and Vavilala, 2019)
Sterols	Anti-24-propylidene-cholest-5-en-3β-ol	*Laurencia papillosa*	Antibacterial activity of algal sterols is due to their ability to interacting with negatively charged phosphate groups in the bacterial membrane. This interaction distorts the membrane and affects the osmoregulation causing disruption of the cell membrane.	*Staphylococcus aureus*, *Bacillus subtilis*, *Escherichia coli*, *Pseudomonas aeruginosa*, *Klebsiella pneumoniae* and *Shigella flexneri* (Alhanout et al., 2010; Kavita et al., 2014)
	–	*Lessonia nigrescens*	–	*Mycobacterium tuberculosis* (Wachter et al., 2001)
Terpenes	Phlorotannins	Brown algae, *Ecklonia kurome*	It may be due to the interaction of phlorotannins with the bacterial proteins and enzymes, which will result in bactericidal action.	Gram-positive bacteria, *S. aureus*, *B. cereus*, and Gram-negative bacteria *C. jejuni*, *E. coli*, *S. enteritidis*, *S. typhimurium*, *V. parahaemolyticus* (Nagayama et al., 2002)

TABLE 1.1 *(Continued)*

Name of the Compound	Structure	Source	Mode of Action	Spectrum
	Sesquiterpenes	*Laurencia luzonesis*	Exact mechanism not known	*Bacillus megaterium*
	Bromoditerpenes	*Sphaerococcus coronopifolius*	Exact mechanism not known	Gram-positive organism *S. aureus* (Rodrigues, 2015)
	–	*Cystoseira mediterranea*, *Enteromorpha linza*, *Ulva rigida*, *Gracilaria gracilis*, and *Ectocarpus siliculosus*	–	*Candida* sp., *Enterococcus faecalis*, *S. aureus*, *Streptococcus epidermidis*, *Pseudomonas aeruginosa*, and *Escherichia coli* (Etahiri, 2001)

rigida valine, leucine, and phenylalanine. Antimicrobial peptides (AMPs) with short-chain peptides are also obtained from algae which are more bioactive than the original proteins and kill a wider spectrum of microbial pathogens. Antibacterial response against MRSA, *Bacillus cereus*, and *Escherichia coli* was demonstrated by proteins extracted from green algae *Tetraselmis suecica*. Protein fraction of *Tetraselmis* has been found to contain AMP, which could be synthesized and modified by substitution of lysine and alanine. These synthesized peptides were more potent as an antimicrobial agent and had low toxicity (Guzmán et al., 2019) in comparison to natural compounds. AMP being amphiphilic, interact with both polar and nonpolar sites on bacterial cellular membrane to create pores which result in cellular leakage resulting in the killing of cells. These AMP have an advantage over antibiotics as they interact with the bacterial membrane and are not affected by antimicrobial resistance by pathogens (Lordan et al., 2011). Alkaline protease and papain hydrolysate from the freshwater alga *Spirulina platensis* resulted in AMP with 18 amino acid peptide called SP-1 which showed antibacterial activity against *Staphylococcus aureus* and *Escherichia coli* (Sun et al., 2018). Enzymatic hydrolysis of proteins extracted from *Saccharina longicruris* resulted in 10 Kda protein hydrolysate which showed antimicrobial activity against *S. aureus* (Beaulieu et al., 2015).

1.2.2 ALKALOIDS

Alkaloids are naturally occurring nitrogenous organic compounds. Hordenine is the first alkaloid isolated from red algae *Phyllophora nervosa* of the black sea (Guven et al., 1970). As many as 35 different types of alkaloids have been isolated from marine algal species, such as phenylethylamine (PEA) alkaloids, indole-derived alkaloids, and halogenated alkaloids. Alkaloid fractions in green algae (*Ulva lactuca, Codium tomentosum*), brown algae (*Dictyopteris membranacea, Sargassum vulgare, Cystoseira barbata*), and red algae (*Gelidium latifolium*) showed wide spectrum antimicrobial activity against gram-positive bacteria like *Bacillus subtilis*, *Bacillus* sp., *Staphylococcus aureus*, and *Staphylococcus epidermidis* and gram-negative bacteria like *Escherichia coli*, *Salmonella typhimurium*, *Pseudomonas aeruginosa*, and *Klebsiella pneumoniae* (Alghazeer et al., 2013). Blue-green algae (BGA) like *Nodularia harveyana* and *Nostoc insulare* contain alkaloids named norharmane (9H-pyrido(3,4-β) indole), which shows antibacterial activity against gram-positive *Staphylococcus aureus*, *Bacillus* sp. and different *Vibrio* species (Zheng et al., 2005; Volk and Furkert, 2006). Antimicrobial potential of *Sargassum wightii* (Brown algae) and *Kappaphycus alwarezii* (Red algae) is also attributed to metabolites which are alkaloids, phenols, and sugars (Arputha et al., 2012). Extracts of *Caulerpa racemosa* and *Caulerpa lentillifera* contained Caulerpin, a bisindole alkaloid which imparted antibacterial property against methicillin-resistant *Staphylococcus aureus* (MRSA) and *Escherichia coli* K1 property (Yap et al., 2019).

1.2.3 BROMOPHENOLS

Bromophenols are organic compounds containing one or more bromine atoms covalently bonded to the phenol backbone. Bromophenols found in some marine algae, are formed by electrophilic bromination of phenol. Red algae *Rhodomela confervoides* has been reported to contain five different types of bromophenols as bis (2,3-dibromo-4,5-dihydroxybenzyl) ether, 3-bromo-4-(2,3-dibromo-4,5-dihydroxyphenyl) methyl-5-(hydroxymethyl)-1,2-benzenediol, 3-bromo-4-(2,3-dibromo-4,5-dihydroxyphenyl) methyl-5-(methoxymethyl)-1,2-benzenediol, 3-bromo-4-(2,3-dibromo-4,5-dihydroxyphenyl) methyl-5-(ethoxymethyl)-1,2-benzenediol and 4,4′-methylenebis (5,6-dibromo-1,2-benzenediol) which show antibacterial effects against different strains of bacteria, such as *Staphylococcus aureus*, *Staphylococcus epidermidis*, *Escherichia coli* and *Pseudomonas aeruginosa* (Xu et al., 2003; Lijun et al., 2005). Bromophenols extracted from red marine algae *Kappaphycus* sp. and *Kappaphycus alvarezii* showed prominent antibacterial activity against *Staphylococcus aureus*, *Pseudomonas fluorescens*, and *Vibrio cholerae* (Rajasulochana et al., 2012) and oral pathogen *Porphyromonas gingivalis*. These bromophenols downregulate virulent proteins (PPAD, Rgp, and HgA) required for pathogenesis of the pathogenic bacteria. Marine algae *Osmundaria colensoi* contain a nitrogenous bromophenol colensolide A and lanosol and its other derivatives that inhibited the growth of *Mycobacterium smegmatis* (Popplewell and Northcote, 2009). Molecular docking score of bromophenols substituted with methyl, glycosyl, and benzene has suggested promising antibacterial potential and pharmacokinetic properties (Cherian, 2019).

1.2.4 CLADOPHORES

Methanolic extracts of green algae *Cladophora socialis* contain many important oligomeric polyphenols (5–12). Compound 4 is 2,3,8,9-tetrahydroxybenzo[c] chromen-6-one is named Cladophorols A which inhibits bacterial DNA gyrase. Novel cladophorols (compound 5) was identified as 4,9,10-trihydroxy-2-[(3,3′,4′-trihydroxy-(1,1′-biphenyl)-4-yl)oxy]benzo[c]chromen-6-one named cladophorol (Weidner-Wells et al., 1998). Another compound identified as α-hydro-ω-[3,4-dihydroxyphenyl] octa[oxy(2-hydroxyphen-4-yl)] a was named cladophorol C. All these compounds showed high antimicrobial activities against MRSA, MDR strains of VRE, *Escherichia coli*, *Klebsiella pneumoniae*, *Pseudomonas aeruginosa*, and *Acinetobacter baumannii*. Cladophorol A and B also inhibited the malaria-causing protozoan *Plasmodium falciparum* and showed no toxicity toward human cell lines (Lavoie et al., 2019). These properties of cladophorols make them a promising class of algal metabolites for further exploration of molecular mechanism of action in their bactericidal property and answer to antibiotic resistance.

1.2.5 CHRYSOPHAENTINS

Marine algae *Chrysophaeum taylori* contain naturally occurring compound called Chrysophaentins which shows a novel mechanism of bactericidal activity. Chrysophaentins inhibit enzymes, bind to the bacterial guanosine triphosphatase (GTPase), and inhibit FtsZ and Z ring formation in bacteria, thus interfering with cell division (Keffer et al., 2013). Chrysophaentins showed high antibacterial action against drug-resistant gram-positive bacteria like VRE, MRSA, and MDR *Staphylococcus aureus*. These compounds with their mechanism of targeting cell division make them promising drug candidates for an answer to drug resistance (Kusuma et al., 2019).

1.2.6 FATTY ACIDS

Almost all prokaryotic and eukaryotic cells have fatty acids as integral membrane components and are necessary for maintenance of membrane and cellular integrity. Free fatty acids found in algal cells inhibit electron transport chain (ETC) and oxidative phosphorylation, the nutrient uptake process and interfere with enzyme activity and auto degradation of cellular products leading to bacterial lysis (Zheng et al., 2005; Smith et al., 2010). These mechanism accounts for the antimicrobial activity of fatty acids. Unsaturated fatty acids, the PUFAs hexadecatrienoic acid (HTA), eicosapentaenoic acid (EPA) and the monounsaturated fatty acid palmitoleic acid (PA) (Desbois et al., 2009) isolated from marine diatom *Phaeodactylum tricornutum* showed high antibacterial activity against gram-positive MRSA, *Staphylococcus aureus*, *Staphylococcus epidermidis*, and *Bacillus weihenstephanensis*, and gram-negative bacteria like *Vibrio* sp. and *Listonella anguillarum*. Free fatty acids, cyclopentane tridecanoic acid, tetradecanoic acid, and α-Cedrene extracted from brown seaweed, *Stoechospermum marginatum* showed high antibacterial activity against drug-resistant *Staphylococcus aureus*, *Escherichia coli* and *Proteus vulgaris* (Anjali et al., 2019). Similarly, extracts of fatty acids like 10,13-octadecadienoic acid and cyclopentaneacetic acid from *Sargassum fusiforme* and *Sargassum vulgare* showed antibacterial activity against drug-resistant *Staphylococcus aureus* and *Klebsiella pneumoniae*, *Scenedesmus obliquus* extract with fatty acids inhibited pathogens such as *Staphylococcus aureus*, *Escherichia coli*, and *Pseudomonas aeruginosa* responsible for food spoilage (Guedes et al., 2011). Algal fatty acids showed a bactericidal effect against peptic ulcers causing bacterium *Helicobacter pylori* (Petschow et al., 1996). Fatty acids like hexadecanoic acid, 9-hexadecenoic acid, 10-octadecenoic acid, 13-octadecenoic acid, eicosanoic acid isolated from red algae *Gracilaria edulis* inhibited fish pathogens *Aeromonas hydrophila* and *Vibrio* spp. (Kasanah et al., 2019). Extracted linoleic acid (LA) from *Chaetomorpha linum* also inhibited pathogens like *Vibrio vulnificus* and *Vibrio ordalii* found in fish. These fatty acids from algae are alternative to the antibiotics for which pathogens have developed resistance.

1.2.7 LECTINS

Marine algal lectins from Rhodophyta, Chlorophyta, and Phaeophyta species inhibited four *Vibrio* bacterial species (*V. neresis*, *V. pelagius*, *Eucheuma serra*, and *Galaxaura marginata*). Aqueous ethanol and saline extracts of red algae *E. serra* and *Pterocladia capillacea* have shown strong antimicrobial activity at low concentrations on the *V. vulnificus*. This antimicrobial action of marine algal lectins makes it a suitable candidate as antibacterials in aquaculture (Liao et al., 2003). Lectins isolated from the red algae *Solieria filiformis* showed antimicrobial activity against pathogens like *Salmonella typhi*, *Klebsiella pneumoniae*, *Pseudomonas aeruginosa*, *Serratia marcescens*, *Enterobacter aerogenes*, and *Proteus* species (Holanda et al., 2005). Similarly, lectins extracted from the red algae *Galaxaura marginata* and *Eucheuma serra* significantly inhibited *Vibrio vulnificus*. Lectin binds to the mannan present in the cell surface of gram-negative bacteria to kill the bacteria (Liao et al., 2003).

1.2.8 LACTONES

Marine organisms contain furanones which are cyclic organics compounds formed by esterification of hydrocarboxylic acids. Halogenated furanones has been extracted from *Delisea pulchra,* a red alga, which is a strong secondary metabolite against highly pathogenic microorganism. Many pathogenic microorganisms communicate through quorum sensing using acyl-homoserine lactones (AHLs), the signaling molecules that bind to the luminescence transcriptional activator (LuxR) regulatory proteins. Halogenated furanone extracts from *Delisea pulchra* interfered with quorum sensing and biofilm formation in *Pseudomonas aeruginosa* and *E. coli* by competed with AHLs for binding to the LuxR receptor and LuxS respectively (Brameyer and Heermann, 2015; Ren et al., 2004). Halogenated lactones, from *Delisea pulchra* and 4-bromo-5-(bromomethylene)-3-(1′-hydroxybutyl)-2(5H)-furanone interfered with the production of carbapenem and other virulent proteins by *Erwinia carotovora.* These secondary metabolites also inhibit carABCDEFGH gene expression required for quorum sensing (Manefield et al., 2001). Furanones from *Delisea pulchra* also kills pathogens like *Campylobacter jejuni* and *Escherichia coli* responsible for food poisoning (Chatterjee et al., 2016).

1.2.9 PHLOROTANNINS

Phlorotannins are found in marine algae-like brown algae such kelps and rockweeds or sargassacean species and red algae. Phlorotannins are produced through acetatemalonate pathway and stored in irregular shaped fucosan vesicles called physodes inside algal cells (Verma et al., 2015). Phlorotannins are polymeric units of phloroglucinol (1,3,5-trihydroxybenzene) having molecular weights ranging from 120–650 KDa. More than 150 compounds of phlorotannins are known which are categorized into six main subgroups: fucols, phlorethols, fucophloretols, fuhalols, and eckols. Few examples of phlorotannins are fucodiphlorethol G from the seaweed *Ecklonia cava,* eckol from *Ecklonia* species or phlorofucofuroeckol-B from *Eisenia arborea.* The antibacterial activity of phlorotannins involves inhibition of oxidative phosphorylation, binding to enzymes, cell membrane and lysis of the cell. The -OH- group and aromatic phenolic compounds (PCs) of phloroglucinol, binds to -NH groups of bacterial proteins by hydrogen and hydrophobic interactions rendering them inactive (Wang et al., 2009). The presence of methyl- or acetyl-vinyl as tertiary structure increases antibacterial activity of phloroglucinol against *Vibrio.* Phloroglucinol from *Fucus vesiculosus* showed significant bacteriostatic effect against pathogenic gram-positive *Staphylococcus aureus* and *Streptococcus pneumoniae* compared to the gram-negative test bacteria (Bogolitsyn et al., 2019). Phlorotannins from *Sargassum thunbergii* caused cell lysis by disputing cell wall of *Vibrio parahaemolyticus* (Wei et al., 2016). Phlorofucofuroeckol-A, from brown algae *Eisenia bicyclis* showed antibacterial activity against antibiotic-resistant *Propionibacterium acnes*, which causes acne (Lee et al., 2014). Diphlorethohydroxycarmalol (DPHC), a phlorotannin extracted from marine brown algae *Ishige okamurae* showed high *in vitro* antibacterial effect against cutaneous pathogens *Staphylococcus aureus* and *Pseudomonas aeruginosa.* DPHC was used as a component of wound dressing hydrogel with polyvinyl alcohol (PVC). A wound dressing prepared with PVC hydrogel and DPHC completely healed the wound in experimental mice and reduced the bacterial load (Kim et al., 2020). Phlorotannins from brown algae *Ecklonia stolonifera, Ecklonia kurome* extracted by silicic acid chromatography and 2,4-Diacetylphloroglucinol collected from the marine red algae *Ceratodictyon spongiosum* has been reported to be effective against MRSA (Lee et al., 2014) and gram-negative bacterium *Vibrio parahaemolyticus.* Phlorotannins suppressed the expression of mecI mecR1 and mecA, the genes of the mec operon which determines the resistance to methicillin in *Staphylococcus aureus* (Eom et al., 2014; Lee et al., 2014). Phlototannins are also effective against *B. cereus*, *Shigella flexneri*, *Streptococcus pyogenes*, *Klebsiella pneumoniae*, *Campylobactor* spp., *C. jejuni*, *E. coli*, *S. enteritidis*, *S. typhimurium*, *V. parahaemolyticus* and *Vibrio* sp. (Nagayam et al., 2002). Studies on mice showed that phlorotannins do not cause any toxicity to the animals and are safe for use (Gopal et al., 2008).

1.2.10 PIGMENTS

Pigments are colorful compounds with fluorescence because they have distinct excitation wavelengths. Pigments commonly found in algae are chlorophylls, carotenoids, and phycobilin. These colorful compounds providing color and a photoprotection to algae. These pigments show both antioxidant and antibacterial properties. Derivatives of chlorophyll α, chlorophyllide-a, and pheophytin-α, extracted from marine alga *Isochrysis galbana* inhibited pathogens like *Brevibacterium* sp., *Micrococcus* sp., *Staphylococcus aureus*, and *Flavobacterium* sp. (Bruce et al., 1967). Chlorophyll pigments (a and b) from *Scenedesmus quadricauda*, *Chlamydomonas reinhardtii*, and *Chlorella vulgaris* showed significant antibacterial effects against gram-positive *Bacillus subtilis*, gram-negative bacteria (Smith et al., 2010). Goud et al. (2007) reported their antimicrobial activity against oral bacteria such as *Fusobacterium nucleatum* and *Porphyromonas gingivalis.* Beta carotene isolated from *Dunaliella salina* and *Chlorococcum humicola* exhibited antimicrobial activity. Similarly, astaxanthin from *Haematococcus pluvialis*, also showed antibacterial activity (Sivathanu and Sumathi, 2010). Pheophorbide found in the extract of *Caulerpa racemosa* antibacterial effects against MRSA and *Escherichia coli* K1 (Yap et al., 2019). Phycocyanin (PC) from *Spirulina platensis*, cyanobacteria *Anabaena* and *Westiellopsis* species contain pigment PC which showed antibacterial effects against pathogenic strains such as *Streptococcus* sp., *Escherichia coli*, *Bacillus* sp., and *Pseudomonas* sp., *Xanthomonas* sp., *Bacillus subtilis* and *Pseudomonas* sp. (Sabarinathan and Ganesan, 2008). Similarly, extracts of brown algae and seaweeds and microalgae *Laurencia obtusa*, *Turbinaria triquetra*, and *Ulva lactuca* containing fucoxanthin showed a significant antimicrobial activity against *Bacillus cereus*, *Bacillus subtilis*, *Klebsiella pneumoniae*, *Klebsiella oxytoca*, *Pseudomonas aeruginosa*, *Escherichia coli*, *Staphylococcus aureus*, *Staphylococcus epidermidis*, and *Streptococcus agalactiae* (Deyab, 2013; Karpiński and Adamczak, 2019).

1.2.11 POLYSACCHARIDES

Marine algae contain sulfated polysaccharides in their cell walls which gives strength and rigidity. Algal polysaccharides with glycoprotein receptors can bind to the molecules present on the bacterial cell wall, membrane, and nucleic acids. This disintegrates the cell wall leading to cell lysis (He et al., 2010). Sulfated polysaccharides, xyloarabinogalactan, from marine alga *Chaetomorpha aerea* show bactericidal effect against the gram-positive *Bacillus subtilis* and *Micrococcus luteus* and drug-resistant *Staphylococcus aureus* (Pierre et al., 2011; Amorim et al., 2012). Fucoidans, a type of sulfated polysaccharide extracted from various marine algae such as marine algae-like *Kjellmaniella crassifolia*, *Polyopes affinis*, and *Cladosiphon okamuranus* are source of fucoidans (a sulfated polysaccharide) which prevent biofilm formation and adhesion of *Helicobactor pylori* gastric mucosa. Laminarin, a polysaccharide from brown marine algae *Laminaria hyperborea* and *Ascophyllum nodosum*, *Cystoseira barbata* exhibited a broad range of antibacterial activity against both gram-positive and gram-negative bacteria, such as *Salmonella typhimurium*, *Escherichia coli*, *Listeria monocytogenes*,

Staphylococcus aureus, *Micrococcus luteus*, *Klebsiella pneumoniae*, *Salmonella typhimurium*, *Salmonella enterica*, and *Pseudomonas aeruginosa* and also showed high antioxidant properties (Kadam et al., 2015). These products have also been used to enhance the immune systems of farmed fishes to combat *Piscirickettsia salmonis* infection. Laminarin-based cream have been found to be very effective in wound healing in rats (Sellimi et al., 2018). *Sargassum swartzii* and *Chlamydomonas reinhardtii* are another source of sulfated polysaccharide inhibited biofilm formation in pathogens like *Bacillus subtilis*, *Streptococcus* sp., *Escherichia coli*, and *Neisseria mucosa* (Vishwakarma and Vavilala, 2019). Polysaccharides extracted from red algae *Grateloupia turuturu*, brown seaweed, *Dictyopteris membranacea*, and red seaweed, *Pterocladia capillacea*, showed a significant reduction in the growth of *Staphylococcus aureus*, *Escherichia coli* and *Pseudomonas fluorescens* (Abou Zeid et al., 2014).

1.2.12 STEROLS

Sterols are lipid molecules having four-ringed structures as core with both polar and nonpolar structure. Sterols are distributed in all the prokaryote and eukaryote cell wall giving it fluidity. Different classes of algae show the presence of different categories of sterols. Brown algae is rich in cholesterol and stigmast-5-en-3-ol, fucosterol, and its derivatives, while red algae contain desmosterol, cholesterol, and its derivatives. Green algae are rich in 24-ethylcholesterol and ergosterol (Sanchez-Machado et al., 2004). Antibacterial activity of algal sterols is due to their ability to interacting with negatively charged phosphate groups in the bacterial membrane. This interaction distorts the membrane and affect the osmoregulation causing disruption of the cell membrane (Alhanout et al., 2010). A sterol anti-24-propylidene-cholest-5-en-3-β-ol found in extracts from *Laurencia papillosa* exhibited antibacterial activity against pathogenic bacteria such as *Staphylococcus aureus*, *Bacillus subtilis*, *Escherichia coli*, *Pseudomonas aeruginosa*, *Klebsiella pneumoniae* and *Shigella flexneri* (Kavita et al., 2014). Saringosterol, and its derivatives isolated from the alga *Lessonia nigrescens*, exhibited inhibition of *Mycobacterium tuberculosis* posing to be promising anti-tuberculosis drugs (Wachter et al., 2001). Similarly, many different unsaturated sterols like cholest-5-24-1,3-(acetyloxy)-3b-ol, ergost-5-en-3-β-ol and 24-oxocholesterol acetate, from *Isochrysis galbana*, inhibited resistant strains of *Mycobacterium tuberculosis* (Prakash et al., 2010).

1.2.13 TERPENES

Terpenes are natural unsaturated hydrocarbons consisting of repeated carbon isoprene units with substituent groups attached to the backbone. There are as many as 30,000 terpenoids known so far. *Laurencia*, a genus of red algae showed the presence of 512 sesquiterpenes and 133 diterpenes of diverse structures (Harizani et al., 2016). Other species of red algae like *Chondrococcus hornemanni*, *Hypnea pannosa*, *Sphaerococcus coronopifolius*, *Plocamium cornutum*, *Plocamium leptophyllum*, and *Portieria hornemannii*, also produce terpenes. Terpenes in algae, *P. purpureum* and *E. australicus* are solely synthesized by the MEP pathway in plastids which is non mevalonic acid pathway. Red algal species that produce volatile sesquiterpenes are reported to possess *MTPSL* genes. PpMTPSL and EaMTPSL2. Bromophycolides J-Q are the novel metabolites found in *Callophycus serratus*, which showed high antibacterial activity against antibiotic-resistant strains of vancomycin-resistant *Enterococcus* (VRE) a MRSA and *Mycobacterium tuberculosis* (Lane et al., 2009). Algae *Sphaerococcus coronopifolius* contains brominated diterpene-based compounds, and sphaerane bromoditerpens and sphaerodactylomelol that inhibit the growth of bacteria such as *Staphylococcus aureus*, *Escherichia coli* and *Pseudomonas aeruginosa* (Etahiri et al., 2001; Rodrigues et al., 2015). Bromosphaerone, sphaerolabdadiene-3,14-diol, and sphaerococcenol A, 12-Shydroxybromosphaerodiol isolated from *Sphaerococcus coronopifolius* showed antibacterial activity against *Staphylococcus aureus* (Etahiri et al., 2001). *Ulva lactuca* showed antibacterial activity against drug-resistant strains of *Staphylococcus aureus*, *Escherichia coli*, and *Klebsiella pneumoniae* attributed to the presence of Neophytadiene, a sesquiterpenoid (Anjali et al., 2019). Sesquiterpenes isolated from red algae species *Laurencia rigida* and *Laurencia luzonesis* contain deschloroelatol, elatol, luzonenone, luzofuran, 3,4-epoxypalisadin, 1,2-dehydro-3,4-epoxypalisadin B, and 15-hydroxypalisadin which show antibacterial action on *Bacillus* (Kuniyoshi et al., 2005).

1.3 CONCLUSION AND FUTURE PROSPECTS

The natural products obtained from various sources, especially marine organisms, for finding answer to existing infectious diseases, has expanded the scope of algae metabolites as therapeutic targets. As oceans cover 70% of the earth surface, it provides a natural habitat to an immense taxonomic diverse range of algae-like Phaeophyta, Rhodophyta, and Chlorophyta, which provide novel metabolic products of various economic importance. Algal metabolites have been studied since the last few decades for their medicinal properties, showing high antibacterial activity against a wide spectrum of bacteria, including MDR strains. As discussed, the mechanism of actions of several algae metabolites is unique, showing no toxicity towards mammalian (mice, humans, monkeys) when administered directly. Also, a multi-disciplinary strategies and the latest technologies like gene transfer technologies, nanotechnology, computational drug delivery methodologies open new opportunities for exploring antimicrobial potentials of novel metabolites from marine algae. Furthermore, the prospects of large-scale production should be looked out, targeting bioactive compounds from microalgae as antimicrobial drug.

KEYWORDS

- **algae natural products**
- **alkaloids**
- **antibacterial activity**
- **bioactive compound**
- **cladophores**
- **drug resistance**
- **mode of action**
- **novel drug**
- **pigments**
- **polysaccharides**
- **terpenes**

REFERENCES

Abou, Z. A. H., Aboutabl, E. A., Sleem, A. A., & El-Rafie, H. M., (2014). Water-soluble polysaccharides extracted from *Pterocladia capillacea* and *Dictyopteris membranacea* and their biological activities. *Carbohydr. Polym.*, *113*, 62–66.

Abu-Ghannam, N., & Rajauria, G., (2013). Antimicrobial activity of compounds isolated from algae. In: Dominguez, H., (ed.), *Functional Ingredients from Algae for Foods and Nutraceuticals* (pp. 287–306). Woodhead Publishing Ltd.: Sawston.

Alghazeer, R., Whida, F., Abduelrhman, E., Gammoudi, F., & Naili, M., (2013). *In vitro* antibacterial activity of alkaloid extracts from green, red, and brown macroalgae from the western coast of Libya. *Afr. J. Biotechnol.*, *12*(51), 7086–7091.

Alhanout, K., Malesinki, S., Vidal, N., Peyrot, V., Rolain, J. M., & Brunel, J. M., (2010). New insights into the antibacterial mechanism of action of squalamine. *J. Antimicrob. Chemother.*, *65*(8), 1688–1693.

Amorim, R. D. D., Rodrigues, J. A. G., Holanda, M. L., et al., (2012). Antimicrobial effect of a crude sulfated polysaccharide from the red seaweed *Gracilaria ornata*. *Braz. Arch. Biol. Technol.*, *55*(2), 171–181.

Anjali, K. P., Sangeetha, B. M., Devi, G., Raghunathan, R., & Dutta, S., (2019). Bioprospecting of seaweeds (*Ulva lactuca* and *Stoechospermum marginatum*): The compound characterization and functional applications in medicine: a comparative study. *J. Photochem. Photobiol. B. Biol.*, *200*, 111622.

Arputha, M. B., Nithya, K., Manikandan, M. S., Selvamani, P., & Latha, S., (2012). Antimicrobial evaluation of the organic extracts of *Sargassum wightii* (brown algae) and *Kappaphycus alvarezii* (red algae) collected from the coast of Meemesal, Tamilnadu. *Int. J. Pharma Bio Sci.*, *2*(4), 439–446.

Beaulieu, L., Bondu, S., Doiron, K., Rioux, L. E., & Turgeon, S. L., (2015). Characterization of antibacterial activity from protein hydrolysates of the macroalga *Saccharina longicruris* and identification of peptides implied in bioactivity. *J. Funct. Food*, *17*, 685–697.

Bogolitsyn, K., Dobrodeeva, L., Druzhinina, A., Ovchinnikov, D., Parshina, A., & Shulgina, E., (2019). Biological activity of a polyphenolic complex of arctic brown algae. *J. App. Phycol.*, *31*(5), 3341–3348.

Brameyer, S., & Heermann, R., (2015). Specificity of signal-binding via non-AHL LuxR-type receptors. *PLoS One*, *10*(4), 1–15.

Bruce, D. L., Duff, D. C., & Antia, N. J., (1967). The identification of two antibacterial products of the marine planktonic alga *Isochrysis galbana*. *J. Gen. Microbiol.*, *48*, 292–298.

Chatterjee, M., Anju, C. P., Biswas, L., Kumar, V. A., Mohan, C. G., & Biswas, R., (2016). Antibiotic resistance in *Pseudomonas aeruginosa* and alternative therapeutic options. *Int. J. Med. Microbiol.*, *306*(1), 48–58.

Cherian, C., Jannet, V. J., & Sharan, L., (2019). Marine bromophenols as an effective inhibitor of virulent proteins (peptidyl arginine deiminase, gingipain R and hemagglutinin A) in *Porphyromas gingivalis*. *Arch. Oral Biol.*, *100*, 119–128.

Chojnacka, K., & Kim, S. K., (2015). Introduction of marine algae extracts. In: *Marine Algae Extracts* (pp. 1–14). Wiley-VCH: Weinheim, Germany.

Desbois, A. P., Mearns-Spragg, A., & Smith, V. J., (2009). A fatty acid from the diatom *Phaeodactylum tricornutum* is antibacterial against diverse bacteria including multi-resistant *Staphylococcus aureus* (MRSA). *Mar. Biotechnol.*, *11*(1), 45–52.

Deyab, M., (2013). Antibacterial activity of some marine algal extracts against most nosocomial bacterial infections. *Egypt. J. Exp. Biol. (Bot.)*, *9*, 281–286.

Eom, S. H., Lee, D. S., Jung, Y. J., Park, et al., (2014). The mechanism of antibacterial activity of phlorofucofuroeckol-A against methicillin-resistant *Staphylococcus aureus*. *Appl. Microbiol. Biotechnol.*, *98*(23), 9795–9804.

Etahiri, S., Bultel-Ponce, V., Caux, C., & Guyot, M., (2001). New bromoditerpenes from the red alga *Sphaerococcus coronopifolius*. *J. Nat. Prod.*, *64*(8), 1024–1027.

Gopal, R., Vijayakumaran, M., Vennkatesan, R., & Kathiroli, S., (2008). Marine organisms in Indian medicine and their prospects. *Nat. Prod. Rad.*, *7*(2), 139–145.

Goud, J. P. M., Seshikala, D., & Charya, M. A. S., (2007). Antibacterial activity and biomolecular composition of certain freshwater micro-algae from river Godavari (India). *Sci. World J.*, *2*(3), 19–23.

Guedes, A. C., Barbosa, C. R., Amaro, H. M., Pereira, C. I., & Malcata, F. X., (2011). Microalgal and cyanobacterial cell extracts for use as natural antibacterial additives against food pathogens. *Int. J. Food Sci. Technol.*, *46*(4), 862–870.

Guven, K. C., Bora, A., & Sunam, G., (1970). Hordenine from the alga *Phyllophora nervosa*. *Phytochemistry*, *9*(8), 1893–1897.

Guzmán, F., Wong, G., Román, T., et al., (2019). Identification of antimicrobial peptides from the microalgae *Tetraselmis suecica* (Kylin) butcher and bactericidal activity improvement. *Marine Drugs*, *17*(8), 453–456.

Harizani, M., Ioannou, E., & Roussis, V., (2016). The Laurencia paradox: An endless source of chemodiversity. *Prog Chem Org Nat Prod.*, *102*, 91–252.

He, F., Yang, Y., Yang, G., & Yu, L. J., (2010). Studies on antibacterial activity and antibacterial mechanism of a novel polysaccharide from *Streptomyces virginia* H03. *Food Control*, *21*(9), 1257–1262.

Holanda, M. L., Melo, V. M. M., Silva, L., Amorim, R. C. N., Pereira, M. G., & Benevides, N. M. B., (2005). Differential activity of a lectin from *Solieria filiformis* against human pathogenic bacteria. *Brazilian J. Med. Biol. Res.*, *38*(12), 1769–1773.

Kadam, S. U., O'Donnell, C. P., Rai, D. K., et al., (2015). Laminarin from Irish brown seaweeds *Ascophyllum nodosum* and *Laminaria hyperborea*: Ultrasound-assisted extraction, characterization and bioactivity. *Mar. Drugs*, *13*(7), 4270–4280.

Karpiński, T. M., & Adamczak, A., (2019). Fucoxanthin - an antibacterial carotenoid. *Antioxidants (Basel)*, *8*(8), 239–264.

Kasanah, N., Amelia, W., Mukminin, A., Triyanto, S., & Isnansetyo, A., (2019). Antibacterial activity of Indonesian red algae *Gracilaria edulis* against bacterial fish pathogens and characterization of active fractions. *Nat. Product Res.*, *33*(22), 3303–3307.

Kavita, K., Singh, V. K., & Jha, B., (2014). 24-Branched Δ5 sterols from *Laurencia papillosa* red seaweed with antibacterial activity against human pathogenic bacteria. *Microbiol. Res.*, *169*(4), 301–306.

Keffer, J. L., Huecas, S., Hammill, J. T., Wipf, P., Andreu, J. M., & Bewley, C. A., (2013). Chrysophaentins are competitive inhibitors of FtsZ and inhibit Z-ring formation in live bacteria. *Bioorg. Med. Chem.*, *21*(18), 5673–5678.

Kim, M. S., Oh, G. W., Jang, Y. M., et al., (2020). Antimicrobial hydrogels based on PVA and diphlorethohydroxycarmalol (DPHC) derived from brown alga *Ishige okamurae*: An *in vitro* and *in vivo* study for wound dressing application. *Mater. Sci. Eng. C.*, *107*, 110352.

Kuniyoshi, M., Wahome, P. G., Miono, T., et al., (2005). Terpenoids from *Laurencia luzonensis*. *J. Nat. Prod.*, *68*(9), 1314–1317.

Kusuma, K. D., Payne, M., Ung, A. T., Bottomley, A. L., & Harry, E. J., (2019). FtsZ as an antibacterial target: Status and guidelines for progressing this avenue. *ACS Infect. Dis.*, *5*(8), 1279–1294.

Lane, A. L., Stout, E. P., Lin, A. S., et al., (2009). Antimalarial bromophycolides J-Q from the Fijian red alga *Callophycus serratus*. *J. Org. Chem.*, *74*(7), 2736–2742.

Lavoie, S., Sweeney, J. A. M., Mojib, N., et al., (2019). Antibacterial oligomeric polyphenols from the green alga *Cladophora socialis*. *J. Org. Chem.*, *84*(9), 5035–5045.

Lee, J. H., Eom, S. H., Lee, E. H., et al., (2014). *In vitro* antibacterial and synergistic effect of phlorotannins isolated from edible brown seaweed *Eisenia bicyclis* against acne-related bacteria. *Algae*, *29*(1), 47–55.
Liao, W. R., Lin, J. Y., Shieh, W. Y., Jeng, W. L., & Huang, R., (2003). Antibiotic activity of lectins from marine algae against marine *Vibrios*. *J. Ind. Microbiol. Biotechnol.*, *30*(7), 433–439.
Lijun, H., Nianjun, X., Jiangong, S., Xiaojun, Y., & Chengkui, Z., (2005). Isolation and pharmacological activities of bromophenols from *Rhodomela confervoides*. *Chin. J. Oceanol. Limnol.*, *23*(2), 226–229.
Lordan, S., Ross, R. P., & Stanton, C., (2011). Marine bioactives as functional food ingredients: Potential to reduce the incidence of chronic diseases. *Mar. Drugs*, *9*(6), 1056–1100.
Manefield, M., Welch, M., Givskov, M., Salmond, G. P. C., & Kjelleberg, S., (2001). Halogenated furanones from the red alga, *Delisea pulchra*, inhibit carbapenem antibiotic synthesis and exoenzyme virulence factor production in the phytopathogen *Erwinia carotovora*. *FEMS Microbiol. Lett.*, *205*(1), 131–138.
Nagayama, K., Iwamura, Y., Shibata, T., Hirayama, I., & Nakamura, T., (2002). Bactericidal activity of phlorotannins from the brown alga *Ecklonia kurome*. *J. Antimicrob. Chemother.*, *50*(6), 889–893.
Petschow, B. W., Batema, R. P., & Ford, L. L., (1996). Susceptibility of *Helicobacter pylori* to bactericidal properties of medium-chain monoglycerides and free fatty acids. *Antimicrob. Agents Chemother.*, *40*(2), 302–306.
Pierre, G., Sopena, V., Juin, C., Mastouri, A., Graber, M., & Maugard, T., (2011). Antibacterial activity of a sulfated galactan extracted from the marine alga *Chaetomorpha aerea* against *Staphylococcus aureus*. *Biotechnol. Bioprocess Eng.*, *16*(5), 937–945.
Popplewell, W. L., & Northcote, P. T., (2009). Colensolide A: A new nitrogenous bromophenol from the New Zealand marine red alga *Osmundaria colensoi*. *Tetrahedron Lett.*, *50*(49), 6814–6817.
Prakash, S., Sasikala, S. L., & Huxley, A. V., (2010). Isolation and identification of MDR-*Mycobacterium tuberculosis* and screening of partially characterized antimycobacterial compounds from chosen marine microalgae. *Asian Pac. J. Trop. Med.*, *3*(8), 655–661.
Pratt, R., Daniels, T. C., Eiler, J. J., et al., (1944). Chlorellin, an antibacterial substance from *Chlorella*. *Science*, *99*, 351, 352.
Rajasulochana, P., Krishnamoorthy, P., & Dhamotharan, R., (2012). Isolation, identification of bromophenol compound and antibacterial activity of *Kappaphycus* sp. *Int. J. Pharm. Bio-Sci.*, *3*(2), 173–186.
Ren, D. C., Bedzyk, L. A., Ye, R. W., Thomas, S. M., & Wood, T. K., (2004). Differential gene expression shows natural brominated furanones interfere with the autoinducer-2 bacterial signaling system of *Escherichia coli*. *Biotechnol. Bioeng.*, *88*(5), 630–642.
Rizzo, L., Fraschetti, S., Alifano, P., Tredici, M. S., & Stabili, L., (2016). Association of *Vibrio* community with the Atlantic Mediterranean invasive alga *Caulerpa cylindracea*. *J. Exp. Mar. Biol. Ecol.*, *475*, 129–136.
Rodrigues, D., Alves, C., Horta, A., et al., (2015). Antitumor and antimicrobial potential of bromoditerpenes isolated from the red alga, *Sphaerococcus coronopifolius*. *Mar. Drugs*, *13*(2), 713–726.
Sabarinathan, K. G., & Ganesan, G., (2008). Antibacterial and toxicity evaluation of C-phycocyanin and cell extract of filamentous freshwater cyanobacterium–*Westiellopsis* sps. *Eur. Rev. Med. Pharmacol. Sci.*, *12*(2), 79–82.

Sanchez-Machado, D. I., Lopez-Hernandez, J., Paseiro-Losada, P., & Lopez-Cervantes, J., (2004). An HPLC method for the quantification of sterols in edible seaweeds. *Biomed. Chromatogr.*, *18*(3), 183–190.
Sellimi, S., Maalej, H., Rekik, D. M., et al., (2018). Antioxidant, antibacterial and *in vivo* wound healing properties of laminaran purified from *Cystoseira barbata* seaweed. *Int. J. Biol. Macromol.*, *119*, 633–644.
Sivathanu, B., & Sumathi, P., (2010). Protective role of p-carotene from *Chlorococcum humicola* against reactive oxygen species and lipid peroxidation in AP induced toxicity. *J. Pharm. Res.*, *1*, 21–35.
Smith, V. J., Desbois, A. P., & Dyrynda, E. A., (2010). Conventional and unconventional antimicrobials from fish, marine invertebrates and micro-algae. *Mar. Drugs*, *8*(4), 1213–1262.
Sun, Y. J., Chang, R., Li, Q. Y., & Li, B. S., (2018). Isolation and characterization of an antibacterial peptide from protein hydrolysates of *Spirulina platensis*. *Eur. Food Res. Technol.*, *244*(8), 1509–1509.
Verma, P., Arun, A., & Sahoo, D., (2015). Algae world. In: Sahoo, D., & Seckbach, J., (eds.), *Algae World* (pp. 177–204). Springer, Dordrecht.
Vishwakarma, J., & Vavilala, S. L., (2019). Evaluating the antibacterial and antibiofilm potential of sulfated polysaccharides extracted from green algae *Chlamydomonas reinhardtii*. *J. App. Microbiol.*, *127*(4), 1004–1017.
Volk, R. B., & Furkert, F. H., (2006). Antialgal, antibacterial and antifungal activity of two metabolites produced and excreted by cyanobacteria during growth. *Microbiol. Res.*, *161*(2), 180–186.
Wachter, G. A., Franzblau, S. G., Montenegro, G., Hoffmann, J. J., Maiese, W. M., & Timmermann, B. N., (2001). Inhibition of *Mycobacterium tuberculosis* growth by saringosterol from *Lessonia nigrescens*. *J. Nat. Prod.*, *64*(11), 1463, 1464.
Wang, Y., Xu, Z., Bach, S. J., & McAllister, T. A., (2009). Sensitivity of *Escherichia coli* to seaweed (*Ascophyllum nodosum*) phlorotannins and terrestrial tannins. *Asian Australas. J. Anim. Sci.*, *22*(2), 238–245.
Wei, Y. X., Liu, Q., Xu, C. J., Yu, J., Zhao, L., & Guo, Q., (2016). Damage to the membrane permeability and cell death of *Vibrio parahaemolyticus* caused by phlorotannins with low molecular weight from *Sargassum thunbergii*. *J. Aquat. Food Prod. Technol.*, *25*(3), 323–333.
Weidner-Wells, M. A., Altom, J., Fernandez, J., et al., (1998). DNA gyrase inhibitory activity of ellagic acid derivatives. *Bioorg. Med. Chem. Lett.*, *8*(1), 97–100.
Xu, N. J., Fan, X., Yan, X. J., et al., (2003). Antibacterial bromophenols from the marine red alga *Rhodomela confervoides*. *Phytochemistry*, *62*(8), 1221–1224.
Yap, W. F., Tay, V., Tan, S. H., Yow, Y. Y., & Chew, J., (2019). Decoding antioxidant and antibacterial potentials of Malaysian green seaweeds: *Caulerpa racemosa* and *Caulerpa lentillifera*. *Antibiotics (Basel)*, *8*(3), 152–160.
Yoo, M. S., Jin, H. J., & Lee, S. Y., (2020). Synergistic antibacterial efficacies of chlorhexidine digluconate or protamine sulfated combined with *Laminaria japonica* or *Rosmarinus officinalis* extracts against *streptococcus mutans*. *Biocontrol Sci.*, *25*(1), 41, 42.
Zheng, L., Chen, H. M., Han, X. T., Lin, W., & Yan, X. J., (2005). Antimicrobial screening and active compound isolation from marine bacterium NJ6-3-1 associated with the sponge *Hymeniacidon perleve*. *World J. Microbiol. Biotechnol.*, *21*(2), 201–206.
Zidar, N., Montalvão, S., Hodnik, Ž., et al., (2014). Antimicrobial activity of the marine alkaloids, clathrodin and oroidin, and their synthetic analogues. *Mar Drugs*, *12*(2), 940–963.

CHAPTER 2

VALUABLE BIO-PRODUCTS FROM ALGAE AND THEIR COMMERCIAL APPLICATIONS

MATHESWARAN YAAMINI,[1]
CHIDAMBARAM KULANDAISAMY VENIL,[1] and LAURENT DUFOSSÉ[2]

[1]*Department of Biotechnology, Anna University, Regional Campus, Coimbatore–641046, Tamil Nadu, India*

[2]*University of Reunion, CHEMBIOPRO Chemistry and Biotechnology of Natural Products, ESIROI Food Department, Sainte-Clotilde F–97490, Reunion Island, Indian Ocean, France*

ABSTRACT

Algae are diverse aquatic organisms and are categorized as macro and micro-algae based on their size. They are classified as red algae (Rhodophyta), brown algae (Phaeophyta), and green algae (Chlorophyta). The recent advances in algal biotechnology are much prospective ones that could extend the spectrum of utilization of algae into new areas. Algae are rich in diversity, producing structurally complex but bioactive compounds that are difficult to produce via synthetic routes. Both macro and microalgae are established natural sources for the commercial production of high-end products for consumption by various stakeholders and industries. The high-value products include pigments, proteins, lipids, carbohydrates, vitamins, antioxidants, etc., with wide applications in nutraceuticals, pharmaceuticals, and cosmetics industries. In addition, algal biomass is potentate enough to be used as a feedstock for producing biofuel, biofertilizers, and other

Algal Metabolites: Biotechnological, Commercial, and Industrial Applications. Jeyabalan Sangeetha, Devarajan Thangadurai, Saher Islam, & Ravichandra Hospet (Eds.)

bio-based products. Exploiting microalgae to treat the wastewater is also one of the interesting disciplines attracting the researchers as this contributes to explore suitable methodologies to sustainably exploit the algal resources and keeping the environment safe. This chapter appraises the general distribution of algae and production of different high-end value-added products from algal sources and their prospective commercial applicability.

2.1 INTRODUCTION

Algae are mainly a unit of aquatic organisms belonging to the kingdom Protista and are varied in nature. Being ordinary plant organisms, they grow in water resources such as freshwater, saltwater, moist soil, and wet rocks. They range from microscopic to macroscopic in sizes and are considered as the building blocks of life. Algae are generally classified as red algae (Rhodophyta), brown algae (Phaeophyta), and green algae (Chlorophyta), and based on their size, and they are grouped as macro and microalgae. Macroalgae, like seaweeds, are multicellular, large in size, and visible to the naked eyes, whereas the microalgae are microscopic, single cells, prokaryotic (cyanobacteria), or eukaryotic (chlorophyte) (Khan et al., 2018). Macro-algae like Rhodophyta and Phaeophyta are used in the production of phycocolloids like agar-agar, alginates, or carrageenans (Kilinc et al., 2013). They are economically significant sources for extracting crude oils, applying in foodstuffs, and for producing a number of pharmaceutical and industrial goods.

The structure of algae, their character, reproductive mechanism, etc., are the main factors to be explored for commercially exploiting them economically and ecofriendly. Liu et al. (2013) report that there are about 200,000 to 800,000 microalgal species, and of them, merely about 35,000 species are documented. These algae survive in extreme climatic conditions like salinity, temperature, nutritional requirements, UV radiations, etc., by acquiring resistance mechanisms and eventually producing novel compounds with biological activities (Morgan-Kiss et al., 2006). Therefore, microalgae are prospective natural sources for commercially producing a range of high-end products. Microalgae possess exalted levels of oils, carbohydrates, sugars, and proteins, and they could be utilized to generate renewable fuels. Moreover, they can also be used as a food source for animals, even for humans. Already, both macro and microalgae are extensively used in food products such as ice creams, to provide consistency, stabilize their features, and preserving the critical nutrients for healthiness (Wells et al., 2017).

Microalgae are capable of growing on wastewaters by utilizing the nutrients from wastewater for food and so they could be employed for remediating wastewater. They are also capable of sequestering excess CO_2 from the atmosphere and industries and thus be a sustainable feedstock for the bio-processing plants (Bhalamurugan et al., 2018). Microalgae can be exploited for their incredible medicinal properties which make them possible candidates for bio-based chemicals in pharma and nutraceutical industries (Pal et al., 2019). Microalgae produce an extensive variety of value supplementary goods, like proteins, lipids, pigments, etc., which could be augmented under strained ecological conditions (Figure 2.1). Microalgae are prospective sources to produce stable isotopes and yet a few species only have been exploited (Milledge, 2011). In this respect till now, this factor has thrown up a better scope for further investigation of these species to expand the perspective and cheap availability of isotopes in the future.

The researchers are expected to understand the algae and their activities under different settings, and then they could choose specific strains for their study on the specified industrial applications. The down streaming processes are concerned in producing an elevated volume of value addition algal products, and the production cost could be lowered by deriving numerous products in a single cycle. This chapter mainly focuses on the potential of high-value bio-products from algae and their significant industrial applications.

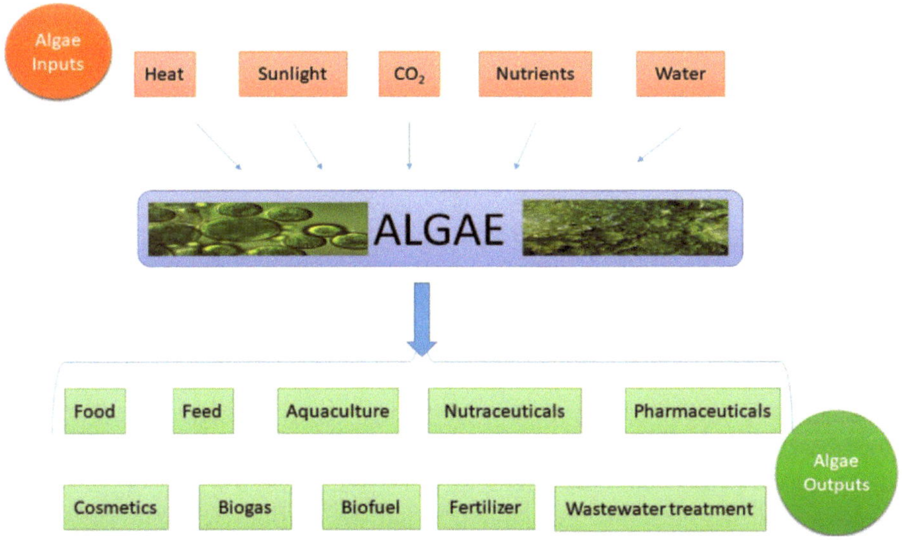

FIGURE 2.1 Applications of algae in different sectors.

2.2 OCCURRENCE AND DISTRIBUTION

Algae being aquatic organisms, breed in low salinity and marine waters, but few algae crop up in high salinity waters as well. They originate from polar stretches to scorching deserts, from mountains to oceans and lakes. Some algae are found allied with other living organisms (e.g., fungi + algae = lichens). Light, temperature, salinity, and nutritional requirements are the main factors that influence their growth in their natural habitats (Metsoviti et al., 2019). Algae are broad-based in distribution and originate normally in soil, rocks, trees, and in the other natural substrates too. The distribution of algae is incredibly uneven, and it depends on the intricate relations of their growth factors. Algae are the most well-known photosynthetic organisms and occur in urban environments too. The microscopic algae, phytoplanktons, live suspended in water and form the food base for most marine food chains and are the most explored group of algae; they float freely and largely in water bodies as colonial and filamentous forms, mainly in freshwater. Cyanobacteria are widespread on the exterior of both contemporary and antique buildings owing to their ability to defy very severe circumstances (Rampelotto, 2013). Both micro and macroalgae are characterized by varied morphology and are omnipresent around the world. They are concerned with mutualistic relations. Microalgae are considered the main aquatic prime producers and shape the foundation of the aquatic food webs, and offer organic resources to bacteria and higher consumers (Chapman, 2013).

2.3 TYPES OF ALGAE

There are seven main types of algae, each with divergent sizes, functions, and colors, which are discussed in subsections.

2.3.1 EUGLENOPHYTA (EUGLENOIDS)

Euglena are bright and brine protists while some of them are autotrophic. Instead of cell wall, they are covered by pellicle, i.e., protein-rich layer. They have chloroplasts and are gifted of photosynthesis. Akin to animal cells, euglenoids are heterotrophic and nourish on carbon-rich materials found in the water and other unicellular organisms. A few of the euglenoids are capable to endure in the darkness for some time. Owing to its photosynthetic capability, *Euglena* has been grouped along with algae in the phylum Euglenophyta. These organisms have this capability due to endo-symbiotic associations with photosynthetic green algae (Krajčovič et al., 2015).

2.3.2 CHRYSOPHYTA (GOLDEN-BROWN ALGAE AND DIATOMS)

Chrysophyta are largely the profuse type of single cellular algae found both in fresh water and in saltwater, accounting for more than 100,000 diverse species; diatoms are a lot more widespread than golden-brown algae; they comprise numerous types of plankton originate in the ocean. Diatoms as an alternative to cell walls enclose by frustule (silica shell). Depending on the species, the shape and structure of the frustule vary. Golden-brown algae are less in number. They are commonly known as nanoplanktons, with cells only 50 micrometers in diameter (Kristiansen, 2009).

2.3.3 PYRROPHYTA (FIRE ALGAE)

These being single cellular algae, that utilize flagellates for movement, are generally originating in oceans and in freshwater sources. Some species of Pyrrophyta are bioluminescent. They are classified into two types, Dinoflagellates, and Cryptomonads. Owing to the very high abundance of Dinoflagellates, ocean or lakes sometimes appear red, and this phenomenon is known as red tides. They are poisonous, which produce neurotoxins that can upset muscle function in humans and in some other organisms. Cryptomonads are alike Dinoflagellates, positive to produce unsafe algal blooms, resulting in the water to appear as red or dark brown (Anderson, 2010).

2.3.4 CHLOROPHYTA (GREEN ALGAE)

They are typically found in freshwater environments, and some of them exist in the oceans too. These algae have cell walls; a few species have one or two flagella and have chloroplasts, which experience photosynthesis. Green algae include thousands of single cellular and multicellular species. In multicellular species, colonies are ranging from four cells to several thousand.

2.3.5 RHODOPHYTA (RED ALGAE)

Red algae generally originate in tropical marine locations and grow on hard surfaces, including tropical reefs or attached to other algae. This eukaryotic cell lacks flagella and centrioles. These algae create a number of different seaweeds (Fong and Paul, 2011).

2.3.6 PHAEOPHYTA (BROWN ALGAE)

These are the leading species of algae, comprising a variety of seaweeds and kelp found in oceanic environment. The life cycle of these protists involves alteration of generations. These algae can grow up to 100 m in length. A few examples of brown algae comprise sargassum weeds, rockweed, and giant kelp (Wehr, 2015).

2.3.7 XANTHOPHYTA (YELLOW-GREEN ALGAE)

These algae characteristically originate in freshwater, but they can also originate in saltwater and damp soil environments too. Yellow-green algae are the least species of algae, with only 450 to 650 species. Their chloroplast lacks a certain pigment, which causes them to appear lighter in color (Salmaso and Tolotti, 2009).

2.4 ALGAE CULTIVATION

Macroalgae are cultivated and harvested all over the world with high nutritional value. They are substantially consumed in Eastern countries. In the remaining parts of the world, macroalgae are used as animal feed, fertilizers, or food additives (Wells et al., 2017). Algal biomass grows in ponds of containers called biofactories or in the cultivated algal production system (CAPS). Inorganic nutrients, carbon dioxide (CO_2), water, and light are provided to algal culture to promote biomass growth. Algae prefer dim light that is not too bright and hence some systems use shading light and diffuse it (Zullaikah et al., 2019). Various species produce biomass best at specific temperature, and consequently, some systems utilize recycled water on the outside of the bio factory to maintain optimum temperature.

While roughly estimating the algal cultivation cost, CO_2 may cost about 5% and this can be minimized by cultivating the algae or locating the algal farm near the manufacturing plant which produces CO_2. Nutrients may be provided from the wastewater, recovered from the algal tank, and harvested. Following the harvest of algal oil, the leftover biomass contains substantial nutrients. Closed system of algal cultivation presents the advantage of recycling the high nutrients water through this system (Li et al., 2019). This stratagem of recycling the nutrient water notably reduces the cost of added nutrients. Moreover, it also minimizes the water loss from evaporation. Algae culture exploits high saline water, so that waste steam from agriculture or brine water could be used for cultivating algae (Behera et al., 2015).

Microalgal growth and the chemical constitution of the culture medium are predominantly controlled by light, temperature, CO_2, nutrients, etc. Growth medium should contain adequate nutrients for microalgal growth. Exploitation of monocultures is important in diverse microalgal applications and controlled cultivation systems. This requisite leads to the improvement of closed photo-bioreactors. Constant light is mandatory for culturing *Botryococcus braunii* and *Chlorella* sp. whereas dark and light photoperiod is required for culturing *Chlorella* sp. (Wang et al., 2019). pH of the medium is vital and typically neutral pH is necessary to avoid the precipitation of the major elements. The pH of the algal cultures is influenced by the medium composition, amount of CO_2 dissolved, temperature, and metabolic activity of algal cells (Panahi et al., 2019). The optimum temperature for microalgae cultivation is between 20°C and 40°C. The algae, *Scenedesmus* sp. grows at a temperature of 15 to 36°C and the chlorophyll χontent is condensed while the levels of carotenoids, saccharides, and lipids are increased (Khan et al., 2018). Researchers are looking for algae with good thermostability. Inorganic and organic compounds are required for cellular function. Many algae need only inorganic nutrients, and such algae could be used as feedstock for biomass creation (Hannon et al., 2010).

2.5 VALUABLE BIO-PRODUCTS FROM ALGAE AND THEIR APPLICATIONS

Microalgae are affluent in diversity and possess nutritionally crucial compounds like proteins, pigments, carbohydrates, polyunsaturated fatty acids (PUFAs), dietary fibers, and bioactive compounds with a broad range of health benefits. Microalgae are in addition broadly utilized as organic colorant, pleasant aroma, and carrier molecules. Algae, both micro and macro species are a significant group of natural resources and ideal niches for producing valuable bio-products like pigments, carbohydrates, sugars, proteins, oils, etc. (Griffiths et al., 2016). Both micro and macro-algae products are widely used in foodstuffs for coloring, providing texture, stabilizing features and to provide important nutrients for health; they are also used for producing renewable energy sources such as biodiesel, biogas, bioethanol, biobutanol, hydrogen, etc. (Anbuchezhian et al., 2015). Besides, they could be used as raw materials for co-firing to produce electricity, liquid fuel production via pyrolysis and thermochemical liquefaction (bio-oil), or biomethane generation through fermentation process (Hochmann et al., 2014). Further, microalgae could be used as essential live feeds and supplements in the aquaculture of larval and juvenile abalone, sunfish larvae, and rotifer.

Algae are used as biofertilizer in agriculture to improve the fertility of soil (Abdel-Raouf et al., 2012). Algal pigments are used in foodstuffs like ice creams, jellies, and confectionary for coloring besides in nutraceuticals and pharmaceuticals products, owing to their medicinal properties (Begum et al., 2015). Adding up, algal species are used in wastewater treatment too because of their low energy requirements and economics; they help in reducing the toxic heavy metals into sludge formation. Their absorption capacity through CO_2 and carbon monoxide emission helps in reducing air pollution; in addition, they release oxygen during photosynthesis. Thus, algae are much valuable for producing various bioproducts and sustaining the environment in a friendly manner. The algal products from high to low market value are shown in Table 2.1.

TABLE 2.1 Algal Products from High to Low-Value Market Potential

High-Value	Medium-High Value	Low to Medium Value
Nutraceuticals: • Astaxanthine • Beta-carotene • Omega-3-fatty acids (DHA and EPA) • Coenzyme Q10	Nutraceuticals and feed additives: • *Spirulina* and *Chlorella* biomasses • Animals (poultry, etc.), and fish feed additives	Fertilizers and animal feed: • Aquaculture feed • Animal feed • Fertilizers

2.5.1 NUTRACEUTICALS

Algal nutraceuticals are dietary supplements obtained from microalgae. Most of the algal nutraceuticals are prepared from microalgae and some from macroalgae (Wells et al., 2017). The majority of the macroalgae are used in food more than in nutraceuticals. Algal nutraceuticals derived from microalgae are categorized into three types: Single-cell proteins, PUFAs, carotenoids, and pigments (Figure 2.2) (Galasso et al., 2019). The major source for single-cell proteins is *Spirulina. Spirulina* is a rich source of proteins, vitamins, amino acids, minerals, fibers, etc., and it is used to fight malnutrition in growing children and infants (Bleakley and Hayes, 2017). *Spirulina* is a popular healthy food and used for instant energy in the form of *Spirulina* tablets providing all required nutrients without excess calories and fats. *Spirulina* also has high amount of PUFAs, 1.5 to 2.0% of 5–6% of lipids. In the genus of cryptophytes, the strain *Storeatula* is high in PUFA production which includes γ-linolenic acid (36% of PUFAs), α-linolenic acid (ALA), linolenic acid (36%), stearidonic acid, eicosapentaenoic acid (EPA), docosahexaenoic acid (DHA) and arachidonic acid (AA) (López et al., 2019).

Carotenoids are a class of more than 1,100 naturally occurring pigments synthesized by plants, algae, and photosynthetic bacteria with a wide range of colors (http://carotenoiddb.jp/). The two primary isomers of carotene: α-carotenoids and β-carotenoids, differ in the position of a double bond (and thus a hydrogen) in the cyclic group at one end (Mezzomo and Ferreira, 2016). α-carotenoids, β-carotenoids, β-cryptoxanthin, lutein, zeaxanthin, and lycopene are the most common dietary carotenoids. α-Carotenoids, β-carotenoids and β-cryptoxanthin are provitamin A, which are converted by the body to retinol. Lutein, zeaxanthin, and lycopene are non-pro-vitamin A. These and other carotenoids like astaxanthin can also act as antioxidants. β-carotene is synthesized by microalgae *Dunaliella salina* whereas astaxanthin, canthaxanthin, and lutein were produced by *Haematococcus pluvialis.* Canthaxanthin, astaxanthin, and β-carotene were produced by *Chlorella vulgaris* whereas lutein and β-carotene were produced by *Scenedesmus almeriensis* (Guedes et al., 2011).

Spirulina, the blue-green algae (BGA), is also known with new taxonomy names *Arthrospira platensis* and *Arthrospira maxima. Arthospira maxima* is cultivated globally and used as dietary supplements and even as whole food. *Spirulina* is used for the production of phycocyanin (PC) and biomass (Lee, 2001; Costa et al., 2003). *Spirulina* has a high amount of protein of about 55 to 70% by dry weight, depending upon the sources. It contains all essential amino acids (EAAs) with a reduced amount of methionine, cystine, and lysine as compared to the standard proteins such as eggs, meats, or milk. However, it is superior to all standard proteins such that as legumes.

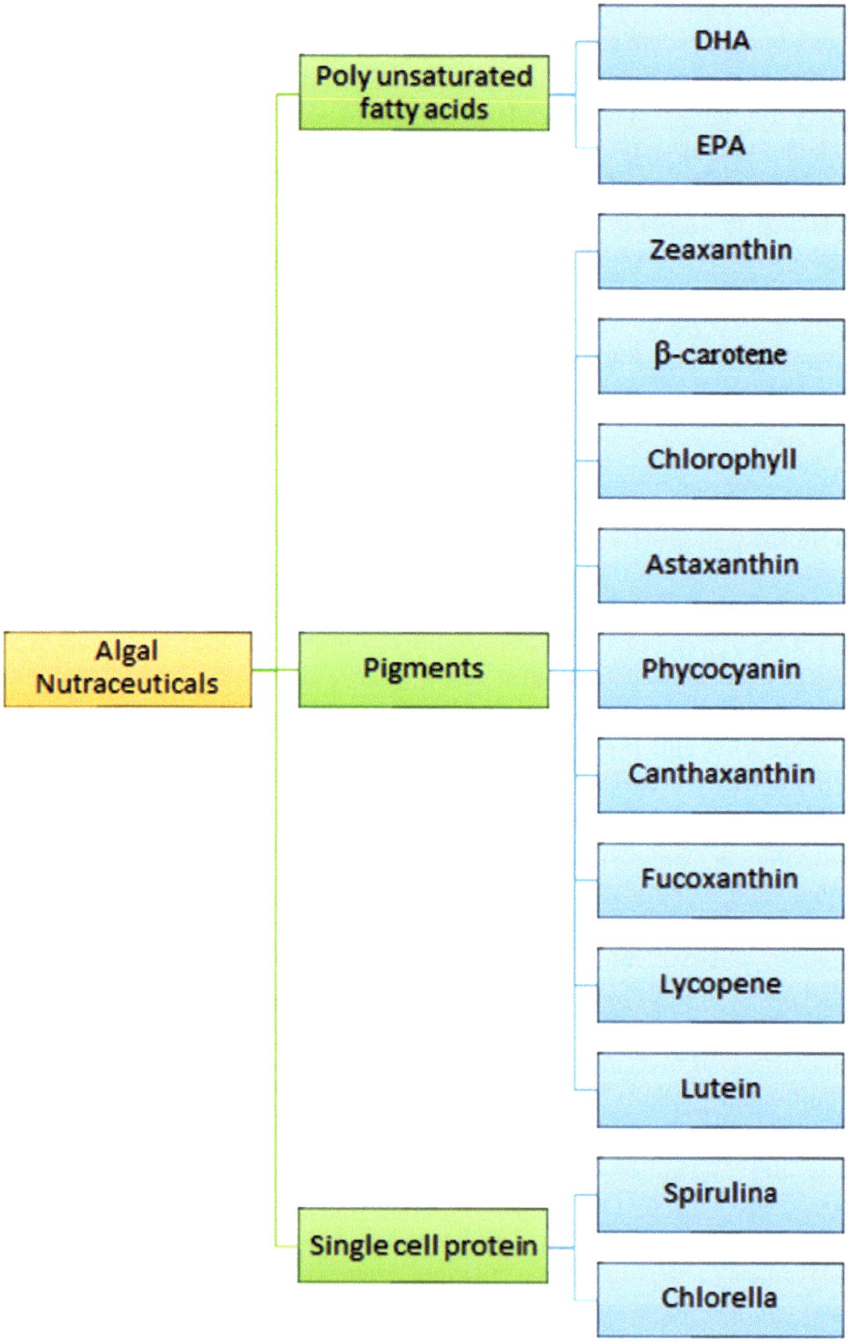

FIGURE 2.2 Algal nutraceuticals from microalgae.

Spirulina also contains vitamin B1 (thiamine), B2 (riboflavin), B3 (nicotinamide), B6 (pyridoxine), B9 (folic acid), B12 (cyanocobalamin), vitamin C, vitamin D and vitamin E (Peltomaa et al., 2017). It is rich in potassium, and also contains calcium, chromium, copper, iron, magnesium, manganese, phosphorous, selenium, sodium, and zinc. Moreover, it contains chlorophyll α, xanthophylls, beta-carotene, echinenone, zeaxanthin, canthaxanthin, diatoxanthin, 3-hydroxyechinenone, beta-cryptoxanthin, oscillaxanthin, plus the phycobiliproteins (PBP) C-phycocyanin and allophycocyanin. Since *Spirulina* is a rich source of proteins, vitamins, amino acids, minerals, crude fibers, etc., it can be used to fight against malnutrition in growing children and infants (Wells et al., 2017).

Chlorella is a single-cell freshwater algae, commonly found in Taiwan and Japan. *Chlorella vulgaris* contains a higher amount of chlorophyll than any other green plants. *Chlorella vulgaris* is often called a 'super food' not only for its chlorophyll χontent but also for its well-balanced nutritional compounds (Koyande et al., 2019). It contains 50 to 60% of protein and all nine EAAs. *Chlorella vulgaris*/Chlorophyta is used to produce biomass (Lee, 2001). When ingested by humans, some carotenoids from *Chlorella vulgaris* are partially converted to vitamin A, the leftover amount of unconverted beta carotene and retinal esters are integrated into the chylomicrons, secreted in lymph, and then transported to the liver. Carotenoids without β-ions ring do not exhibit provitamin activity (Eroglu and Harrison, 2013). Algal-based beta-carotene contains almost equal amount of *trans* and *cis* geometrical isomers of beta-carotene and in particularly the 9-*cis* isomer. Thus, the chemically synthesized beta-carotene can contain only *trans*-isomers. All through the heating and digesting of food, *cis* isomers will be formed in natural food. Here, 9 *cis*-beta carotene acts as a superior antioxidant when compared to all other *trans*-isomers. It helps in preventing post-exercise symptoms. It also acts as a nutrition solution to protect sunlight. Thus, this minimizes the damages by UV rays from the sun, such as sunburns, skin damage, etc. It also provides nutrition to the skin. It is a potent antioxidant, and it is a strong singlet oxygen quencher.

PBP are water-soluble, intensely fluorescent proteins that function as components in the photosynthesis apparatus in cyanobacteria and in certain eukaryotic algae such as rhodophytes, cryptomonads, and glaucocystophytes. Phycocyanin and phycoerythrin are compounds present in red algae or rhodophyta. Phycobilisomes are light-absorbing antenna of photosystem II in cyanobacteria, red algae, and glaucophytes (Singh et al., 2015). These photosynthetic accessory pigments, collectively known as PBP, are deeply colored (red or blue), water-soluble, proteinaceous, and complex compounds. R-phycoerythrin is only isolated from red algae, not from other taxa. It is the major light-harvesting pigment of red algae and cyanobacteria and is widely used. Visually, phycoerythrin appears red, phycocyanins (PCs) range from purple (phycoerythrocyanin, R-phycocyanin) to deep-blue (C-phycocyanin) and all phycocyanin are blue with a hint of green. Phycobilins are linear tetrapyrrole prosthetic groups linked to proteins (bilins), their functional state being covalently linked to specific cysteine residues of the proteins. For producing phycocyanin, *Spirulina* is used (Lee, 2001; Costa et al., 2003). Phycobilins or PBP are water-soluble accessory pigments. Phycobilins are also used in food colorants and cosmetics products. For example, a blue phycobilin from *Arthrospira*, is used to color cosmetics and food in Japan (Metting, 1996). Allophycocyanin absorbs and emits red light (650–660 nm) and it is found in cyanobacterium and red algae. These compounds act at a longer wavelength than phycoerythrin and phycocyanin (Sonani et al., 2016).

2.5.2 ANIMAL FEED

Microalgae is an essential food source for marine bivalve mollusks (clams, oysters, and scallops), the larval stage of marine gastropods and larvae of several marine fish species. They are nutritionally rich in proteins, vitamins, and non-saturated fats. *Spirulina* and *Dunaliella* are marketable healthy foods and are exploited globally. Other species used as animal feed are *Chlorella*, *Dunaliella*, *Nostoc*, and *Aphanizomenon* (Hochmann and Zibbermann, 2014). Algal products can be used for animal/livestock and other value-added products. Microalgae are used as essential live feeds and supplements in the aquaculture of larval and juvenile abalone, sunfish larvae, and rotifer. *Pavlova*, a golden-brown flagellate alga similar to *Isochrysis* that has high DHA profile and is an excellent food source for zooplankton. *Tetraselmis* is a large green flagellate with a high lipid and amino acid contents and feeds marine animals (Sathasivam et al., 2019). It is an excellent feeding source for larval shrimp. Similarly, *Nanochloropsis* is used for feeding zooplanktons in aquaculture industries. *Thalassiosira weissflogii* is used as a feed for shrimp and shellfish.

The incorporation of microalgal biomass into feeds brings vitamins, EAAs, polysaccharides, mono-, and n-3 and n-6 PUFAs, minerals, and pigments (Priyadarshani and Rath, 2012). Many nutritional and toxicological assessments have shown that microalgal biomass may be used as feed supplements (Becker, 2004). The most important microalgae species used as feed in fish industry are *Chlorella* sp., *Isochrysis* sp., *Pavlova* sp., *Phaeodactylum* sp., *Chaetoceros* sp., *Nannochloropsis* sp., *Skeletonema* sp., *Thalassiosira* sp., *Haematococcus* sp. and *Tetraselmis* sp. (Divya and Aanand, 2020). Exploiting the small amounts of microalgae biomass in feed can benefit animals' physiology by improving immune response, disease resistance, antiviral, and antibacterial action, gut function, and probiotic colonization stimulation. All these aspects combined together result in animal weight control, growth promotion, improved feed conversion ratio, and reproductive performance (Harel and Clayton, 2004). The microalgae *Schizochytrium* sp. is mainly applied in animal feed production as a feed supplement under the form of DHA-Gold extract. Although not consensual, this extract may impact positively on broiler productive performance by leading to an improvement in final body weight and body weight gain (Ribeiro et al., 2014).

2.5.3 COSMETICS

When skin is exposed to sunlight, the radiation is absorbed by melanin, a complex pigment which imparts color to human skin and acts as a protective barrier for human skin cells. Radiation from the sun helps synthesize tyrosinase and helps in the formation of melanosomes maturing melanin into keratinocytes (Maddodi et al., 2012). The pigment from algae, fucoxanthin from *Laminaria japonica*, *Alaria* sp., *Chorda* sp. and *Macrocystis* sp. helps to decrease the activity of tyrosinase and melanogenesis. The antioxidant properties of β-carotene from green and red algae aid in to fight against skin aging (Joshi et al., 2018). Many algal species like *Turbinaria ornata*, *Ahnfeltiopsis*, *Colpomenia*, *Gracilaria*, *Halymenia*, *Hydroclathrus*, *Polysiphonia*, etc., are used as anti-aging agents. Mycosporine-like amino acids (MAA) from *Porphyra umbilicalis* protect against UV which causes skin damage and premature skin aging (Kelman et al., 2012). The polysaccharides from *S. japonica*, *Chondrus crispus*, and *Codium tomentosum* help in the absorption of water and gives a soothing effect, and aid in proper moisture circulation by keeping the skin moisture in extremely hot and dry environments (Wang et al., 2013). Polysaccharides such as alginates, carrageenan derived from *Phaeophyceae* and *Rhodophyceae* act as gelling agents and are used in shampoos, lotions, etc. (Joshi et al., 2018). The microalgal species like *Spirulina*, *Chlorella* plays an important role in the cosmetics industry and the metabolites from these organisms enrich the beauty products. The proteins, amino acids from these algal species give natural moisturizing ability and keep the skin hydrated. The algal species, such as *Spirulina*, *Dunaliella* sp., *Chlamydomonas* sp. produce valuable metabolites like astaxanthin, proteins, MAA, and give nourishment to skin and hair. The cleansing properties of red algae refine the skin and give overall health benefits to skin cells (Pangestuti et al., 2018).

2.5.4 BIOFUEL

The algae play a central role for biofuel extraction in the world because they are safer and non-competitive organisms used for biodiesel production. Dickinson et al. (2017) reported that these kinds of algae could be grown devoid of much care on leftover nutrients. And so, such kinds of algae are measured as better source for biodiesel production. Algae contain 80% energy substance as that possessed by petroleum (Chisti, 2013) and 30% lipid content (Kligerman and Bouwer, 2015). The microalgae possess higher lipid content up to 85% and *Botryococcus braunii* have approximately 30–40% of hydrocarbon content that could be harvested easily (Mirza et al., 2008). The green algae, *Chlamydomonas reinhardtii*, *Dunaliella salina*, *Chlorella* sp. and *Botryococcus braunii* contain over 60% of lipid, which is secreted into the cell wall (Liu et al., 2016). Other algal groups include the diatoms *Phaeodactylum tricornutum* and *Thalassiosira pseudonana*, and other heterokonts including *Nannochloropsis* sp. and *Isochrysis* sp. for lipid secretion (Alishah-Aratboni et al., 2019).

There are many challenges for producing biodiesel from algae. For biodiesel production, algae oils are extracted and converted into biodiesels. For increasing the productivity of biofuels, the algae are cultivated in photobioreactors, and thus it is easier to maintain axenic conditions in the reactor. Depending upon the type of algae, different oil contents with varied compositions have been identified. Microalgae that would grow even in less nutrient content should be chosen for cultivation (Khan et al., 2018).

The microalgae *Chlorella vulgaris*, *Spirulina maxima*, *Nannochloropsis* sp., *Neochloris oleoabundans*, *Scenedesmus obliquus*, and *Dunaliella tertiolecta* were screened for quality and quantity in biofuel production. *N. oleoabundans* and *Nannochloropsis* sp. proved to be suitable raw materials for biofuel production with high oil content (Gouveia, 2009). The algal species *Oedogonium* sp. and *Spirogyra* sp. were comparatively studied by Sharif et al. (2008) for biodiesel production and found that algal oil and biodiesel (ester) were higher in *Oedogonium* sp. than *Spirogyra* sp.

2.5.5 BIOPLASTICS

Plastics are referred to as "green plastics" if they exhibit properties like: source renewability, biodegradability/compostability afterlife, and environmental friendliness (Rasul et al., 2017). Bioplastics made from algae are referred to as a blue bioplastics. Algae are excellent feedstock for plastics production and are referred to as "green plastics." Rahman and Miller (2017) reported that the microalgal biomass could be used for bioplastics production by utilizing the biomass directly or its secondary metabolites. Algae bioplastic mainly evolves as a by-product of algal biofuel production. These plastics are made by adding denatured algae biomass to petroleum-based plastics like polyurethane and polyethylene as filters. By using algae, it reduces the amount of petroleum used per unit of plastic and contain the property of biodegradability. Filamentous algae of order Cladophorales are claimed to be well suited for use in hybrid plastics (Ross et al., 2017).

2.5.6 WASTEWATER TREATMENT

Algae may be unicellular or multicellular organisms used in the treatment of wastewater. Both micro- and macro-algae can be used to sequester, remove, or transform pollutants including heavy metals, nutrients, and xenobiotics from wastewater or CO_2 from exhaust (Hochmann and Zibermann, 2014). Treating wastewater by bacteria requires constant oxygen supply that are expensive and tedious whereas microalgae release oxygen during photosynthesis and remove nitrogen and phosphorous, which can be used as biofuel and feedstocks (Hochmann et al., 2014).

Algae and bacteria are in a symbiotic relationship in the conversion of organic matter in wastewater. Bacteria metabolize organic wastes to produce new biomass and release carbon dioxide, and then algae utilize carbon dioxide for photosynthesis, and it releases oxygen for aerobic bacterial activity. *Chlorella* reduces nitrogen and phosphorous entering the nearby coastal water, thus preventing the eutrophication problem, which results in depletion of oxygen in water followed by fish death.

2.5.7 BIO-FERTILIZERS

Algae biofertilizers are natural, organic, and water energy sources that could retain essential nutrients and water in the soil which are required for the proper growth of the plants. Powdered and liquid form of algae can be effectively used as bio-fertilizers. Blue-green algae such as *Nostoc* sp., *Anabaena* sp., *Tolypothrix* sp., etc., synthesize, and liberate plant growth-promoting substances such as auxin and amino compounds which stimulate plant growth (Singh et al., 2016). Algal biofertilizers like the BGA such as *Nostoc* sp., *Anabaena* sp., *Tolypothrix* sp., *Aulosira* sp. etc., have the potential to fix atmospheric nitrogen and are used in paddy fields (Chittora et al., 2020).

The blue-green algae, cyanobacteria can photosynthesize and fix nitrogen and adapt to different soil types. The important roles played by cyanobacteria in agriculture include: enhancement in soil porosity, release of growth-promoting hormones, vitamins, and amino acids, increase in water holding capacity, decrease in soil salinity, preventing weeds, increase in soil phosphate, etc. The cyanobacteria serve as the inexpensive nitrogen source to plants besides increasing the crop yield and makes the soil fertile (Abinandan et al., 2019).

2.6 CONCLUSIONS

The current trend is exploring the ways and means to exploit the natural resources in a sustainable manner without harming our living environment and ecology. Already the academia and industry are in this direction, and various bio-products that are harmless to human health and environmentally degradable in nature are in the pipeline. Such bio-products are being sourced from natural sources like plants, microorganisms, algae, etc. Of these, algae are comparatively prospective wealthy natural source for producing various high-end value-added products and for other environment friendly applications. Moreover, useful exploitation of algal sources is a gateway for million-dollar business in the years to come as it is ecologically safe with high commercial value. However, presently the algal industry is at infancy which needs boosting both in research and development. As appraised in this chapter, algal bio-products have proved their mettle and worthiness in the fields of pharmaceutical, agriculture, cosmetics, foodstuff, remediation of wastewater, bio-plastics, bio-fuel, animal feed, nutraceuticals, coloring pigments, etc. Further researches and development are imperative for the gradual and continuous exploitation of algal sources for eco-friendly production and application of this viable natural source.

ACKNOWLEDGMENTS

Dr. C.K. Venil thanks the UGC for awarding the Dr. D.S. Kothari Postdoctoral Fellowship (BL/17-18/0479). Also, the authors thank Anna University, Regional Campus – Coimbatore, for providing the necessary facilities to carry out the project work. Professor Laurent Dufossé deeply thank the Conseil Reunion Regional, Réunion island, Indian Ocean, for continuous financial support of research activities dedicated to microbial pigments.

CONFLICTS OF INTEREST

The authors declare no conflict of interest.

FUNDING

This work was supported and funded by the University Grants Commission (UGC), New Delhi, under Dr. D.S. Kothari Post-doctoral Fellowship (BL/17-18/0479) dated 25th September 2018.

KEYWORDS

- **algae**
- **derivatives**
- **eco-friendly**
- **metabolites**
- **nutraceuticals**
- **pharmaceuticals**

REFERENCES

Abdel-Raouf, N., Al-Homaidan, A. A., & Ibraheem, I. B. M., (2012). Agricultural importance of algae. *African Journal of Biotechnology*, *11*, 11648–11658.

Abinandan, S., Subashchandrabose, S. R., Venkateswarlu, K., & Megharaj, M., (2019). Soil microalgae and cyanobacteria: The biotechnological potential in the maintenance of soil fertility and health. *Critical Reviews in Biotechnology*, *39*, 981–998.

Alishah, A. H., Rafiei, N., Garcia-Granados, R., Alemzadeh, A., & Morones-Ramírez, J. R., (2019). Biomass and lipid induction strategies in microalgae for biofuel production and other applications. *Microbial Cell Factories*, *18*(1), 178.

Anbuchezhian, R., Karuppiah, V., & Li, Z., (2015). Prospect of marine algae for production of industrially important chemicals. In: Das, D., (ed.), *Algal Biorefinery: An Integrated Approach.* Springer, Cham.

Andreson, D. K. K., (2010). Dinoflagellates: A remarkable evolutionary experiment. *American Journal of Botany*, *91*, 1523–1534.

Becker, W., (2004). Microalgae in human and animal nutrition. In: Richmond, A., (ed.), *Microalgal Culture Handbook* (pp. 312–351). Blackwell, Oxford.

Begum, H., Yusoff, F. M., Banerjee, S., Khatoon, H., & Shariff, M., (2015). Availability and utilization of pigments from microalgae. *Critical Reviews in Food Science and Nutrition*, *56*, 2209–2222.

Behera, S., Singh, R., Arora, R., Sharma, N. K., Shukla, M., & Kumar, S., (2015). Scope of algae as third-generation biofuels. *Frontiers in Bioengineering and Biotechnology*, *2*, 90.

Bhalamurugan, G. L., Valerie, O., & Mark, L., (2018). Valuable bioproducts obtained from microalgal biomass and their commercial applications: A review. *Environmental Engineering Research*, *23*(3), 229–241.

Bleakley, S., & Hayes, M., (2017). Algal proteins: Extraction, application, and challenges concerning production. *Foods (Basel, Switzerland)*, *6*(5), 33.

Chapman, R. L., (2013). Algae: The world's most important "plants" - an introduction. *Mitigation and Adaptation Strategies for Global Change*, *18*, 5–12.

Chisti, Y., (2013). Constraints to commercialization of algal fuels. *Journal of Biotechnology*, *167*(3), 201–214.

Chittora, D., Meena, M., Barupal, T., & Swapnil, P., (2020). Cyanobacteria as a source of biofertilizers for sustainable agriculture. *Biochemistry and Biophysics Reports*, *22*, 100737.

Costa, J. A. V., Colla, L. M., & Duarte, P., (2003). *Spirulina platensis* growth in open raceway ponds using freshwater supplemented with carbon-nitrogen and metal ions. *Zeits Nstuforsch C-C J. Bioscience*, *58*, 76–80.

Dickinson, S., Mientus, M., Frey, D., Amini-Hajibashi, A., Ozturk, S., Shaikh, F., Sengupta, D., & El-Halwagi, M. M., (2017). A review of biodiesel production from microalgae. *Clean Technologies Environmental Policy*, *19*, 637–668.

Divya, M., & Aanand, S., (2020). Microalgae: A boon for larviculture of aquatic organisms. *International Journal of Applied Research*, *6*, 138–143.

Eroglu, A., & Harrision, E. H., (2013). Carotenoid metabolism in mammals, including man: Formation, occurrence and function of apocarotenoids. *Journal of Lipid Research*, *54*, 1719–1730.

Fong, P., & Paul, V. J., (2011). Coral reef algae. In: Dubinsky, Z., & Stambler, N., (eds.), *Coral Reefs: An Ecosystem in Transition.* Springer, Dordrecht.

Galasso, C., Gentile, A., Orefice, I., Ianora, A., Bruno, A., Noonan, D. M., Sansone, C., Albini, A., & Brunet, C., (2019). Microalgal derivatives as potential nutraceutical and food supplements for human health: A focus on cancer prevention and interception. *Nutrients*, *11*(6), 1226.

Gouveia, L. A. C. O., (2009). Microalgae as a raw material for biofuels production. *Journal of Industrial Microbiology and Biotechnology*, *36*, 269–274.

Griffiths, M., Harrison, S. T. L., Smit, M., & Maharajh, D., (2016). Major commercial products from micro- and macroalgae. In: Bux, F., & Chisti, Y., (eds.), *Algae Biotechnology: Green Energy and Technology*. Springer, Cham.
Guedes, A. C., Amaro, H. M., & Malcata, F. X., (2011). Microalgae as sources of carotenoids. *Marine Drugs*, *9*(4), 625–644.
Hannon, M., Gimpel, J., Tran, M., Rasala, B., & Mayfield, S., (2010). Biofuels from algae: Challenges and potential. *Biofuels*, *1*(5), 763–784.
Harel, M., & Clayton, D., (2004). *Feed Formulation for Terrestrial and Aquatic Animals*. US Patent 20070082008 (WO/2004.080196).
Hochman, G., & Ziberman, D., (2014). Algal farming and its bio-products. In: McCann et al., (ed.), *Plants and Bioenergy: Advances in Plant Biology*, (49–64). Springer Science + Business Media, New York.
Hochman, G., Trachtenberg, M. C., & Zilberman, D., (2014). Algae crops: Co-production of algae biofuels. In: Dierig, D., & Cruz, V. M., (eds.), *Industrial Crops: Breeding for Bioenergy and Bioproducts.* Springer Science, New York.
Joshi, S., Kumari, R., & Upasani, V. N., (2018). Applications of algae in cosmetics: An overview. *International Journal of Innovative Research in Science, Engineering and Technology*, *7*, 1269–1278.
Kelman, D., Posner, E. K., McDermid, K. J., Tabandera, N. K., Wright, P. R., & Wright, A. D., (2012). Antioxidant activity of Hawaiian marine algae. *Marine Drugs*, *10*, 403–416.
Khan, M. I., Shin, J. H., & Kim, J. D., (2018). The promising future of microalgae: Current status, challenges, and optimization of a sustainable and renewable industry for biofuels, feed, and other products. *Microbial Cell Factories*, *17*(1), 36.
Kilinc, B., Cirik, S., Turan, G., Tekogul, H., & Koru, E., (2013). Seaweeds for food and industrial applications. In: Muzzalupo, I., (ed.), *Food Industry*. IntechOpen, London.
Kligerman, D. C., & Bouwer, E. J., (2015). Prospects for biodiesel production from algae-based wastewater treatment in Brazil: A review. *Renewable and Sustainable Energy Reviews*, *52*, 1834–1846.
Koyande, A. K., Chew, K. W., Rambabu, K., Tao, Y., Chu, D. T., & Show, P. L., (2019). Microalgae: A potential alternative to health supplementation for humans. *Food Science and Human Wellness*, *8*, 16–24.
Krajčovič, J., Vesteg, M., & Schwartzbach, S. D., (2015). Euglenoid flagellates: A multifaceted biotechnology platform. *Journal of Biotechnology*, *202*, 135–145.
Kristiansen, (2009). Chrysophytes-golden algae. *Encyclopedia of Inland Waters*, 123–129.
Lee, Y. K., (2001). Microalgal mass culture systems and methods: Their limitation and potential. *Journal of Applied Phycology*, *13*, 306–315.
Li, K., Liu, Q., Fang, F., Luo, R., Lu, Q., Zhou, W., Huo, S., et al., (2019). Microalgae based wastewater treatment for nutrients recovery: A review. *Bioresource Technology*, *291*, 121934.
Liu, L., Pohnert, G., & Wei, D., (2016). Extracellular metabolites from industrial microalgae and their biotechnological potential. *Marine Drugs*, *14*(10), 191.
Liu, Z., Zhang, F., & Chen, F., (2013). High throughput screening of CO_2-tolerating microalgae using GasPak bags. *Aquatic Biosystems*, *9*(1), 23.
López, G., Yate, C., Ramos, F. A., Cala, M. P., Restrepo, S., & Baena, S., (2019). Production of polyunsaturated fatty acids and lipids from autotrophic, mixotrophic and heterotrophic cultivation of *Galdieria* sp. strain USBA-GBX-832. *Scientific Reports*, *9*, 10791.
Maddodi, N., Jayanthy, A., & Setaluri, V., (2012). Shining light on skin pigmentation: The darker and the brighter side of effects of UV radiation. *Photochemistry and Photobiology.*, *88*(5), 1075–1082.

Metsoviti, M. N., Papapolymerou, G., Karapanagiotidis, I. T., & Katsoulas, N., (2019). Comparison of growth rate and nutrient content of five microalgae species cultivated in greenhouses. *Plants (Basel, Switzerland)*, *8*(8), 279.
Metting, F. B., (1996). Biodiversity and application of microalgae. *Journal of Industrial Microbiology and Biotechnology*, *17*, 477–489.
Mezzomo, N., & Ferreira, S. R. S., (2016). Carotenoids functionality, sources and processing by supercritical technology: A review. *Journal of Chemistry*, *2016*, 3164312.
Milledge, J. M., (2011). Commercial application of microalgae other than as biofuels: A brief review. *Reviews in Environmental Science and Biotechnology*, *10*, 31–41.
Mirza, U. K., Ahmad, N., & Majeed, T., (2008). An overview of biomass energy utilization in Pakistan. *Renewable and Sustainable Energy Reviews*, *12*, 1988–1996.
Morgan-Kiss, R. M., Priscu, J. C., Pocock, T., Gudynaite-Savitch, L., & Huner, N. P., (2006). Adaptation and acclimation of photosynthetic microorganisms to permanently cold environments. *Microbiology and Molecular Biology Reviews*, *70*(1), 222–252.
Pal, P., Chew, K. W., Yen, H. W., Lim, J. W., Lam, M. K., & Show, P. L., (2019). Cultivation of oily microalgae for the production of third generation biofuels. *Sustainability*, *11*, 5424.
Panahi, Y., Yari, K. A., Sahebkar, A., & Heidari, H. R., (2019). Impact of cultivation condition and media content on *Chlorella vulgaris* composition. *Advanced Pharmaceutical Bulletin*, *9*(2), 182–194.
Pangestuti, R., Siahaan, E. A., & Kim, S. K., (2018). Photoprotective substances derived from marine algae. *Marine Drugs*, *16*(11), 399.
Peltomaa, E., Johnson, M. D., & Taipale, S. J., (2017). Marine cryptophytes are great sources of EPA and DHA. *Marine Drugs*, *16*(1), 3.
Priyadarshani, I., & Rath, B., (2012). Bioactive compounds from microalgae and cyanobacteria: Utility and applications. *International Journal of Pharmaceutical Scientific Research*, *3*(10), 4123–4130.
Rahman, A., & Miller, C. D., (2017). Microalgae as a source of bioplastics. *Algal Green Chemistry: Recent Progress in Biotechnology*, 121–138.
Rampelotto, P. H., (2013). Extremophiles and extreme environments. *Life (Basel, Switzerland)*, *3*(3), 482–485.
Rasul, I., Farrukh, A., Myuhammed, H., Siddique, Saima, M., Azhar, R., Anam, M., Muhammed, A., et al., (2017). *Algal Biotechnology: A Green Light for Engineered Algae*, 301–334.
Ribeiro, F., Gallego-Urrea, J. A., Jurkschat, K., Crossley, A., Hassellov, M., Taylor, C., Soares, M. V. M., & Loureiro, S., (2014). Silver nanoparticles and silver nitrate induce high toxicity to *Pseudokirchneriella subcapitata*, *Daphnia magna* and *Danio rerio*. *Science of the Total Environment*, *466*, *467*, 232–241.
Ross, M. E., Stanley, M. S., Day, J. G., & Semião, A., (2017). A comparison of methods for the non-destructive fresh weight determination of filamentous algae for growth rate analysis and dry weight estimation. *Journal of Applied Phycology*, *29*(6), 2925–2936.
Salmaso, N., & Tolotti, M., (2009). Other Phytoflagellates and groups of lesser importance. *Encyclopedia of Inland Waters*, 174–183.
Sathasivam, R., Radhakrishnan, R., Hashem, A., & Abd, A. E. F., (2019). Microalgae metabolites: A rich source for food and medicine. *Saudi Journal of Biological Sciences*, *26*(4), 709–722.
Sharif, H. A. B. M., Salleh, A., Boyce, A. N., Chowdhury, P., & Naqiuddin, M., (2008). Biodiesel fuel production from algae as renewable energy. *American Journal of Biochemistry and Biotechnology*, *4*(3), 250–254.

Singh, J. S., Kumar, A., Rai, A. N., & Singh, D. P., (2016). Cyanobacteria: A precious bio-resource in agriculture, ecosystem, and environmental sustainability. *Frontiers in Microbiology*, *7*, 529.

Singh, N. K., Sonani, R. R., Rastogi, R. P., & Madamwar, D., (2015). The phycobilisomes: An early requisite for efficient photosynthesis in cyanobacteria. *EXCLI Journal*, *14*, 268–289.

Sonani, R. R., Rastogi, R. P., Patel, R., & Madamwar, D., (2016). Recent advances in production, purification and applications of phycobiliproteins. *World Journal of Biological Chemistry*, *7*(1), 100–109.

Wang, J., Jin, W., Hou, Y., Niu, X., Zhang, H., & Zhang, Q., (2013). Chemical composition and moisture-absorption/retention ability of polysaccharides extracted from five algae. *International Journal of Biological Macromolecules*, *57*, 26–29.

Wang, S. K., Guo, C., Wu, W., Sui, K. Y., & Liu, C. Z., (2019). Effects of incident light intensity and light path length on cell growth and oil accumulation in *Botryococcus braunii*. *Engineering in Life Sciences*, *19*, 104–111.

Wehr, J. D., (2015). Brown algae. Freshwater algae of North America. *Aquatic Ecology*, 851–871.

Wells, M. L., Potin, P., Craigie, J. S., Raven, J. A., Merchant, S. S., Helliwell, K. E., Smith, A. G., et al., (2017). Algae as nutritional and functional food sources: Revisiting our understanding. *Journal of Applied Phycology*, *29*(2), 949–982.

Zullaikah, S., Utomo, A. T., Yasmin, M., Ong, L. K., & Ju, Y. H., (2019). Ecofuel conversion technology of inedible lipid feedstocks to renewable fuel. *Advances in Eco fuels for a Sustainable Environment*, 237–276.

CHAPTER 3

HALIDE METABOLITES FROM THE ALGAL BIOMASS: A PROMISING THERAPEUTIC FOR TARGETING RARE DISEASES

SUSHIL KUMAR,[1] GOURAV CHOUDHIR,[2] and ANKIT SINGH[2]

[1]*Department of Botany, Shaheed Mangal Pandey Government Girls PG College, Madhavpuram, Meerut–250002, Uttar Pradesh, India*

[2]*Center for Rural Development and Technology, Indian Institute of Technology Delhi, Hauz Khas–110016, New Delhi, India*

ABSTRACT

Secondary metabolites constitute a wide and varied group of organic compounds. All photoautotrophic organisms, from algae to higher plants, synthesize and accumulate different kinds and amounts of these compounds. Marine habitat is full of surprises as it harbors a vast variety of micro and macro-organisms equipped with a rich library of secondary metabolites. Marine algae exhibit antibacterial, antioxidant, anti-inflammatory, antidiabetic, anticancer antiobese, antimalarial, antihyperlipidemia, anticoagulant, hepatoprotective, antiangiogenic, and other activities. Halogenated secondary metabolites are good sources of a large number of pharmaceuticals. Numerous studies are ongoing to find and identify pharmaceutically active products which may be an alternative to synthetic drugs in the near future. The synthetic drugs are expensive along with several side effects on the patient's body. Macroalgae are comparatively more explored than microalgae for halogenated secondary metabolites. Different categories of halides from red algae are sesquiterpenes,

Algal Metabolites: Biotechnological, Commercial, and Industrial Applications. Jeyabalan Sangeetha, Devarajan Thangadurai, Saher Islam, & Ravichandra Hospet (Eds.)

diterpene, triterpenes, C15 acetogenins, indoles, and steroids. In the present chapter, we will briefly go through different marine alga for their halogenated natural products. This is an established fact that halogenation increase bioactivity of secondary metabolites.

3.1 INTRODUCTION

Algae are photoautotrophic, thalloid, and non-embryophytes. Algae commonly inhabit aquatic habitats. A number of terrestrial algal types grow on moist and damp soils. Aquatic species may be either freshwater or marine according to the water they live. Due to the short life span algae multiply rapidly and fix CO_2 efficiently to convert it into biomass. The rate of biomass accumulation is much more than cereal crops on the land. Quantitatively marine representatives are more than freshwater ones. Various species have been utilized by man for different purposes since long back. From prehistoric periods, these are being used for food and fodder. With the dawn of modern civilizations, algae became the inseparable part of life in certain areas of the Globe. Recently, scientists turned their attention towards a vast array of medicinally bioactive natural products produced by this comparatively less explored group of organisms.

Microalgae as well as macroalgae are the producers of many valuable products of industrial commercialization but presently are in their early stages of development. The industry using marine algal products has the potential to become a multibillion industry in the near future. A big part of marine resources is still not even touched or least explored. The majority of algae have been proved to synthesize valuable bioactive organic compounds that bring them in biotechnological and pharmaceutical focus. The research is now aimed at new bioactive metabolites identification, isolation as well as to find their roles in the treatment of various diseases. The market of natural products of pharmaceutical and nutraceutical significance is increasing at a fast pace globally. Algae are rising as a wellspring of promising undiscovered pharmaceutical production. The estimation of algal natural products is high in the market but very low amount produced naturally, hence biotechnological approaches now used to maximize it. Many algal products are commercialized and others are under consideration (Bhattacharjee, 2016). Drug discovery from marine sources is relatively newer than from land sources. The former began with SCUBA (Self-Contained Underwater Breathing Apparatus) diving and new sampling methods in the 1940s. The number of pharmaceutically potential compounds is now more than 28,000 and many are in progress to be discovered per year (Blunt et al., 2015). Regardless of a great number of compounds have been isolated and characterized from marine organisms, only a few members are approved and are in pipeline around the world. Microalgae are poorly explored for their pharmacological activities, but now the attention turned to them to search biotechnological and commercial potentials of microalgae (Mimouni et al., 2012; Guedes et al., 2013; Samarakoon et al., 2013). Brown seaweeds consist of carotenoid, fucoxanthin along with compounds from diatoms have antioxidant, anti-inflammatory, antidiabetic, anticancer, antiobese, antimalarial, and antiangiogenic activities (Peng et al., 2011). Numerous natural products of great economic importance have been obtained from algae. These belong to different categories, e.g., fibers, minerals, polysaccharides, proteins, pigments, vitamins, steroids, lectins, and halogenated compounds (Mariya and Kurup, 2016; Mehta et al., 2016; Oramary et al., 2016). Marine macroalgae as well as microalgae are pharmaceutically more beneficial (Jacob et al., 2013; Raja, 2014). The world market is dominated by pharmaceutical products by 70% to 80%. Algae remained a good source of antibiotics, antiviral substances, antifungal agents, therapeutic proteins, and other drugs for many years (Tomar, 2012; Silva et al., 2013).

Blue-green algae (BGA) are prokaryotic photoautotrophs and are rich in pharmaceutical compounds. *Spirulina* is among the most studied genus from this group and reported to be used in the treatment of diabetes, obesity, cancer, hypertension, HIV, and hyperglycemia. *Spirulina* is known to boost the immunity, especially in response to heavy metal toxicity and drugs. It reduces sugar and lipids levels in blood serum (Colla et al., 2007, 2008). *Nostoc* has been accounted for its nutritional and pharmaceutical properties used in the treatment of fistula and cancer. *Nostoc* reported to enhance immunity and control blood pressure. Bioactive ingredients are proteins and PUFAs. *Chlorella*, a green alga in addition to its nutritional value being rich in proteins, minerals, and vitamins also have antioxidant, anticancer, antibacterial antihyperlipidemia and anticoagulant activities. *Chlorella* is effectively used as hepatoprotective and immunity stimulating activities. *Dunaliella*, a unicellular green alga is a significant antifungal and antibacterial activity.

Recently the interest in marine natural products resulting in the isolation, structure elucidation and sorted for their pharmacological properties. From time to time new compounds have been discovered and tested to evaluate its efficiency in pharmacological tests. Many compounds have been extracted from algae with different activities responsible for medicinal use.

Phycobillins, polyphenols, and vitamins present in algae are the basis of antioxidant properties. These compounds bear free radical and free oxygen species scavenging abilities. The antioxidant properties are well known to prevent mutation in the genetic material and in turn prevent cancer in the premalignant state (Bisen, 2016). Green algae mostly filamentous have been reported for their great antioxidant activities. Many marine genera with this property are now used commercially (Mutafova et al., 2016).

3.2 RARE DISEASE

The World Health Organization (WHO) identifies rare diseases (RD) at a prevalence rate of 0.6%–1%, which can be enfeebling or life-threatening. With ever increasing awareness about RD, it was felt that there is need to establish a centralized body which can strengthen, support, and improve international collaboration on RD and therefore in 2011 the International Rare Diseases Research Consortium (IRDiRC) was established. RD is believed to affect about 10% of the population globally, but the definition and categorization may vary depending on the region. For example, the prevalence rate for the United States, European Union (EU), Australia, and Japan is 0.75%, 0.5%, 0.11%, and 0.4%, respectively. The early-onset and fatality rate of 8% leads to the death of 50% of children effected by RD while 50% die before the age of 5 (Song et al., 2012). As per the definition by the United States under the Rare Disease Act (2002), rare disease is a collective term for diseases that affects less than 200,000 people among the entire resident population. It is estimated that about 30,000,000 patients are currently living with a rare disease in the United States of America (Whicher et al., 2018) with almost equivalent number in the EU (European Medicines Agency, 2018). RD lead to adverse effect on the physical function, cognitive, and communication impairments along with emotional and social isolation, which affects the quality of life (QoL) of the person affected by it. The patients become dependent on families and caregivers for management of their everyday life due to the challenges associated with the diseases. When we consider the effects of the disease on people associated with the patients, the number of people directly affected by these diseases rises drastically by a multiplier of 3–4 people. The strong public stigma associated with health-related issues creates the fear of rejection among people associated with it and a sense of secrecy. This adversely affects psychological, emotional, and physical development of patients in the absence of peer group interaction (Zhu et al., 2017). Although the probability of some diseases is higher in certain populations or some countries, but most of them are not recognized by geographical, historical, or political borders. Since many disorders are covered under the umbrella term of RD, often, it is felt that there is a lack of information about some of the diseases. In the absence of clearly defined symptoms of rare disease, many of them go unidentified. It was found that around 7,000 RD affect the human in different ways that depend on the age and gender of the patients. The patients with RD suffer with consequences of the conditions which could go utterly unrecognized due to the lack of knowledgeable physicians or face very high costs for disease-specific medication. In an estimated patient population of 100,000 US adults with AATD disease which is a condition associated with severe deficiency of alpha-1 antitrypsin, fewer than 10,000 patients have been correctly identified till now (Aboussouan and Stoller, 2009; Stoller and Brantly, 2013). This kind of anonymity for RD is a global phenomenon wherein patients visit multiple doctors with compounding symptoms and delayed diagnosis. Reportedly, 43% of patients reported consulting at least 3 doctors before initial diagnosis, while 12% of patients visited between 6 and 12 doctors (Aboussouan and Stoller, 2009). Treatment of the rare disease is very costly because of a lack of information and medication. For example, the annual cost of treatment of AATD disease falls between US$128,856 to US$135,408, as four argumentation therapy drugs currently available cost between $0.59 to $0.62/mg (Stoller, 2018). The same is the case of patients with LAM disease on regular 2 mg per day dosage of rapamycin with one tablet costing about $31.50, which takes the annual cost up to US$11,497.50. In United States, patients without medical insurance cannot afford treatment due to high cost associated with it while the claims for reimbursement even for medication by patients on high deductible insurance plans takes away large part of deductibles (Stoller, 2018). The patients in Germany, the United Kingdom, France, Switzerland, and several countries of the Scandinavian region have easy reach to many medicines in a short time. However, there is a huge disparity in drug availability and expenditure between different regions of the EU. The lesser developed countries such as Hungry, Greece, and the Czech Republic face difficulties in availing these medicines to their residents while Germany, Switzerland, and France invest heavily on medicines for RD (Deticek et al., 2018). Orphan drug is the collective term used to categorize medicines which are needed for treatment RD. Most of the pharmaceutical companies shy away from investing money in the discovery of these drugs due to several challenges associated with it. The main deterrent is the need for huge investment and risks involved in the formulation of medicines catering to the small potential market. There is insufficient knowledge of pathophysiological mechanisms due to the rarity of the disease and lack of funding. Thus, the target sites for potential diagnosis or treatment drugs are not known, which proves to be another hindrance. Etiological studies require continued, multicentric collaboration with surveys covering large demographic samples, which in the case of RD, prove to be challenging due to small patient populations. The existing model for drug development might not be most suitable for potential medicine for RD. It can be emphasized from the fact that in the last 10–15 years of study, which cost more than 1 billion US Dollars, only 5 of 5,000 experimental compounds could pass through pre-clinical trials, and only 1 got FDA approval for human usage (Groft, 2013) (Table 3.1).

A global effort to generate products that can prevent, identify, and treat rare diseases (RD) is the need of the hour. One of the revitalizing strategies would be cross-linking information acquired from studies on RD to improve our understanding of common diseases. A faster drug development plan could be set up that can access resources from a wide array of databases which would help to strengthen various collaborative efforts. Another countermeasure to deal with issues related to lack of funding would be allotment of funds and support systems which are earmarked for rare disease studies. These funds could be directly linked to the organization working on RD such as Alpha-1 Foundation, HHT Foundation, Hermansky-Pudlak Syndrome Network, Cystic Fibrosis Foundation, LAM Foundation and National Organization for Rare Disorders. Although there are several challenges associated with RD, collaboration, innovation, and resilience of patients, government, support groups, scientists, and commercial stakeholders can go a long way in overcoming these odds.

As per the experts (Austin et al., 2018) in the field of RD, the near future (2017–2027) objectives of IRDiRC should be to:

- Detection of all susceptible cases of RD within a year of notification to the existing medical system. Engagement of currently undiagnosable patients within a coordinated diagnostic and research scenario functioning at a global scale.

  The main logistical hindrances for completing this objective are lack of awareness among public and general physicians, ineffective referral system within medical community, and challenges associated with the exchange of knowledge about diagnostic tools and data between practitioners and researchers worldwide. Scientists have been able to sequence the coding region of the genome which has revealed information about several known diseases. However, there is lack of information about the functioning of non-coding regions and

TABLE 3.1 Important Rare Diseases, Their Causes, Size of Population Affected and Common Symptoms

Name of Disease	Factors Responsible	Population Affect	Symptoms
Hutchinson-Gilford progeria	Gene mutation LMNA locus	1 person among 4 million births	Postnatal growth retardation, midface hypoplasia, micro-gnathia, premature atherosclerosis and coronary artery disease, absence of subcutaneous fat, alopecia, generalized osteodysplasia with osteolysis and pathologic fractures
Fields' disease	–	1 in 3.75 billion	Neuromuscular disease muscular degeneration
Fibrodysplasia ossificans progressiva	Mutation in activin receptor IA/ activin-like kinase-2 (ACVR1/ALK2)	1 in 2 million	Skeletal muscle and connective tissue, tendons, and ligaments
Microcephaly	Deletion of multiple gene or deletion of a single gene	2 to 12 babies per 10,000	Less develop the brain, reduced head size
Von Hippel-Lindau	A germline mutation of the VHL tumor suppressor gene	1 in 36,000 births	Arising of the tumor at multiple organs
Paraneoplastic pemphigus	Somatic gene mutation in B-lymphocytes	Data not available	An autoimmune disorder stemming from an underlying tumor that leads to blistering of the skin and mucous membranes
Morgellons	Sores with fibrous material	Data not available	Presence of fibers underneath, embedded in, and erupting from unbroken skin or slow-healing sores

structural variants related to disease associated with them (Boycott et al., 2017). The results of whole-exome sequencing (WES) have revealed that there is an array of disease genes and variants have yet not been discovered and more efforts are required. We need methods that are complementary to WES, such as whole-genome sequencing (WGS), long read technologies and customized transcriptomes that can more effectively bind to targeted non-coding regions and structural variants. The extended time gap between primary detection and the start of treatment can be shortened by easing the availability of detailed information about sub-specialist medical support and diagnostic laboratories along with efforts to improve the affordability of diseases sequencing and analysis costs. There is a need for several studies which can effectively analyze robust data collected about diagnostic services, clinical services, and their cost-effectiveness to facilitate reimbursement of sequencing-based diagnosis by insurance companies (Deticek et al., 2018; Stoller, 2018). We need a global scale co-operation of clinical and research laboratories to overcome issues of undiagnosed diseases. The patients should be informed about the need/importance of medical data sharing related to rare diseases, and their consent should be taken from the outset of the clinical testing process. There is a need to establish biorepositories where samples which were collected for sequencing and other genome analysis can be stored for future use. Some examples of these models can be undiagnosed diseases program (UDP) which was designed by National Institute of Health (NIH) in the United States, followed by Undiagnosed Diseases Network International (UDNI) on international level (Taruscio et al., 2015). More collaboration is needed among various national programs, such as the Japanese Initiative on Rare and Undiagnosed Diseases (IRUD) which can be based on data sharing to the global network for cases where no diagnosis could be done after initial sequencing. This data can be analyzed by global experts and results could be shared with the aim of facilitating diagnoses of RD for patients all over the world. It is also felt that there is need of educating medical expertise about RD to encourage better engagement with patients. They can be educated through various courses about RD and new diagnostic methods which are available for various targeted end-users such as doctors, patients, etc., with varying level of information at the outset. Patients and their caregivers should also be informed about scientific aspects of the disease by experts in the field such researchers and clinicians which would help in better disease management on their part. All the information about newly reported diseases and genes responsible for them should be maintained in databases such as online mendelian inheritance in man (OMIM) and Orphanet. Surveys by specialty doctors, diagnostic labs, supporting organization and more targeted sampling based on information from RD network would help to shorten the time period between diagnosis and treatment initiation.

- Increases the number of available therapies by 1,000 which would be focused on RD without approved options.

Since 2010, IRDiRC has kept an eye on number of orphan medicinal products (OMPs) which have received approval in EU and the United States and reportedly more than 220 OMPs have received their first approval for a new indication. Although these numbers are significant, we cannot ignore the fact that thousands of patients suffering with one or another form of rare disease are still waiting for approval of a therapy which specifically caters to their needs. In order to exponentially increase the number of approved therapies globally, we need redesigned innovative methodologies for clinical trials, collection of data/specimen and its storage, better maintenance of natural history studies, more co-ordination between various players involved and informed clinical endpoints. Most of the drugs developed in recent times were often clustered around similar technologies or therapeutic approaches which might soon be utilized to their maximum capacity. There are chances that a market gap would be created in the future when these easily developable indicators addressed by traditional approaches have been fully explored, leaving behind more complicated drug targets and technologies that have yet not been investigated. It is suggested that IRDiRC objective of developing 1,000 new therapies within the next decade can be reached only via significant increment in R&D efforts which should triple the current rate of production with compounded annual growth rate of about or above 10%. Significant advancements can be achieved in the therapeutic development process by adopting radically new approaches which are based on common standards across distinct research fields, exchange of knowledge about best practices, creation of sustainable business models and modifications in regulatory law governing the therapies. We need new methodologies which can help to streamline drug development which encompasses early-stage improvements such as increased efficiency of data collection and sharing, improvements in our knowledge about progression of disease and phenotypic expression of related genes, better approaches for small scale clinical trials and improved safety/efficacy in methods for pre-clinical assessment. Additional steps during late-stage advancements can include better definition of endpoints which is more suitable for large scale analyses of benefits for patients; provide medical relevance, generation of regulatory benefits/risks evidences and quantification of commercial importance of product for buyer, companies involved and society as whole. To develop a full understanding of the disease and better reorganization of endpoints associated with it, better integration of patients during clinical research, drug development and its evaluation is important. The importance of this collective knowledge, endowment, encouragement, and participation cannot be underestimated. Equalized contribution from all regulators through interaction is essential in ensuring that the developed product meets all the regulatory criteria which in turn would improve the market outreach of the product. This could be achieved through regular interaction between regulators and help with protocols for product development. As per the predictions, with the increase in a number of applications seeking assistance with protocol a number of changes would have to be implemented such as streamlining the approval process, more collaboration between regulators functioning in various jurisdictional areas leading to creation of uniform review process, improved human resource and training programs and updated knowledge of regulatory laws. Apart from all the measures mentioned above, it is essential to encourage changes in the current drug development scenario such as development of a new model of sharing risk and incentives between all the partners involved, i.e., public and private partners, redesigning of existing agents and development of more flexible regulatory laws.

- Development of methodologies to analyze the effects of diagnoses and argumentative therapies on patients.

The methods currently being used for impact assessment of newly developed diagnosis and therapies are highly substandard irrespective of the organization responsible for it. Therefore, efforts are required to develop improved impact assessment methodologies and tools. It can be done by more engagement with patients from under-developed areas in the world which would expand the global outreach of the program and improve inclusion of underrepresented demographics. The use of pre-existing databases such as NIH Genetic Testing Registry and RARE-Best practices could go a long way in developing rigorous processes and qualitative markers for evaluation. More investment is required on the part of funders which can help to tackle the complexities associated with the measurement of health outcomes.

3.3 HALIDES COMPOUND FROM ALGAE

This is well-known fact that secondary metabolites are not involved in normal vital activities but frequently contribute to the survival of marine algae equally to that of the primary metabolites. Certain secondary metabolites are equally important as the primary metabolites in algae. Halogenated secondary metabolites or halides are the major players of medicinal properties exhibited by marine algae. Halogenation of secondary metabolites increases their biological activity leading them to effective drugs. An enormous number of organohalides metabolites with commercial potential have been produced by marine algae. The chemical nature of halogenated compounds isolated showed great variation from simple to complex organic molecules. These molecules range from simple aliphatic linear chain compounds to cyclic and aromatic molecules. Pharmaceutical utilization of halogenated mixes have been begun in the ongoing past. Although, other potentials like food explored more. Antifouling properties well explored from the long past. In the last few decades many halogenated compounds have been discovered with their potential as new drugs is an open area of research. A great number of marine algal species are poorly explored. Macroalgae are comparatively more explored than microalgae or phytoplanktons for halogenated secondary metabolites. Phytoplanktons make the first trophic level, i.e., producers in the marine food chain and promising materials as they exhibit quick adaptation in the changed environment. Due to frequent changes in their surroundings, they produce an array of secondary metabolites to adapt fast (Cabrita et al., 2010).

Biosynthesis of organohalogens is an inherent property of various organisms more than 5,000 such compounds have been discovered (Gribble, 2018). These belong to different categories of organic compounds, viz. simple phenolic or aliphatic, polyketides, and oligopeptides. Correspondingly the biological activities are also diverse. Numerous terrestrial life forms have been accounted for to produce a number of halogenated compounds. Halogenated secondary metabolites from terrestrial organisms are known to be chlorinated, while marine organisms produce largely brominated secondary metabolites irrespective of the amount of chlorides (559 mM) as compared to bromides (0.86 mM) in marine habitats. This is due to comparatively easy bromide ions oxidation of into bromonium ions. Bromonium ions undergo electrophilic addition to unsaturated hydrocarbons, i.e., alkenes and aromatic ring skeletons. Fluorine containing organohalides are rarely present (Faulkner, 1995; Gribble, 2003; Anderson and Chapman, 2006; Naismith, 2006; Vetter, 2006; Kulik et al., 2009). Almost biologically produced organohalogens are derived from marine organisms. Probably serve as defense agents, e.g., antibacterial and antifeedant (Gribble, 2003). Among marine seaweeds red algae contain the highest degree of organohalogens synthesis and storage. The variety of halogenated metabolites was earlier used as signatures for chemotaxonomic studies to establish relationships among species. Later on, this was reported that a seasonal variation occurred in the chemical composition and restricted the chemotaxonomic value of organohalogens.

Iodinated secondary metabolites are more common in brown algae. Bromine or chlorine-containing halides in brown, green, and red algae are less than 1%, 7%, and 90% of their secondary metabolites respectively (Harper et al., 2001). Current research has been focused in searching, isolation, and characterization of novel organohalides from marine algal genera. This interest mainly evolved probably due to the potential of halogenated compounds in drug discovery and biotechnological applications. Halogenated natural compounds are not considered good signatures taxonomically due to seasonal and geographical variations in the same taxa (Fenical and Norris, 1975).

3.4 STRUCTURAL AND FUNCTIONAL DIVERSITY OF HALIDES COMPOUND FROM ALGAL BIOMASS

Marine algae produce a collection of many simple and complex halides, probably for defense purposes. Simple organohalides like chloromethane, bromomethane, and iodomethane have been discovered from phytoplanktons (Scarratt and Moore, 1996). This is believed that the peculiar odor of ocean water is largely because of the presence of organohalogens secreted by marine algae. Major examples of such simple organohalogens from different algal genera can be illustrated as $CHBr_3$, CH_2Cl_2, $CHCl_3$, CH_3Br, CH_3I, CH_2Br, CH_2BrI, CH_2I_2, CHI_3, $CHCl_2Br$, $CHBr_2I$, $CHClBr_2$, $CHClBrI$, $CHBrI_2$, CH_3CH_2I, CH_3CH_2Br, $CH_3CH_2CH_2Br$, $CH_3CH_2CH_2I$, $(CH_3)_2CHI$, etc. (Gribble, 2018).

Although, the number of halogenated secondary metabolites from brown algae is relatively lesser than red algae but show great variety. Halogenated alkanes reported from *Cystoseira barbata* may be monosubstituted (CH_3I and CH_3Br), poly-hetero substituted ($CHBr_3$ and $CHBr_2I$) (Nightingale et al., 1995; Milkova et al., 1997). In Laminariales about 90% out of the total iodine in the form of hormone-like tyrosine derivatives monoiodotyrosine and diiodotyrosine (Hou et al., 1997). Halogenated phenolics are of two types phloroglucinols and phlorotannins. From *Eisenia arborea* bromophloroglucinol and iodophloroglucinol have been reported (Glombitza, 1985). Monomeric bromophenols have been reported to be released by Laminariale *Ecklonia kurome* and *Eisenia bicyclis* (Shibata et al., 2006).

A phlorotannin bromotriphlorethol has been isolated from *Cystophora congesta* (Koch and Grigson, 1984). At least 17 brominated, chlorinated, and iodinated phlorotannins were isolated and characterized from *Cystophora retroflexa* (Sailor and Glombitza, 1999) and *Carpophyllum angustifolium* reported to have chlorophloroglucinol, iodophlorethol, chlorodifucol, and chlorobifuhalol (Glombitza and Schmidt, 1999). Phloroglucinols and their derivatives have also been isolated from *Eisenia arborea*, *Analipus japonicus* and *Colpomenia sinuosa* by different workers. Brown algae *Egregia menziesii* yields three oxylipins which are derived from 18 and 20 carbon PUFAs (Todd et al., 1993). From *Padina tetrastromatica*, a halogenated nor-sesquiterpene and an iodine-containing meroterpene from *Ascophyllum nodosum* are important halogenated terpenes belonging to brown algae (Parameswaran et al., 1994; Arizumi et al., 1994). Halogenated meroterpenoid, fallachromenoic acid has been isolated from *Sargassum fallax* (Reddy et al., 2009). Two novel selinane brominated sesquiterpenes, 9-bromoselin-4(14),11-diene and 1-bromoselin-4(14),11-diene have been extracted from *Dictyopteris divaricata* (Kladi, 2008). Halogenated secondary metabolites are reported from a variety of blue green algae. *Lyngbya majuscula* and *Nostoc linckia* are the most prolific genera (Sailler and Glombitza, 1999).

Genus *Laurencia* belongs to the order Ceramiales and family Rhodomelaceae. It includes about 146 taxonomic species and inhabited temperate, subtropical, and tropical coastal waters. *Laurencia* have been extensively explored for their halogenated natural products categorized as sesquiterpenes, diterpene, triterpenes, and C15 acetogenins. Sesquiterpenes are more prevalent from different species of *Laurencia*. Besides, there are various halogenated metabolites represented by lesser numbers including aromatic compounds, indoles, and steroids (Wright, 2003; Suzuki et al., 2009). Due to the luxuriant growth of red algae in tropical and temperate regions, these are more rapidly investigated for their natural products, including halogenated secondary metabolites. Red algae are represented by over 4,000 species all around the world (Rovirosa et al., 2013). Red algae are in fact the prominent producers of organohalides out of all marine algal groups. *Polysiphonia*, *Rhodomela*, and *Symphyocladia* are the genera from the same family reported halogenated phenols/aromatics (Wang et al., 2013). *Laurencia* is being extensively screened; new halogenated compounds remain adding day by day. Mainly terpenes have been isolated and screened for their pharmaceutical effects. The extent of halogenation is relatively high in *Laurencia* (Hill, 2007). From *Lauenrecia glandulifera* five chlorinated C15 acetogenins were isolated and characterized while a halogenated sesquiterpene alcohol, elatol was first time reported from *Laurencia microcladia* which shows potent antibacterial activity (Kladi et al., 2009; Lhullier et al., 2009). *Laurencia glandulifera* collected from the island of Crete in South Greece has reported new glandulaurencianol A and B are brominated diterpenes with a new related non-brominated analogue (Kladi et al., 2014). Ji et al. (2009) reported, "*Laurencia saitoi* has been reported to produce halide sesquiterpenes 2,10-dibromochamigra-2,7-dien-9-ol, 10-bromo-3-chloro-2,7-epoxychamigr-9-en-8-ol, (9*R*)-2-bromo-3-chloro-6,9-epoxybisabola-7(14),10-diene and (9*S*)-2-bromo-3-chloro-6,9-epoxy-bisabola-7(14),10-diene." Same species reported novel nor-sesquiterpenes and three sesquiterpenes named as 2-bromo-ionone, 2-hydroxyluzofuranone, 2-hydroxyluzofuranone B and 4-hydroxypalisadin C. Furthermore, other brominated compounds were at that point revealed for various macroalgal species, in particular aplysistatin, 5-acetoxypalisadin B, palisadin A, palisadin B and 2,3,5,6-tetrabromoindole (Su et al., 2009). Two new halogenated secondary metabolites iso-obtusol and elatol were reported from *Laurencia majuscula.* Both screened for their antibacterial activities (Vairappan, 2003).

Literary survey indicates that numerous *Laurencia* species are known to commonly produce chamigranes, a group of sesquiterpenes and similar metabolites (Harizani, 2016). Four species namely *Laurencia nipponica, L. nidifica, L. obtusa* and *L. majuscula* are rich sources of halochamigranes. In number most of the halochamigrane sesquiterpenes reported are bromochlorinated followed by brominated and a few are chlorinated. *Laurencia nipponica* is reported to produce halogenated secondary metabolites. Chamigrane type sesquiterpenoids, for example, halochamigrene epoxide, prepacifenol, and C15 bromoethers such as "laurencin, laureatin, isoprelaurefucin, epilaurallene, and kumausallene." These seven sorts of secondary metabolite production continued as before in the wild and under different culture conditions (Masuda, 1997). If combined, lauranes, cyclolauranes, and other similar halogenated secondary metabolites make the second largest category of sesquiterpenes characterized in *Laurencia* species. These compounds may accumulate in the grazers and can be isolated from them. Bisabolanes and related metabolites were primarily reported from higher plants and made a large class of sesquiterpenes from marine algal forms. Until now, 72 bisabolanes and similar compounds have been identified from *Laurencia* and their consumers. Perforanes, cycloperforanes are a fairly large class of sesquiterpenes. Brasilane skeleton containing sesquiterpenes have also been reported from different species of *Laurencia*. Cuparane sesquiterpenes constitute a small group of sesquiterpenes and reported from the following five species of *Laurencia: L. implicata, L. glandulifera, L. majuscula, L. okamurai* and *L. microcladia.* About 14 sesquiterpenes cuparane have been separated from the genus *Laurencia*. In 1975, *Laurencia glandulifera* was reported to produce α-bromocuparene and its isomer α-isobromocuparene. These two were the first cuparanes sesquiterpenes to be reported from marine algal forms. Eudesmanes also called selinanes and cycloeudesmanes are common occurrences in terrestrial plants. Till date, 21 such secondary metabolites isolated from *Laurencia* spp. (Harizani, 2016).

Diterpenes are less prevalent as compared to sesquiterpenes. Different *Laurencia* species have been reported to produce 133 diterpenes. Around 33 among them are pargueranes, isopargueranes, and similar diterpenes. Second frequently found diterpene groups include the halogenated secondary metabolites such as labdanes, pimaranes, and similar organic compounds. Several *Laurencia* species has been reported to produce about 23 diterpenes so far. All were found brominated at C-3. Laukarlaol and neoconcinndiol hydroperoxide are the only exceptions. Irieanes and neoirieanes make the third category of diterpenes reported from *Laurencia* species and their grazers. Around 21 such diterpenes have been isolated so far. The chemical structure of irieane and neoirieane are not reported from other sources; in this way, these are unique to *Laurencia* and their feeders. A halogenated diterpene, dactylomelol was isolated from *A. dactylomela* (Estrada et al., 1989). Bicyclic obtusane skeleton was first reported in obtusadiol isolated from *Laurencia obtusa* (Howard and Fenical, 1978). New halogenated acetogenins isolated from *Laurencia obtusa* containing two tetrahydropyrans each connected by a methylene to a tetrahydrofuran ring (Esselin, 2018).

Blunt et al. (1978) collected *Laurencia thyrsifera* from Seal Reef-Kaikoura, New Zealand and isolated thyrsiferol. Since then, many compounds with similar structure and properties have been reported by different researchers. Specifically, 50 triterpenes and other related metabolites with a 2,7-dioxabicyclodecane group have been reported from *Laurencia* species and the ocean bunnies devouring them up until now. The bromochlorinated intricatetraol, at first reported from *Laurencia intricata* gathered at various locations close to Hokkaido in Japan, is the main cause of a triterpene with a C-2 balanced structure (Suzuki, 1993). *L. tristicha* collected from Taiwan, studied for its halogenated natural products nine new along with an equal number of previously known sesquiterpenes have been isolated and characterized. The sesquiterpenes belongs to chamigrane-type, cuparane-type, and halogenated chamigrene derivatives (Chen, 2016).

Red algae under genus *Laurencia* are the source of non-terpenoid C15 acetogenins, polyketides emerging most likely from a C16 unsaturated fatty acid. Various chemical researches on different species of *Laurencia* revealed the presence of 244 acetogenins and chemically related compounds. Certain mollusks feeding on these algae also found to accumulate such acetogenins. By and large, these are brominated, much of the time bromochlorinated or dibrominated, while it isn't unprecedented to have at least three halogen molecules, and just only here and there do they need halogen atoms. Presently, 65 acetogenin compounds are known to bear five-member cyclic ethers moieties have been reported from *Laurencia* species (Harizani, 2016). It is significant that the structures of elatenyne, investigated from *Laurencia elata* and *Laurencia decumbens* (Hall and Reiss, 1986; Ji et al., 2007), *L. decumbens* reported the presence of laurendecumbenyne (Ji et al., 2007) and compound containing a 2,7-dioxabicyclo[4.4.0]decane skeleton was obtained from *Laurencia majuscula* (Wright, 1993). The class of halogenated secondary metabolites represented maneonenes and isomaneonenes. The first compounds of this category were reported from *Laurencia nidifica* by Erickson et al. (1976). The metabolites include (Z)-maneonenes A, B, and C, isomaneonenes A and B and (E)-maneonene B. Maneonenes and isomaneonenes differ structurally as Maneonenes have a chlorine and bromine atoms at C-5 and C-13 respectively, while isomaneonenes consists of two bromine atoms at C-12 and C-13 (Sun et al., 1976; Waraszkiewicz et al., 1978). *L. microcladia* collected from the French coast was reported to produce microcladallenes A, B, and C. All the three microcladallenes acetogenins have a bromoallene group at the end and a bromine atom at C-12 (Kennedy et al., 1984). A novel tricyclic acetogeninpoitediene with a cis-dibromo-olefin functional group at the end was reported from *Laurencia poitei* (Wright et al., 1983).

Sun et al. (2013) isolated similisines A and B from *L. similis*. Similisines A and B are enantiomers and polybrominated spiro-trisindoles. Kamada and Vairappan (2015) collected *L. nangii* Lohok Butun (Malaysia) and isolated a novel sesquiterpene named neolaurene with five already known compounds. Previously known compounds were 2,10-dibromo-3-chloro-α-chamigrene, neolaurallene, deoxyprepacifenol, intricatetraol, and cycloelatanene B (Kamada and Vairappan, 2015). Marine red alga *Laurencia nangii* isolated and characterized two novels halogenated nonterpenoids C15 acetogenin snangallenes A and B along with two already isolated halogenated secondary metabolites named 2,10-dibromo-3-chloro-α-chamigrene and itomanallene A (Kamada et al., 2019b). *Laurencia majuscula* investigated for halogenated secondary metabolites and resulted in the identification of three novel tricyclic sesquiterpenes rhodolaurenones B-C, omphalaurediol along with nine previously isolated such as rhodolaureol, rhodolaurenone A, isorhodolaureol, (−)-laurencenone D, cartilagineol, elatol, (+)-deschloroelatol, 2-chloro-3-hydroxy-α-chamigren-9-one and (+)-laurencenone B (Kamada et al., 2019a). Seco-laurokamurone A has a new carbon skeleton which may arise due to the oxidative cleavage of seco-laurokamurone which may arise by oxidative cleavage of laurokamurane, an already isolated compound (Yu et al., 2014).

Various members of family Rhodomelaceae investigated for organohalogens and results, but up to now, even a single halogenated monoterpene is not reported from this family. Plocamiaceae and Rhizophyllidaceae are the two families of red algae to which halogenated monoterpenes are reported (Coll and Wright, 1987). *Plocamium cornutum* is rich in halogenated monoterpenes. *Plocamium cornutum* collected from South Africa yielded five halogenated monoterpenes including "(1Z,3E,5R,6S)-1-bromo-5,6-dichloro-2,6-dimethyl-octa-1,3,7-triene and (1Z,3E,5S,6S)-1-bromo-5,6-dichloro-2,6-dimethyl-octa-1,3,7-triene" (Afolayan et al., 2009). *Plocamium hamatum* and *Plocamium costatum* collected from separate locations in Australia have been reported in eight novel halogenated monoterpenes (Motti et al., 2014). *Plocamium cartilagineum*, a red alga from the South African coast reported a novel halogenated monoterpene designated (-)-(5E,7Z)-348-trichloro-7-dichloromethyl-3-methyl-1,5,7-octatriene has been isolated together with three already identified halogenated compounds (Sabry et al., 2017). Three novel monoterpenes plocoralides A–C are polyhalogenated and were isolated and identified along three previously known compounds in the organic extract of *Plocamium corallorhiza*. The structures of the novel compounds were named as 4,8-dibromo-1,1-dichloro-3,7-dimethyl-2*E*,6*E*-octadiene, 4,8-dibromo-1,1,7-trichloro-3,7-dimethyl-2*E*,5*Z*-octadiene,and4,6-dibromo-1,1-dichloro-3,7-dimethyl-2*E*,7-octadiene (Knott et al., 2005). A temperate species *Plocamium costatum* havebeenreportedanovel 1,4-dibromo-2,3,6-trichloro-3,7-dimethyl-7-octene along with three already identified polyhalogenated monoterpenes (Konig et al., 1999). Red macroalga *Plocamium suhrii* reported five previously known halogenated monoterpenes along with two (1*Z*,3*R*,4*S*,5*E*,7*Z*)-1-bromo-3,4,8-trichloro-7-(dichloromethyl)-3-methylocta-1,5,7-triene and

(3*R*,4*S*)-3,4,6,7-tetrachloro-3,7-dimethyl-octen-1-ene (Entunes, 2011). Chemical studies on *Ptilonia australasica* reported four novel polybrominated natural compounds including three ptilones A-C and australasol A. Such carbon skeletons were never reported in algal halogenated secondary metabolites (Tran et al., 2016).

Rhodophyllis membranacea has been reported to bear 11 new tetra halogenated indoles; out of them, four have bromine, chlorine, and iodine. This was supposed to be the first report of such kinds of compounds. In addition, four pentahalogenated indoles and uncharacterized tribromotrichloroindole were isolated (Woolner, 2017). The genus *Asparagopsis* belongs to the family Bonnemaisoniaceae of red algae. *Asparagopsis taxiformis* edible seaweed is reported to have more than 100 organohalogens. The dominant organohalide found in this seaweed is bromoform (Moore, 1977; McConnell and Moore, 1977). Low molecular weight halogenated compounds found to be produced by *Asparagopsis*. *Asparagopsis taxiformis* famous as 'limu kohu' is an edible seaweed that is a delicacy in Hawaiians. This is a source of over 100 organohalogens with the new mahorone and 5-bromomahorone. The samples were collected from the Indian Ocean off the coast of Mayotte. Mahorone and 5-bromomahorone are first reported natural compounds of their type 2,3-dibromocyclopentenones (Greff, 2014). Two brominated cyclopentenones were identified from *A. taxiformis* in a more recent study (Maximo, 2018).

Lane et al. (2009) and Lin et al. (2010) isolated eight and four bromophycolides from Fijian red alga *Callophycus serratus*. Six halogenated meroditerpenoids callophycoic acid I, callophycol C, iodocallophycols E, iodocallophycols F, callophycoic acid J and iodocallophycoic acid B. Out of them iodocallophycols E and iodocallophycoic acid B are new organohalogens added to the meroditerpenoids family of compounds produced by *Callophycus serratus* (Woolner et al., 2018). Brominated heptanones which are supposed to be the precursors of bromoform isolated from other red algae, *Bonnemaisonia hamifera* (McConnell and Fenical, 1980). Rhodomelin A, a novel ureidobromophenol was isolated and characterized from *Rhodomela confervoides*. This is the first natural ureidopyrrolidone alkaloid with a gamma-aminobutyric acid unit (Li et al., 2018). Bromophenols are known for their anti-inflammatory actions and various are reported from different algal species a novel one is bis(3-bromo-4,5-dihydroxybenzyl) ether isolated from *Polysiphonia morrowii*, a red alga, although its physiological functions are yet unknown (Choi, 2018). Four new macrocyclic 12-membered rings with an embedded tetrahydrofuran ring containing halogenated acetogenins reported from *Laurenciella marilzae* (Table 3.2 and Figure 3.1).

TABLE 3.2 A Few Halogenated Secondary Metabolites and Their Important Medicinal Properties

Category	Halogenated Secondary Metabolite	Name of Algae	Medicinal Properties	References
Brominated	Rhodomelin A	*Rhodomela confervoides*	Potent scavenging activity	Li et al. (2018)
	Bis-benzyl ether	*Polysiphonia morrowii*	DPPH, ABTS radicals, and Anti-inflammatory	Choi et al. (2018)
	Aplysin	*Laurencia tristicha*	Protect from alcohol-induced liver damage	Ge et al. (2018)
	α-Bromocuparane and α-isobromocuparane	*Laurencia johnstonii*	Antiamoeboid activity	Wada et al. 92018)
	Bis(2,3-dibromo-4,5-dihydroxybenzyl) ether (BDDE)	*Odonthalia corymbifera*	Antidiabetic	Xu et al. (2016)
	Bromoanaindolone	*Anabaena constricta*	Antibacterial activity against *Bacillus cereus*	Volk et al. (2009)
	Bromophycocolides	*Callophycus serratus*	Stops growth of *Lindra thalassiae*, a marine pathogenic fungus and antibacterial activity against *S. aureus* and *E. faecium*	Lane et al. (2009)
	Lyngbyastatin 10	*Lyngbya semiplena*	Inhibit porcine pancreatic elastase	Kwan et al. (2009)
	Aplysistatin	*Laurencia saitoi*	Antileukemic agent	Pettit et al. (1977)
	Pentabromopropen-2-yl tribromo and dibromoacetate	*Asparogopsis taxiformis*	–	Sugano et al. (1990)
	10-acetoxyangasiol, aplysiadiol, cupalaurenol, 1-methyl-2,3,5-tribromoindole and chamigrane epoxide	*Laurencia* sp.	Inhibit bovine lens aldose reductase and treat diabetes	Vairappan et al. (2010)

TABLE 3.2 *(Continued)*

Category	**Halogenated Secondary Metabolite**	**Name of Algae**	**Medicinal Properties**	**References**
	Bromoditerpenes	*Sphaerococcus coronopifolius*	Antibacterial activity against pathogenic bacteria; Antitumor and antimicrobial activity	Rodrigues et al. (2015)
Chlorinated	Ambiguine K and M, Ambiguine A	*Fischerella ambigua*	Against *Mycobacterium tuberculosis*	Mo et al. (2009)
	Fallachromenoic acid (meroditerpenoid)	*Sargassum fallax*	Against *Bacillus anthracis*	Reddy et al. (2009)
	Halogenated monoterpenes	*Plocamium cornutum*	Antitumor activities against a P388 Murine Leukaemia cell line	Afolayan et al. (2009)
	C15 acetogeninen-ynes	*Laurencia glandulifera*	Antiplasmodial, antistaphylococcal activity	Kladi et al. (2008)
Bromo-chlorinated	Obtusol (halogenated sesquiterpene)	*Laurencia dendroidea*	Larvicidal against the dengue fever mosquito *Aedes aegypti*	Salvador-Neto et al. (2016)
	Elatol	*Laurencia microcladia*	Antibacterial and antifeedant	Lhullier et al. (2009)
Iodinated	Monoiodotyrosine (MIT) and diiodotyrosine (DIT)	*Laminaria* sp.	Precursor of thyroid hormone	Heyland and Moroz (2005)

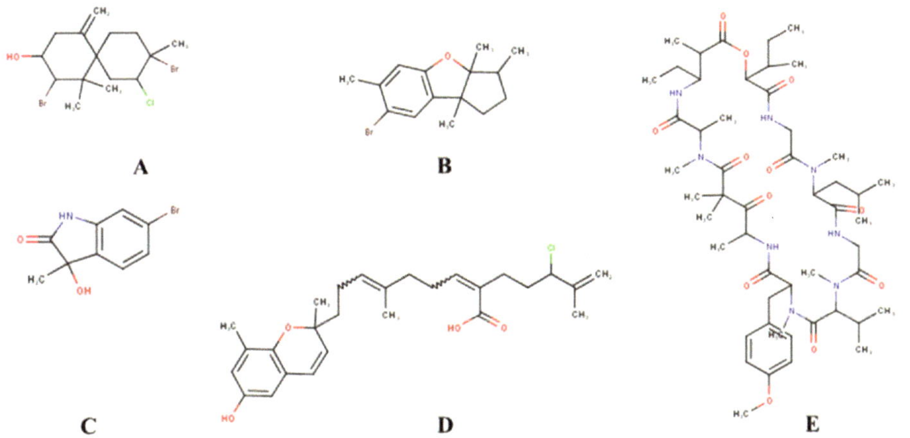

FIGURE 3.1 Medicinally important halide compounds from the natural sources. (A) obtusol; (B) aplysin; (C) bromoanaindolone; (D) fallachromenoic acid; and (E) lyngbyastatin 10.

3.5 POTENTIAL ROLE OF HALIDE COMPOUNDS IN THE CURRENT DRUG DISCOVERY

Halides are made of halides functional groups with a distinct class of chemical moieties. Mainly chlorine, fluorine, iodine, and bromine are important halides. Halogen bonding is directional in large molecules where halides act as the electron acceptor. The organic halogen has anisotropic charge distribution where negative charges form an equatorial ring along with positively charged Σ-hole region and a C-X bond protrusion. The electrostatic interaction between abundant electron molecules and halogen Σ-hole leads to the formation of linear halogen bonds. Halogenation can contribute significantly to future strategies of optimizing drug development. Currently, high-throughput screening consists of about half of the halogenated molecules. Although halogen substituent is widely used as inhibitors and drugs, there is a dire need for more research on the chemical and structural mechanisms involved in inhibitor affinity and identification. This lack of information limits their current use in rational drug design. In the past, halogen ligand bonding mainly focused on only their steric and lipophilic contributions. However, recent studies have identified the stabilizing bond formation capabilities of halogen in biological macromolecules through the involvement of halogen bonds, hydrogen bonds and multipolar linkages (Brammer et al., 2001; Paulini et al., 2005; Lu et al., 2009). A plethora of studies in recent times have improved our knowledge about the importance of halogen bonding within biological systems and the scope of their utilization in rational drug designing. The perturbations in halide structure due to multiple geometries associated with it make halide bonds ideal candidate for modern drug discovery (Wilcken et al., 2016). In recent years, halides have played essential roles in the optimization of drugs from the hit to lead molecules due to improving the drug binding affinity as well as enhances the pharmacodynamics and pharmacodynamics properties of drugs (Xu et al., 2014). Jiang et al. (2016) reported that the position of halide bonds leads to perturbation of the binding affinity of substrates with the enzymes. Lu et al. (2012) reviewed that multipolar interaction, halogen, and hydrogen bonding potentials of halide compounds impart those capabilities to form diverse bonds with targets. Halogen bonding considers the most substantial interaction in the binding of target molecules compared to other chemical interactions. The modification of any functional group with the halides leads to the pharmacokinetics and pharmacodynamics properties. Artemisinin is a drug that has lower bioavailability due to the C-H bonds and the higher rate of oxidation. However, the perturbation of C-H bonds with the C-F leads to the bioavailability of these drugs by the lowering of oxidation (Gillis et al., 2015).

3.6 CONCLUSION AND FUTURE PERSPECTIVES

Our improved knowledge of marine resources and huge investments by governments all over the world has made seaweeds a lucrative target for scientific studies. Halogenation modulates the bioactivity of secondary metabolites and leads to pharmaceutical significance in future. Marine algae especially red algae followed by brown and blue-green algae are rich sources of halogenated secondary metabolites. These halogenated products can be developed in multibillion-dollar pharmaceutically valuable products. Till now, only little is discovered from the natural treasure of the ocean and more research should be undertaken in this field. The patients of RD and their caregivers go through severe emotional trauma and bear huge economic burden while concurrently facing several challenges associated with the disease. As per reports from IRDiRC, there is lack of argumentative therapies in the market and new approaches for drug development are much needed. Obviously, it is well established that most of the pharmaceutical companies are hesitant in developing and discovering orphan drugs used for the treatment of RD due to their limited scope in the market. The other challenges associated with it are low sample size, non-availability of volunteers for clinical trials and the rigorous laws imposed by governments.

The discovery of new molecules having medicinal properties may lead to the development of cost-effective drugs. Halogenated marine natural products have a great hope in this regard. Several studies mentioned above indicate that the efficacy of drugs can be improved by halogenation. More research is needed which can directly link halogenated seaweed products and their derivatives with treatment of RD. There is also a dire need for more research on the chemical and structural mechanisms involved in inhibitor affinity and identification of RD.

KEYWORDS

- **halides**
- **halogenated secondary metabolites**
- **macroalgae**
- **marine habitat**
- **microalgae**
- **organohalogens**
- **orphan drugs**
- **rare diseases**
- **terpene acetogenins**

REFERENCES

Aboussouan, L. S., & Stoller, J. K., (2009). Detection of alpha-1 antitrypsin deficiency: A review. *Respiratory Medicine*, *103*(3), 335–341. https://doi.org/10.1016/j.rmed.2008.10.006.

Afolayan, A. F., Mann, M. G. A., Lategan, C. A., Smith, P. J., Bolton, J. J., Denzil, R., & Beukes, D. R., (2009). Antiplasmodial halogenated monoterpenes from the marine red alga *Plocamium cornutum*. *Phytochemistry*, *70*, 597–600.

Anderson, J. L. R., & Chapman, S. K., (2006). Molecular mechanisms of enzyme-catalyzed halogenation. *Molecular BioSystems*, *2*, 350–357.

Arizumi, H., Hata, K., & Shimizu, N., (1994). Isolation and determination of structure of antioxidant chromanol derivative from *Ascophyllum nodosum*. *Jpn. Kokkai Tokyo Koho*, *122*, 160312.

Austin, C. P., Cutillo, C. M., Lau, L. P. L., et al., (2018). Future of rare diseases research 2017–2027: An IRDiRC perspective. *Clinical and Translational Science*, *11*(1), 21–27. https://doi.org/10.1111/cts.12500.

Bhattacharjee, M., (2016). Pharmaceutically valuable bioactive compounds of algae. *Asian Journal of Pharmaceutical and Clinical Research*, *9*(6), 43–47.

Bisen, P. S., (2016). Nutritional therapy as a potent alternative to chemotherapy against cancer. *Journal of Cancer Science and Therapy*, *8*(6), 168, 169.

Blunt, J. W., Copp, B. R., Keyzers, R. A., Munro, M. H. G., & Prinsep, M. R., (2015). Marine natural products. *Natural Product Reports*, *32*, 116–211. doi: 10.1039/C4NP00144C.

Blunt, J. W., Hartshorn, M. P., McLennan, T. J., Munro, M. H. G., Robinson, W. T., & Yorke, S. C., (1978). Thyrsiferol, a squalene-derived metabolite of *Laurencia thyrsifera*. *Tetrahedron Letters*, *19*(1), 69–72.

Boycott, K. M., Rath, A., Chong, J. X., et al., (2017). International cooperation to enable the diagnosis of all rare genetic diseases. *American Journal of Human Genetics*, *100*(5), 695–705. https://doi.org/10.1016/j.ajhg.2017.04.003.

Brammer, L., Bruton, E. A., & Sherwood, P., (2001). Understanding the behavior of halogens as hydrogen bond acceptors. *Crystal Growth and Design*, *1*(4), 277–290.

Cabrita, M. T., Vale, C., & Rauter, A. P., (2010). Halogenated compounds from marine algae. *Marine Drugs*, *8*, 2301–2317.

Chen, J. Y., Huang, C. Y., Lin, Y. S., et al., (2016). Halogenated sesquiterpenoids from the red alga *Laurencia tristicha* collected in Taiwan. *Journal of Natural Products*, *79*(9), 2315–2323.

Choi, Y. K., Ye, B. K., Kim, E. A., et al., (2018). Bis(3-bromo-4,5-dihydroxybenzyl) ether, a novel bromophenol from the marine red alga *Polysiphonia morrowii* that suppresses LPS-induced inflammatory response by inhibiting ROS-mediated ERK signaling pathway in RAW 264.7 macrophages. *Biomedicine and Pharmacotherapy*, *103*, 1170–1177. doi: 10.1016/j.biopha.2018.04.121.

Coll, J. C., & Wright, A. D., (1987). Tropical marine algae I. New halogenated monoterpenes from *Chondrococcus hornemannii* (*Rhodophyta*, *Gigartinales*, *Rhizophyllidaceae*). *Australian Journal of Chemistry*, *40*, 1893–1900.

Colla, L. M., Muccillo-Baisch, A. L., & Vieira, C. J. A., (2008). *Spirulina platensis* effects on the levels of total cholesterol, HDL and triacylglycerols in rabbits fed with a hypercholesterolemic diet. *Brazilian Archives of Biology and Technology*, *51*(2), 405–411.

Colla, L. M., Oliveira, R. C., Reichert, C., & Costa, J. A., (2007). Production of biomass and nutraceutical compounds by *Spirulina platensis* under different temperature and nitrogen regimes. *Bioresource Technology*, *98*(7), 1489–1493.

Deticek, A., Locatelli, I., & Kos, M., (2018). Patient access to medicines for rare diseases in European countries. *Value in Health*, *21*(5), 553–560. https://doi.org/10.1016/j.jval.2018.01.007.

Entunes, E. M., Afolayan, A. F., Chiwakata, M. T., et al., (2011). Identification and *in vitro* anti-esophageal cancer activity of a series of halogenated monoterpenes isolated from the South African seaweeds *Plocamium suhrii* and *Plocamium cornutum*. *Phytochemistry*, *72*(8), 769–777.

Esselin, H., Tomi, F., Bighelli, A., & Sutour, S., (2018). New metabolites isolated from a *Laurencia obtusa* population collected in Corsica. *Molecules*, *23*, 720, doi: 10.3390/molecules23040720.

Estrada, D. M., Ravelo, J. L., Ruiz-Perez, C., & Martın, J. D., (1989). Dactylomelol, a new class of diterpene from the sea hare *Aplysia dactylomela*. *Tetrahedron Letters*, *30*, 6219.

European Medicines Agency, (2018). *Rare Diseases, Orphan Medicines: Getting the Facts Straight.* http://www.ema.europa.eu/docs/en_GB/document_library/Other/2018/02/WC500244578.pdf (accessed on 7 April 2022).

Faulkner, D. J., (1995). Chemical riches from the oceans. *Chemistry in Britain*, *31*(9), 680–684.
Fenical, W., & Norris, J. N., (1975). Chemotaxonomy in marine algae: Chemical separation of some *Laurencia* species (Rhodophyta) from the Gulf of California. *Journal of Phycology*, *11*, 104–108.
Ge, N., Liang, H., Zhao, Y. Y., Liu, Y., Gong, A. J., & Zhang, W. L., (2018). Aplysin protects against alcohol-induced liver injury via alleviating oxidative damage and modulating endogenous apoptosis-related genes expression in rats. *Journal of Food Science*, *83*(10), 2612–2621. doi, 10.1111/1750-3841.14320.
Gillis, E. P., Eastman, K. J., Hill, M. D., Donnelly, D. J., & Meanwell, N. A., (2015). Applications of fluorine in medicinal chemistry. *Journal of Medicinal Chemistry*, *58*(21), 8315–8359.
Glombitza, K. W., & Gerstberger, G., (1985). Phlorotannins with dibenzodioxin structural elements from the brown alga *Eisenia arborea*. *Phytochemistry*, *24*, 543–551.
Glombitza, K. W., & Schmidt, A., (1999). Nonhalogenated and halogenated phlorotannins from the brown alga *Carpophyllum angustifolium*. *Journal of Natural Products*, *62*, 1238–1240.
Greff, S., Zubia, M., Jouve, G. G., Massi, L., Perez, T., & Thomas, O. P., (2014). Mahorones, highly brominated cyclopentenones from the red alga *Asparagopsis taxiformis*. *Journal of Natural Products*, *77*, 1150. doi: 10.1021/NP401094HC.
Gribble, G. W., (2003). The diversity of naturally produced organohalogens. *Chemosphere*, *52*(2), 289–297.
Gribble, G. W., (2018). Newly discovered naturally occurring organohalogens. *Archive for Organic Chemistry Part I*, 372–410.
Groft, S. C., (2013). Rare diseases research: Expanding collaborative translational research opportunities. *Chest*, *144*(1), 16–23. https://doi.org/10.1378/chest.13-0606.
Guedes, A. C., Giao, M. S., Seabra, R., Ferreira, A. C. S., Tamagnini, P., & Moradas-Ferreira, P., (2013). Evaluation of the antioxidant activity of cell extracts from microalgae. *Marine Drugs*, *11*, 1256–1270. doi: 10.3390/md11041256.
Hall, J. G., & Reiss, J. A., (1986). Elatenyne - a pyrano[3,2-β]pyranyl vinyl acetylene from the red alga *Laurencia elata*. *Australian Journal of Chemistry*, *39*, 1401–1409.
Harizani, M., Ioannou, E., & Roussis, (2016). The *Laurencia* paradox: An endless source of chemodiversity. In: Kinghorn, A. D., Falk, H., Gibbons, S., & Kobayashi, J., (eds.), *Progress in the Chemistry of Organic Natural Products* (Vol. 102, pp. 91–153). Springer International Publishing, Switzerland.
Harper, M. K., Bugni, T. S., Copp, B. R., et al., (2001). Introduction to the chemical ecology of marine natural products. In: McClintock, J. B., & Baker, B. J., (eds.), *Marine Chemical Ecology* (pp. 3–71). CRC: Boca Raton, FL, USA.
Heyland, A., & Moroz, L. L., (2005). Cross-kingdom hormonal signaling: An insight from thyroid hormone functions in marine larvae. *Journal of Experimental Biology*, *208*(Pt 23), 4355–4361. doi:10.1242/jeb.01877.
Hill, R. A., (2007). Marine natural products. *Annual Reports on the Progress of Chemistry Section B: Organic Chemistry*, *103*, 125–139.
Hou, X., Chai, C., Qian, Q., Yan, X., & Fan, X., (1997). Determination of chemical species of Iodine in some seaweeds (I). *Science of the Total Environment*, *204*, 215–221.
Howard, B. M., & Fenical, W., (1978). Obtusadiol, a unique bromoditerpenoid from the marine red alga *Laurencia obtusa*. *Tetrahedron Letters*, *18*, 2453. https://doi.org/10.1021/cg015522k.
Jakob, G., Wolf, J., Bul, T. V. L., et al., (2013). Surveying a diverse pool of microalgae as a bioresource for future biotechnological applications. *Journal of Petroleum and Environmental Biotechnology*, *4*, 153. doi, 10.4172/2157-7463.1000153.

Ji, N. Y., Li, X. M., Li, K., & Wang, B. G., (2007). Laurendecumallenes A-B and laurendecumenynes A-B, halogenated nonterpenoid C15 acetogenins from the marine red alga *Laurencia decumbens*. *Journal of Natural Products*, *70*, 1499–1502.

Ji, N. Y., Li, X. M., Li, K., & Wang, B. G., (2009). Halogenated sesquiterpenes from the marine red alga *Laurencia saitoi* (Rhodomelaceae). *Helvetica Chimica Acta*, *92*, 1873–1879.

Jiang, S., Zhang, L., Cui, D., et al., (2016). The important role of halogen bond in substrate selectivity of enzymatic catalysis. *Scientific Reports*, *6*, 34750. doi: 10.1038/srep34750.

Kamada, T., & Vairappan, C. S., (2015). New Laurene-type sesquiterpene from Bornean *Laurencia nangii*. *Natural Product Communications*, *10*, 1, 2.

Kamada, T., Phan, C. S., & Vairappan, C. S., (2019a). New antibacterial halogenated tricyclic sesquiterpenes from Bornean *Laurencia majuscula* (Harvey) Lucas. *Natural Product Research*, *33*(4), 464–471.

Kamada, T., Phan, C. S., & Vairappan, C. S., (2019b). Nangallenes A and B, halogenated nonterpenoid C_{15}-acetogenins from the Bornean red alga *Laurencia nangii*. *Journal of Asian Natural Products Research*, *21*(3), 241–247.

Kennedy, D. J., Selby, I. A., Cowe, H. J., Cox, P. J., & Thomson, R. H., (1984). Bromoallenes from the alga *Laurencia microcladia*. *Journal of the Chemical Society, Chemical Communications*, *3*, 153–155.

Kladi, M., Ntountaniotis, D., Zervou, Vagias, M., Ioannou, E., & Roussis, V., (2014). Glandulaurencianols A–C, brominated diterpenes from the red alga, *Laurencia glandulifera* and the sea hare, *Aplysia punctata*. *Tetrahedron Letters*, *55*, 2835–2837. doi: 10.1016/J.TETLET.2014.03.083.

Kladi, M., Vagias, C., Papazafiri, P., Brogi, S., Tafi, A., & Roussis, V., (2009). Tetrahydrofuran acetogenins from *Laurencia glandulifera*. *Journal of Natural Products*, *72*, 190–193.

Kladi, M., Vagias, C., Stavri, M., Rahman, M., Gibbons, S., & Roussis, V., (2008). C_{15} acetogenins with antistaphylococcal activity from the red alga *Laurencia glandulifera*. *Phytochemistry Letters*, *1*, 31–36.

Knott, G. M., Catherine, H. M., Andrese, E., Hendricks, D. T., Bolton, J. J., & Beukes, D. R., (2005). Plocoralides A–C, polyhalogenated monoterpenes from the marine alga *Plocamium corallorhiza*. *Phytochemistry*, *66*(10), 1108–1112.

Koch, M. L., & Gregson, R. P., (1984). Brominated phlorethols and non-halogenated phlorotannins from the brown alga *Cystophora congesta*. *Phytochemistry*, *13*, 2633–2637.

Konig, G. M., Wright, A. D., & De Nys, R., (1999). Halogenated monoterpenes from *Plocamium costatum* and their biological activity. *Journal of Natural Products*, *62*(2), 383–385. doi: 10.1021/np980408y.

Kulik, H. J., Blasiak, L. C., Marzari, N., & Drennan, C. L., (2009). First-principles study of non-heme Fe (II) halogenase SyrB2 reactivity. *Journal of the American Chemical Society*, *131*(40), 14426–14433.

Kwan, J. C., Taori, K., Paul, V. J., & Luesch, H., (2009). Lyngbyastatins 8–10, elastase inhibitors with cyclic depsipeptide scaffolds isolated from the marine cyanobacterium *Lyngbya semiplena*. *Marine Drugs*, *7*, 528–538. doi: 10.3390/md7040528.

Lane, A. L., Stout, E. P., Lin, A. S., et al., (2009). Antimalarial bromophycolides J-Q from the Fijian red alga *Callophycus serratus*. *The Journal of Organic Chemistry.*, *74*, 2736–2742.

Lhullier, C., Donnangelo, A., Caro, et al., (2009). Isolation of elatol from *Laurencia microcladia* and its palatability to the sea urchin *Echinometra lucunter*. *Biochemical Systematics and Ecology*, *37*(4), 254–259.

Li, K., Wang, Y. F., Li, X. M., et al., (2018). Isolation, synthesis, and radical-scavenging activity of rhodomelin A, a ureidobromophenol from the marine red alga *Rhodomela confervoides*. *Organic Letters*, *20*(2), 417–420. doi: 10.1021/acs.orglett.7b03716.
Lin, A. S., Stout, E. P., Prudhomme, J., et al., (2010). Bioactive bromophycolides R-U from the Fijian red alga *Callophycus serratus*. *Journal of Natural Products*, *73*(2), 275–278. doi: 10.1021/np900686w.
Lu, Y., Liu, Y., Xu, Z., Li, H., Liu, H., & Zhu, W., (2012). Halogen bonding for rational drug design and new drug discovery. *Expert Opinion on Drug Discovery*, *7*(5), 375–383.
Lu, Y., Wang, Y., Xu, Z., et al., (2009). C–X–H contacts in biomolecular systems: How they contribute to protein-ligand binding affinity. *Journal of Physical Chemistry B*, *113*(37), 12615–12621.https://doi.org/10.1021/jp906352e.
Mariya, J. G., & Kurup, M. G., (2016). *In vitro* antioxidant properties of edible marine algae *Sargassum swartzii*, *Ulva fasciata* and *Chaetomorpha antennina* of Kerala coast. *Journal of Pharmaceutical Bioprocessing*, *4*(6), 100–108.
Masuda, M., Abe, T., Sato, S., Suzuki, T., & Suzuki, M., (1997). Diversity of halogenated secondary metabolites in the red alga *Laurencia nipponica* (*Rhodomelaceae*, *Ceramiales*). *Journal of Phycology*, *33*(2), 196–208.
Maximo, P., Ferreira, L. M., Branco, P., Lima, P., & Lourenço, A., (2018). Secondary metabolites and biological activity of invasive macroalgae of Southern Europe. *Marine Drugs*, *16*(8), 265. doi, 10.3390/md16080265.
Mc Connell, O., & Fenical, W., (1977). Halogen chemistry of the red alga *Asparagopsis*. *Phytochemistry*, *16*, 367.
Mehta, P., Shepard, J., Rouse, K., et al., (2016). Growth and tolerability of healthy term infants fed a new formula supplemented with DHA from *Schizochytrium* sp. microalgae. *Journal of Vascular Medicine and Surgery*, *4*, 267. doi: 10.4172/2329-6925.1000267.
Milkova, T., Talev, G., Christov, R., Dimitrikova-Konaklieva, S., & Popov, S., (2007). Sterols and volatiles in *Cystoseira barbata* and *Cystoseira crinita* from the black sea. *Phytochemistry*, *45*, 93–95.
Mimouni, V., Ulmann, L., Pasquet, V., Mathieu, M., Picot, L., & Bougaran, G., (2012). The potential of microalgae for the production of bioactive molecules of pharmaceutical interest. *Current Pharmaceutical Biotechnology*, *13*, 2733–2750. doi: 10.2174/138920112804724828.
Mo, S., Krunic, A., Chilipala, G., & Orjala, J., (2009). Antimicrobial ambiguine isonitriles from the cyanobacterium *Fischerella ambigua*. *Journal of Natural Products*, *72*(5), 894–899.
Moore, R. E., (1977). Volatile compounds from marine algae. *Accounts of Chemical Research*, *10*(2), 40–47.
Motti, A. P., Hall, P. T., Hagiwara, K. A., Simmons, C. J., Willis, R., & Wright, A. D., (2014). Accelerated identification of halogenated monoterpenes from Australian specimens of the red algae *Plocamium hamatum* and *Plocamium costatum*. *Journal of Natural Products*, *77*, 1193. doi: 10.1021/NP500059H.
Mutafova, B., Mutafov, S., Fernandes, P., & Berkov, S., (2016). Microbial transformations of plant origin compounds as a step in preparation of highly valuable pharmaceuticals. *Journal of Drug Metabolism and Toxicology*, *7*(2). doi: 10.4172/2157-7609.1000204.
Naismith, J. H., (2006). Inferring the chemical mechanism from structures of enzymes. *Chemical Society Reviews*, *35*, 763–770.
Nightingale, P. D., Malin, G., & Liss, P. S., (1995). Production of chloroform and other low-molecular-weight halocarbons by some species of macroalgae. *Limnology and Oceanography*, *40*, 680–689.

Oramary, S. O. M., Koramarky, D. M. I., Salih, S. A., & Mustafa, A. A., (2016). Feeding common carp fish (*Cyprinus carpio*) on natural foods (Algae, phytoplankton, zooplankton and others) on Tigris River in Mosul dam/Duhok, Kurdistan Region of Iraq. *Journal of Aquaculture Research and Development*, *7*, 413. doi: 10.4172/2155-9546.1000413.

Parameswaran, P. S., Bhat, K. L., Das, B., Kamat, S. Y., & Harnos, S., (1994). Halogenated terpenoids from the brown alga *Padina tetrastromatica* (Hauck). *Indian Journal of Chemistry Section–B*, *33*, 1006–1008.

Paulini, R., Müller, K., & Diederich, F., (2005). Orthogonal multipolar interactions in structural chemistry and biology. *Angewandte Chemie - International Edition*, *44*(12), 1788–1805. https://doi.org/10.1002/anie.200462213.

Peng, J., Yuan, J. P., Wu, C. F., & Wang, J. H., (2011). Fucoxanthin, a marine carotenoid present in brown seaweeds and diatoms: Metabolism and bioactivities relevant to human health. *Marine Drugs*, *9*, 1806–1828. doi: 10.3390/md9101806.

Pettit, G. R., Herald, C. L., Allen, M. S., et al., (1977). The isolation and structure of aplysistatin. *Journal of the American Chemical Society*, *99*, 262, 263.

Raja, R., Shanmugam, H., Ganesan, V., & Carvalho, I. S., (2014). Biomass from microalgae: An overview. *Oceanography*, *2*, 118. doi: 10.4172/2332-2632.1000118.

Reddy, P., & Urban, S., (2009). Meroditerpenoids from the Southern Australian marine brown alga *Sargassum fallax*. *Phytochemistry*, *70*, 250–255.

Rodrigues, D., Alves, C., & Horta, A., (2015). Antitumor and antimicrobial potential of bromoditerpenes isolated from the red alga. *Marine Drugs*, *13*, 713–726. doi, 10.3390/md13020713.

Rovirosa, J., Soler, A., Blanc, V., Leon, R., & San-Martin, A., (2013). Bioactive monoterpenes from Antarctic *Plocamium cartilagineum*. *Journal of the Chilean Chemical Society*, *58*, 2025, 2026.

Sabry, O. M. M., Goegera, D. E., Valeriotec, A. F., & Gerwick, W. H., (2017). Cytotoxic halogenated monoterpenes from *Plocamium cartilagineum*. *Natural Product Research*, *31*(3), 261–267.

Sailler, B., & Glombitza, K. W., (1999). Halogenated phlorethols and fucophlorethols from the brown alga *Cystophora retroflexa*. *Natural Toxins*, *7*, 57.

Salvador-Neto, O., Gomes, S. A., Soares, A. R., et al., (2016). Larvicidal potential of the halogenated sesquiterpene (+)-obtusol, isolated from the alga *Laurencia dendroidea* J. Agardh (Ceramiales: Rhodomelaceae), against the dengue vector mosquito *Aedes aegypti* (Linnaeus) (Diptera: Culicidae). *Marine Drugs*, *14*(2), 20. doi: 10.3390/md14020020.

Samarakoon, K. W., Ko, J. Y., Shah, M. M. R., et al., (2013). *In vitro* studies of anti-inflammatory and anticancer activities of organic solvent extracts from cultured marine microalgae. *Algae*, *28*, 111–119. doi: 10.4490/algae.2013.28.1.111.

Scarratt, M. G., & Moore, R. M., (1996). Production of methyl chloride and methyl bromide in laboratory cultures of marine phytoplankton. *Marine Chemistry*, *54*(3, 4), 263–272.

Shibata, T., Hama, Y., Miyasaki, T., Ito, M., & Nakamura, T., (2006). Extracellular secretion of phenolic substances from living brown algae. *Journal of Applied Phycology*, *18*, 787–794.

Silva, M., Vieira, L. M. M., Almeida, A. P., et al., (2013). Chemical study and biological activity evaluation of two Azorean macroalgae: *Ulva rigida* and *Gelidium microdon*. *Oceanography*, *1*, 102. doi: 10.4172/2332- 2632.1000102.

Song, P., Gao, J., Inagaki, Y., Kokudo, N., & Tang, W., (2012). Rare diseases, orphan drugs, and their regulation in Asia: Current status and future perspectives. *Intractable and Rare Diseases Research*, *1*(1), 3–9. https://doi.org/10.5582/irdr.2012.v1.1.3.

Stoller, J. K., & Brantly, M., (2013). The challenge of detecting alpha-1 antitrypsin deficiency. *COPD: Journal of Chronic Obstructive Pulmonary Disease*, *10*(S1), 26–34. https://doi.org/10.3109/15412555.2013.763782.

Stoller, J. K., (2018). The challenge of rare diseases. *Chest*, *153*(6), 1309–1314. https://doi.org/10.1016/j.chest.2017.12.018.

Su, H., Yu, Z. H., Li, J., et al., (2009). Two new bromoindoles from red alga *Laurencia similis*. *Chinese Chemical Letters*, *20*(4), 456–458.

Sugano, M., Sato, A., Nagak, H., Yoshiok, S., Shiraki, T., & Horikoshi, H., (1990). Aldose reductase inhibitors from the red alga, *Asparagopsis taxiformis*. *Tetrahedron Letters*, *31*(48), 7015–7016.

Sun, H. H., Waraszkiewicz, S. M., & Erickson, K. L., (1976). C15 halogenated compounds from the Hawaiian marine alga *Laurencia nidifica*. VI. the isomaneonenes. *Tetrahedron Letters*, *47*, 4227–4230.

Sun, W. S., Su, S., Zhu, R. X., Tu, et al., (2013). A pair of unprecedented spiro-trisindole enantiomers fused through a five-member ring from *Laurencia similis*. *Tetrahedron Letters*, *54*, 3617.

Suzuki, M., Matsuo, Y., Takeda, S., & Suzuki, T., (1993). Intricatetraol, a halogenated triterpene alcohol from the red alga *Laurencia intricata*. *Phytochemistry*, *33*(3), 651–656.

Suzuki, M., Takahashi, Y., Nakano, S., et al., (2009). An experimental approach to study the biosynthesis of brominated metabolites by the red algal genus *Laurencia*. *Phytochemistry*, *70*(11, 12), 1410–1415.

Taruscio, D., Groft, S. C., Cederroth, H., et al., (2015). Undiagnosed diseases network international (UDNI): White paper for global actions to meet patient needs. *Molecular Genetics and Metabolism*, *116*(4), 223–225. https://doi.org/10.1016/j.ymgme.2015.11.003.

Todd, J. S., Proteau, P. J., & Gerwick, W. H., (1993). Egregiachlorides A-C: New chlorinated oxylipins from the brown marine alga *Egregia menziesii*. *Tetrahedron Letters*, *34*, 7689–7692.

Tomar, V., (2012). Raman spectroscopy of algae: A review. *Journal of Nanomedicine and Nanotechnology*, *3*(2), 131. doi:10.4172/2157-7439.1000131.

Tran, T. D., Pham, N. B., & Quinn, R. J., (2016). Unique polybrominated hydrocarbons from the Australian endemic red alga *Ptilonia australasica*. *Journal of Natural Products*, *79*, 570–577. doi: 10.1021/acs.jnatprod.5b00989.

Vairappan, C. S., (2003). Potent antibacterial activity of halogenated metabolites from Malaysian red algae, *Laurencia majuscula* (*Rhodomelaceae*, *Ceramiales*). *Biomolecular Engineering*, *20*(4–6), 255–259.

Vairappan, C. S., Ishii, T., Lee, T. K., Suzuki, M., & Zhaoqi, Z., (2010). Antibacterial activities of a new brominated diterpene from Borneon *Laurencia* spp. *Marine Drugs*, *8*, 1743–1749. doi, 10.3390/md8061743.

Vetter, W., (2006). Marine halogenated natural products. *Reviews of Environmental Contamination and Toxicology*, *188*, 1–57.

Volk, R. B., Girreser, U., Al-Rafai, M., & Laatsch, H., (2009). Bromoanaindolone, a novel antimicrobial exometabolite from the cyanobacterium *Anabena constricta*. *Natural Product Research*, *23*(7), 607–612.

Wada, H., Matsumoto, T., Suzuki, K., et al., (2018). Differences and similarities between disseminated intravascular coagulation and thrombotic microangiopathy. *Thrombosis Journal*, *16*, 14. https://doi.org/10.1186/s12959-018-0168-2.

Wang, B. N., Gloer, J. B., Ji, N. Y., & Zhao, J. C., (2013). Halogenated organic molecules of *Rhodomelaceae* origin: Chemistry and biology. *Chemical Reviews*, *113*(5), 3632–3685. doi: 10.1021/cr9002215.

Waraszkiewicz, S. M., Sun, H. H., Erickson, K. L., Finer, J., & Clardy, J., (1978). C15 halogenated compounds from the Hawaiian marine alga *Laurencia nidifica*, maneonenes and isomaneonenes. *Journal of Organic Chemistry*, *43*(16), 3194–3204.

Whicher, D., Philbin, S., & Aronson, N., (2018). An overview of the impact of rare disease characteristics on research methodology. *Orphanet Journal of Rare Diseases*, *13*, 14. doi: 10.1186/s13023-017-0755-5.

Wilcken, R., Zimmermann, M. O., Lange, A., Joerger, A. C., & Boeckler, F. M., (2013). Principles and applications of halogen bonding in medicinal chemistry and chemical biology. *Journal of Medicinal Chemistry*, *56*(4), 1363–1388.

Woolner, V. H., (2017). *New Halogenated Secondary Metabolites from Red Algae of the South Pacific*. PhD Thesis, Victoria University of Wellington.

Woolner, V. H., Gordon, R. M. A., Miller, et al., (2018). Halogenated meroditerpenoids from a South Pacific collection of the red alga *Callophycus serratus*. *Journal of Natural Products*, *81*(11), 2446–2454.

Wright, A. D., Goclik, E., & König, G. M., (2003). Three new sesquiterpenes from the red alga *Laurencia perforate*. *Journal of Natural Products*, *66*, 435–437.

Wright, A. D., Konig, G. M., De Nys, R., & Sticher, O., (1993). Seven new metabolites from the marine red alga *Laurencia majuscula*. *Journal of Natural Products*, *56*, 394–401.

Wright, A. E., Wing, R. M., & Sims, J. J., (1983). Poitediene a new metabolite from the marine red alga *Laurencia poitei*. *Tetrahedron Letters*, *24*(43), 4649–4652.

Xu, Z., Yang, Z., Liu, Y., Lu, Y., Chen, K., & Zhu, W., (2014). Halogen bond: Its role beyond drug-target binding affinity for drug discovery and development. *Journal of Chemical Information and Modeling*, *54*(1), 69–78.

Yu, X. Q., He, W. F., Liu, D. Q., et al., (2014). A *seco*-laurane sesquiterpene and related laurane derivatives from the red alga *Laurencia okamurai* Yamada. *Phytochemistry*, *103*, 162–170. doi, 10.1016/j.phytochem.2014.03.021.

Zhu, X., Smith, R. A., & Parrott, R. L., (2017). Living with a rare health condition: The influence of a support community and public stigma on communication, stress, and available support. *Journal of Applied Communication Research*, *45*(2), 179–198. https://doi.org/10.1080/00909882.2017.1288292.

CHAPTER 4

METABOLITE PROFILING OF ALGAL BIOMASS AND ITS INDUSTRIAL AND BIOLOGICAL APPLICATIONS

A. AMALA LOURTHURAJ,[1] MARIADHAS VALAN ARASU,[2] R. LEKSHMI,[3] R. RAJAKRISHNAN,[2] AMEER KHUSRO,[3] NAIF ABDULLAH AL-DHABI,[1] PAUL AGASTIAN,[4] and M. A. RATHI[5]

[1]*Department of Biotechnology, Guru Nanak College (Autonomous), Velachery, Chennai–600042, Tamil Nadu, India*

[2]*Department of Botany and Microbiology, College of Science, King Saud University, P.O. Box 2455, Riyadh–11451, Saudi Arabia*

[3]*Department of Botany and Biotechnology, MSM College, Kayamkulam, Alappuzha, Kerala, India*

[4]*Research Department of Plant Biology and Biotechnology, Loyola College, Nungambakkam, Chennai, Tamil Nadu, India*

[5]*Department of Biochemistry, Sree Narayana Guru College, Coimbatore–641105, Tamil Nadu, India*

ABSTRACT

Algae have highly complex photosynthetic mechanisms to channelize absorbed solar energy into other forms of energy for the production of metabolites. These algae are the sources of promising catalysts and can be effectively used for the sustainable production of metabolites, food, and green energy sources such as biodiesel. Algae harbor various metabolites, including biocides (insecticides, herbicides, and algaecides), cyanotoxins,

Algal Metabolites: Biotechnological, Commercial, and Industrial Applications. Jeyabalan Sangeetha, Devarajan Thangadurai, Saher Islam, & Ravichandra Hospet (Eds.)

phytohormones, photoprotective compounds (halogenated compounds, carotenoids, polysaccharides, scytonemin, and halogenated compounds), free fatty acids, sterols, phytols, terpenoids, phytoene, and phenolics. Algae metabolites are used as anti-inflammatory agents, antiviral, anticancer, immunosuppressant, and antibiotics. The secondary metabolites, photoprotective agents are widely used in various skincare products like regenerants, anti-inflammatory drugs, antioxidants, anti-irritant, and anti-aging creams. Algae and cyanobacteria are highly useful in the field of agriculture. Cyanotoxins, derived from cyanobacteria showed inhibitory response on viruses, bacteria, fungi, bivalves, and crustaceans. Algae produce large quantities of polyunsaturated fatty acids and contributing in manufacturing of oils and fats in large quantities. Macroalgae yield carrageenan and used in food industry as stabilizer, emulsers due to their suspension and thickening forming properties in various foods, mainly milk-based food products such as jams, evaporated milk, chocolate milk and ice creams. In this chapter, we described the functions of various metabolites from marine alga and their applications. Besides, we have also discussed the role of algae in the production of biofuel and biomedicine.

4.1 INTRODUCTION

Algae are an umbrella term that includes a diverse group of organisms, mostly associated with wet and moist habitats, available in different colors and capable of producing their own food by photosynthesis. Every form of life on earth requires water for its survival. Algae, not just requires, but they actually live in water of many kinds with an additional quality of producing food on their own by the process called photosynthesis. Every alga thus holds two of these determined characteristics, i.e., algae are aquatic and auto-trophic. Hence, algae are the primary producers of food for most of the aquatic lives and also, they are the chief contributors of producing the atmospheric air we breathe. For this reason, we can generally say that, "In the kingdom of producers, algae are the pioneers." Notably, something else is also even more interesting about algae. Since algae are existing in the extreme kind of ecological conditions, it has to develop a wide range of tolerance against all sort of biotic and abiotic stresses. Thus, to attain and maintain its sustainability, algae produce a variety of active chemical substances as the primary and secondary metabolites. Such kinds of chemical substances of algal origin, with considerable and promising mechanism of action are called the algal metabolites.

Algae have their own set of lovers worldwide, in all parts of the geographical time scale, for a variety of reasons. In the past few decades, the potentials of algae and algal metabolites, as a good quality based supplementary food, as a natural colorant for food additives, as a food preservative, as a source for bio-fuel, as a fertilizer, as a tool to control pollution, as a chemical substance in the biological experiments, as an ingredient in paints, and as a potential therapeutic agent for various life-threatening diseases with antibacterial, antiviral, anti-fungal, anti-protozoan, anti-inflammatory, antioxidant, anticancer properties, were reported enormously (Al-Dhabi and Arasu, 2016; Kumaresan et al., 2017; Sannasimuthu et al., 2020; Velayutham et al., 2021). Even now, definitely in the near future, also thereafter, our comprehensive scientific community's modernistic thirst thrives towards algae and algal metabolites. Such a scientific drive cannot happen as such. It requires acceleration and a perfect control, thus comes the hero biotechnology in action. The biotechnological approach focuses on two different aspects. In one way, it engineers the qualities of algae, in other way it explores the applications of algal metabolites in a right way. Taking everything in account, this chapter summarizes various kinds of algal metabolites and their biotechnological applications.

4.2 GENERAL CHARACTERISTICS

Algae are aquatic, simple, and plant-like organisms: They live in freshwater (rivers, lakes, ponds, rocks, or solid supports where there is the flow of water) or in marine water. Algae are autotrophic: The presence of pigments, especially chlorophyll enables them to produce their own food by photosynthesis. Algae are beautifully colored. The presence of the pigments chlorophylls, carotenoids, and phycobiliproteins (PBP) makes them look green, brown, or red in color. Algae may be unicellular or multicellular: Most of the algae are unicellular and can be seen only with the help of a microscope (microalgae). Some are multicellular and macroscopic in size and are as large as 50 or more meters in length (macroalgae). Algae may be eukaryotic or prokaryotic: Algae are eukaryotic with a defined nucleus. Cyanobacteria or blue-green algae (*Spirulina plantensis*, *Nostoc commune*, *Anabaena sphaerica*) are prokaryotic in nature. Considering the above characteristics, algae are broadly expressed in two important aspects. The first approach is based on the color, which includes green algae Chlorophyceae, brown algae Phaeophyceae and red algae Rhodophyceae. The second approach is based on the size. The algae which are unicellular and can be visualized only with the aid of microscope are called microalgae (*Chlorella vulgaris*, *Haematococcus pluvialis*, *Thalassiosira pseudonana*, etc.), and the large-sized groups can be seen even through the naked eye are called macroalgae (*Codium adhaerens*, *Ulva lactusa* and *Sargassum vulgare*).

4.3 CLASSIFICATION

Classification of algae is the most important aspect in the study of algal biology. Starting from Carolus Linnaeus, who is lovably known as the father of modern taxonomy, many other eminent researchers in various time periods, classified the algae based on their distinct characteristics like the type of pigments, mode of nutrition, morphological features, structural architecture, methods of reproduction, etc. It will definitely take a very accountable time to study the methods of classification of algae. Baweja and Sahoo (2015) summarized almost all of the classification methods instantly and ingeniously. Harwey's classification (1836) was connected with the thallus color and possessed four groups chlorospermae (green), melanospermae (brown), rhodospermae (red) and diatamacea (diatoms). Eichler's classification (1883) grouped fungi and algae into five groups together, cyanophyceae (blue-green algae (BGA)), diatomeae (diatoms), chlorophyceae (green algae), phaeophyceae (brown algae) and rhodophyceae (red algae). Engler and Prantle's classification (1887) included algae and fungi in six groups, schizophyta (cyanobacteria), flagellatae, dianoflagellatae (flagellated protists), bacillariales (diatoms), euphyceae (algae) and eumycetes (fungi) under Euthallophyta. West's classification (1916) was based on the flagella, the algae were classified into isokontae (equal size flagella), akontae (no flagella), stephanokontae (crowned flagella) and heterokontae (unequal flagella). Pascher (1932) classified algae in to eight divisions, chrysophyta, phaeophyta, pyrrophyta, euglenophyta, chlorophyta, charophyta, rhodophyta, and cyanophyta. Tilden's classification was connected with pigments and the five classes were myxophyceae (blue-green) rhodophyceae (red), phaeophyceae (brown), chrysophyceae (yellow-green) and chlorophyceae (green).

In 1935, Fritsch, the father of phycology, classified algae in to 11 classes, chlorophyceae (green), xanthophyceae (yellow-green), chrysophyceae (orange), bacillariophyceae (yellow or golden brown), cryptophyceae (light brown), dinophyceae (brown or dark yellow), chloromonadineae (bright green), euglenophyceae (grass green), phaeophyceae (brown), rhodophyceae (red) and myxophyceae (blue-green). Smith (1955) classified algae in to seven divisions and 13 classes, chlorophyta (chlorophyceae and charophyceae), euglenophyta (euglenophyceae), pyrrophycophyta (desmophyceae and dinophyceae), chrysophyta (chrysophyceae, xanthophyceae, and bacilliriophyceae), phaeophyta (isogenerateae, heterogenerateae, and cyclosporeae), cyanophyta (myxophyceae) and rhodophyta (rhodophyceae). Papenfus (1955) used phylogenetic relationships for his classification with seven divisions and 12 classes, chlophycophyta (chlorophyceae), charophycophyta (charophyceae), euglenophycophyta (euglenophyceae), chrysophycophyta (xanthophyceae, chryosophyceae, and bacilliriophyceae), pyophycophyta (dianophyceae, chryptophyceae, and chloromonadophyceae), phaeophycophyta (phaeophyceae) and rhodophycophyta (rhodophycophyceae). Chapman's classification (1962) included four divisions and nine classes, euphycophyta (chlorophyceae, phaeophyceae, and rhodophyceae), myxophycophyta (myxophyceae), chrysophycophyta (chrysophyceae, xanthophyceae, and babilliriophyceae) and pyrophycophyta (chryptophyceae and dinophyceae). Christensen (1964) included the division cyanophyta in prokaryota. The other 10 divisions are included in eukaryota, rhodophyta (non-motile), chlorophyta, charophyta, euglenophyta, xanthophyta, chrysophyta, bacillariophyta, pyrrophyta, cryptophyta, and phaeophyta (all are motile). Prescott's classification (1969) included nine phyla and various classes, chlorophyta (chlorophyceae and charophyceae), eugleophyta, chrysophyta (chrysophyceae, bacillariophyceae, and xanthophyceae), pyrrophyta (desmophyceae and dinophyceae), phaeophyta (isogeneratae, heterogeneratae, and cyclosporae), rhodophyta (bangioideae and florideae), cyanophyta (coccogoneae and harmogoneae), cryptophyta, and chloromonadophyta. Round (1973) classified algae in to two groups and 12 phyla. He included the phylum cyanophyta in prokaryota. Eukaryota included 11 more phyla, euglenophyta, chlorophyta, charophyta, prasinophyta, xanthophyta, haptophyta, dinophyta, babillariophyta, chrysophyta, phaeophyta, and rhodophyta. Chapman and Chapman (1973) classified alga in to two groups, 11 divisions and 15 classes. The division cyanophyta included in the group prokaryota and the other 10 divisions, rhodophyta, chlorophyta, euglenophyta, chloromonadophyta, xanthophyta, bacillariophyta, chrysophyta, phaeophyta, pyrrophyta, and cryptophyta are included in the group eukaryota. Bold and Wynne (1978), used phyco prefix to phyta as suggested by Papenfuss (1955). The nine divisions of algae in their classification are cyanochloronta, chlorophycophyta, charophyta, euglenophycophyta, phaeophycophyta, chrysophycophyta, pyrrophycophyta, cryptophycophyta, and rhodophycophyta. Parker (1982) considered the presence of well-defined membrane bound organelles and classified algae in to two groups, six divisions and 16 classes. Two of the six divisions (Cyanophycota and Prochlorophycota) in prokaryota and the other four divisions (Rhodophycota, Chromophycota, Euglenophycota, and Chlorophycota) in Eukaryota. Lee (2008) grouped the single division Cyanophyta in Prokaryota and based on the membrane structural organization of chloroplast, the group eukaryota was added with eight more divisions, Glaucophyta, Rhodophyta, Chlophyta, Euglenophyta, Dinophyta, Cryptophyta, Prymnesiophyta, and Heterokontophyta.

4.4 IDENTIFICATION

Since, algae include a diverse group of divisions and classes and are occurring abundantly in a variety of ecological niches, proper recognition and identification of the genus and species becomes more essential for all sort of phycological analysis. Rippka (1988) discussed about various methods employed in the recognition and identification of cyanobacteria. The tentative identification was based on the common names, ecological distribution, and respective phenotypic characters such as color, shape, and forms. Traditional approaches for identification include analysis of structural properties of the cells or filaments, color of the living forms, nature of colonial forms, analysis of botanical code using herbarium specimens, botanical descriptions, more importantly physiological, biochemical, and genetic analysis of the samples. The technical identification was based on critical examination under microscope, oxygen evolution experiments, pigment content analysis and pure culture techniques. Estep and MacIntyre (1989) used system-based image analysis (IA). The collected samples were processed and counted using a microscope. The cells were also photographed and studied with improved image quality. Temraleeva et al. (2016) clearly explained the applications of modern techniques in the recognition and identification of specific strains of algal species including the phylogenetic relationships. The samples were collected and cultured using different methods of mixed culture, plate culture on buried soils and water-soil cultures under defined conditions using special culture mediums like blue-green (BG) 11, BG11o, Bold 3N and cyanophycean. The DNA sample from the cultured species was isolated, amplified using polymerase chain reaction (PCR), electrophoresed using 1% agarose gel, then the DNA fragments were purified and sequenced. The 18S rRNA gene was sequenced and analyzed for the determination of phylogenetic relationships. The sporopollenin content was isolated by centrifugation and it was qualitatively analyzed using hot acetolysis method.

They also reported the presence of *Hemiflagellochloresis kazakhstanica*, *Chlorella mirabilis*, and *Muriella terestris*.

4.5 COLLECTION

Francoeur et al. (2013) summarized different types of qualitative and quantitative methods for the collection of algal samples. The qualitative collection methods were considered as simple and cost effective. The quantitative methods were useful for analyzing the expression of abundance, biomass content and metabolic activity of algae collected from a defined volume of water or surface area of sample collection. The methods were also discussed based on the type of algal aquatic habitats. The free-floating phytoplankton in shallow surface water was collected using a jar or bottle. A rigid acrylic tube was used to collect depth integrated phytoplankton. Kemmerer or Van Dorn Bottle was used to collect the samples from the deeper layer of water. Plankton nets with a mesh of defined sizes were used to collect oligotrophic water. The filamentous metaphyton samples were collected by scooping the material into a jar or collecting with meshes attached with dip nets or suctioning by syringes or by handpicking. The benthic algae samples present at the bottom of the water bodies were collected based on the nature of the substratum where the algal species were growing. The sample was scrapped out using rubber spatulas, spoons, scalpels, razor blades, tooth brushes, etc. The samples were collected along with the substratum in plastic bags. Coring tubes were used for the collection of benthal algal species in a defined area for quantitative purposes. The gravimetric aluminum foil technique was used to collect samples from irregularly defined areas such as rock surfaces. In this technique, the samples were collected from a particular place. The aluminum foil was dipped into the sample removed surface and trimmed to match the surface area of the sample removed region. The ratio of the masses was calculated to measure the area of the sample collected region using the following formula: Size of unknown area (cm^2) = Foil mass from unknown area (mg) × Size of known area (cm^2)/Foil mass from known area (mg).

4.6 ISOLATION

Alasalli et al. (2013) described various methods employed in the isolation and characterization of chemical constituents of the algae. Proteins and other lipid molecules were isolated by Soxhlet extraction and characterized using elemental analysis, Kjeldhal analysis and nuclear magnetic resonance (NMR) analysis. Carotenoids were isolated using liquid-liquid extraction, phenols were isolated using sonication, and all were analyzed by high performance liquid chromatography (HPLC). Antioxidants and other value-added compounds were isolated using pressurized liquid extraction (PLE) and analyzed by capillary electrophoresis and gas chromatography (GC). Fatty acids were extracted by esterification and analyzed by GC. Sulfates and other sugars were isolated using autoclaving and characterized by barium chloride-gelatin method, phenol-sulfuric acid method and HPLC. Supercritical fluid extraction (SFE) was comparatively best among all other methods and reported as an effective method for the isolation of carbohydrates, pigments, carotenoids, lipids, and other active algal biomolecules. Gas chromatography-mass spectrometry (GC-MS), HPLC, and spectroscopy methods were employed for the analysis and characterization of a variety of algal metabolite.

4.7 ALGAL METABOLITES

A variety of chemical substances, produced by the diverse groups of algae, may be required for their development, metabolism, survival, or protection, directly or indirectly for the sustainability, can be collectively called algal metabolites. These primary or secondary algal metabolites include pigments, lipids, alkaloids, phenolic compounds (PCs), photoreceptive substances, phycocolloids, vitamins, nutraceuticals, cosmetics, food, and feed additives.

4.7.1 PIGMENTS

Pigments isolated from the natural sources like marine algae are far better in many aspects compared to the pigments which are chemically synthesized. The simple isolation methods using organic solvents with increased stability of the common algal pigments chlorophylls (*Chlorella*) β-carotene (*Dunaliella salina*), fucoxanthin (phaeophytes), peridin (*Amphidinium carterar*), phycoerythrin (*Porphyridium cruentum*) and phycobilins (*Spirulina plantensis*) pigments and their significant uses were reported (Alan et al., 2018). Chlorophyll α, chlorophyll β and carotenoids were isolated from the freshwater algae *Cladophora glomerata*, *Cladophora rivularis* and *Ulva flexuosa*. Supercritical fluid extraction method was used for extraction of the pigments. The estimated content of chlorophylls was comparatively higher than the total content of carotenoids by this method (Fabrowska et al., 2017).

In a study carried out by Serive et al. (2017), 37 species of microalgae were used for the community analysis of pigment patterns. All of the species were collected and cultured. They employed a novel extraction procedure for isolation of pigments from the cultures. HPLC was applied for the analysis of the isolated pigments. The HPLC analysis signified the presence of 124 algal pigments. The common names of the pure pigments are listed below. Alloxanthin, antheraxanthin, astaxanthin, canthaxanthin, chlorophyll α, chlorophyllide α, chlorophyll β, chlorophyll χ2, chlorophyll χ3, chlorophyll α and b allomers, crocoxanthin, β,ε-carotene, β,β-carotene, diadinoxanthin, diatoxanthin, dinoxanthin, fucoxanthin, 19'-hexanoylfucoxanthin, 19'-but-fucoxanthin, 19'-hex-fucoxanthin, lutein, 9'-cis-neoxanthin, prasinoxanthin, pheophorbide a methyl ester, pheophytin a, peridinin isomer, violaxanthin, and zeaxanthin.

4.7.2 LIPIDS

The lipid molecules present in algae were classified into phospholipids (PLs), glycerolipids, neutral lipids, betaine lipids, etc. Phospholipids: phosphatidylglycerol (PG), phosphatidylcholine (PC), phosphatidyl-ethanolamine (PE), phosphatidylinositol (PI), phosphatidic acid (PA), and phosphatidylserine (PS). Glycolipids (GL): Sulfolipid, monogalactosyldiacylglycerol (MGDG) and digalactosyl diacylglycerol (DGDG). Non polar glycerolipids: Monoacylglycerol, diacylglycerol and triacylglycerol. Betaine lipids: 1,2-diacylglyceryl-3-O-4'-(N,N,N-trimethyl)-homoserine(DGTS),1,2-diacylglyceryl-3-O-2'-(hydroxymethyl)-(N,N,N-trimethyl)-β-alanine (DGTA) and 1,2-diacylglyceryl-3-O-carboxy-(hydroxymethyl)-choline (DGCC)). Fatty acids: Monounsaturated fatty acids (MUFA), polyunsaturated fatty acids (PUFA) and oxylipins (Oxygenated derivatives of PUFA). Sterols: Isofucosterol, Clionasterol, poriferasterol, campesterol, stigmasterol, desmosterol, ergosterol, β-sitosterol, clerosterol, codisterol, cholesterol, 24-methylenecholesterol, brassicasterol, 22-(E)-dehydrocholesterol, etc. (Kumari et al., 2013).

The presence of fatty acids such as α-linolenic acid (ALA), eicosapentanoic acid, arachidonic acid (AA), palmitic acid, lauric acid, stearic acid, oleic acid, myristic acid, hexadecatrienoic acid (HTA), acrylic acid and eicosanoids were reported. The following list of Sterols; Cholesterol, clinoasterol, ALA, γ-linolenic acid, GL, 5-dehydroavenasterol, fucosterol, myristic acid, desmosterol, 5-α-24(S)-ethyl-cholesterane-3-β-6-β-diol, oleic acid, linoleic acid (LA), prostaglandin A_2, Prostaglandin E_2, phytol, and leukotriene B_4. Terpenoids: Diterpenes, secospantane diterpene, acetogenins, neomangicol A, neomangicol B and neomangocol C were also reported (Chauhan and Kasture, 2014).

4.7.3 ALKALOIDS

There are three groups of alkaloids were found in marine algae. These are, phenylethylamine (PEA) alkaloids (phenylethylamine, tyramine, hordenine, and dopamine), indole group alkaloids (almazotone, caulersin, cauleprin, denticine, fragilamide, martensine, and marterfragine), and halogenated indole alkaloids (bromoindole, sulfur-containing bromo alkaloids, polyhalogenated indoles, bromobisindole, and lophocladines) (Guven et al., 2010).

4.7.4 PHENOLIC COMPOUNDS (PCs)

Phenolic compounds (PCs) with bioactive properties were reported by Freile-Pelegrin and Robledo (2013). It included tannins (hydrolyzabe tannins, flavonoid based condensed tannins, and phlorotannins), phenolic antioxidants (phloroglucinol, eckol, dieckol, 2-phloroeckol, 6,6'-bieckol and phlorofucofuroeckol B) and also some other classes of PCs (bromophenols, terphenoids, tichocorpols, meroditerphenoids, coumarines, and vanillic acid).

4.7.5 PHOTORECEPTIVE SUBSTANCES

The photoreceptive substances were extracted and the chief activity of the compounds was discussed (Pangestuti et al., 2018). The active groups are sulfonated polysaccharides (fucoidan and carrageenan), carotenoids (carotenes and xanthophylls, mycosporine like amino acids (MAAs–shinorine, palythine, porphyra-344, asterina 330 and microsporine glycine) and other polyphenols (triphlorethol–A, phloroglucinol, fucofloroeckol–A and dieckol).

4.7.6 POLYSACCHARIDES

The isolation methods and biological activities of the algal polysaccharides (carrageenan, porphyran, sulfated galactan, sulfated manan, fucoidan, laminaran, ulvan, rhamnan sulfate, heterofucan, and s-fucan) were summarized (de Jesus-Raposo et al., 2015; Xu et al., 2017).

4.7.7 PHYCOCOLLOIDS

Phycocolloids include agar, carrageenan, and alginates are employed in science laboratories and food industries for their special gel-forming or solidifying properties. Pangestuti et al. (2015) described the methods for extracting and purifying the phycocolloids from various species of agarophytes, carageenophytes, and alginophytes. The bioactive potentials of these compounds were also discussed. Alginates enriched with 80% of guluronic acid compared to 20% of mannuronic acid was extracted and characterized from five important algal species (Larsen et al., 2003). Agar a well-known gelling agent was isolated in good quality and quantity from the red algae *Gracilaria gracilis*.

4.8 GLOBAL BURDEN OF DISEASES AND MANAGEMENT OF HUMAN HEALTH

The top 15 diseases in humans, by 2030, based on the current prevalence of occurrence were reported by the world health organization (WHO). The reported diseases are ischemic heart disease, cerebrovascular disease, HIV/AIDS, chronic obstructive pulmonary disease, lower respiratory infections, trachea, bronchus, and lung cancer, diabetes mellitus, perinatal conditions, stomach cancer, hypersensitive heart disease, nephritis, and nephrosis, liver, colon, and rectum cancers respectively (Mathers and Loncar, 2006). Heart diseases, infectious diseases, diabetes mellitus and different types of cancer are most common among the above listed diseases. Infectious diseases are caused by bacteria, viruses, fungi, and parasites like protozoa, helminths, and ectoparasites (mosquitoes). They have the tendency to spread through food, water, soil, air, or personal contacts from one individual to others. Depending upon the nature and rate of spreading, such diseases can also be classified as endemics (common in certain groups or places), epidemics (significant spread of disease to many people or many places) and pandemics (rapid and global spread of disease). Cancer was reported as the first or second leading factor of premature deaths in many countries. Lung cancer, breast cancer, colon-rectum cancer, prostate cancer, stomach cancer and cervix-uterus cancer are most common in humans. It is an important agenda of the United Nations to reduce the mortality rate within the next 10 years (WHO Cancer Report, 2020). Diabetes mellitus has become a very common disease in humans. It was estimated that, more than 422 million people were reported with diabetes. In simple 1 in every 11 people were diabetic (WHO Diabetes Report, 2018). Early diagnosis, prevention, control, and cure of such disease in humans remain as a promising challenge for the scientific community worldwide. This is very crucial for not alone the management of good health, also it is important for the development of every nation's economy and for sustainability of the environment too.

4.9 BIOTECHNOLOGICAL APPLICATIONS OF ALGAL METABOLITES

Since, the double-helical structure of DNA was discovered by Watson and Crick (1953), the field of biotechnology boomed magnificently. The r-DNA technology developed by Cohen et al. (1973) enhanced the chances for manipulation of the organism at their genetic level, almost as we desire. Algal biotechnology is a separate field of biotechnology specially emphasizes on understanding the ecology of algae, identification of new algal species, development of advanced culture techniques, screening of a wide range of chemical constituents of algal origin, enhancement of production of algal metabolites under controlled conditions, characterization of alga derived metabolites and determination of biological potentials of algal metabolites. The recent developments in algal genome sequencing and production of transgenic algae accelerate the potentials of algae and algae-based products (Armin, 2015). Hence, the search on algae and applications of algal metabolites in the fields of medicine, agriculture, and the environment becomes enormous. The antibacterial, antiviral, antifungal, antidiabetic, antioxidant, and anti-tumor properties of the algal metabolites are widely investigated and the bioactive potentials are demonstrated extensively for the benefit of mankind. Such a wide range of discoveries with huge applications significantly improved the global market values of the algal metabolites.

Nivedita and Poonam (2017) summarized the biotechnological applications of diverse groups of algae and algae derived bioactive compounds. Broad ecological distribution, high biomass production, diverse range of chemical composition, easy collection and cultivation techniques, sophisticated research facilities, altogether made algae as a challenging source for the research and development sectors. The presence of enriched contents of the organic molecules like carbohydrates, proteins, and lipids served as a good source for the production of biofuels. The algal groups of *Spirulina*, *Dunaliella*, *Scenedesmus*, *Desmodesmus*, *Laminaria*, *Oedogonium*, *Cladophora* were highly used in the production of algal oils. Biodiesel production from algae becomes feasible because of the presence of high content of fatty acids. *Chlorella*, *Euglena*, *Spirogyra*, *Scenedesmus*, *Desmodesmus*, *Pseudokirchneriella*, *Phormidium*, *Nitzschia* were capable of producing bio diesel. The hydrogen-based compounds could be produced from algae in large concentrations by different kinds of biotechnological and biochemical approaches like manipulating the environmental and growth conditions, direct, and indirect photolysis processes, altered ATP driven pathways. Thermal pretreatment of algae, especially *Chlorella*, was found useful for producing biomethane. The increased content of carbohydrates made algae as a major source for producing bioethanol by anaerobic fermentation processes. *Chlorella*, *Dunaliella*, *Spirulina*, *Arthrospira*, *Sargassum*, *Gracilaria*, *Scenedesmus*, *Chlorococcum littorale*, *Prymnesium parvum*, *Euglena gracilis*, *Ulva lactuca*, *Lamineria digita*, *Laminaria hyperborean* were widely involved in bioethanol production. Starch, sugars, and cellulose content of algae enhanced the production of biobutanol. *Ulva lactuca* and *Saccharina* were used because of the high concentration of hexoses and pentoses, resulted in the production of commercially valuable organic solvents acetone, butanol, and ethanol.

Algal metabolites are intensively analyzed for their therapeutic potentials. Algal drugs can be cost effective when compared to the drugs produced from plant or animal sources. Hence, algae are widely examined for the presence of proteins, polypeptides, polysaccharides, phytochemicals, pigments, vitamins, precursor molecules, etc., with therapeutic properties. Many commercially viable bioactive compounds with promising antibacterial properties were isolated from *Chlorella vulgaris* and *Chlamydomonas pyrenoidosa*. Some algal toxins isolated from *Ochromonas* sp. and *Prymnesium parvum* exhibited different types of bioactive potentials. Different types of metabolites isolated from cyanobacteria actively expressed a wide range of antifungal, antibacterial, and antiviral properties. A cyanobacterial metabolite cyanovirin-N was proved effective against HIV. Cyanobacteria extracts were also exhibited antineoplasmic activity. The active metabolites isolated from *Spirulina plantensis* exhibited their antiviral potentials against human cytomegalovirus, herpex simplex virus and measles. Some of the sulfur containing polysaccharides *Porphyridium* sp. was very useful against HSV-1 and HSV-2. Chlorosulfite isolated from *Poterioochromonas malhamensis* showed inhibition property against protein tyrosine kinase. Cryptophycin isolated from *Nostoc* sp., exhibited effective anticancer property. Scytonemin, a protein serine inhibitor isolated from *Stigonema* sp., showed antiproliferative and anti-inflammatory properties. The algal metabolites mycosporine, sporopollenin, and scytonemin isolated from different types of algae such as *Ankistrodesmus spiralis*, *Chlorella minutissima*, *Chlorella sorokiniana*, *Characium terrestre*, *Allexandrium catenella*, etc., are widely used in cosmetic products as an active formulation in antiaging creams, sun protectants, rejuvenating, and hair care products. More specifically the *Spirulina* based products such as *Spirulina* firming masks, *Spirulina* whitening facial masks, *Spirulina* facial scrub, etc., were widely marketed by many cosmetic companies. Phormiskin Bioprotech G, a *Phormidium persicium* based product was marketed for its special photo-protective property.

Algae are extensively used for the extraction of natural pigments such as β-carotene, leutin, zeazanthin, astaxanthin, canthaxanthin, etc. β-carotene isolated from *Dunaliella salina*, exhibited a very effective antioxidant property, hence, it was used in many anticancer drug formulations. β-carotene serves as a precursor for vitamin A and used in nutraceutical formulations. It was also used as natural food colorant, cholesterol controlling agent in managing heart diseases. Astaxanthin, isolated from *Haematococcus pluvialis*, used as natural food colorant, nutraceuticals ingredient, active drug candidate for neurogenerative disorders, metabolic disorders, diabetes mellitus, and various types of cancer. The PBP such as phycocyanin (PC), allophycocyanin, and phycoerythrin, isolated mainly from *Spirulina plantensis* and *Phorphyridium*, was used as natural dye, food colorant, fluorescent marker, cosmetic ingredient, and diagnostic tool.

The polyunsaturated fatty acids such as linolenic acid, eicosapentaenoic acid (EPA), docosapentaenoic acid (DPA) and docosahexaenoic acid (DHA), isolated from *Chlorella minutissima*, *Schizochytrium*, *Spirulina*, *Dunaliella*, etc., was used in the prevention of asthma, arthritis, cancer, depression, schizophrenia, skin, kidney, and cardiovascular disorders. Some species of algae, especially *Chlorella* and *Spirulina* are widely used as food additives for the presence of proteins, poly unsaturated fatty acids, pigments, vitamins, carbohydrates, phenolics, etc. They are also added to various food substances like cookies, noodles, pastas, gums, snacks, and beverages. They are also packed and sold as capsules, tablets, powders, and syrups. The cyanobacteria *Nostoc*, *Anabaena*, and *Tolypothrix* are the sources of the plant growth-promoting factors such as vitamin B12, Auxin like indole-3-acetic acid (IAA) and indole-3-propionic acid. Most of the algal leftover biomass is mostly used as natural fertilizers. The polysaccharides like agar, alginates, and carrageenan were used as gels and thickeners. The uses of agar as gels in laboratories for biomolecules separation, carrageenan as emulsifiers and stabilizers in food industries and alginates uses in textile industries are reported. Carrageenan is widely studied for its antiviral, anticoagulant, and antitumor properties. *Chlorella*, *Nannochloropsis*, *Pavlova*, *Phaeodactylum*, etc., were enriched with protein and other sources, thus used as fish feed in aquaculture. The algal based products are successfully marketed worldwide.

Alassali et al. (2013) described the biological applications of the secondary metabolites including lectins, sterols, polysaccharide complex molecules, beta carotene, astraxanthin, polyunsaturated fatty acids, phenolics, sesquiterpenes, phlorotannins, bromoditerpenes, halogenated furanones, lectins, sulfated polysaccharides, fucoidans, laminarans, porphyran, alginic acid, carrageenan, ulvan, bioactive peptides and their bioactive properties as natural colorants, paint additives, vitamins, minerals, antibiotics, antiviral, antifungal, antifouling, antioxidant, anticoagulant, and anticancer compounds, drugs for degenerative and cardiovascular diseases (CVDs) were reported. These compounds were isolated using different type of methods from nine species of microalgae (*Spirulina*, *Chlorella*, *Muriellopsis*, *Dunaliella*, etc.), and more than 22 species of brown algae (*Halidrys cerenoides*, *Euklohia kurome*, etc.), red algae (*Laurencia rigida*, *Pelisea pulchra*, etc.), and green algae (*Enteromorphalina*, *Ulva rigida*, etc.). The hyperglycemic effects of bromophenols, the antimicrobial properties of pacifenol and styphotriol against *Pseudomonas aeruginosa* and *Streptococcus enteritidis* were specified. The organic solvent extracts of *Cystoseria tamariscifolia* were rich in sulfated polysaccharides and poly sulfates, were tested and proved to be effective against herpes simplex viruses (HSV-1 and HSV-2), cytomegalovirus, respiratory syncytial virus (RSV), etc. The cyanobacterial and *Fucus vesiculosus* extracts exhibited antiviral properties against HIV. Algal phlorotannins and hydrophilic compounds were reported for their anti-allergic, anti-HIV, anti-diabetic, and anticancer properties. Seaweed carotenoids fucoxanthin, neoxanthin, canthaxanthin, peridin, and a metabolite mythyl jasmonate induced apoptosis in many cancer cells. Stypolactone and aromatic acid were effective against human lung and colon carcinoma cells. 14-keto-stypodiol diacetate, monogalactosyl diacylglycerol, DGDG and polyphosphotidylglycerol were reported as anti-inflammatory and anti-thrombic agents. Epitaondiol diacetate was characterized for its positive effect on rat cardiovascular system.

Michalak and Chojnacka (2015) summarized various categories of bioactive substances of algal origins. The potentials of bioactive substances like alginate, cellulose, fucoidan, laminarin, agar, carrageenan, cellulose, furcellaran, mannan, porphytan, xylan, amylose, amylopectin, cellulose, inulin, pectin, ulvan, etc. (polysaccharides), three classes of pigments phycobilins (phycocyanin and phycoerythrin), carotenoids (α-carotene, β-carotene, and lycopene), xanthophylls (fucoxanthin, zeaxanthin, and lutein) and chlorophylls (chlorophyll α, chlorophyll β and chlorophyll χ). The algal extracts were classified based on their active biological properties. Proteins, polyphenols, polysaccharides, pigments, and poly unsaturated fatty acids were categorized for their antibacterial properties. The polyunsaturated fatty acids, pigments, terpenes, and phenols were categorized for their antifungal properties. Proteins, microsporine like amino acids, glutathione (GSH), polyphenols, polysaccharides, poly unsaturated fatty acids, carotenoids, tocopherol, and ascorbate were categorized for their antioxidant property. Proteins, carotenoids, polysaccharides, sterols, polysterols, and porphyrin derivatives were categorized for their anti-inflammatory properties. Polyphenols, carotenoids, and polysaccharides were categorized for their antitumor properties. Proteins, diterpenes, polyphenols, polysaccharides, and carotenoids were categorized for their antiviral properties. β-glucan, fucoidan, laminarin, and astaxanthin were reported for their antimicrobial and antioxidant properties. Carotenoids, astaxanthin, and other photoreceptive compounds such as mycosporine glycine, palythine, asterina-330, palythinol, porphyra-334, shinorine polythene, GSH, phycobilin, and other micosporine like amino acids were widely used as an active ingredient in cosmetic and health care product formulations. Jigna and Avani (2013) reviewed the bioactive potentials of various algal metabolites. Zoanthamine alkaloids were highlighted for their antibacterial, anti-osteoporotic, and cytotoxic activities. Sphingosine derivatives were reported for their antiviral property. Stypoldine, monoterpen, and nalomon were reported for their cytotoxic activities.

4.10 THE NEED FOR DEVELOPMENT OF NEW ANTIBIOTICS AGAINST ANTIBIOTIC RESISTANCE BACTERIA

WHO developed a global priority pathogens list (Global PPL) by the application of a technique called muti-criteria decision analysis (MCDA). The major criteria considered for the analysis are all-cause mortality, transmitability, preventability, treatability, health care and community burdens, prevalence of resistance, 10 years more trends of resistance and current pipeline. The Global PPL was finalized in the WHO meeting held at Geneva in January 2017. Based on the threats and demand for the discovery and development of new antibiotics, the antibiotic-resistant bacteria were classified into three tiers: Priority 1: CRITICAL category includes, *Acinetobacter baumannii* (carbapenem-resistant), *Pseudomonas aeruginosa* (carbapenem-resistant) and *Enterobacteriae* (carbapenem-resistant and *Sephalosporin* resistant species of *Klebsiella pneumoniae, Escherichia coli, Enterobacter* spp., *Serratia* spp., *Proteus* spp., *Providencia* spp., and *Morganella* spp.). Priority 2: HIGH category includes *Enterococcus faecium* (vancomycin-resistant), *Staphylococcus aureus* (methicillin-resistant and vancomycin-resistant), *Helicobacter pyroli* (clarithromycin resistant), *Campylobacter* (fluoroquinolone-resistant), *Salmonella* spp. (fluoroquinolone resistant) and *Neisseria gonorrhoeae* (third-generation cephalosporin-resistant and fluoroquinolone-resistant). Priority 3: MEDIUM category includes *Streptococcus pneumoniae* (penicillin non-susceptible), *Haemophilus influenzae* (ampicillin resistant) and *Shigella* spp. (fluoroquinolone resistant). Mycobacteria (*Mycobacterium tuberculosis*) was already listed in global priority category and there still exists a great need of potential antibiotics. The panel also insisted the research and development sectors for the development of antibiotics for pediatric populations and production of oral formulations for community diseases like typhoid and extended-spectrum beta-lactamase (ESBL) producing bacterial infections.

4.10.1 ALGAL METABOLITES WITH ANTIBACTERIAL ACTIVITY

El-Shafay et al. (2016) studied the antibacterial effects of bioactive compounds isolated from the algae *Sargassum fusiforme* and *Surgassum vulgare* against the multi-drug resistant (bacteria (MDR), *Pseudomonas aeruginosa* (PA 1 and PA 2), *Staphylococcus aureus*, *Klebsiella pneumoniae*, *Shigella flexneri* and *Escherichia coli*. In the preliminary antibiotic susceptibility analysis, the diethyl ether extract of *S. fusiforme* and ethanolic extract of *S. vulgare* exhibited good antibacterial activity against the pathogens, especially against *Staphylococcus aureus* and *Klebsiella pneumoniae*. So, the diethyl ether extract was subjected to GC-MS analysis. The GC-MS report of diethyl ether extract revealed the presence of the following bioactive compounds: 10,13-octadeadienoic acid, tritriacontane, eicosane, 2-butyloctanol, pentacosane, dotriacontate, octatriacontyl pentafluoropropionate, nonacosane, hentriacontane, eicosane, tricosane, dicosane, and pentatriacontane. The GC-MS report of ethanol extract of *S. vulgare* showed the presence of di-n-acetyl phthalate (48%) and other compounds (cyclopentaneacetic acid, tonalid, tetratetraconate, octacosane, heneicosane, pentacosane, and heptacosane). The phytochemical analysis indicated the presence of fatty acids, acetogenins, terpenes, indoles, halogenated hydrocarbons, and PCs.

Rajivgandhi et al. (2018) isolated around 50 types of endophytic actinomycetes from the green algae *Caulerpa taxifolia* collected from various parts of India. The effects of the isolates were checked against the MDR urinary tract infection-causing pathogens, *Escherichia coli*, *Proteus mirabilis*, *Pseudomonas aeruginosa*, *Klebsiella pneumoniae* and *Enterobacter* sp. Among the 50 isolates, 20 of them showed activity against most of these pathogens. The top five isolates were named as DMS 1, DMS 2, DMS 3, DMS 4 and DMS 5. DMS 3 which showed excellent activity against most of the pathogens, hence, subjected to liquid-liquid extraction using ethyl acetate for the isolation of bioactive compounds. The ethyl acetate fractions exhibited a variety of algal secondary metabolites. They suggested for further research on these secondary metabolites as a potential target for developing new antibiotics against MDR uropathogens.

Kasanah et al. (2015) reviewed and summarized more than 40 molecules isolated from different species of red algae and the antibacterial activity against many pathogenic bacterial species were discussed. The algal metabolites such as laurenterol, halomon, callicladol, paclifenol, irienol, laurepinnacin, elatol, 5-acetoxypalisadin, palisadin B, pannosane, pannosonal, chlorofucin, acetoxyangasiol, aplysidiol, cupalaurenol, 1-methyl-2,3,5-trobromoindole, chamgrane epoxide, cholest-4-ene-3,6,dione, 5a-cholkestane-3,6-dione, cholest-4-en-3-on, stigmasterol, betastigmasterol, brassicasterol, campesterol, Beta-cryptoxanthin, bosseopentaenoic acid, 10-hydroxyectadectetraeonoic acid, 13-S-hydroxyoctadecatrienoic acid, 13-ketoocta-dienoic acid, 14-methyl-10-oxopentadecanoic acid, octadec-9-eroic acid, octadecanoic acid, chlorobenzene, bromophycolide P, bromophycolide Q, bromophycoic acid A, bromophycoic acid C, bromophycoic acid, callophycoic acid G, callophycoic acvid A, neuromenolides N, Neromenolide B, etc., were determined. All of these algal metabolites were isolated from different algal species of *Laurencia glandulifera*, *L. microladia*, *L. saitoi*, *L. pannosa*, *Asparagopsis taxiformis*, *Graciela gracius* and *Neuromenia fraxinifolia*. The target pathogens reported are, *Escherichia coli*, *Pseudomonas aeruginosa*, Methicillin Resistant *Staphylococcus aureus* (MRSA), and Vancomycin-Resistant *Enterococcus faecium*.

Jena and Subudhi (2019) reported the antimicrobial potentials of algal extracts and algal compounds. Ambiguine I isonitrile isolated from *Fischrella* sp. was effective against *Staphylococcus albus* and *Bacillus subtilis*. Carbamidocyclophanes extracted from *Nostoc* sp. and Norbietane diterpenoid was effective against *Staphylococcus aureus*. Gamma-lactone malyngolide 14 was effective against *Lyngbya majuscula*, *Mycobacterium smegmatis* and *Streptococcus pyogenes*. Noscomin isolated from *Nostoc commune* was effective against *Bacillus cereus*, *Staphylococcus epidermidis* and *Escherichia coli.* Cycloeudesmol determined from *Chondria oppostíclada* was effective against *Staphylococcus aureus* and *Candida albicans*. Hapalindole characterized from *Fischerella* sp. was effective against *Staphylococcus aureus*, *Pseudomonas aeruginosa*, *Salmonella typhi* and *Escherichia coli.* Chlorellin isolated from *Chlorella vulgaris* and *C. pyrenoidosa* was effective against *Bacillus subtilis*, *Streptococcus pyogenes*, *Staphylococcus aureus*, *Escherichia coli* and *Pseudomonas aeruginosa*. α-linolenic acid was effective against *Bacillus subtilis*, *B. cereus*, *Staphylococcus aureus*, *Enterobacteter aerogenes* and methicillin-resistant *Staphylococcus aureus* (MRSA) (Pratt, 1948).

Polyphenols extracted from different species of marine algae were active against *Pseudomonas aeruginosa*, *Listeria monocytogenes*, *Enterococcus faecalis*, *Salmonella abony*, *Bacillus subtilis*, *Staphylococcus aureus*, *Escherichia coli*, *Aeromonas hydrophila*, *Proteus vulgaris*, *Klebsiella pneumoniae*, *Shigella flexneri*, *Vibrio vulnificans* and *Salmonella typhi.* Sulfonated polysaccharides showed activity against *Bacillus subtilis*, *Staphylococcus aureus*, *Escherichia coli*, *Enterobacter aerogens*, *Pseudomonas aeruginisa*, *Salmonella choleraesuis* and *Salmonella typhi.* Sulfonated galactan isolated from *Chaetomorpha aerea* showed antibacterial activity against *Staphylococcus aureus*, *Salmonella enteritidis*, *Pseudomonas aeruginosa*, *Enterococcus faecalis*, *Bacillus subtilis* and *Micrococcus luteus.* Labdane diterpenoids isolated from *Ulva fasciata* showed activity against *Vibrio parahaemolyticus*, *Vibrio alginolyticus* and *Vibrio vulnificus.* Fatty acids, alkanes, and polyphenols isolated from *Haematococcus pluvialis* showed activity against *Staphylococcus* aureus and *Escherichia coli.* Bromophenols isolated from *Odonthalia corymbifera* showed activity against *Micrococcus luteus*, *Bacillus subtilis*, *Staphylococcus aureus*, *Proteus vulgaris*, *Escherichia coli* and *Salmonella typhimurium*. Indolic compounds, Poly Unsaturated Fatty Acids (PUFAs), b-iodine, and neophytadiene showed activity against *Candida albicans*, *Punaliella salina*, *Staphylococcus aureus* and *Escherichia coli* (Abu-Ghannam and Rajauria, 2013). Three types of fatty acids, oleic acid, palmitic acid, and stearic acid were isolated and characterized from *Ulva rigidia.* These compounds were active against *Salmonella typhimurium*, *Aeromonas salmonicida*, *A. hydrophila*, *Streptococcus agalactiae*, *Staphylococcus aureus* and *Enterococcus faecalis* (Ismail et al., 2018).

Four different strains of *Dunaliella* strains were analyzed for the production of bioactive compounds and their antibacterial activity was tested against different strains of *Escherichia coli*, *Brochothrix thermosphacta*, *Enterobacter cloacae* and *Staphylococcus aureus*. Lutein among the 13 other pigments produced, stearic acid among the other 11 fatty acids produced and a phenol ferulic acid exhibited maximum antibacterial activity (Kilic et al., 2019).

4.10.2 ALGAL METABOLITES WITH ANTIVIRAL ACTIVITY

David Baltimore (1971) classified viruses into seven classes based on the concept of central dogma proposed by Francis Crick (1958). Viruses, though they are non-living and submicroscopic in size, they can easily cause severe ill effects in other living organisms. Compared to all other diseases, viral outbreaks are unpredictable, unmanageable, and most dangerous to the mankind. The WHO's Report on, "Managing Epidemics (2018)" listed top 15 deadliest diseases: Ebola virus disease, Lassa fever, Crimean Congo hemorrhagic fever, yellow fever, zika, chikungunya, avian, and other zoonotic influenza, and pandemic influenza, middle-east respiratory syndrome, monkeypox, cholera, plague, leptospirosis, meningococcal meningitis and demonstrated about the causes, mode of transmission, signs, and symptoms, complications, distribution, and management processes. Among these, viral diseases were most in number and were found extremely difficult in management. Recently in March, 2020, Hardling and Lanese, had reported about the mortality effects of the top 12 deadliest viruses. It included Marburg virus, Ebola virus, rabies virus, human immunodeficiency virus (HIV), Hantavirus, influenza virus, dengue virus, Rotavirus, severe acute respiratory syndrome – coronaviruses (SARS-CoV, SARS-CoV-2) and middle east respiratory syndrome coronavirus (MERS-CoV). COVID-19 pandemic, caused by coronavirus emerged as the deadliest viral disease in 2020. All over the World, by April 2020, more than 3.5 million people were affected, and around 2.5 lakhs of deaths were recorded. Such kind of pandemics deeply demanded the necessity for the discovery and development of new vaccines and antiviral therapeutic compounds. Braney and Graham (2013) discussed about the status of patented antiviral vaccines and insisted on various approaches regarding the development of new vaccines through the biotechnological approaches like analyzing the genome of viruses, understating the structure, function, and properties of antigens and antibodies, determining the mechanism of action of viruses and viral components, understanding the immunological processes, isolating, and analyzing human monoclonal antibodies. De Clarcq and Li (2016) described about 90 approved antiviral drugs of different origin and were effective against nine human viruses: HIV, hepatitis C virus (HCV), influenza virus, RSV, hepatitis B virus (HBV), human papillomavirus (HPV), human cytomegalovirus (HCV), herpes simplex virus (HSV) and varicella-zoster virus (VZV).

The antiviral drugs were supposed to be formulated with multilevel therapeutic potentials such as anti-oxidant, anti-inflammatory, and anti-immunomodulatory properties. Since, algae were naturally evolved with a diverse group of primary and secondary metabolites, they could be the definite source for the antiviral compounds. The drug used in anti-viral therapy mostly exhibited the following mode of actions. The nucleoside analogs showed inhibition of viral replication by replacing the natural nucleotides. The non-nucleosides slowed down the reverse transcription process by blocking nucleoside binding site. Protein inhibitory property of the antiviral agents resulted in the production of immature viruses. Blocking of the receptor prevented entry of the virus into the host. The algal metabolites lectins (cyanovirin-N, microvirin, scytovirin, grifithin, etc.), sulfated polysaccharides, fucoidans, lamarins, and phenolic compounds exhibited all such property together in a study of HIV therapy (Natalaya et al., 2019).

The algal metabolites zidovudine, idoxuridine, trifluridine, virabaine, bromovinyldozyuridine, fluoriodoaracystosine, ganciclovir, acyclovir, sulfo-quinovosyl diacylglycerol (SQDG), scytovirin, and spirulin were effective against HIV-1, HIV-2 and HSV (Micheal et al., 1990; Sayda et al., 2012; Susana et al., 2012). *Haematococcus pluvialis* and *Dunaliella salina* extracts were tested against HIV. The GCMS analysis of the extracts revealed the presence of fatty acids such as propionic acid, butanoin acid, palmitic acid, ALA, β-ionone, neophytadiene, and phytol (Susana et al., 2012). Lectins showed a very good antiviral property against HIV and influenza virus (Ana et al., 2019). Cyanovirin-N, scytovirin, and grifithsin were reported for their antiviral properties against HCV (Takebe et al., 2013).

The algal polysaccharides carrageenan (IVA, DENV, HIV, HRV, HPV, HSV-1 and HSV-2), galactan (HSV-1, HSV-2, HIV-1, HIV-2, DENV, and HAV), alginate (HIV, IAV, and HBV), fucan (HSV-1, HSV-2, HCMV, VSV, HIV-1 and sindbis virus), laminarin (HIV), naviculan (HSV-1 and HSV-2), p-KG03 (EMCV and IVA), A1 and A2 (IVA, IVB, RSV-A, and RSV-B), calcium spirulan (HSV-1 MV, MuV IVA, Polio, Coxsackie HW-1 and HCMV), nostaflan (HSV-1, HSV-2, IVA, and HCMV) were reported and used widely for their antiviral properties against many human viruses.

4.10.3 ALGAL METABOLITES WITH ANTIFUNGAL ACTIVITY

Many of the fungi identified were associated with marine or freshwater algae. The algal fungal relationship was critical to explain unless properly investigated. The parasitic fungal complications could be neutralized with natural bioactive constituents of algae. Such algae and algal components were investigated widely and used in commercial pharmaceutical sectors (Nagale and Wedge, 1990). The crude extracts of *Dilsea carnosa*, *Laurencia pinnatifida* and *Odanthalia dentata* were tested against *Microsporum canis* and *Trichophyton veruucosum* (Tariq, 1991). Steroids, alcoholic, and nitrogen-based compounds of various algae-like *U. prolifera*, *A. spicifera* and *L. paniculata* exhibited activity against *Candida albicans*, *Aspergillus niger*, *Mucor* sp., and *Paecilomyces* sp. The GCMS analysis of the *L. paniculata* revealed the presence of bioactive compounds aristolene, phytol, cholesta-3,5-diene, cholest-α-en-3-one, trans-2-α-bisabolene epoxide, etc. (Mickymaray et al., 2018). The cyclic polypeptides lazaphylin A, B, C, D, and E were isolated from *Anabena laxa* showed fungicidal activity against *Aspergillus oryzae* and *Candida albicans*. They also exhibited antitumor activity against the human epidermal carcinoma and human colorectal adenocarcinoma cell lines. The extracts of *U. lactuca*, *U. fasciculata*, *E. compressa*, *P. capillacea*, *H. musciformis* and *P. paronica* were tested against *Fusarium solani*, *Fusarium oxysporum*, *Trichoderma hamatum*, *Aspergillus flavipes* and *Candida albicans*. The GCMS analysis of the algal extracts revealed the presence of a variety of bioactive algal compounds like palmitic acid, ethyl ester, methyl ester, LA, n-heptacosane, trichloromethyloxirane, 11-actadecenoic acid, myristic acid, etc. The GCMS analysis of different algal extracts like *Bostrychia tenella* exhibited various bioactive compounds with antifungal, antibacterial, and anti leishmanicidal properties. The chemical profile of algae included different classes like alkanes, alkenes, alcohols, aldehydes, aromatics, esters, amides, carboxylic acids, fatty acids, cholesterol derivatives, phenols methyl esters, etc. The metabolites, encalptol, phytol, benzoic acid, 3,4, dihydroxy benzaldehyde, 4-hydroxy-benzaldehyde, phytadiene, neophytadiene, phy-2-ene, hexahydrofernesyl acetone, 7-β-aminomethyl cholesterol, 7-α-β-aminocholesterol, etc., were determined from algae. Antifungal scytopycins and hassallidins were isolated from *Anabeena* sp., and *Nostoc* sp. (Shishido et al., 2015). Two potential antifungal compounds, butylated hydroxytoluena (BHT) and hexadecanoic acid methyl ester were isolated by TLC method from *Microcystis aeruginosa* and the chemical structures where characterized by GCMS and NMR (Marrez and Sultan, 2016). The endometabolites and exometabolites of microalgae, lobocyclamide A, B, C, cryptophycin, anachelin H and scytoscalarol were isolated for their fungicidal properties along with other bioactive compounds of algae (Maheep, 2014).

4.10.4 ALGAL DRUGS WITH ANTIPARASITIC ACTIVITY

Malaria (*Plasmodium* spp.), African trypanosomiasis (*Trypanosoma* spp.), Chagas disease (American trypanosomiasis), Leishmaniasis (*Leishmania* spp.), Toxoplasmosis (*Toxoplasma* spp.) and cryptosporidiosis (*Cryptosporidium* spp.) are the most common parasitic infection in humans. The repurposing category of are widely used in the management of such parasitic infections. Some of the parasites showed drug resistance against drugs. Hence, an innovative approach is mandatory to treat parasitic infection (Katherine et al., 2014). Human African trypanosomiasis (HAT) and Chagas disease were caused by the *Trypanosoma parasites*, *Trypanosoma brucei rhodesience*, and *Trypanosoma cruzi.* Marine based drugs, including some of the algal metabolites, and their synthetic analogues were mostly involved in the management of tryanosomatid diseases (Jones et al., 2013). Leishmaniasis could be easily transmitted in humans through the sand flies, Phlebotomous, and Lutzomiya. Bifurcatriol, atomaric acid, pachydictyol A, dolabelladienetriol, fucossterol, elatol, obstural, fucoidan, silphiperfol-5-en-3-ol, etc., belonging to diterpene, triterpene, sesquiterpene, and sulfated polysaccharides were reported very effective in controlling *Leishmania* spp. (Yamthe et al., 2017). Sulfoquinovosyldiacyl glycerols (SQDG) isolated from the chloroform extract of brown algae, *Lobophrora variegata* exhibited definite antiprotozoan property against *Gierdia intentinalis*, *Entamoeba histolytica* and *Trichomonas vaginalis* (Cantillo-ciaeu et al., 2013). A linear diterene, bifurcatriol isolated and characterized from *Bifurcaria bifurcata* was proved very effective against *Plasmodium falciparam*, *Trypanosoma brucei rhodesiense*, *Tryoanosoma cruzi* and *Leishmania donovani* (Smyrniotopoulus et al., 2017). The halogenated terpenes, brominated meroterpenes, tocopherol derivatives and carotenoid like compounds isolated from *Cytoseira tamariscifolia*, *Cytoseira baccata* and *Ulva lactuca* were able to inhibit the growth of *Plasmodium* sp. in liver stage (LS) (Jasmine et al. (2013).

4.10.5 ALGAL METABOLITES WITH ANTI-DIABETIC ACTIVITY

The presence of secondary algal metabolites phlorotannins, the polysaccharides fucoids, alginates, laminarans, alginic acid, and the pigment fucoxanthin made algae as a rich source of therapeutic substances with various bioactive potentials. Eckol, dieckol, 2-phloroeckol, 7-phloroeckol, 6,6'-eckol fucodiphloroeckol, phlorofucoeckol, dioxinodehydroeckol, fhlorofucofuroeckol-A, fucodiphloroethol G, fucosterol, floroglucinol, diphlorethohydroxycarmalol (DPHC), sargahydroquinoic acid, pheophorbide-A, pheophytin-α, etc., were the active molecules responsible for antidiabetic property. The ability of these molecules to inhibit the enzyme-like amylase, glucosidase, aldolase, peptidase, protein tyrosine phosphatase, etc., could play a significant role in controlling diabetes mellitus (Thilia, 2020). The phytochemical screening of a marine seaweed, *Halymenia durvillei* extracts revealed rich content of bioactive compounds, sterols, phenols, saponins, hydroquinones, and also for their nutritional values, it was suggested as a good nutraceutical source for diabetic people. In an *in-silico* analysis, astaxanthin, AA, brassicasterol, β-stigmasterol, β-carotene, canthaxanthin, DHA, EPA, fucoxanthin, γ-aminobutyric acid, γ-linolenic acid, lutein lycopene, microcolin A, okadoic acid, zeaxanthin was used as ligands and fructose 1,6 bisphosphatase, glucokinase, human multidrug-resistant protein, glycogen synthase kinase, peroxisome proliferator-activated receptor, and cytochrome P450 were used as receptors. The high binding efficiency of the algal pigments fucoxanthin, zeaxanthin, and lutein were high with most of the enzymes, suggested their antidiabetic property. In an *in vitro* enzyme inhibitory analysis, the extracts of *Sargassum* spp. were used. The GCMS analysis of the algal extracts revealed the presence of fucosterol with effective antidiabetic property (Unnikrishnan et al., 2015). *In vitro* analysis of *Ishige okamurae* revealed the presence of phlorotannins composed of phloroglucinol functional units and suggested their significant anti-diabetic and anti-obesity properties (Hye-Won et al., 2019).

4.10.6 ALGAL METABOLITES WITH ANTICANCER PROPERTY

The anticancer potentials of polysaccharides extracted from various natural sources were discussed with their possible mechanism of action (Fedorov, 2013). The polysaccharides, fucanoidans, phycocyanin, chlorophyll, pheophytin, β-carotene, fucoxanthin, siphonaxanthin, benzochromenone, stypodiol diacetate, glycoprotein, meroditerpenoids, styposterol triacetate, elatol, bis (2,3 dibromo-4-5 dihydroxybenzyl ether (BDDE), sargachromanol E, cannabinoids, yessotoxins, monoterpenes, riphenylated toluquinines and toluhydroquinones were analyzed extensively for their anticancer potentials. The biological properties of these algal metabolites and their significant role in anticancer therapy are very prominent (Table 4.1).

TABLE 4.1 Anticancer Activities from Algal Metabolites*

Algal Metabolites	Types of Cancer/Cell Lines
Apratoxin A	Osteocarcinoma U2O6, lung cancer NCI-H460, Oral epidermoid cancer KB, mammary adenocarcinoma MDA-MB-435, colon cancer, etc.
Apratoxin B & C	Oral epidermoid cancer KB
Apratoxin D	Lung cancer H460
Apratoxin E	Osteocarcinoma U2O6, Colon adenocarcinoma HT29, Epithelial carcinoma Hela
Apratoxin F & H	Colorecteral cancer cells HCT-116
Aurilide B	Lung cancer H460
Aurilide C	Lung cancer NCL-H460
Belamide A	Colon carcinoma HCT-116
Bisebromoamide	Epithelial carcinoma Hela
Biselyngbyaside	Epithelial carcinoma Hela, central nervous system SNB-78, lung cancer NCT H522
Borophycin	Human epidermoid carcinoma, Human adenocolorectal adenocarcinoma
Calothrixin A & B	Epithelial carcinoma Hela, leukemia CEM, Jurka T cells
Carotenoids	Colon carcinoma HCT-116
Caylobolide A & B	Colorecteral cancer cells HCT-116, colon adenocarcinoma HT29, epithelial carcinoma Hela
Chrysolaminaran	Colorectal adenocarcinoma HT 29, DLD-1
Coibamide A	Lung cancer NCI-H460, breast cancer MDA-MB-231 melanoma LOX IMVI, leukemia HL-60, astrocytoma SNB75, mammary adenocarcinoma MDA-MB-435, ovarian carcinoma SKOV3
Cryptophycin 1 & 8	Human solid tumors and tumor cell lines
Curacin A	Colon adenocarcinoma HT29, Hep G2 Cells
Dolostatins	Breast carcinoma MX-1, lung carcinoma LX
Draganomide	Lung carcinoma A549, colorectal adenocarcinoma HT 29, melanoma MEL-28
Eicosapentaenoic acid	Breast carcinoma BT 20
Ethyl homonate A	Lung carcinoma
Fucoidan	Lewin lung adenocarcinoma
Fucoxanthin	Human leukemia HL-60, colon adenocarcinoma Caco-2, HT-29, DLD-1, prostate cancer PC – #, DU 145 and LNCaP
Grassystatins	Lung carcinoma
Hectochlarin	Lung carcinoma
Hermitabides A & B	Lung carcinoma
Hoiamide A & B	Lung carcinoma

TABLE 4.1 *(Continued)*

Algal Metabolites	Types of Cancer/Cell Lines
Homodolastatin 16	Esophageal cancer WHCO1 & WHCO6, cervical cancer cells ME180
Isomalyngamides	Breast cancer MCF-7
Jamaicamides	Lung carcinoma
Kalkitoxin	Colon carcinoma HCT-116
Kempokektins	Colon carcinoma
Lagunamide C	Lung carcinoma A549, prostate carcinoma PC3, ileocecal colorectal carcinoma HCT8, ovarian carcinoma SK-OV
Lyngbyabellins	Nasopharyngeal cancer KB, colon adenocarcinoma LoVo
Lyngbyastatins	Lung carcinoma
Largazole	Human mammary epithelial carcinoma MDA-MB-21
Majasculomide C	Glioblastoma SF-295, lung carcinoma NCI-H460
Melevimide D	Lung carcinoma A549, melanoma MEL-28, colon carcinoma HT-29
Malyngamides	Lung carcinoma H460
Nostocyclo peptides A1 & A2	Oral epidermoid carcinoma KB, colon carcinoma LoVo
Obyanamide	Cervical carcinoma, lung carcinoma, gastero carcinoma BGC
Palauamide A	Lung carcinoma
Palmysolidc	Lung carcinoma
Pitipeptolides	Colon carcinoma
L-Phycocyanin	Human mammary epithelial carcinoma MDA-MB-21
2-Pentadecanone	Breast cancer cell lines MCF 7
Polysaccharide aldehydes	Colon adenocarcinoma Caco-2, COLO 205, lung carcinoma A549
Pseuododysidenin	Lung adenocarcinoma A549, colon adenocarcinoma HT-29, melanoma MEL-28
Scytonemin	Jurkat T cells
Somocysstinamide A	Mouse neuroblastoma cells
Stigmasterol	Liver hepatocellular carcinoma Hep G2
Symplostatin	Mice colon carcinoma
Tasiamide	Colon carcinoma
Tasipeptins	Oral epidermoid cancer KB
Tiglicamides	Oral epidermoid cancer KB
Ulongapeptin	Oral epidermoid cancer KB
Violoxanthin	Breast adenocarcinoma MCF-7
Wewakazole	Lung carcinoma H460
Wewakpeptins	Lung carcinoma H460

*Vijayakumar and Menaka (2015); Martinez et al. (2018); Abd El-Hack et al. (2019).

4.11 CONCLUSIONS

Since the starting of the human civilization, bioactive compounds, which are obtained from various algae have been explored for human kinds. Algae are rich sources of essential lipids, proteins, polysaccharides, polyketides, vitamins, antioxidants, steroids, fibers, lectins, and pigments. Therefore, algae are widely in various countries due to their pharmaceuticals as well as nutraceuticals. The secondary metabolites derived from algae possess various pharmaceutical uses such as antimicrobial, anticancer, and antiviral activities. In recent years, our understanding in the field of algal metabolites has improved significantly, however there are still several steps we have to move. Definitely, by disclosing novel functions of various available algal metabolites, a new era will appear with unique reflection to humanity. This chapter has emphasized that algae are promising sources of bioactive compounds such as polyphenolics, lipids, polysaccharides, steroids, alkaloids, and terpenes which have various utilities in many industries. Nevertheless, further studies are required for investigating the profile of secondary metabolites of algae in order to make them more useful for the welfare of human. Further study will no doubt investigate new functions for secondary metabolites and future research in this field will be promising.

KEYWORDS

- **algal biomass**
- **algal metabolites**
- **alkaloids**
- **antibacterial activity**
- **anticancer property**
- **anti-diabetic activity**
- **antifungal activity**
- **antiparasitic activity**
- **antiviral activity**
- **biological applications**
- **human health**
- **lipids**
- **metabolite profiling**
- **phenolic compounds**
- **photoreceptive substances**
- **phycocolloids**
- **pigments**
- **polysaccharides**

REFERENCES

Abd El-Hack, M. E., Abdelnour, S., Alagawany, M., Abdo, M., Sakr, M. A., Khafaga, A. F., & Gebriel, M. G., (2019). Microalgae in modern cancer therapy: Current knowledge. *Biomed. Pharmacother.*, *111*, 42–50.

Abdo, S. M., Mona, H. H., Waleed, M. E., Rawheya, A. S. E. D., & Gamila, H. A., (2012). Antiviral activity of freshwater algae. *J. Appl. Pharm. Sci.*, *2*, 21–25.

Abu-Ghannam, N., & Rajauria, G., (2013). Antimicrobial activity of compounds isolated from algae. In: Domínguez, H., (ed.), *Functional Ingredients from Algae for Foods and Nutraceuticals* (pp. 287–306). Woodhead Publishing: Cambridge, UK.

Alassali, A., Cybulska, I., Brudecki, G. P., Farzanah, R., & Thomsen, M. H., (2016). Methods for upstream extraction and chemical characterization of secondary metabolites from algae biomass. *Adv. Tech. Biol. Med.*, 4.

Al-Dhabi, N. A., & Arasu, M. V., (2016). Quantification of phytochemicals from commercial *Spirulina* products and their antioxidant activities. *Evi.-Bas. Comp. Alt. Med.*

Andrews, K. T., Fisher, G., & Skinner-Adams, T. S., (2014). Drug repurposing and human parasitic protozoan diseases. *International Journal for Parasitology: Drugs and Drug Resistance*, *2*, 95–111.

Armin, H., (2015). Algae biotechnology – green cell-factories on the rise. *Curr. Biotechnol.*, *4*, 389–415.

Baltimore, D., (1971). Expression of animal virus genomes. *Bacteriol. Rev.*, *35*, 235–241.

Baweja, P., & Sahoo, D., (2015). *Classification of Algae* (Vol. 26, pp. 31–55). Springer, New York.

Besednova, N. N., Zvyagintseva, T. N., Kuznetsova, T. A., Makarenkova, I. D., Smolina, T. P., Fedyanina, L. N., Kryzhanovsky, S. P., & Zaporozhets, T. S., (2019). Marine algae metabolites as promising therapeutics for the prevention and treatment of HIV/AIDS. *Metabolites*, *9*, 87.

Bold, H. C., & Wynne, M. J., (1978). *Introduction to Algae: Structure and Reproduction* (p. 720). Prentice-Hall of India, New Delhi.

Cantillo-Ciau, Z., Moo-Puc, R., Quijano, L., & Freile-Pelegrín, Y., (2010). The tropical brown alga *Lobophora variegata:* A source of antiprotozoal compounds. *Mar. Drugs*, *8*, 1292–1304.

Chapman, D. J., & Chapman, V. J., (1973). *The Algae* (2nd edn., p. 497). Macmillan, London.

Chapman, V. J., (1962). *The Algae* (p. 492). Macmillan Publishers, London.

Chauhan, J., & Kasture, A., (2014). Antimicrobial compounds of marine algae from Indian coast. *Int. J. Curr. Microbiol. Appl. Sci.*, *7*, 526–532.

Christensen, T., (1964). The gross classification of the algae. In: Jackson, D. F., (ed.), *Algae and Man* (pp. 59–64). Plenum Press, New York.

Clercq, D. E., & Li, G., (2016). Approved antiviral drugs over the past 50 years. *Clin. Microbiol. Rev.*, *29*, 695–747.

Cohen, S. N., Annie, C. Y. C., Boyer, H. W., & Helling, R. B., (1973). Construction of biologically functional bacterial plasmids *in vitro*. *Proceedings of National Academy of Sciences of the United States of America*, *11*, 3240–3244.

Crick, F. H. C., (1958). On protein synthesis. *Symp. Soc. Exp. Biol.*, *12*, 138–163.

El Shafay, S. M., Ali, S. S., & El-Sheekh, M. M., (2016). Antimicrobial activity of some seaweeds species from Red sea, against multi-drug resistant bacteria. *Egypt. J. Aquat. Res.*, *42*, 65–74.

Estep, K. W., & MacIntyre, F., (1989). Counting, sizing, and identification of algae using image analysis. *Sarsia*, *74*, 261–268.

Fabrowska, J., Messyasz, B., Szyling, J., Walkowiak, J., & Łęska, B., (2017). Isolation of chlorophylls and carotenoids from freshwater algae using different extraction methods. *Phycol. Res.*, *66*, 52–57.

Fedorov, S. N., Ermakova, S. P., Zvyagintseva, T. N., & Stonik, V. A., (2013). Anticancer and cancer preventive properties of marine polysaccharides: Some results and prospects. *Mar. Drugs*, *11*, 4876–4901.

Francoeur, S. N., Rier, S. T., & Whorley, S. B., (2013). Methods for sampling and analyzing wetland algae. In: Anderson, J., & Davis, C., (eds.), *Wetland Techniques* (pp. 1–58). Springer, Dordrecht.

Freile-Pelegrin, Y., & Robledo, D., (2013). Bioactive phenolic compounds from algae. In: Hernández-Ledesma, B., & Herrero, M., (eds.), *Bioactive Compounds from Marine Foods: Plant and Animal Sources* (pp. 113–129). Wiley, Hoboken.

Fritsch, F. E., (1935). *The Structure and Reproduction of Algae* (Vol. 1). Cambridge University Press, London.

Gondim, A. C. S., Silva, S. R., Mathys, L., Noppen, S., Liekens, S., Sampaio, A. H., Nagano, C. S., et al., (2019). Potent antiviral activity of carbohydrate-specific algal and leguminous lectins from the Brazilian biodiversity. *Med. Chem. Comm.*, *10*, 390–398.

Govindan, R., Govindan, R., Muthuchamy, M., Subramaniyan, S., Natesan, M., et al., (2018). Antibacterial effect of endophytic actinomycetes from marine algae against multi drug resistant gram negative bacteria. *Examines Mar. Biol. Oceanogr.*, *1*(4), EIMBO.000522.2018, doi: 10.31031/EIMBO.2018.01.000522.

Graham, B. S., (2013). Advances in antiviral vaccine development. *Immunol. Rev.*, *255*, 230–242.

Güven, K. C., Percot, A., & Sezik, E., (2010). Alkaloids in marine algae. *Mar. Drugs*, *8*, 269–284.

Harding, A., & Lanese, N., (2020). *The 12 Deadliest Viruses on Earth*. https://www.livescience.com/56598-deadliest-viruses-on-earth.html (accessed on 7 April 2022).

Ismail, A., Ktari, L., Romdhane, Y. B. R., Aoun, B., Sadok, S., Boudabous, A., & El Bour, M., (2018). Antimicrobial fatty acids from green alga *Ulva rigida* (Chlorophyta). *BioMed Res. Int.*, 3069595.

Jena, J., & Subudhi, E., (2019). Microalgae: An untapped resource for natural antimicrobials. In: Sukla, L., Subudhi, E., & Pradhan, D., (eds.), *The Role of Microalgae in Wastewater Treatment*. Springer, Singapore.

Jones, A. J., Grkovic, T., Melissa, L., & Sykes, V. M., (2013). Trypanocidal activity of marine natural products. *Mar. Drugs*, *11*, 4058–4082.

Kasanah, N., Hadisaputro, T., Seto, D. S., Amelia, W., & Isnansetyo, A., (2015). Antibacterial compounds from red seaweeds (*Rhodophyta*). *Indonesia. J. Chem.*, *15*, 201–209.

Kilic, N. K., Erdem, K., & Donmez, G., (2019). Bioactive compounds produced by *Dunaliella* species, antimicrobial effects and optimization of the efficiency. *Turkish J. Fisheries Aquatic Sci.*, *19*, 923–933.

Kumaresan, V., Nizam, F., Ravichandran, G., Viswanathan, K., Palanisamy, R., Bhatt, P., Arasu, M. V., et al., (2017). Transcriptome changes of blue-green algae, *Arthrospira* sp. in response to sulfate stress. *Algal Research*, *23*, 96–103.

Kumari, P., Bijo, A. J., Mantri, V. A., Reddy, C. R. K., & Jha, B., (2013). Fatty acid profiling of tropical marine macroalgae: An analysis from chemotaxonomic and nutritional perspectives. *Phytochemistry*, *86*, 44–56.

Larsen, B., Salem, D. M. S. A., Sallan, M. A. E., Mishrikey, M. M., & Beltagy, A. I., (2003). Characterization of the alginates from algae harvested at the Egyptian red sea coast. *Carbohy. Res.*, *338*, 2325–2336.

Lauritano, C., & Ianora, A., (2016). Marine organisms with anti-diabetes properties. *Mar. Drugs*, *14*, 220.

Lee, R. E., (2008). *Phycology* (4th edn., p. 561). Cambridge University Press, London.

Maheep, K., (2014). Harvesting of valuable eno- and exo-metabolites form cyanobacteria: A potential source. *Asian J. Pharm. Clin. Res.*, *7*, 24–28.
Mahomoodally, M. F., Lobine, D., Rengasamy, K. R. R., Gowrishankar, S., Tewari, D., Zengin, G., Kim, D. H., & Sivanesan, I., (2019). Marine algae: A potential resource of anti-HSV molecules. *Processes*, *7*, 887.
Managing epidemics, (2018). *Key Facts About Major Deadly Diseases*. Geneva: World Health Organization.
Marrez, D., & Sultan, Y., (2016). Antifungal activity of the cyanobacterium *Microcystis aeruginosa* against mycotoxigenic fungi. *J. Appl. Pharm. Sci.*, *6*, 191–198.
Martínez, A. K. A., Lauritano, C., Romano, G., & Ianora, A., (2018). Marine microalgae with anti-cancer properties. *Mar. Drugs*, *16*, 165.
Mathers, C. D., & Loncar, D., (2006). Projections of global mortality and burden of disease from 2002 to 2030. *PLoS Med.*, *3*, e442.
Michalak, I., & Chojnacka, K., (2015) Algae as production systems of bioactive compounds. *Eng. Life Sci.*, *15*, 160–176.
Nagle, D. G., & Wedge, D. E., (2002). Antifungal properties of cyanobacteria and algae: Ecological and agricultural implications. In: Mallik, A. U., & Inderjit, (eds.), *Chemical Ecology of Plants: Allelopathy in Aquatic and Terrestrial Ecosystems*. Birkhäuser, Basel.
Najam, L., & Al-Harras, A., (2018). Extraction of natural pigments from marine algae. *J. Agri. Mar. Sci.*, *23*, 81–91.
Neushul, M., (1990). Antiviral carbohydrates from marine red algae. In: Lindstrom, S. C., & Gabrielson, P. W., (eds.), *Thirteenth International Seaweed Symposium. Developments in Hydrobiology*, (Vol. 58). Springer, Dordrecht.
Pangestuti, R., Siahaan, E. A., & Kim, S. K., (2018). Photoprotective substances derived from marine algae. *Mar. Drugs*, *16*, 399.
Papenfuss, G. F., (1955). Classification of algae. In: *A Century of Progress in the Natural Sciences* (pp 115–224). Proceedings of California Academic Sciences, San Francisco.
Parker, S. P., (1982). *Synopsis and Classification of Living Organisms* (Vol. 1, 2, p. 2424) McGraw Hill, New York.
Pascher, A., (1914). Uber flagellaten and algen. *Ber Deutsch Bot Ges.*, *32*, 136–160.
Pascher, A., (1931). Systematischeubersicht uber die mit flagellaten in zusamennhang stehenden algentriehen und versuch einer einreihung dieser algenstame in die stamme des pflanzen–reiches. *Beibefte Botanischen Centralblatt*, *48*, 317–332.
Pratt, R., (1948). Studies on *Chlorella vulgaris*. XI. Relation between surface tension and accumulation of *Chlorella*. *Am. J. Bot.*, *35*, 634–637.
Prescott, G. W., (1969). *The Algae: A Review*. Thomas Nelson and Sons, London.
Raposo, M. F. J., Morais, A. M. B., & Morais, R. M. S. C., (2015). Marine polysaccharides from algae with potential biomedical applications. *Mar Drugs.*, *13*, 2967–3028.
Rippka, R., (1988). Recognition and identification of cyanobacteria. In: Packer, L., & Glazer, A. N., (eds.), *Cyanobacteria: Methods in Enzymology* (Vol. 167, pp. 28–67) Academic Press, New York.
Round, F. E., (1973). *The Biology of the Algae* (p. 585). Edward Arnold Publishers, London.
Sahoo, D., & Seckbach, J., (2015). The algae world. *Cellular Origin, Life in Extreme Habitats and Astrobiology*.
Sannasimuthu, A., Ramani, M., Pasupuleti, M., Saraswathi, N. T., Arasu, M. V., Al-Dhabi, N. A., Arshad, A., Mala, K., & Arockiaraj, J., (2020). Peroxiredoxin of *Arthrospira platensis*

derived short molecule YT12 influenced in antioxidant and anticancer activity. *Cell Biology International*, *44*, 2232–2242.
Serive, B., Nicolau, E., Bérard, J. B., Kaas, R., Pasquet, V., Picot, L., & Cadoret, J. P., (2017). Community analysis of pigment patterns from 37 microalgae strains reveals new carotenoids and porphyrins characteristic of distinct strains and taxonomic groups. *PLoS One*, *12*, e0171872.
Sharma, N., & Sharma, P., (2017). Industrial and biotechnological applications of algae: A review. *J. Adv. Plant Biol.*, *1*, 1–26.
Singh, R., Parihar, P., Singh, M., Bajguz, A., Kumar, J., Singh, S., Singh, V. P., & Prasad, S. M., (2017). Uncovering potential applications of cyanobacteria and algal metabolites in biology, agriculture and medicine: Current status and future prospects. *Front. Microbiol.*, *8*, 515.
Smith, G. M., (1955). *Cryptogamic Botany* (Vol. 1). Algae and fungi. McGraw-Hill Book Company, New York.
Smyrniotopoulos, V., Merten, C., Kaiser, M., & Tasdemir, D., (2017). Bifurcatriol, a new antiprotozoal acyclic diterpene from the brown alga *Bifurcaria bifurcata*. *Mar. Drugs*, *15*, 245.
Spavieri, J., Allmendinger, A., Kaiser, M., Itoe, M. A., Blunden, G., Mota, M. M., & Tasdemir, D., (2013). Assessment of dual life stage antiplasmodial activity of British seaweeds. *Mar. Drugs*, *11*, 4019–4034.
Susana, S., Laura, J., Merichel, P., Miguel, H., Irene, R., Elena, I., & Guillermo, R., (2012). Antiviral compounds obtained from microalgae are commonly used as carotenoid sources. *J. Appl. Phycol.*, *24*, 731–774.
Takebe, Y., Saucedo, C. J., Lund, G., Uenishi, R., Hase, S., et al., (2013). Antiviral lectins from red and blue-green algae show potent *in vitro* and *in vivo* activity against hepatitis C virus. *PLoS One*, *8*, e64449.
Tânia, S., Anu, H., Jouni, J., Liwei, L., Matti, W., Anisha, T., David, F., Perttu, P., Ana, A., Marl, F., & Kaarina, S., (2015). Antifungal compounds from cyanobacteria. *Mar. Drugs*, *13*, 2124–2140.
Tarik, V. N., (1999). Antifungal activity in crude extracts of marine red algae. *Mycol. Res.*, *95*, 1433–1440.
Temraleeva, A. D., Dronova, S. A., Moskalenko, S. V., et al., (2016). Modern methods for isolation, purification, and cultivation of soil cyanobacteria. *Microbiol.*, *85*, 389–399.
Tilden, J. E., (1933). The algae and their life relations. *Fundamentals of Phycology* (1st edn.) Hofner Publishing Company, New York.
Unnikrishnan, P. S., Suthindhiran, K., & Jayasri, M. A., (2015). Alpha-amylase inhibition and antioxidant activity of marine green algae and its possible role in diabetes management. *Pharmacognosy Magazine*, *11*, S511–S515.
Velayutham, M., Ajay, G., Arasu, M. V., Al-Dhabi, N. A., Ki Choon, C., Elumalai, P., Harikrishnan, R., et al., (2021). GR15 peptide of S-adenosylmethionine synthase (SAMe) from *Arthrospira platensis* demonstrated antioxidant mechanism against H_2O_2 induced oxidative stress in in-vitro MDCK cells and in-vivo zebrafish larvae model. *J. Biotec.*, *342*, 79–91.
Vijayakumar, S., & Menakha, M., (2015). Pharmaceutical applications of cyanobacteria: A review. *Journal of Acute Medicine*, *5*, 15–23.
Watson, J. D., & Crick, F. H. C., (1953). A structure for deoxyribose nucleic acid. *Nature*, *171*, 737–738.

West, G. S., (1916). *A treatise on the British freshwater algae.* Cambridge University Press, London.

Wild, C. P., Weiderpass, E., & Stewart, B. W., (2020). *World Cancer Report: Cancer Research for Cancer Prevention.* Lyon, France: International Agency for Research on Cancer.

World Health Organization, (2020). *Coronavirus Disease (COVID-19).* Situation Report – 107.

World Health Organization, (2018). *Diabetes Report.* Infographic.

World Health Organization. (2018). *Global Priority List of Antibiotic-Resistant Bacteria to Guide Research, Discovery and Development of New Antibiotics.*

Xu, S. Y., Huang, X., & Cheong, K. L., (2017). Recent advances in marine algae polysaccharides: Isolation, structure, and activities. *Mar. Drugs, 15*, 388.

Yamthe, L. R., Appiah-Opong, R., Tsouh, F. P. V., Tsabang, N., Fekam, B. F., Nyarko, A. K., & Wilson, M. D., (2017). Marine algae as source of novel antileishmanial drugs: A review. *Mar. Drugs, 15*, 323.

Yang, H. W., Fernando, K., Oh, J. Y., Li, X., Jeon, Y. J., & Ryu, B., (2019). Anti-obesity and anti-diabetic effects of *Ishige okamurae*. *Mar. Drugs, 17*, 202.

PART II

MICROALGAL METABOLITES: SOURCES, BIOSYNTHESIS, AND APPLICATIONS

CHAPTER 5

MICROALGAE: A VALUABLE RESOURCE FOR RECOVERY OF BIOLOGICALLY ACTIVE ALGAL METABOLITES

JAYASEELAN ARUN,[1] KANNAPPAN PANCHAMOORTHY GOPINATH,[2] SIVAKUMAR SHRI VIGNESHWAR,[2] and AUTHILINGAM SWETHA[2]

[1]*Center for Waste Management–'International Research Center,' Sathyabama Institute of Science and Technology, Jeppiaar Nagar (OMR), Chennai–600119, Tamil Nadu, India*

[2]*Department of Chemical Engineering, SSN College of Engineering, Kalavakkam–603110, Tamil Nadu, India*

ABSTRACT

Microalgae is a bio-resource that can be easily cultivated in wastewater with the help of carbon dioxide (CO_2) sequestration from the atmosphere. Microalgae provide economical support to civilians as valuable bio-products are produced from algae biomass. Microalgae serve as platform to synthesis of bioactive metabolites, biofuels, and platform chemicals. Most common economical products recovered from microalgae are biofuels, antioxidants, biopolymers, and bio-chemicals. Genetic improvement of microalgae will enhance the quantity and quality of desired product formation. Microalgae need to be put forward in real time applications by overcoming the hindrances in the commercialization process. The circular economy concept is easy to achieve in correlation with attaining sustainable development goals for the next generation.

Algal Metabolites: Biotechnological, Commercial, and Industrial Applications. Jeyabalan Sangeetha, Devarajan Thangadurai, Saher Islam, & Ravichandra Hospet (Eds.)

5.1 INTRODUCTION

Algae are photosynthetic multicellular and unicellular organisms which produce valuable metabolites during their growth period. Algae are mainly into two types such as macroalgae and microalgae. Microalgae serve as a valuable food resource for decades. Macroalgae are used as cattle feed for decades due to their carbohydrate content. Microalgae can survive in hard conditions as they have acclimatizing tendency. This regulates them to produce valuable intracellular and extracellular compounds.

A metabolite is an intermediate or product of metabolism reactions and usually confined to smaller size molecules. Metabolites provide structure, acts as stimulators, signaling, and defense by interaction with other organisms. Metabolites are classified into two types, namely: (1) Primary metabolite (e.g., ethanol and lactic acid) and (2) Secondary metabolites (e.g., pigments and antibiotics). The differences between the primary and secondary metabolites are elaborated in Table 5.1. Chemical compounds isolated from a different class of algae serve a great application in a variety of fields. Over 15,000 chemical compounds have been separated so far. Microalgae have many applications in the field of pharmaceuticals, food, agriculture, and aquafarming.

5.2 MICROALGAE

Microalgae are the earliest formed life on earth. Microalgae on photosynthesis produce proteins, carbohydrates, and lipids. They have the ability to grow rapidly and survive in extreme conditions due to their simple structure (Falkowski et al., 2004). It has been estimated that about 50,000 species of microalgae are present and only about 30,000 have been studied so far (Richmond, 2008). Humans have used microalgae as food for thousands of years (Milledge, 2011). Solar energy gets converted into chemical energy by microalgae through CO_2 sequestration, and its efficiency is 10 times that of a terrestrial plant with many technical advantages (SundarRajan et al., 2020). Around 5,000 tons/day of dry matter of microalgae are produced commercially (Raja et al., 2008). In the Asia-Pacific region, approximately 110 commercial producers of microalgae are present, with capacities ranging from 2 to 500 tons/year. Taiwan, China, and India are having a high number of commercial microalgae producers in Asia. Commercially important microalgae are *Chlorella*, *Spirulina*, *Haematococcus*, *Botryococcus*, *Dunaliella*, *Porphyridium*, *Phaeodactylum*, *Crypthecodinium*,

TABLE 5.1 Differences between Primary and Secondary Metabolites

Criteria	Primary Metabolites	Secondary Metabolites
Definition	The primary metabolites are obtained as product of metabolism reactions which are produced during the growth phase of an organism to perform physiological functions and also provide support for the overall development of the cell	The end products of primary metabolism that are produced after the growth phase has been completed and are important in ecological and other activities of the cell are known as secondary metabolites
Other names	Trophophase	Idiophase
Produced at	Growth phase	Stationary phase
Production	Produced in large quantities, and their extraction is easy.	Produced in small quantities, and their extraction is difficult.
Occurrence	Same in all species	Unique to different species.
Importance	These products have industrial applications.	Secondary metabolites such as antibiotics, pigments are also important.
	Primary products play an important role in reproduction, development, and cell growth.	They also indirectly support the cell, by sustaining their life for a longer duration.
Examples	Ethanol, lactic acid, nucleotides, and vitamins	Pigments, antibiotics, and drugs

Nannochloris, Chaetoceros, Isochrysis, Tetraselmis, Nitzschia, Skeletonema, and Schizochytrium. A wide range of highly valuable products such as carotenoids, polyunsaturated fatty acid (PUFA), polysaccharides, phycobiliprotein, and phycotoxin are being produced by microalgae. Some examples of microalgae are *Chlorella vulgaris*, *Dunaliella salina*, *Isochrysis galbana*, *Botryococcus braunii* and *Arthrospira platensis*.

Microalgae are the appropriate bio asset due to their multiple applications and adaptability. They can be cultivated in brackish water as they absorb nutrients from the wastewater as a source for growth. It requires less water than earthly plants (Umamaheswari and Shanthakumar, 2016). Further, microalgae can be harvested through bio-flocculation method (Gerardo et al., 2015). There were various studies devoted to understanding the relevance of microalgae in the treatment of wastewater and in addition, its capability to create biofuel or biomass and CO_2 sequestration (Caporgno et al., 2015; Arun et al., 2020).

5.2.1 IMPORTANCE OF MICROALGAE IN DIET

Due to the modern lifestyle, most people consume high caloric food products that result in many health issues such as diabetes, obesity, and heart diseases. To remain healthy, a properly balanced diet is necessary, and this diet should contain minerals, vitamins, and PUFA's. Microalgae are one of the remarkable natural sources for a healthy diet but not well explored. Proteins, carbohydrates, lipids, many nutritionally valuable compounds are found to be rich in several species of microalgae (Arun et al., 2017a, 2019). An important fact of cyanobacteria is that their cell wall lacks polysaccharides. Because of this, they are easily digested and thus more suitable for human consumption. At present, microalgae production and marketing were commercialized and available in market in the form of capsules, tablets, and liquids. The microalgae products were consumed when mixed with noodles, beverages, and even in breakfast cereals (Liang et al., 2004). Protein, carbohydrate, and lipid content of few microalgae are mentioned in Table 5.2. Many microalgae species are being consumed as healthy food and some among them with high nutritive value and rich protein contents are *Aphanizomenon flos-aquae*, *Chlorella* sp., *Dunaliella salina* (*D. salina*), *Dunaliella tertiolecta* (*D. tertiolecta*), and *Spirulina plantensis* (*S. plantensis*). Amongst all these microalgae, *Chlorella* and *Spirulina* are currently at the peak of the microalgal market.

TABLE 5.2 Biochemical Composition of Various Microalgae

Algae	Protein (%)	Carbohydrate (%)	Lipid (%)	References
Spirulina platensis	50–65	8–14	4–9	Becker (1994)
Euglena sp.	39–61	14–18	14–20	
Catenella repens	8	29	5	Chakraborty and Santra (2008)
Ulva lactuca	8	35	4	
Porphyridium sp.	28–39	50–57	–	Becker (1994)
Spirulina maxima	60–71	13–16	6–7	
Tetraselmis	52	15	16–45	Brown (1991)
Chlamydomonas sp.	43–56	2/9/17	14–22	Renaud et al. (1994)
Chlorella sp.	51–58	12–17	14–22	Arun et al. (2017b)

5.3 BIOLOGICALLY ACTIVE METABOLITES

5.3.1 FATTY ACIDS

Fatty acids are essentially the structural components of the major lipids compounds. Different microalgae are composed of varying amounts and types of fatty acids. For healthy growth, our diet should contain some essential fatty acids (EFAs) such as arachidonic, linoleic, and linolenic acids. Microalgae comprises of all these EFAs. Humans are not good enough to produce these fatty acids fast enough to meet the needs. Thus, consuming microalgae is the best supplement for EFAs. At present, many researchers have been interested in the microalgae due to their large-scale production, including the marine algae. This is due to the fact that microalgae contain many health benefits of mono and polyunsaturated fatty acids (MUFA and PUFA). Besides, PUFA plays an important key role in tissue and cellular metabolism, oxygen, and electron transport, regulation of membrane fluidity, and thermal adaptation (Funk, 2001).

5.3.2 EICOSAPENTAENOIC ACID

Among the PUFA, a typical family of PUFA named eicosapentaenoic acid (EPA) is currently attracting research interests. EPA is a 20 carbons length fatty acid with five double bonds from the carboxy (Δ) terminus or the last double bond at the third carbon of the methyl (ω) terminus (Nettleton, 1995). The EPA is being biosynthesized through a series of reactions and they can be classified into two distinct ones. The first one is the de novo synthesis of oleic acid (18:1 ω9) from acetate, which followed by the conversion of linoleic acid (LA; 18:2 ω-6) and α-linolenic acid (ALA; 18:3 ω-3). Then ω-3 PUFA formed by the subsequent stepwise dehydration and elongation. The EPA is normally esterified within the cell by cyclooxygenase and lipoxygenase activities to form a complex lipid molecule. In higher animals and humans, the EPA are the precursors of a group of eicosanoids, a hormone-like substance such as thromboxane, leukotrienes, and prostaglandins that are very important in maintaining regulatory and developmental physiology. Many microalgal classes have been found to be containing EPA. Nevertheless, only some of these species have the potential for industrial production. This is because of the fact that when grown under conventional photoautotrophic conditions, the majority of the microalgal species exhibit very low specific growth rates and lower cell densities (Wen and Chen, 2003). We must also consider that some microalgae are capable of heterotrophic growth with one or more substances as their carbon and energy sources. This method of growth of microalgae eliminates the necessity of light and therefore gives the great possibility of increasing their cell density and growth rates (Cardozo et al., 2002). This method is also cost-efficient when compared to the photoautotrophic method. In the heterotrophic production of EPA, some important points to be considered are the ability to metabolize and divide in the dark, growth in inexpensive and readily available media and the fast adaptation to new environments. As many microalgae have been found to be capable of producing EPA heterotrophically, these can be used for the large-scale production of EPA in an inexpensive method when compared to that of the photoautotrophic method. Even though some microalgae are grown as a source of these fatty acids, transgenic algae are being engineered to produce EPA, as if transgenic oilseeds crop could provide an alternative sustainable source for human consumption (Tan and Johns, 1996). At present, this method of releasing transgenic organisms that are nutritionally enhanced with EPA is limited by the continued customer animosity to transgenic food products. One best alternative would be to use the EPA produced from transgenic algae as a high potential food source in aquaculture. By following this method, the health benefits of these fatty acids can be delivered to the human diet.

5.3.3 CAROTENOIDS

Carotenoids can be defined as a natural pigment obtained from five-carbon isoprene units, which are polymerized enzymatically to form a regular, highly conjugated, 40-carbon structure (maximum of 15 conjugated double bonds). Either one or both the ends of carbon skeleton might undergo a cyclization reaction to produce β-ionone group, which is additionally substituted by hydroxy, oxo or epoxy groups at various positions to form various xanthophylls (Takaichi, 2011). All the photosynthetic organisms produce carotenoids to protect themselves from photodamage and to support photosynthesis. It has been found that a minimum of 600 different carotenoids are performing important biological functions in algae, bacteria, animals, and plants. As animals do not have the ability to produce carotenoids on their own, they obtain it through their diet. In the process of photosynthesis, carotenoids are found in the pigment-protein antenna complexes and they do the following important function: (i) In photosystems, as structural stabilizers for protein assembly, (ii) additional pigment for harvesting light during photosynthesis, and (iii) Inhibiting free or photo radical oxidation that occurs due to excess exposure to light. In the natural algal carotenoids, it has been found that there have been several modifications in the basic structural law of carotenoids (Cardozo et al., 2002).

5.3.3.1 β-CAROTENE

One of the most important carotenoids is the β-carotene and this is because they contain an active form of provitamin A. Provitamin A is an additive to health food products and multivitamin preparation. Our body can easily absorb the natural form of this pigment than the synthetic one and has a strong effect on us. β-carotene is found to have several applications in the feeds, food, cosmetic, and pharmaceutical industries. As the interest for extracting β-carotene from the various natural sources has increased, the demand for the natural carotenoids have also increased. High salinity, extreme temperature, high light intensity, and nutrient supplement plays a vital role in increasing the production of β-carotene (Raja et al., 2004). At present, due to optimum environmental conditions, the production of β-carotene and both 9-cis isomers and all-trans. As of vitamin formation is considered, β-carotene is found to be twice as active as cis form. A large amount of this metabolite is being produced in Australia when compared to the USA, China, Australia, and Israel (García-González et al., 2005). In both mice and humans, β-carotene produced from algae shows protecting effect against atherosclerosis. Many epidemiological studies state that humans tend to maintain their serum level and reduce their level of occurrence of several types of cancer and degenerative diseases when they are fed with a β-carotene-rich diet. β-carotene also has many other health benefits such as protection against oxidative damage and UV radiation, the prevention of harmful oxy-radicals, and UV-induced erythema (Heinrich et al., 2003).

5.3.3.2 HEALTH BENEFITS OF CAROTENOIDS

The majority of the carotenoids have therapeutic values, including anticancer and anti-inflammatory activities. This property has largely been ascribed to the organism's strong antioxidant effect which protects the organism from oxidative stress (Lindqvist and Andersson, 2002). For the nutritional purpose of the humans, provitamin A is offered by some carotenoids and is catalyzed by the intestinal enzyme β-carotene to get converted to the retina. In the last decades, the main problem that has been concerning the developing countries is the deficiency of vitamin A. High intake of carotenoids and health benefits has been a favorable correlation for the past few days. A healthy meal as accepted by the health association is the daily intake of fruits and vegetables, providing about 6 mg of carotenoids. Seafood, pink-fleshed fishes, fruits, vegetables, and watercress are found to contain an appreciable amount of dietary carotenoids.

5.3.4 ANTIOXIDANTS

Antioxidants are artificial or natural substances that delay or prevent cell damage. It delays cells from damaging by preventing chain reactions occurred due to the chemical reaction (mainly oxidation) that leads to damage the cell of the organism. Substances for instance ascorbic acid terminates such reaction. Antioxidants are available in vegetables and fruits. Recent studies have shown that taking supplements along with the antioxidants could be efficient in preventing disease. The major examples of antioxidants include vitamin C and E, selenium, and carotenoids. Carotenoids include beta-carotene, lutein, lycopene, and zeaxanthin.

5.3.4.1 MICRO ALGAL ANTIOXIDANTS FOR HUMAN HEALTH

Microalgae biomass is a multi-component antioxidant system. It is more effective due to various interactions among different antioxidant components. They have the ability to accumulate vitamin E (alpha-/beta-tocopherol/

alpha-tocotrienol). Microalgae are rich in vitamins. Vitamin E has a wide range of importance in human health, helpful in treating cancer, heart disease, Alzheimer's disease (AD), Parkinson's disease and other medical conditions (Pham-Huy et al., 2008). In the food industry, harvested microalgae are used as a preservative and for photoprotection in skin creams (Alberts et al., 1996). For example, microalgae such as *D. tertiolecta* and *Tetraselmis sueccica*, increases the production of vitamin E when cultured under nitrogen deprived environment (Durmaz, 2007). *Euglena gracilis* produces a high concentration of alpha-tocopherol (Rodríguez-Zavala et al., 2010). Some of the microalgae can produce water-soluble vitamin. Ascorbic acid can be accumulated by some algae which is a water-soluble vitamin.

Spirulina sp. can synthesize a large amount of vitamin B12. Ascorbic acid (vitamin C) is an essential antioxidant for carnitine, collagen, and neurotransmitter synthesis. It is also used as a food additive. Vitamin C also fights for cancer and atherosclerosis prevention and also as an immunomodulator. Lipophilic antioxidants such as butylated hydroxyl toluene (BHT) accumulated by *Chlorella* sp. and *Dunaliella* sp. (Barbosa et al., 2005). Microalga *B. braunii* accumulates a high amount of BHT. BHT is used as an antioxidant source or in a food additive. *Dunaliella* sp. produces high amounts of glutathione (GSH). GSH is a non-protein thiol antioxidant that prevents cellular damage of various disease conditions and has an enzymatic property that scavenges the ROS (O^{2-} and H_2O_2). GSH is used in the cosmetic, food, and pharmaceutical industry.

5.3.4.2 *ANTIOXIDANT ACTIVITY*

Microalgae and plants through photosynthesis, photorespiration, respiration, and other metabolic processes, yield hydrogen peroxide as a product from the enzymatic activity of glycolate oxidase and urate oxidase. It is suspected that those antioxidants are the reason for some side effects such as liver damage and carcinogenesis. Recently many studies are undergoing to find any natural antioxidants having both safety and effectiveness, which can be used as an alternative for antioxidants such as BHA and BHT. Thus, microalgae is the best source of natural antioxidants. Algae containing enzymatic antioxidant are required to maintain the concentration of ROS to protect cells from damage. Lipids (peroxidation of PUFAs in membranes), proteins (denaturation), carbohydrates, and nucleic acids are highly damaged due to ROS. SOD, catalase (CAT) and glutathione peroxidase (GPX), and peroxiredoxin (PrxR) are primary scavenging enzymatic defense systems.

The hydrophilic antioxidants AA and GSH effectively scavenge oxygen radicals. ROS is removed directly from the pigment bed by Carotenoids and TOH. Microalgae like *Nostoc muscorum*, *Anabaena flos-aquae*, *Anabaena oryzae*, *Nostoc humifusum*, *Oscillatoria* sp., *Spirulina platensis*, *Phormedium fragile*, *Wollea saccata* and *Chlorella vulgaris* also exhibit antioxidant properties.

5.3.5 PHYCOCOLLOIDS

Phycocolloids are polymers of sugar units with high-molecular-weight molecules. They constitute the main structural compound in the cell wall of seaweeds and pathogens. Mostly extracted polysaccharides from seaweeds are agar, alginate, and carrageenan because of its significance in food and cosmetic industries. Unlike the other alginate is a binary polyuronide extracted from Phaeophyceae, which contains mannuronic and guluronic acid. While agar and carrageenan are extracted from Rhodophyceae, are sulfated polysaccharides. Their properties such as gelling, viscosifying, and emulsifying have great commercial and scientific advantages.

5.3.5.1 AGAR

Seaweed galactans containing α (1–4)-3,6-anhydro-L-galactose and β (1–3)-D-galactose residues and little amount of sulfate esterification is called Agar. It is assumed that chains of alternating D- and L-galactopyranosyl residues are assembled in the Golgi apparatus. In the Golgi apparatus at an early stage of biosynthesis Sulfation of L-galactopyranosyl residues occurs, while ring closure and methylation may occur afterward. Further migration out of the Golgi into the cell-wall matrix occurs and modification of the agar polysaccharides begins (Hemmingson et al., 1996). Depending on specific physicochemical characteristics, quality, and quantity vary. Moreover, a parameter such as environment, growth, and reproductive cycle has an adverse effect (Daugherty and Bird, 1988). For example, different compositions of agar extracted from two species of *Gracilaria* (Marinho-Soriano, 2001).

Agar is used in food products as well as in industrial products and processes. Agar is also used as an alternative for gel in biological culture media. They serve their purpose in the medical/pharmaceutical field as well, such as bulking agents, laxatives, suppositories, capsules, tablets, and anticoagulants. The highly purified Agar (the neutral fractions called agarose) is used for separation in molecular biology. It is used in medicinal and pharmaceutical areas, such as in a therapy against cancer cells since it can induce the apoptosis of these cells *in vitro* (Chen et al., 2004).

5.3.5.2 CARRAGEENAN

Water-soluble sulfated galactans with an alternating backbone consisting of α (1–4)-3,6-anhydro-D-galactose and β (1–3)-D-galactose is called carrageenan (Gonçalves et al., 2002). Natural carrageenans have distinct compositions. The three commercially important carrageenans are known to be λ-, ι- and κ-carrageenans. Two other types are μ- and ν-carrageenans. To supply carrageenan to the food industries; macroalga *Kappaphycus alvarezii* is majorly grown. The μ- and ν-carrageenans are the biological precursors of κ and ι-carrageenans, respectively. Because of their thickening and suspension properties in milk-based products, carrageenans (especially κ- and ι-carrageenans) are more widely used.

5.3.5.3 ALGINATE

Alginic acid or alginate is the general name of a family of linear polysaccharides containing 1,4-linked β-D-mannuronic and α-L-guluronic acid residues (Andrade et al., 2004). In the textile industry, alginate is used for sizing cotton yarn and it has an advantage for its solution properties and as a gelling agent. To form a highly viscous solution, alginate produced by brown seaweed is used in the food and pharmaceutical industries (Yamasaki et al., 2005).

5.3.6 HALOGENATED COMPOUND

Red algae and brown algae produce a halogenated compound (Butler and Carter-Franklin, 2004). Some types of halogenated compounds are indoles, terpenes, acetogenins, phenols, fatty acids, and halogenated compounds. In the majority of the cases, they possess antibacterial, antitumor, and feeding defendant properties. Genus *Laurencia* (Rhodophyta) is a major producer of halogenated compounds (Wright et al., 2003). These compounds comprise mainly of derivatives of sesquiterpenes, diterpenes, triterpenes, acetogenins, fatty acids, and brominated indoles. For example, polyhalogenated monoterpenes obtained from red algae possess antimicrobial, antitubercular, and anticancer activity. *Bonnemaisonia nootkana*, *Bonnemaisonia hamifera* and *Trailliella intricata* synthesizes brominated fatty acids. The presence of halogen atoms in the fatty acid chain has an adverse effect such as it could change the conformation of biological membranes (Dembitsky and Srebnik, 2002). The pathways for halogenation of fatty acids are well known. It proceeds via interhalogen addition to unsaturated sites followed by molecular rearrangements. Halogenated compounds in algae are in profuse amounts; these compounds serve their purpose in marine natural products. The presence of peroxides and halide ions haloperoxidase catalyzes the halogenation of organic compounds. It is also used in carrying out oxidation, epoxidation, and sulfoxidation of organic compounds. Thus, on the whole, halogenation helps to increase or induce the biological activity.

5.4 ALGAL METABOLITES: WORLDWIDE SCENARIO

Food, cosmetic, and pharmaceutical industry has an increasing market for more than a period of time and algal products have been used in these fields. Although the challenge, of growing algae on a larger scale without bringing damage to the marine environment, is currently studied. Micro and macro-algae are necessary micronutrients for the development of aquaculture, which includes vitamins, sterols, and fatty acids. In 2000, 45.71 million tons (by weight) of aquaculture was produced. Mollusk production was one of the important proportions of total global aquaculture production. This accounts for 20% of global aquaculture production.

Carotenoids serve their use in food coloring agents and also in nutritional supplements, with their global market of US 935 million dollars. Investment in fish farming and avian culture has been increased such that astaxanthin-canthaxanthin accounts for US 150 million dollars per year (Leeson and Caston, 2004). For example, salmon farming is a multibillion-dollar industry with growing demands. Known to its predominant emerging market, astaxanthin production has changed the preferences from synthetic pigments to biotechnology-derived products. By optimizing the fermentation of yeast and microalgae condition has been enhanced for better yield of astaxanthin production (Bhosale, 2004).

From species of the red algal families *Gelidiaceae* and *Gracilariaceae* about 10,000 tons of Agar, valued at $200 million, are produced worldwide. Agar is an expensive product in the global market, due to the scarcity of exploitable agar-producing seaweeds. Growing at a rate of 5% per year for the past 25 years, the market for carrageenan has a remarkable functionality.

25,000 tons of carrageenan worthing $200 million are produced globally. Carrageenan from farm-raised and natural stands of *Eucheuma cottonii* and *Eucheuma spinosum* exported from Philippines. Nearly, 50,000 wet tons of *Chondrus crispusin* harvested per year in the Maritime Provinces of Canada.

Due to fear of getting skin cancer, sunscreen use has been increased significantly in the last decade. According to reports in the past two decades, occurrences of non-melanoma skin cancer (NMSC) increased sharply. Therefore, health care professionals recommend daily use of sunscreen. NMSC has been reduced significantly by the use of sunscreens simultaneously avoiding exposure to UV radiation. Therefore, sunscreen is the first line of defense against skin cancer. Due to this reason, cosmetic industries grow greatly by the use of mycosporine-like amino acids, which is a highly efficient UV blocker.

5.5 FUTURE SCOPE

To produce microalgae products as a cost-effective and sustainable methodology for the bio-refinery aspect. Utilization of biomass for high value products is necessary; achieved through imparting abiotic stress, which has a great impact on biomass. Genetic engineering of microalgae helps in achieving greater productivity of microalga also with additional uses under stress conditions. For example, stress-tolerant genes from other organisms could be injected into the host. In another point of view, the world today needs a continuous and increasing demand for fuel energy. It is blatant that fossil fuel is alarmingly depleted thus biofuels serves its purpose in as many fields as possible. Microalgae serve as bio-resource for bioenergy (bio-oil, bio-char, and syngas) production.

5.6 CONCLUSION

Algae are responsible for more than 50% of the oxygen production and it reduces global warming. Nutritional value of microalgae varies based on shape, size, digestibility, and toxicity of the species. The primary determinant for edible application was the composition of microalgae. Drug discovery process was an evergreen growing field in the pharmaceutical sector. Moreover, algae with enormous nutritional values have been explored in recent years by various researchers. Algae with high biological activity, production of metabolites with high complexity, and diverse biological properties are enormous.

KEYWORDS

- **active compounds**
- **antioxidants**
- **carotenoids**
- **fatty acids**
- **metabolites**
- **microalgae**
- **wastewater**

REFERENCES

Alberts, D. S., Goldman, R., Xu, M. J., Dorr, R. T., Quinn, J., Welch, K., et al., (1996). Disposition and metabolism of topically administered α-tocopherol acetate: A common ingredient of commercially available sunscreens and cosmetics. *Nutrition and Cancer, 26*(2), 193–201.

Andrade, L. R., Salgado, L. T., Farina, M., Pereira, M. S., Mourao, P. A., & Amado, F. G. M., (2004). Ultrastructure of acidic polysaccharides from the cell walls of brown algae. *Journal of Structural Biology, 145*(3), 216–225.

Arun, J., Avinash, U., Arun, K. B., Pandimadevi, M., & Gopinath, K., (2017a). Ultrasound assisted enhanced extraction of lutein (β, ε-carotene-3, 3'-diol) from microalga (*Chlorella pyrenoidosa*) grown in wastewater: Optimization through response surface methodology. *Global Nest Journal, 19*(4), 574–583.

Arun, J., Gopinath, K. P., SundarRajan, P., Malolan, R., Adithya, S., Jayaraman, R. S., et al., (2020). Hydrothermal liquefaction of *Scenedesmus obliquus* using a novel catalyst derived from clamshells: Solid residue as catalyst for hydrogen production. *Bioresource Technology, 310*, 123443.

Arun, J., Gopinath, K., Shreekanth, S., Sahana, R., Raghavi, M., & Gnanaprakash, D., (2019). Effects of process parameters on hydrothermal liquefaction of microalgae biomass grown in municipal wastewater. *Petroleum Chemistry, 59*(2), 194–200.

Arun, J., Shreekanth, S. J., Sahana, R., Raghavi, M. S., Gopinath, K. P., & Gnanaprakash, D., (2017b). Studies on the influence of process parameters on hydrothermal catalytic liquefaction of microalgae (*Chlorella vulgaris*) biomass grown in wastewater. *Bioresource Technology, 244*, 963–968.

Barbosa, M. J., Zijffers, J. W., Nisworo, A., Vaes, W., Van, S. J., & Wijffels, R. H., (2005). Optimization of biomass, vitamins, and carotenoid yield on light energy in a flat-panel reactor using the A-stat technique. *Biotechnology and Bioengineering, 89*(2), 233–242.

Becker, E. W., (1994). *Microalgae: Biotechnology and Microbiology*. Cambridge University Press, UK.

Bhosale, P., (2004). Environmental and cultural stimulants in the production of carotenoids from microorganisms. *Applied Microbiology and Biotechnology, 63*(4), 351–361.

Brown, M. R., (1991). The amino-acid and sugar composition of 16 species of microalgae used in mariculture. *Journal of Experimental Marine Biology and Ecology, 145*(1), 79–99.
Butler, A., & Carter-Franklin, J. N., (2004). The role of vanadium bromoperoxidase in the biosynthesis of halogenated marine natural products. *Natural Product Reports, 21*(1), 180–188.
Caporgno, M., Taleb, A., Olkiewicz, M., Font, J., Pruvost, J., Legrand, J., et al., (2015). Microalgae cultivation in urban wastewater: Nutrient removal and biomass production for biodiesel and methane. *Algal Research, 10*, 232–239.
Cardozo, K. H., Oliveira, M. A., Tavares, M. F., Colepicolo, P., & Pinto, E., (2002). Daily oscillation of fatty acids and malondialdehyde in the dinoflagellate *Lingulodinium polyedrum*. *Biological Rhythm Research, 33*(4), 371–382.
Chakraborty, S., & Santra, S. C., (2008). Biochemical composition of eight benthic algae collected from Sunderban. *Indian Journal of Geo-Marine Sciences, 37*(3), 329–332.
Chen, Y. H., Tu, C. J., & Wu, H. T., (2004). Growth-inhibitory effects of the red alga *Gelidium amansii* on cultured cells. *Biological and Pharmaceutical Bulletin, 27*(2), 180–184.
Daugherty, B. K., & Bird, K. T., (1988). Salinity and temperature effects on agar production from *Gracilaria verrucosa* strain G-16. *Aquaculture, 75*(1, 2), 105–113.
Dembitsky, V. M., & Srebnik, M., (2002). Natural halogenated fatty acids: Their analogues and derivatives. *Progress in Lipid Research, 41*(4), 315–367.
Durmaz, Y., (2007). Vitamin E (α-tocopherol) production by the marine microalgae *Nannochloropsis oculata* (*Eustigmatophyceae*) in nitrogen limitation. *Aquaculture, 272*(1–4), 717–722.
Falkowski, P. G., Katz, M. E., Knoll, A. H., Quigg, A., Raven, J. A., Schofield, O., et al., (2004). The evolution of modern eukaryotic phytoplankton. *Science, 305*(5682), 354–360.
Funk, C. D., (2001). Prostaglandins and leukotrienes: Advances in eicosanoid biology. *Science, 294*(5548), 1871–1875.
García-González, M., Moreno, J., Manzano, J. C., Florencio, F. J., & Guerrero, M. G., (2005). Production of *Dunaliella salina* biomass rich in 9-cis-β-carotene and lutein in a closed tubular photobioreactor. *Journal of Biotechnology, 115*(1), 81–90.
Gerardo, M. L., Van, D. H. S., Vervaeren, H., Coward, T., & Skill, S. C., (2015). Harvesting of microalgae within a biorefinery approach: A review of the developments and case studies from pilot plants. *Algal Research, 11*, 248–262.
Gonçalves, A. G., Ducatti, D. R., Duarte, M. E. R., & Noseda, M. D., (2002). Sulfated and pyruvylated disaccharide alditols obtained from a red seaweed galactan: ESIMS and NMR approaches. *Carbohydrate Research, 337*(24), 2443–2453.
Heinrich, U., Gärtner, C., Wiebusch, M., Eichler, O., Sies, H., Tronnier, H., et al., (2003). Supplementation with β-carotene or a similar amount of mixed carotenoids protects humans from UV-induced erythema. *The Journal of Nutrition, 133*(1), 98–101.
Hemmingson, J. A., Furneaux, R. H., & Murray-Brown, V. H., (1996). Biosynthesis of agar polysaccharides in *Gracilaria chilensis* Bird, McLachlan et Oliveira. *Carbohydrate Research, 287*(1), 101–115.
Leeson, S., & Caston, L., (2004). Enrichment of eggs with lutein. *Poultry Science, 83*(10), 1709–1712.
Liang, S., Liu, X., Chen, F., & Chen, Z., (2004). Current microalgal health food R & D activities in China. In: *Asian Pacific Phycology in the 21st Century: Prospects and Challenges* (pp. 45–48). Springer.
Lindqvist, A., & Andersson, S., (2002). Biochemical properties of purified recombinant human β-carotene 15, 15′-monooxygenase. *Journal of Biological Chemistry, 277*(26), 23942–23948.

Marinho-Soriano, E., (2001). Agar polysaccharides from *Gracilaria* species (*Rhodophyta, Gracilariaceae*). *Journal of Biotechnology, 89*(1), 81–84.
Milledge, J. J., (2011). Commercial application of microalgae other than as biofuels: A brief review. *Reviews in Environmental Science and Bio/Technology, 10*(1), 31–41.
Nettleton, J. A., (1995). Introduction to fatty acids. In: *Omega-3 Fatty Acids and Health* (pp. 1–63). Springer. Boston, MA.
Pham-Huy, L. A., He, H., & Pham-Huy, C., (2008). Free radicals, antioxidants in disease and health. *International Journal of Biomedical Science, 4*(2), 89.
Raja, R., Anbazhagan, C., Lakshmi, D., & Rengasamy, R., (2004). Nutritional studies on *Dunaliella salina* (Volvocales, *Chlorophyta*) under laboratory conditions. *Seaweed Resources Utilization, 26*, 127–146.
Raja, R., Hemaiswarya, S., Kumar, N. A., Sridhar, S., & Rengasamy, R., (2008). A perspective on the biotechnological potential of microalgae. *Critical Reviews in Microbiology, 34*(2), 77–88.
Renaud, S. M., Parry, D. L., & Thinh, L. V., (1994). Microalgae for use in tropical aquaculture I: Gross chemical and fatty acid composition of twelve species of microalgae from the Northern Territory, Australia. *Journal of Applied Phycology, 6*(3), 337–345.
Richmond, A., (2008). *Handbook of Microalgal Culture: Biotechnology and Applied Phycology*. John Wiley & Sons, UK.
Rodríguez-Zavala, J., Ortiz-Cruz, M., Mendoza-Hernández, G., & Moreno-Sánchez, R., (2010). Increased synthesis of α-tocopherol, paramylon and tyrosine by *Euglena gracilis* under conditions of high biomass production. *Journal of Applied Microbiology, 109*(6), 2160–2172.
SundarRajan, P., Gopinath, K. P., Arun, J., GracePavithra, K., Pavendan, K., & AdithyaJoseph, A., (2020). An insight into carbon balance of product streams from hydrothermal liquefaction of *Scenedesmus abundans* biomass. *Renewable Energy, 151*, 79–87.
Takaichi, S., (2011). Carotenoids in algae: distributions, biosyntheses and functions. *Marine Drugs, 9*(6), 1101–1118.
Tan, C. K., & Johns, M. R., (1996). Screening of diatoms for heterotrophic eicosapentaenoic acid production. *Journal of Applied Phycology, 8*(1), 59–64.
Umamaheswari, J., & Shanthakumar, S., (2016). Efficacy of microalgae for industrial wastewater treatment: A review on operating conditions, treatment efficiency and biomass productivity. *Reviews in Environmental Science and Bio/Technology, 15*(2), 265–284.
Wen, Z. Y., & Chen, F., (2003). Heterotrophic production of eicosapentaenoic acid by microalgae. *Biotechnology Advances, 21*(4), 273–294.
Wright, A. D., Goclik, E., & König, G. M., (2003). Three new sesquiterpenes from the red alga *Laurencia perforata*. *Journal of Natural Products, 66*(3), 435–437.
Yamasaki, M., Ogura, K., Hashimoto, W., Mikami, B., & Murata, K., (2005). A structural basis for depolymerization of alginate by polysaccharide lyase family-7. *Journal of Molecular Biology, 352*(1), 11–21.

CHAPTER 6

BIOTECHNOLOGICAL APPLICATIONS OF SECONDARY METABOLITES FROM MICROALGAE

AMEET KUMAR,[1] ANEELA NAWAZ,[1] USMAN ALI CHAUDHRY,[2] AAMER ALI SHAH,[1] MALIK BADSHAH,[1] and SAMIULLAH KHAN[1]

[1]*Department of Microbiology, Faculty of Biological Sciences, Quaid-i-Azam University, Islamabad–45320, Pakistan*

[2]*Infection Control and Disease Prevention Center, Ministry of Health, Tabuk, Kingdom of Saudi Arabia*

ABSTRACT

Microalgae are photoautotrophic microorganisms which are proficient producers of biologically active metabolites, food, and feed supplement, chemicals, and biofuels. These microorganisms have the ability to promote health and reduce the risk of the development of degenerative diseases. They are capable of forming a vast variety of bioactive chemicals. These bioactive compounds are secondary metabolites which include carbohydrates, amino acids, peptides, vitamins, antibiotics, enzymes, and organic acid. Through microalgae biomass, a large variety of different high-value compounds are obtained, such as polyunsaturated fatty acids (PUFA), phytosterols, polyphenols, pigments such as carotenoids, phycobiliproteins (PBP). Microalgae metabolites show a vast range of biological activities such as antibacterial, anticancer, anti-inflammatory, antioxidant, antiobesity, antiviral, and immunomodulatory. Microalgal biotechnology has become a subject of study for various fields, due to the varied byproducts that can be

Algal Metabolites: Biotechnological, Commercial, and Industrial Applications. Jeyabalan Sangeetha, Devarajan Thangadurai, Saher Islam, & Ravichandra Hospet (Eds.)

obtained from these microorganisms, which are further applied for various biotechnological applications. Microalgae metabolites have a wide range of applications in pharmaceutical, nutraceuticals, food, cosmetics, and petrochemical industries. In this context, the aim is to discuss bioactive metabolites produced by microalgae for possible applications in the life sciences.

6.1 INTRODUCTION

Microalgae are unicellular, aquatic, photosynthetic organisms and play a vital role in the ecosystem. Microalgae are autotrophic organisms that use light energy and inorganic nutrients and are capable of creating biomass from carbon dioxide (CO_2), nutrients, and solar energy for their survival and form bioactive compounds that have high aggregated therapeutic functions and nutritional value. The size of picoplanktonic microalgae ranges from 0.2 to 2 μm and size of filamentous forms of microalgae is of 100 μm or higher (Tyagi et al., 2010). Most of the algae dwell in marine and freshwater environment. Algae range from small unicellular microalgae like diatoms and cyanobacteria to giant, multicellular macroalgae such as giant kelp (Pratt, 1992). Algae contribute about 40% of global photosynthesis. These microorganisms have the aptness to turnout different biologically active compounds (De Morais et al., 2015).

The biomass constitutes different primary metabolites such as oils, lipids, and sugar which are further processed by different existing pathways for manufacturing of high-value products that include animal and human feed supplements, industrial chemicals, transport fuels, and pharmaceutical products. Algae-derived compounds and biomass have a potential role in different applications such as food supplements, pharmaceuticals, human nutrition, and in cosmetics. Some algae are reservoir of natural antioxidants (Faten and Emad, 2009).

Metabolites are substances that form during the metabolic process. It will be the end product or intermediate of metabolism and are small in size. Two types of metabolites are formed during the metabolic process; i.e., primary metabolites and secondary metabolites. Primary metabolites are those substances that are directly involved in the development, reproduction, and normal growth. Secondary metabolites are those chemical substances that are not necessary or directly involved in the growth, reproduction, or development of organisms. Secondary metabolites have a major role in ecological function. The initiation of secondary metabolism is correlated with special environmental conditions or in some cases it is linked with developmental changes (Cardozo et al., 2005).

Algae have the potential to combat different environmental conditions. The EPS secretion helps them to live under harsh conditions. Those algae which don't have external defense mechanism can survive by developing defense strategies by the production of secondary metabolites with repellent or antibiotic effects. Most of the algae contain special types of bioactive compounds and metabolites having activity against different animal and human pathogens. These organisms also possess the capability of producing other molecules such as proteins, pigments, antioxidants, polyunsaturated fatty acids (PUFAs), which have a potential function in therapeutics against human disorders. These molecules have a wide range of applications in cosmetics, food, and pharmaceuticals industries (Lipton, 2003). Different microalgae have been screened for finding out bioactive agents which shows positive response as anti-inflammatory, antifungal, anticancer, and antimicrobial.

Microalgal biotechnology is now becoming a vast subject and opens the door for various fields. Microalgae are a reservoir of different bioactive compounds. Bioactive compounds are active substances with functional properties in the human body. These can be produced directly from primary metabolism such as vitamins, pigments, proteins, and fatty acids. Different compounds are also produced during secondary metabolism such as linolenic acid, oleic acid, cyanovirin, vitamin E, palmitoleic acid (PA), lutein, phycocyanin (PC), zeaxanthin, β-carotene and vitamin B12. Bioactive compounds are also accumulated in biomass in most microalgae and some exometabolites are excreted into medium. These compounds include phycocyanins, carotenoids, fatty acids, polyunsaturated compounds and phycocyanins (De Morais et al., 2015).

Bioactive compounds have a profound effect on health. These compounds show antiviral, antifungal, antienzymatic, antimicrobial, and anticancer activity. Microalgae are a natural reservoir of many biologically active compounds and have a potential role in different life sciences applications. Algae have a major role in the biotechnological exploitation, such as the production of valuable products of pharmaceuticals importance, cosmetics, and products for food industries (Harun et al., 2010). Due to the biotechnological application microalgae should continue to be regarded as a priority area of research (Chu, 2012).

This chapter deals with the application of secondary metabolites of microalgae in different biotechnological applications.

6.2 ALGAL METABOLITES OF PHARMACEUTICAL IMPORTANCE

Algae are a rich source of biologically active metabolites, including primary and secondary metabolites and are potential candidates for the pharmaceutical sector (Abedin et al., 2008; Sharma and Sharma, 2017). Immense bioactive metabolites from microalgae are utilized for commercial applications. Algae contain a vast range of proteins and pharmaceutical products that otherwise are not present or very expensive to be obtained from plant and animal sources (Bhattacharjee, 2016; Sharma and Sharma, 2017). The microalgae contain high value pharmaceutical products but their commercialization is still in infancy. There is great genetic potential present in microalgae for several bioactive agents. Hence the proven ability for production of these bioactive compounds from microalgae has placed these microorganisms in the biotechnological spotlight for application and commercialization in the pharmaceutical industry. From algae, large ranges of products are being derived that include, antimicrobials, antivirals, therapeutic proteins, drugs, antifungal compounds, and many more (Bhattacharjee, 2016; Sharma and Sharma, 2017).

6.2.1 MICROALGAL COMPOUNDS WITH ANTIBACTERIAL ACTIVITY

The discovery of new compounds to test their antimicrobial activity is necessary because the constant use of present antibiotics has developed antibiotic resistance in human pathogen (Prakash et al., 2011; De Morais et al., 2015). Microalgae are rich source of different antimicrobial with a potent antibacterial activity. The antibacterial activity shown by these microorganisms is due to the ability of synthesizing different compounds. The compounds synthesized by these microorganisms are acrylic acids, fatty acids, terpenoids, halogenated aliphatic compounds, sulfur containing heterocyclic compounds, sterols, phenols, acetogenins, and carbohydrate (De Morais et al., 2015).

Extracts from cell and growth media of unicellular algae-like *Chlamydomonas spyrenoidosa*, *Chlorella vulgaris* have antibacterial effect, on both Gram-positive and Gram-negative bacteria (Sharma and Sharma, 2017). The activity of these extracts largely depends on its lipid composition. The antimicrobial activity of microalgae is due to the potential of producing active compounds such as β-cyclocitral, α- and β-ionone, phytol, and neophytadiene (Amaro et al., 2011; De Morais et al., 2015).

The antimicrobial activity of microalgae against human pathogens such as *Pseudomonas aeruginosa*, *Staphylococcus epidermidis*, and *Staphylococcus aureus* has been imputed to eicosapentaenoic acid (EPA), γ-linolenic acid, hexadecatrienoic acid (HTA), docosahexaenoic acid (DHA), lauric acid, oleic acid, PA, lactic acid, and arachidonic acid (AA) (Smith et al., 2010; Amaro et al., 2011; De Morais et al., 2015).

The mechanism of antimicrobial activity is related to affecting different cell structures. The most affected structure is the cell membrane. Damage to the membrane results in the loss of internal substances from the infected cell and results in the imbalances in nutrient absorption which further inhibit cellular respiration (Amaro et al., 2011; De Morais et al., 2015).

Wide range of cyanobacterial species produces extracellular and intracellular metabolites having different biological activities (Noaman et al., 2014; Sharma and Sharma, 2017). *Spirulina*, a microalga produces volatile components such as tetradecane and heptadecane which have antibacterial activity (Plaza et al., 2009; De Morais et al., 2015).

6.2.2 MICROALGAL COMPOUNDS WITH ANTIVIRAL ACTIVITY

Microalgae contain different metabolites which have a functional role against viruses and possess major antiviral activity. There are different antiviral compounds which can be obtained from microalgae. Cyanobacteria contain a compound called cyanovirin-N which shows strong antiviral against HIV by blocking the interaction of viral glycoprotein gp120 with CD4 (Patterson et al., 1993; Sharma and Sharma, 2017).

Microalgae produce extracellular sulfated polysaccharide (EPS) with therapeutic potential and use as a therapeutic agent. Some species of microalgae have highly sulfated antiviral polysaccharides mainly composed of glucose, galactose, and xylose. The sulfate content in the polysaccharide determines the strength of the antiviral activities. The higher the sulfate content in the polysaccharide, the higher will be its antiviral activity. The mechanism of action of polysaccharides is because of viral interactions or due to the positive charges on the surface of cell which inhibit the entry of virus to the cell. The *Spirulina* a cyanobacterium have a tendency to produce sulfated polysaccharide that have a major role both *in vitro* and *in vivo* as an antiviral agent (Amaro et al., 2011; De Jesus et al., 2014; De Morais et al., 2015).

Viruses such as encephalomyocarditis virus, human immunodeficiency virus (HIV), Herpes simplex virus type 1 and 2 (HSV1, HSV2), varicella virus, swine fever virus, hemorrhagic septicemia in salmonid virus can cause life-threatening infections (Smelcerovic et al., 2008). These infections can be treated by compounds produced by algae such as carrageenan, a sulfated polysaccharide that binds directly to the human papilloma virus, which inhibits the process of adsorption and virus uncoating step (De Jesus et al., 2014).

Alginate obtained from red algae has a potential role against HIV, IAV, and HBV. Fucan, a polysaccharide obtained from brown algae such as *Ascophyllum nodosum* and *Macrocystis pyrifera* is appeared active against HIV, IAV, and HBV. Nostaflan obtained from blue-green algae (BGA), *Nostoc flagelliforme* have a potential activity against HSV-1, HSV-2, human cytomegalovirus, and influenza A virus (Aditya et al., 2016).

6.2.3 MICROALGAL COMPOUNDS WITH ANTICANCER ACTIVITY

Microalgae are a source of various bioactive compounds that have potential anticancer activity. *Poteriochromonas malhamensis* produces novel compound called chlorosulfolipid which act as tyrosine kinase inhibitor (Gerwick et al., 1994; Sharma and Sharma, 2017). Various species of cyanobacteria have the efficiency of producing bioactive compounds which have the potential of destroying cancer cells by the process of apoptosis, affecting cell signaling by activation of signaling enzyme of protein kinase C family (Boopathy and Kathiresan, 2010; Sharma and Sharma, 2017). These compounds obtained from microalgae are effective against lung carcinoma and implanted lymphocytic leukemia (Furasawa et al., 1994; Sharma and Sharma, 2017).

A bioactive metabolite cryptophycin obtained from *Nostoc* ATCC 53789 showed higher anticancer activity. Another metabolite scytonemin isolated from *Stigonema* have inhibitory action against protein serine and provide a good choice of drug with anti-inflammatory and anti-proliferative activities (Stevenson et al., 2002; Sharma and Sharma, 2017). The carotenoid fucoxanthin from diatoms has anticancer activity and is responsible for enhancing the apoptosis process in human cancer cells (Maeda et al., 2005; De Morais et al., 2015).

6.2.4 MICROALGAL COMPOUNDS WITH ANTI-INFLAMMATORY ACTIVITY

Inflammation is a quick reaction to a tissue or cell injury caused by detrimental stimuli, such as pathogens and toxins. In such circumstances, the body responds in such a way that it recognizes the agents which is responsible for the tissue or cell injury and helps to neutralize them as soon as possible. At the site of infection, swelling, redness, pain, and heat occur due to inflammation. By ingesting anti-inflammatory compounds, it helps in building the immune response of the body and prevents the body from diseases and enhances the healing process from the infection (De Morais et al., 2015).

Different species of microalgae produce different anti-inflammatory compounds in their biomass and possess a protective function, that's why they are used in food and pharmaceuticals. The important microalgal compounds having these properties are long-chain polyunsaturated fatty acids (PUFAs), sulfurized polysaccharide and pigments (Bhat and Madyastha, 2001; Matsui et al., 2003; Khan et al., 2007).

Microalgal polysaccharide contains the capability to change the immune system through the initiation of reactive oxygen species (ROS), activation of macrophage functions, induction of nitric oxide (NO) and various types of cytokines/chemokines. Chemocytokines serve as signals for immune and inflammatory reactions (Schepetkin and Quinn, 2006; De Morais et al., 2015). Sulfate polysaccharides have anti-inflammatory activity and a role in skin treatments by inhibiting the adhesion and migration of polymorphonuclear leukocytes (Matsui et al., 2003). The PUFAs such as omega-3 and omega-6, docosahexaenoic (DHA), arachidonic (AA) acids and eicosapentaenoic (EPA), were used in the treatment of chronic inflammation such as skin diseases and rheumatism. Different microalgae including *Nannochloropsis*, *Isochrysis*, *Thalassiosira*, *Phaeodactylum*, and *Pavlova*, are the major source of these PUFA's (Ryckebosch et al., 2014; De Morais et al., 2015). Different pigments are also responsible for the anti-inflammatory activity. The carotenoid fucoxanthin is found in diatoms. It is responsible for the apoptosis. Phycocyanin, a pigment from cyanobacteria also showed anti-inflammatory activity by inhibiting the release of histamine (Estrada et al., 2001; Maeda et al., 2005; De Morais et al., 2015).

6.2.5 MICROALGAL METABOLITES WITH ANTI-OXIDANT ACTIVITY

Change in the environmental conditions results in ROS production, which affects the cells by the process of generating free radicals. Photosynthetic organisms have gained different strategies to counter the negative consequences of ROS. PUFAs have a positive role in providing protection to those cells which are affected by the process of oxidation. Phycocyanobilins are considered as efficient scavengers of different oxygen derivatives and have the potential to protect the different living cells against the oxidative stress (Singh et al., 2017). Also, mycosporine like amino acid (MAA) are a family of intracellular compounds which functions in protecting the cell by scavenging the oxygen derivatives and superoxide anions (Suh et al., 2003).

6.2.6 MICROALGAL COMPOUNDS WITH EVENTUALITY OVER DEGENERATIVE DISEASES

Reactive oxidation species (ROS) are responsible for causing irreversible damage to the cell. In the human body, oxidation reaction occurs by ROS which cause unrepairable damage to the cell and its constituents like lipids, DNA, and proteins and finally leads to the several syndromes like degenerative diseases of aging, cardiovascular diseases (CVDs) and some cancers (De Morais et al., 2015).

The natural pigments obtained from the microalgae shows neuroprotective properties and contain valuable functional ingredients in pharmaceuticals and food products which are used in the treatment or prevention of neurodegenerative diseases. Vitamin E obtained from microalgae have potential against many diseases such as heart diseases, atherosclerosis, and neurodegenerative diseases (Pangestuti and Kim, 2011). Carotenoids from microalgae have potential benefits and have positive consequences on human health. The degenerative diseases such as cataract development and macular degeneration are cured by carotenoids. These compounds functions as an antioxidant which decreases the oxidative damage caused by ROS. Phenols ingestion reduces the occurrence of degenerative diseases. Different reports show that phenolic compounds (PCs) obtained from the microalgae have a potential role in fighting against free radicals (El-Baky et al., 2009; De Morais et al., 2015). The natural source of β-carotene is *Dunaliella salina* which is effective against degenerative diseases. Lutein is also obtained from microalgae and is effective against various diseases such as macular degeneration, cataracts, and early stage of atherosclerosis. The extracts obtained from *Chlorella* species having lutein and β-carotene have a potential effect on preventing cognitive disability. Some reports show that lutein extracts from *Chlorella* have a preventive role in cancer. Carotenoids extracts from *Chlorella vulgaris* and *Chlorella ellipsoidea* have a potential role in inhibiting the growth and multiplication of colon cancer. The compound lycopene extracted from *Chlorella marina* helps in reducing the proliferation of prostate cancer in mice and reducing the cholesterol level and low-density lipoprotein level and also improves rheumatoid arthritis. Increased level of carotenoids associated with provitamin A activity containing α-carotene, β-cryptoxanthin, and β-carotene have been related to lowering the risk of angina pectoris. Low level of zeaxanthin and lutein is associated with macular degeneration, which is the vital cause of irreversible vision loss (Renju et al., 2014; De Morais et al., 2015). Scientific reports have shown the use of astaxanthin for multimodal intervention. It is used in many forms of degenerative diseases such as metabolic syndrome, cancer, cardiovascular disease, stomach, and ocular diseases, age related immune dysfunction, cognitive impairment, and skin damage (Vega et al., 2011). Lycopene levels in the body also regulate the coronary heart diseases, risk of atherosclerosis and myocardial infarction (Renju et al., 2014).

6.2.7 MICROALGAL COMPOUNDS WITH ANTIOBESITY ACTIVITY

Weight loss remains the most potential approach for obesity and lowering the risk of related diseases. Achieving and maintaining the weight loss is difficult (El-Baz et al., 2020). A microalgal compound fucoxanthin is a marine carotenoid obtained from diatom, *Phaeodactylum tricornutum* (Kim et al., 2012). Fucoxanthin, a microalgal compound, has a major role in the treatment of obesity and is a promising weight loss and dietary supplement. Scientific studies showed the effect of xanthigen which is a fucoxanthin-based antiobesity supplement (De Morais et al., 2015). Fucoxanthin is also useful in the treatment and prevention of bone diseases such as rheumatoid arthritis and osteoporosis. Some reports also showed the use of fucoxanthin in the treatment of diabetes, and hyperglycemia (Pangestuti and Kim, 2011).

6.3 MICROALGAE AS A SOURCE OF HIGH-VALUE PRODUCTS

Microalgae contain a vast variety of different products which are used for various biotechnological purposes. These high-value products include carotenoids, phycobiliproteins (PBP), phycotoxins, and polysaccharides (Table 6.1). The main advantage of using microalgae for bioactive compounds is their easy cultivation on a large scale for the production of high-value chemicals (Chu, 2012).

TABLE 6.1 Biotechnological Applications of Products from Microalgae (Chu, 2012)

Product	Applications	Microalgal Producers
	Polyunsaturated Fatty Acids (PUFA)	
Eicosapentaenoic acid (EPA)	Nutritional supplements, aquaculture feed	*Pavlova*, *Nannochloropsis*, *Monodus*, and *Phaeodactylum*
Docohexaenoic acid (DHA)	Infant formula, nutritional supplements, aquaculture feed	*Crypthecodinium* and *Schizochytrium*
Gamma-linolenic acid (GLA)	Nutritional supplements	*Spirulina*
Arachidonic acid (AA)	Nutritional supplements	*Porphyridium*
Phycobiliproteins		
Phycocyanin	Natural dye for health food and cosmetics (lipsticks and eyeliners), antioxidant	*Spirulina platensis*
Phycoerythrin	Tool for biomedical research, fluorescent agent, diagnostic tool	Red algae (e.g., *Porphyridium cruentum*)
Carotenoids		
β-carotene	Food colorant, antioxidant, cancer-preventive properties	*Dunaliella salina*
Astaxanthin	Pigmenter for salmon, antioxidant	*Haematococcus pluvialis*
Mycosporine-like amino acids (MAA)	UV-screening agent, sunscreen	*Aphanizomenon flos-aquae*
Polysaccharides	Lubricants and flocculants for industrial applications, viscosifiers, antiviral agent	*Porphyridium cruentum*
	Phycotoxins	
Okadaic acid, gonyautoxins, and yessotoxins	Experimental tools for investigations on neurodegenerative diseases	Dinoflagellates such as *Amphidinium*, *Prorocentrum*, and *Dinophysis*
	Lipids	
Triglycerides and hydrocarbons	Biofuels	*Chlorella protothecoides*, *Botryococcus braunii*

6.4 MICROALGAE AS A SOURCE OF BIOACTIVE COMPOUNDS

As microalgae have the ability to cope with stress, they are also capable of forming a vast variety of bioactive chemicals. These bioactive compounds are secondary metabolites which include carbohydrates, amino acids, peptides, vitamins, antibiotics, enzymes, and organic acid. These metabolites show a vast range of biological activities such as anticancer, antioxidant, antiviral, and immunomodulatory effects (Chu, 2012). From various groups of microalgae, the most prominent is the cyanobacteria, which are the main sources of bioactive molecules used for the biotechnological purpose. These compounds are mainly produced by cyanobacteria and dinoflagellates. These compounds are particularly produced from those species which are responsible for causing algal bloom in fresh or marine water environment (Chu, 2012).

Different toxins can cause major health problems to humans, wild life and domestic animals with different toxicological effects, including hepatotoxicity, neurotoxicity, dermotoxicity, and cytotoxicity (Carmichael, 2001; Chu, 2012). Algal toxins such as microcystins, saxitoxins, homo-, and anatoxin-a are produced by cyanobacteria such as *Anabaena*, *Microcystis*, *Nostoc*, and *Oscillatoria* species. The dinoflagellates which produce toxins include *Alexandrium*, *Karenia*, *Dinophysis*, and *Gymnodinium* species (Chu, 2012). Phycotoxins possess a vast range of biological activities including, antitumor, cytotoxic, antifungal, antibiotic, neurotoxic, and immunosuppressant activities, which provide a great potential for biotechnological exploitation (Camacho et al., 2007; Chu, 2012).

6.5 MICROALGAE ROLE IN DIETARY AND FOOD SUPPLEMENTS

Microalgae constitute a high nutritional value. Microalgal metabolites are used globally in nutraceutical applications. Algae species such as *Spirulina*, *Chlorella*, *Dunaliella*, and *Haematococcus* are major sources of secondary metabolites for the development of healthier food products and can produce different metabolites such as flavonoids, terpenoids, alkaloids, glycosides, and phenazines. Along with these metabolites, algae contains carbohydrates, lipids, proteins, polysaccharides, PBP, and polyols. These secondary metabolites are enormously used in the food and health sectors (Chu, 2012; Wells et al., 2012).

Chlorella a microalga is used in the health food. These microalgae are used in the form of pills and powder. *Chlorella* contains a large amount of proteins, vitamins, minerals, polysaccharides, amino acids, chlorophyll, oligo-elements, minerals, and vitamin B complex (B12). Vitamin B12 has a vital role in the regeneration and formation of blood cells. FDA approves the use of *Chlorella* as a food prior to the conditions of cultivation of micro-algae in proper environment with suitable hygiene and good manufacturing practices (Singh et al., 2017). *Chlorella* also produced different by-products that are used as preservatives in food products such as vegetables and fruits (Priyadarshani and Rath, 2012).

Spirulina is also cultured on an enormous scale for commercial production of algal biomass as dietary and food supplement in various countries. *Spirulina* contains high content of proteins, vitamins, gamma-linolenic acid (GLA) and serve as a nutritious food (Chu, 2012). *Spirulina* is used for therapeutic implications, including for health problems such as arthritis, diabetes, CVDs, anemia, and cancer. *Spirulina* is also used as a functional ingredient and is incorporated into various food products in order to enhance their nutritional qualities. *Spirulina* is used as a therapeutic management of chronic disorders such as hypertension, diabetes, and heart disease. *Spirulina* is also well-known for its antioxidant compounds such as vitamin E and phycocyanin (Mani, 2007; Chu, 2011; Joventino et al., 2012).

Another algae *Nostoc* is regarded as a healthy food due to its low-fat content and high protein and pigment content. Different species of *Nostoc* are used as herbal and food ingredients. The species *Nostoc flagelliforme* contains different amino acid which is needed for human nutrition (Chu, 2012).

6.5.1 MICROALGAL COMPOUNDS AS FOOD COLORANT

Microalgae are used as natural food colorant and added to snacks, food, pasta, drinks, and are essential as food supplements. Pigments from micro-algae are commercially used as natural food coloring agent, carotene from microalgae have a major role in coloring different food materials. Yellow color to margarine is a major application of the β-carotene pigment obtained from microalgae. β-carotene is also used as a food additive for enhancing the colors of the flesh of fish, eggs yolk. β-carotene also have a potential role in improving health and fertility rate of grain fed cattle (Priyadarshani and Rath, 2012).

The application and use of microalgae as a food colorant is limited because it is not photostable and with cooking the colors tends to bleach. Nevertheless, the application of microalgae food coloring is vast and applied to many food products. *Dunaliella salina* is a major source of photosynthetic pigment, beta carotene which is used as a dye and vitamin C supplement (Priyadarshani and Rath, 2012).

6.6 MICROALGAE AND ITS APPLICATIONS IN COSMETICS

Different components of algae species are used in cosmetics as antioxidants, thickening agents, and water-binding agents. *Chlorella* and *Arthrospira* are used in skincare products (Stolz and Obermayer, 2005), cosmetics, and toiletries products (Priyadarshani and Rath, 2012). Extracts from microalgae are found in face and skincare products like refreshing or regenerator care products, anti-aging cream, anti-irritant in peelers and emollient. Extracts from microalgae are also used in different products of sun protection and hair care products. The species that are used for the cosmetic products are *Dunaliella salina*, *Chondrus crispus*, *Alaria esculenta*, *Ascophyllum nodosum*, *Spirulina platensis*, *Nannochloropsis oculata*, *Chlorella vulgaris*, and *Mastocarpus stellatus* (Priyadarshani and Rath, 2012).

Organic metabolites are also produced by certain microalgae such as scytonemin, MAAs and sporopollenin, which are helpful in protecting them from UV radiation and allowing visible radiation that are involved in the process of photosynthesis (Priyadarshani and Rath, 2012).

From the last two decades, the cases of skin cancer increase such as non-melanoma skin cancer (NMSC), so the use of sunscreen is considered beneficial by skincare professionals in order to counter the effect of skin cancer. Scytonemin from cyanobacteria is considered as an efficient and natural UV blockers and is used in these formulations. Scytonemin blocks 90% of UV radiation from entering to the cell. Studies show that scytonemin has a kinase activity and is helpful in curing of disorders like inflammation and proliferation. This scytonemin has an antioxidant activity and acts as a radical scavenger and prevent the cellular damage resulting from the radical oxidizing species (ROS) due to the exposure to UV radiations (Matsui et al., 2012; Rastogi et al., 2015; Singh et al., 2017).

Another photoprotective compounds are carotenoids, beta carotene helps in protecting skin from UV-induced photooxidation. Ketocarotenoid astaxanthin has a major role in protection from pathological damages in human-like inflammation, photooxidation, aging, ulcers, prostate, and mammary carcinogenesis (Gurein et al., 2003; Cardozo et al., 2007; Singh et al., 2017).

In Asia, a common tradition is the whitening of skin. The logic behind whitening is that it is a parameter of beauty and beauty can be judged by skin which is white in color in Asian culture (Li et al., 2008). For this purpose, the most common approach for skin whitening is the use of tyrosinase inhibitors. Fucoxanthin extracted from *Laminaria japonica* suppresses tyrosinase activity as this enzyme catalyzes the rate-limiting step of pigmentation. Fucoxanthin has the ability to negatively regulate the melanogenesis at the transcriptional level (Thomas and Kim, 2013). Another brown algal secondary metabolite that is Phloroglucinol also have tyrosinase inhibitory activity due to their ability to chelate copper (Babitha and Kim, 2011). At industrial scale, they may be widely used in drugs, food additives, and cosmetics (Jha and Zi-Rong, 2004).

Different algae components are used as thickening agents, antioxidants, and water-binding agent. Different species extracts are used in skin and face care products. The species like *Ascophyllum nodosum* (brown algae), *Chondrus crispus* (red algae), *Alaria esculenta* (brown algae), *Nannochloropsis oculata* (algae), *Spirulina platensis* (BGA), *Dunaliella salina* (green algae), *Chlorella vulgaris* (green algae) and *Mastocarpus stellatus* (red algae) have occupied major role in cosmetics and important position in the skincare market (Stolz and Obermayer, 2005; Sharma and Sharma, 2017).

6.7 CONCLUSION

Algae are vast and diverse group of microorganisms having a capacious range of biochemical and physiological characteristics. Microalgae are considered as an important raw material for the production of vitamins, bioactive compounds, amino acids, and pigments. The bioactive metabolites obtained from microalgae are of great interest for the production and development of new products for the food, pharmaceuticals, and cosmetic industries. Microalgae are widely used throughout the world because of its potential application in pharmaceuticals and nutraceuticals. Other applications of secondary metabolites are their activities as antibacterial, antiviral, and anticancer compounds. There is also a hidden potential present in microalgae that will be exploited further in the near future for the discovery of new drugs. Algae have the potential to provide novel bioactive compounds.

KEYWORDS

- **antibacterial**
- **anti-inflammatory**
- **antioxidant**
- **antiviral**
- **bioactive compounds**
- **biotechnology**
- **cosmetics**
- **metabolites**
- **microalgae**
- **pharmaceuticals**

REFERENCES

Abedin, R. M., & Taha, H. M., (2008). Antibacterial and antifungal activity of cyanobacteria and green microalgae. Evaluation of medium components by Plackett-Burman design for antimicrobial activity of *Spirulina platensis*. *Global Journal of Biotechnology and Biochemistry*, *3*(1), 22–31.

Aditya, T., Bitu, G., & MercyEleanor, G., (2016). The role of algae in pharmaceutical development. *Research and Reviews: Journal of Pharmaceutics and Nanotechnology*, *2016*, 82–89.

Amaro, H. M., Guedes, A. C., & Malcata, F. X., (2011). Antimicrobial activities of microalgae: An invited review. In: Méndez-Vilas, A., (ed.), *Science Against Microbial Pathogens: Communicating Current Research and Technological Advances* (Vol. 3, pp. 1272–1284). Spain: Formatex.

Babitha, S., & Kim, E. K., (2011). Effect of marine cosmeceuticals on the pigmentation of skin. In: Kim, S., (ed.), *Marine Cosmeceuticals: Trends and Prospects* (pp. 63–66). Boca Raton, FL, USA: CRC Press.

Bhat, V. B., & Madyastha, K. M., (2001). Scavenging of peroxynitrite by phycocyanin and phycocyanobilin from *Spirulina platensis*: Protection against oxidative damage to DNA. *Biochemical and Biophysical Research Communications*, *285*(2), 262–266.

Bhattacharjee, M., (2016). Pharmaceutically valuable bioactive compounds of algae. *Asian J. Pharm. Clin. Res.*, *9*, 43–47.

Boopathy, N. S., & Kathiresan, K., (2010). Anticancer drugs from marine flora: An overview. *Journal of Oncology*, *2010*, 214186.

Camacho, F. G., Rodríguez, J. G., Mirón, A. S., et al., (2007). Biotechnological significance of toxic marine dinoflagellates. *Biotechnology Advances*, *25*(2), 176–194.

Cardozo, K. H., Guaratini, T., Barros, M. P., et al., (2007). Metabolites from algae with economical impact. *Comparative Biochemistry and Physiology Part C: Toxicology and Pharmacology*, *146*(1, 2), 60–78.

Carmichael, W. W., (2001). Health effects of toxin-producing cyanobacteria: "TheCyano HABs". *Human and Ecological Risk Assessment: An International Journal*, *7*(5), 1393–1407.

Chu, W. L., (2011). Potential applications of antioxidant compounds derived from algae. *Current Topics in Nutraceuticals Research*, *9*(3), 83.

Chu, W. L., (2012). Biotechnological applications of microalgae. *IeJSME*, *6*(1), S24–S37.

De Jesus, R. M. F., De Morais, A. M. M. B., & De Morais, R. M. S. C., (2014). Influence of sulphate on the composition and antibacterial and antiviral properties of the exopolysaccharide from *Porphyridiumcruentum*. *Life Sciences*, *101*(1, 2), 56–63.

De Morais, M. G., Vaz, B. D. S., De Morais, E. G., & Costa, J. A. V., (2015). Biologically active metabolites synthesized by microalgae. *BioMed Research International*, *2015*, 835761. https://doi.org/10.1155/2015/835761.

El-Baky, H. H. A., El Baz, F. K., & El-Baroty, G. S., (2009). Production of phenolic compounds from *Spirulina maxima* microalgae and its protective effects. *African Journal of Biotechnology*, *8*(24).

El-Baz, F. K., Aly, H. F., & Abd-Alla, H. I., (2020). The ameliorating effect of carotenoid rich fraction extracted from *Dunaliella salina* microalga against inflammation-associated cardiac dysfunction in obese rats. *Toxicology Reports*, *7*, 118–124.

Estrada, J. P., Bescós, P. B., & Del, F. A. V., (2001). Antioxidant activity of different fractions of *Spirulina platensis* protean extract. *Ilfarmaco*, *56*(5–7), 497–500.

Faten, M., & Emad, A. S., (2009). Antioxidant activity of extract and semi-purified fractions of marine red macroalga, *Gracilaria verrucosa*. *Aust. J. Basic App. Sci.*, *3*(4), 3179–3185.

Furusawa, E., Moore, R. E., Mynderse, J. S., Norton, T. R., & Patterson, G. M. L., (1994). New purified culture of *Scytonema pseudohofmanni* ATCC 53141 is used to produce Scytophycins A, B, C, D and E, which are potent cytotoxins and antineoplastic agents. USA Patent, 5281533.

Gerwick, W. H., Roberts, M. A., Proteau, P. J., & Chen, J. L., (1994). Screening cultured marine microalgae for anticancer-type activity. *Journal of Applied Phycology*, *6*(2), 143–149.

Guerin, M., Huntley, M. E., & Olaizola, M., (2003). Haematococcus astaxanthin applications for human health and nutrition. *Trends in Biotechnology*, *21*(5), 210–216.

Guzman, S., Gato, A., Lamela, M., Freire-Garabal, M., & Calleja, J. M., (2003). Anti-inflammatory and immunomodulatory activities of polysaccharide from *Chlorella stigmatophora* and *Phaeodactylum tricornutum*. *Phytotherapy Research*, *17*(6), 665–670.

Harun, R., Singh, M., Forde, G. M., & Danquah, M. K., (2010). Bioprocess engineering of microalgae to produce a variety of consumer products. *Renewable and Sustainable Energy Reviews*, *14*(3), 1037–1047.

Jha, R. K., & Zi-Rong, X., (2004). Biomedical compounds from marine organisms. *Marine Drugs*, *2*(3), 123–146.

Joventino, I. P., Alves, H. G., Neves, L. C., et al., (2012). The microalga *Spirulina platensis* presents anti-inflammatory action as well as hypoglycemic and hypolipidemic properties in diabetic rats. *Journal of Complementary and Integrative Medicine*, *9*(1).

Khan, M. N. A., Cho, J. Y., Lee, M. C., et al., (2007). Isolation of two anti-inflammatory and one pro-inflammatory polyunsaturated fatty acids from the brown seaweed *Undaria pinnatifida*. *Journal of Agricultural and Food Chemistry*, *55*(17), 6984–6988.

Kim, S. M., Jung, Y. J., Kwon, O. N., et al., (2012). A potential commercial source of fucoxanthin extracted from the microalga *Phaeodactylum tricornutum*. *Applied Biochemistry and Biotechnology*, *166*(7), 1843–1855.

Li, E. P., Min, H. J., Belk, R. W., Kimura, J., & Bahl, S., (2008). Skin lightening and beauty in four Asian cultures. In: Angela, Y. L., & Dilip, S., (eds.), *NA – Advances in Consumer Research* (Vol. 35, pp. 444–449). Duluth, MN: Association for Consumer Research.

Lipton, A. P., (2003). Marine bioactive compounds and their applications in mariculture marine ecosystem. *Curr. Sci.*, *2*, 4–16.

Maeda, H., Hosokawa, M., Sashima, T., Funayama, K., & Miyashita, K., (2005). Fucoxanthin from edible seaweed, *Undaria pinnatifida*, shows antiobesity effect through UCP1 expression in white adipose tissues. *Biochemical and Biophysical Research Communications*, *332*(2), 392–397.

Mani, I. U., (2007). Spirulina and its therapeutic implications as a food product. In: Gershwin, M. E., & Amha, B., (eds.), *Spirulina in Human Nutrition and Health* (pp. 65–84). Boca Raton, FL, USA: CRC Press.

Matsui, K., Nazifi, E., Hirai, Y., Wada, N., Matsugo, S., & Sakamoto, T., (2012). The cyanobacterial UV-absorbing pigment scytonemin displays radical-scavenging activity. *The Journal of General and Applied Microbiology*, *58*(2), 137–144.

Matsui, M. S., Muizzuddin, N., Arad, S., & Marenus, K., (2003). Sulfated polysaccharides from red microalgae have anti-inflammatory properties *in vitro* and *in vivo*. *Applied Biochemistry and Biotechnology*, *104*(1), 13–22.

Noaman, N. H., Fattah, A., Khaleafa, M., & Zaky, S. H., (2004). Factors affecting antimicrobial activity of *Synechococcus leopoliensis*. *Microbiological Research*, *159*(4), 395–402.

Pangestuti, R., & Kim, S. K., (2011). Biological activities and health benefit effects of natural pigments derived from marine algae. *Journal of Functional Foods*, *3*(4), 255–266.

Patterson, G. M., Baker, K. K., Baldwin, C. L., et al., (1993). Antiviral activity of cultured blue-green algae (*Cyanophyta*). *Journal of Phycology*, *29*(1), 125–130.

Plaza, M., Herrero, M., Cifuentes, A., & Ibanez, E., (2009). Innovative natural functional ingredients from microalgae. *Journal of Agricultural and Food Chemistry*, *57*(16), 7159–7170.

Prakash, J. W., Marimuthu, J., & Jeeva, S., (2011). Antimicrobial activity of certain freshwater microalgae from Thamirabarani River, Tamil Nadu, South India. *Asian Pacific Journal of Tropical Biomedicine*, *1*(2), S170–S173.

Pratt, D. E., (1992). Natural antioxidants from plant material. In: Hang, M., Ho, C., & Lee, C. Y., (eds.), *Phenolic Compounds in Food and Their Effects on Health II: Antioxidants and Cancer Prevention* (pp. 54–71). Washington DC: American Chemical Society.

Priyadarshani, I., & Rath, B., (2012). Commercial and industrial applications of microalgae: A review. *Journal of Algal Biomass Utilization*, *3*(4), 89–100.

Rastogi, R. P., Sonani, R. R., & Madamwar, D., (2015). Cyanobacterial sunscreen scytonemin: Role in photoprotection and biomedical research. *Applied Biochemistry and Biotechnology*, *176*(6), 1551–1563.

Renju, G. L., Kurup, G. M., & Kumari, C. S., (2014). Effect of lycopene from *Chlorella marina* on high cholesterol-induced oxidative damage and inflammation in rats. *Inflammopharmacology*, *22*(1), 45–54.

Ryckebosch, E., Bruneel, C., Termote-Verhalle, R., Goiris, K., Muylaert, K., & Foubert, I., (2014). Nutritional evaluation of microalgae oils rich in omega-3 long-chain polyunsaturated fatty acids as an alternative for fish oil. *Food Chemistry*, *160*, 393–400.

Schepetkin, I. A., & Quinn, M. T., (2006). Botanical polysaccharides: Macrophage immunomodulation and therapeutic potential. *International Immunopharmacology*, *6*(3), 317–333.

Sharma, P., & Sharma, N., (2017). Industrial and biotechnological applications of algae: A review. *Journal of Advances in Plant Biology*, *1*(1), 01.

Singh, R., Parihar, P., Singh, M., et al., (2017). Uncovering potential applications of cyanobacteria and algal metabolites in biology, agriculture and medicine: Current status and future prospects. *Frontiers in Microbiology*, *8*, 515.

Smelcerovic, A., Knezevic-Jugovic, Z., & Petronijevic, Z., (2008). Microbial polysaccharides and their derivatives as current and prospective pharmaceuticals. *Current Pharmaceutical Design*, *14*(29), 3168–3195.

Smith, V. J., Desbois, A. P., & Dyrynda, E. A., (2010). Conventional and unconventional antimicrobials from fish, marine invertebrates and micro-algae. *Marine Drugs*, *8*(4), 1213–1262.

Stevenson, C. S., Capper, E. A., Roshak, A. K., et al., (2002). The identification and characterization of the marine natural product scytonemin as a novel antiproliferative pharmacophore. *Journal of Pharmacology and Experimental Therapeutics*, *303*(2), 858–866.

Stolz, P., & Obermayer, B., (2005). Manufacturing microalgae for skincare. *Cosmetics and Toiletries*, *120*(3), 99–106.

Suh, H. J., Lee, H. W., & Jung, J., (2003). Mycosporine glycine protects biological systems against photodynamic damage by quenching singlet oxygen with a high efficiency. *Photochemistry and Photobiology*, *78*(2), 109–113.

Thomas, N. V., & Kim, S. K., (2013). Beneficial effects of marine algal compounds in cosmeceuticals. *Marine Drugs*, *11*(1), 146–164.

Tyagi, S., Singh, G., Sharma, A., & Aggarwal, G., (2010). Phytochemicals as candidate therapeutics: An overview. *International Journal of Pharmaceutical Sciences Review and Research*, *3*(1), 53–55.

Vega, P. J. M., VílchezLobato, C., Bedmar, F., Cuaresma, F. M., & Forjan, L. E., (2011). Marine carotenoids: Biological functions and commercial applications. *Marine Drugs*, *9*(3), 319–333.

Wells, M. L., Potin, P., Craigie, J. S., et al., (2017). Algae as nutritional and functional food sources: revisiting our understanding. *Journal of Applied Phycology*, *29*(2), 949–982.

CHAPTER 7

MICROALGAL CAROTENOIDS: BIOACCESSIBILITY AND BIOAVAILABILITY FOR BIOTECHNOLOGICAL APPLICATIONS

MARCELE L. NÖRNBERG, PRICILA P. NASS, EDUARDO JACOB-LOPES, and LEILA Q. ZEPKA

Department of Food Science and Technology, Federal University of Santa Maria (UFSM), Roraima Avenue, 1000, 97105-900, Santa Maria, RS, Brazil

ABSTRACT

The growing demand for natural inputs has boosted the biotechnological exploration of microalgal metabolites, with emphasis on carotenoids for being excellent sources of biocompounds applied in various areas of the industry. Carotenoids can behave as active ingredients, minimizing the risk of developing cardiovascular disease, cancer, and age-related macular degeneration (AMD) due to its antioxidant potential. Despite the beneficial properties of carotenoids, their effectiveness at preventing or treating of diseases depends on their bioaccessibility and bioavailability. Thus, models that simulate the digestion process have been described, seeking to contribute to a better understanding of the potential availability of that compound. In that context, this chapter provides a comprehensive description of carotenoids from microalgae, remarkable biological properties, and the bioaccessibility and bioavailability of these metabolites.

Algal Metabolites: Biotechnological, Commercial, and Industrial Applications. Jeyabalan Sangeetha, Devarajan Thangadurai, Saher Islam, & Ravichandra Hospet (Eds.)

7.1 INTRODUCTION

Microalgae are a source of valuable natural ingredients due to their content of bioactive compounds (Jacob-Lopes et al., 2019). Of particular interest are the carotenoids, which represent a model of success in terms of commercial carotenoid production, through the cultivation of *Dunaliella salina* and *Haematococcus pluvialis*, with a focus on β-carotene and astaxanthin. Besides, the global carotenoid market is estimated to be valued at USD 1.7 Billion in 2022 (Rammuni et al., 2018; Fernandes et al., 2020).

In addition, carotenoid metabolites have also gained interest, due to exceptional bioactive properties, including antioxidant activity, pro-vitamin A precursors, immune activators, anti-neurodegenerative, anti-cancer, anti-inflammatory agents, bone health, and protection of macular degeneration (Patias et al., 2017; El-Akabawy and El-Sherif, 2019; Saini and Keum, 2020).

However, the bioactive properties of carotenoids are extremely dependent on their bioaccessibility and bioavailability. Numerous factors affect the carotenoid bioaccessibility and bioavailability, the matrix and its release, changes during digestion, uptake, metabolism, and biodistribution to target tissues following ingestion (Bohn, 2018; Kopec and Failla, 2018; Xavier and Mercadante, 2019).

Thus, this chapter provides an overview of the microalgal carotenoid profile, its importance in human health, and finally, recent advances in the bioaccessibility and bioavailability of these structures.

7.2 CHARACTERISTICS OF MICROALGAL CAROTENOIDS

Carotenoids are responsible for the yellow, orange, and red tones of foods. This effect stems from its basic structure, a polyene backbone that contains a variable number of conjugated double bonds, called a light-absorbing chromophore, requiring seven conjugated double bonds to be colored. The color change occurs as the number of double bonds increases, as there is a shift in the absorption spectrum of the molecule, which characterizes each carotenoid (Rodriguez-Amaya, 2001, 2015; Maldonade et al., 2008; Irazusta et al., 2013; Jomova and Valko, 2013).

The structure of these compounds are generally tetraterpenes (C40) consisting of 8 isoprenoid units (C5) can be classified as xanthophylls are carotenoids that contain oxygen as a functional group such as epoxy (violaxanthin and fucoxanthin), hydroxy (lutein and zeaxanthin), and keto (astaxanthin and canthaxanthin). Carotenes, which contain only by carbon and hydrogen atoms (α-carotene and β-carotene) (Gul et al., 2015; Zaghdoudi et al., 2015; Magosso et al., 2016; Gille et al., 2018; Lafarga et al., 2020). The structures of xanthophylls and carotenes are shown in Figure 7.1(a).

Microalgae can synthesize large quantities of carotenoids that have shown different biotechnological explorations, in addition to physiological properties, discussed in Table 7.1. Furthermore, they have become an attractive option as a natural source of inputs for the industry due to the emerging of health safety issues caused by synthetic products such as allergic reactions and hyperactivity in humans (Khoo et al., 2019). The main commercial sources of natural carotenoids with β-carotene and astaxanthin are *Dunaliella salina* and *Haematococcus pluvialis* and represent a successful model (up to 7%–14% in dry weight) (Rammuni et al., 2018; Fernandes et al., 2020).

The *Dunaliella salina* produces β-carotene in quantities that represent about 10–14% of its dry mass, being the world leader in the production of β-carotene (Borowitzka, 2013; Mulders et al., 2014; Khan et al., 2018). The carotenoids provitamin-A of *D. salina* widely used in pharmaceutical products and as a food color (Guedes et al., 2011; Chen et al., 2015; Jacob-Lopes et al., 2019).

Moreover, astaxanthin produced by *Haematococcus pluvialis* applications are important in the pharmaceutical, nutraceutical, cosmetic, food industries (dye, food additive or dietary supplement), and in animal feed as a source of pigmentation (Vílchez et al., 2011; Xia et al., 2013; Zhang et al., 2014; Panis and Carreon, 2016; Shah et al., 2016; Sathasivam and Ki, 2018; Jacob-Lopes et al., 2019; Hamidi et al., 2020).

Besides, other microalgae species represent a potential dietary source of carotenoids as lutein and its structural isomer zeaxanthin. The *Muriellopsis* sp. produces high lutein content and a high growth rate; therefore, it has been exploited for commercial production (Guedes et al., 2011; Sathasivam et al., 2019). Lutein can be used as a food additive in the nutraceutical field and as cosmetics. It also has been used as a pigment source for birds. In addition, it has been marketed as a food supplement for its beneficial effects on vision (Mehariya et al., 2019; Sathasivam et al., 2019).

The microalgae used for the production of zeaxanthin are the *Chlorella vulgaris*, *Phaeodactylum tricornutum*, *Scenedesmus almeriensis*, and *Nannochloropsis oculata*. Zeaxanthin is mainly used in pharmaceutical products, applications in the cosmetics and food industry, also used in aquaculture and poultry as a pigment source (Mulders et al., 2014; Chen et al., 2015; Dufossé, 2016; Gille et al., 2018; Sathasivam and Ki, 2018; Singh et al., 2020).

TABLE 7.1 Some Carotenoid-Producing Microalgae, Their Biotechnological Explorations, and its Physiological Properties

Microalgae Source	Main Carotenoid	Biotechnological Exploration	Physiological Properties	References
Botryococcus braunii, *Chlorella vulgaris*, *Chlorella zofingiensis*, *Dunaliella salina*, *Galdieria sulfuraria*, *Haematococcus pluvialis*	Astaxanthin	• Aquaculture • Pharmaceuticals • Nutrition • Cosmetics • Health food • Feed additives • Farming of salmon and trout • Nutraceuticals • Food ingredients • Food coloring agent • Food supplement	• Benign prostatic hyperplasia and prostate and liver tumors • Anti-inflammatory properties • Active against liver neoplasms • Strong antioxidant property • Cardiovascular health • Hepatoprotective effect • Anti-nephropathy diabetic • Anti-hypertension • Anti-stroke • Anti-cancer • Anti-neurodegenerative diseases • Photoprotection of eyes and skin	Naguib (2000); Guedes et al. (2011); Vílchez et al. (2011); Cuellar-Bermudez et al. (2014); Koller et al. (2014); Mulders et al. (2014); Zhang et al. (2014); Dufossé (2016); Shah et al. (2016); Yeh et al. (2016); Sathasivam and Ki (2018); Goff et al. (2019); Sathasivam et al. (2019); Hamidi et al. (2020)
Botryococcus braunii, *Chlorella pyrenoidosa*, *Chlorella saccharophila*, *Chlorella zofingiensis*, *Chlorella vulgaris*, *Coccomyxa acidófila*, *Dunaliella bardawil*, *Dunaliella salina*, *Dunaliella tertiolecta*, *Haloferax alexandrinus*, *Halorubrum* sp., *Phormidium autumnale*,	β-carotene	• Aquaculture • Pharmaceuticals • Nutrition • Cosmetics • Health food • Food supplement • Feed • Feed additive	• Provitamin A • Colorectal cancer • Prevention of acute and chronic coronary syndromes • Photoprotection of skin against UV light • Antioxidant property • Prevention of liver fibrosis • Night blindness prevention	Guedes et al. (2011); Vílchez et al. (2011); Cuellar-Bermudez et al. (2014); Koller et al. (2014); Mulders et al. (2014); Zhang et al. (2014); Rodrigues et al. (2015); Dufossé (2016); Patias et al. (2017); Gille et al. (2018); Khan et al. (2018); Jacob-Lopes et al. (2019);
Porphyridium cruentum, *Scenedesmus almeriensis*, *Scenedesmus obliquus*, *Spirulina maxima*, *Spirulina platensis*		• Food colorant	• Anti-neurodegenerative diseases • Immune function • Embryonic development	Maroneze et al. (2019); Nascimento et al. (2019, 2020); Hamidi et al. (2020)

TABLE 7.1 *(Continued)*

Microalgae Source	Main Carotenoid	Biotechnological Exploration	Physiological Properties	References
Chlorella vulgaris, *Chlorella zofingiensis*, *Dietzia natronolimnaea*, *Gordonia jacobaea*, *Haematococcus pluvialis*, *Haloferax alexandrinus*, *Micrococcus roseus*, *Nannochloropsis gladitana*, *Nannochloropsis oculata*, *Nannochloropsis salina*, *Phormidium autumnale*	Canthaxanthin	• Pharmaceuticals • Nutrition • Cosmetics • Health food • Food additive • Farming of salmonids and chicken • Tanning pills • Food coloring agent • Animal feed additive	• Antioxidant property • Prevention of cardiovascular diseases • Antitumoral activity • Anti-cancer • Immune system stimulation • Anti-inflammatory • Neuroprotective effects	Guedes et al. (2011); Cuellar-Bermudez et al. (2014); Koller et al. (2014); Mulders et al. (2014); Rodrigues et al. (2015); Sathasivam and Ki (2018); Sathasivam et al. (2019); Gao et al. (2020); Hamidi et al. (2020)
Chaetoceros gracilis, *Cylindrotheca closterium*, *Isochrysis galbana*, *Nitzschia closterium*, *Nitzschia* sp., *Odontella aurita*, *Phaeodactylum tricornutum*	Fucoxanthin	• Pharmaceuticals • Nutrition • Cosmetics • Animal nutrition • Baby food	• Anti-obesity • Antioxidant property • Anti-cancer • Bone-protective effects • Anti-inflammatory effects • Anti-hepatotoxicity effects • Anti-diabetic • Antihypertensive	Xia et al. (2013); Cuellar-Bermudez et al. (2014); Zhang et al. (2014); Sathasivam and Ki (2018); Hamidi et al. (2020)

TABLE 7.1 *(Continued)*

Microalgae Source	Main Carotenoid	Biotechnological Exploration	Physiological Properties	References
Botryococcus braunii, *Chlorella fusca*, *Chlorella protothecoides*, *Chlorella pyrenoidosa*, *Chlorella sorokiniana*, *Chlorella vulgaris*, *Chlorella zofingiensis*, *Chlorococcum citriforme*, *Coccomyxa acidofila*, *Coccomyxa onubensis*, *Coelastrum proboscideum*, *Dunaliella salina*, *Galdieria sulfuraria*, *Haematococcus pluvialis*, *Muriella aurantiaca*, *Muriella decolor*, *Muriellopsis* sp., *Phormidium autumnale*, *Porphyridium cruentum*, *Scenedesmus almeriensis*, *Scenedesmus obliquus*, *Tetracystis aplanosporum*, *Tetracystis intermedium*, *Tetracystis tetrasporum*	Lutein	• Aquaculture • Pharmaceuticals • Nutrition • Health food • Feed additives • Food supplement • Cosmetics • Pigmentation of animal tissues • Food coloring agent poultry feeding	• Prevention of acute and chronic coronary syndromes and stroke • Helps to maintain a normal visual function • Cataract and age-related macular degeneration prevention • Prevention of retinitis • To avoid gastric infection by *H. pylori* • Anti-cancer • Antioxidant properties • Ocular protective effects • Anti-diabetic retinopathy • Prevention of cardiovascular diseases	Guedes et al. (2011); Vílchez et al. (2011); Xia et al. (2013); Cuellar-Bermudez et al. (2014); Koller et al. (2014); Mulders et al. (2014); Zhang et al. (2014); Rodrigues et al. (2015); Dufossé (2016); Patias et al. (2017); Sathasivam and Ki (2018); Mehariya et al. (2019); Nascimento et al. (2019, 2020); Hamidi et al. (2020)
Botryococcus braunii, *Chlorella pyrenoidosa*, *Chlorella saccharophila*, *Dunaliella salina*, *Gramella oceani*, *Gramella planctonica*, *Kordia*	Zeaxanthin	• Aquaculture • Food additive • Animal feed • Pharmaceutical • Food coloring agents	• Active against liver neoplasms • Prevention of acute and chronic coronary syndromes • Helps maintain normal visual function	Vílchez et al. (2011); Koller et al. (2014); Mulders et al. (2014); Zhang et al. (2014); Rodrigues et al. (2015); Dufossé (2016);

TABLE 7.1 *(Continued)*

Microalgae Source	Main Carotenoid	Biotechnological Exploration	Physiological Properties	References
aquimaris, *Mesoflavibacter zeaxanthinifaciens*, *Muricauda lutaonensis*, *Nannochloropsis gladitana*, *Nannochloropsis oculata*, *Phormidium autumnale*, *Porphyridium cruentum*, *Scenedesmus almeriensi*		• Poultry feeding • Feed • Cosmetics • Food supplement	• Cataract prevention • Prevents age-associated macular degeneration • Anti-colon cancer • Antioxidant properties • Anti-diabetic retinopathy • Prevention of cardiovascular diseases	Sathasivam and Ki (2018); Maroneze et al. (2019); Hamidi et al. (2020); Nascimento et al. (2020); Singh et al. (2020)

Some carotenoids are found mainly in microalgae, such as crocoxanthin, fucoxanthin, echinenone, and canthaxanthin (Figure 7.1(b)). Taking into account the structures, suggest a higher antioxidant activity for microalgal carotenoids about conventional sources, due to the presence of exclusive carotenoids, in which acetylenic bond and the greater number of conjugated double bonds are characteristic associated with the high antioxidant action of microalgae carotenoids (Uenojo et al., 2007; Klassen and Foght, 2011; Patias et al., 2017; Sathasivam and Ki, 2018; Nascimento et al., 2019).

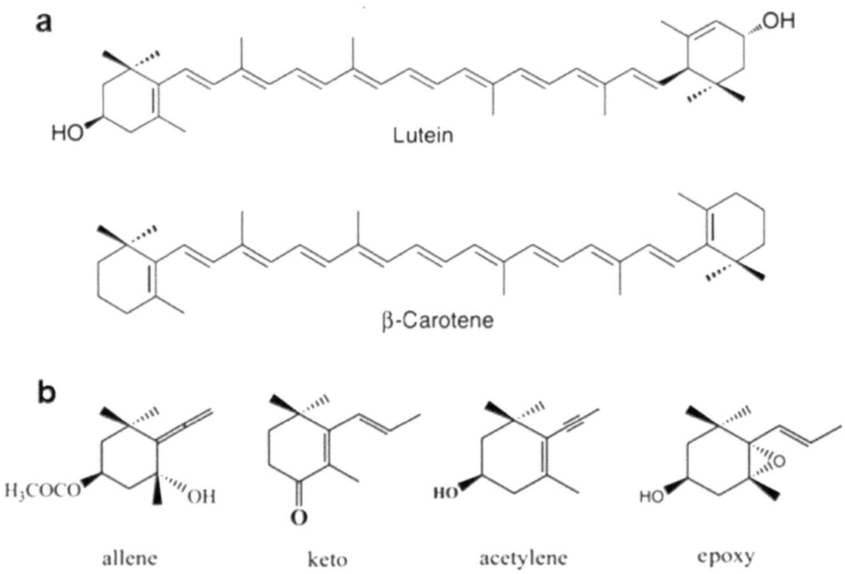

FIGURE 7.1 Chemical structures: (a) xanthophylls and carotenes; (b) unique cyclic end groups from microalgae.

These compounds are highly active against reactive oxygen species (ROS) and free radicals and gained great interest (Nascimento et al., 2020). Thus, microalgae represent natural resources of antioxidants due to their vast biodiversity. On the other hand, not all groups of microalgae can be applied as sources of antioxidants due to their widely diversified product content and to unfavorable growth, among other factors. The main types of microalgae in industrial production are *Arthrospira* sp., *Chlorella* sp., *Nitzschia* sp., *Porphyridium cruentum*, *Nannochloropsis* sp., *Crypthecodinium cohnii*, *Phaeodactylum tricornutum*, and *Schizochytrium* sp., sources of antioxidants and natural bioactive compounds that exhibit beneficial effects on human health (Khan et al., 2018; Hamidi et al., 2020).

7.3 BIOACTIVE CAROTENOIDS AND THEIR IMPORTANCE IN HEALTH

Scientific evidence has shown that microalgae to their ability both to produce bioactive metabolites, including a wide range of different carotenoids that can provide health benefits and to their functioning as bioactive and nutraceutical agents, as they have antioxidant, anti-carcinogenic, anti-inflammatory effects, among many other beneficial health effects (Table 7.1) (Guedes et al., 2011; Zhang et al., 2014; Sathasivam and Ki, 2018; Jacob-Lopes et al., 2019; Hamidi et al., 2020).

The protection of bioactive compounds present in carotenoids against some diseases is associated especially with its high antioxidant potential, the ability to sequester singlet oxygen and react with free radicals, as these can cause harmful effects, such as the triggering of many diseases, including cancer, coronary artery disease, obesity, diabetes, ischemic stroke, Alzheimer's disease (AD), among others (Trejo-Solís et al., 2013; Khan et al., 2018).

7.3.1 *β-CAROTENE*

The β-carotene, in addition to its antioxidant properties, has other benefits to human health, such as prevention of liver fibrosis, prevention of night blindness, anti-neurodegenerative properties, photoprotection of the skin against ultraviolet (UV) light, prevention of acute coronary syndrome and chronic, as well as some types of cancer. It also acts on vision, on increasing the performance of the immune system, on embryonic growth and development, due to its activity as provitamin A (Maldonade et al., 2008; Gul et al., 2015; Rodriguez-Amaya, 2015; Magosso et al., 2016; Gille et al., 2018; Khan et al., 2018; Hamidi et al., 2020).

Studies *in vitro* and *in vivo* showed that β-carotene is involved in reducing angiogenesis by suppressing cell proliferation and migration (Goff et al., 2019). Research with mice showed that a biomass-based diet of powdered microalgae *Dunaliella*, containing 0.6% β-carotene, can reduce cholesterol levels in blood plasma and inhibit the progression of atherosclerosis in diets rich in fat (Harari et al., 2008).

β-carotene have substantial effects on the enzymatic antioxidant defense system, preventing oxidative stress by eliminating free radicals, protecting lipids from peroxidation, factors linked to severe and lethal diseases, such as cardiovascular, Parkinson, atherosclerosis, and cancer (Khan et al., 2018; Goff et al., 2019). The anti-cancer activity of these molecules involves several mechanisms, including the induction of cell apoptosis and suppression of cell proliferation. In particular, an *in vivo* study showed that β-carotene helps to reduce liver neoplasms (Sathasivam and Ki, 2018).

Serum β-carotene levels are inversely correlated with glycated hemoglobin levels A1c (HbA1c) associated with decreased insulin sensitivity (Goff et al., 2019). Sluijs et al. (2015) investigated the association of a carotenoid-rich diet with type 2 diabetes. The carotenoids included in the diet were: β-carotene, α-carotene, β-cryptoxanthin, lycopene, lutein, and zeaxanthin; and the sum of all these carotenoids. According to this study, intake of β-carotene and α-carotene reduced type 2 diabetes, both in men and women.

7.3.2 ASTAXANTHIN

In a comparative study of the antioxidant activities of some carotenoids, Naguib (2000) reported that astaxanthin has higher activity when compared to α-carotene, β-carotene, lutein, and lycopene, in the elimination of ROS, which makes it a powerful antioxidant and anti-inflammatory (Yeh et al., 2016).

The oxidative stress is a causal factor or at least an aid in the pathogenesis of the main neurodegenerative diseases (Alzheimer's, Huntington's, Parkinson's, and amyotrophic lateral sclerosis (ALS). Natural astaxanthin can pass the blood-brain barrier and thus extend its antioxidant effect. Therefore, astaxanthin can work by reducing the effects of Alzheimer's and other neurological diseases (Shah et al., 2016).

Anti-inflammatory effects, enhanced by astaxanthin, preserve essential fatty acids (EFAs) and lymphocyte proteins. Astaxanthin works by inducing actions of the enzymes superoxide dismutase and catalase (CAT). Other studies have shown that astaxanthin protects against inducing liver damage (CCl_4), inhibiting lipid peroxidation, stimulating the cellular antioxidant system, and modulating the inflammatory process (Vílchez et al., 2011), reduces the occurrence of cancer, antiproliferative activity, and reduction of metastases (induced by stress) (Goff et al., 2019).

In addition to the anti-cancer effect, astaxanthin can be used in the prevention and treatment of chronic diseases like cardiovascular disorders, diabetes, and diabetic nephropathy (Zhang et al., 2014), as well as can prevent inflammation in biological systems. It can fight ulcerative diseases by *Helicobacter pylori*, protecting against gastric lesions (ulcers), and improving gastrointestinal health and discomfort (Shah et al., 2016).

Studies have shown that astaxanthin is effective and safe in controlling body weight, being considered as an anti-obesity agent. The administration of astaxanthin, in addition to significantly reducing body weight and adipose tissue, can also reduce hepatic, plasma, and total cholesterol triglycerides (Zhang et al., 2014).

Astaxanthin supplementation may be beneficial for people at increased risk for heart attacks. It is carried by VLDL, LDL, and HDL in the blood and protects LDL cholesterol against oxidation (Shah et al., 2016). In addition, another study carried out in humans, aged 25 to 60 years, showed that 12 weeks of administration of astaxanthin significantly decreased serum triglyceride levels, while significantly increasing high-density lipoprotein (HDL) (Sathasivam and Ki, 2018).

Research has also reported anti-hypertensive effects of astaxanthin in spontaneously hypertensive rats (SHRs). In the study, oral administration of astaxanthin, at a concentration of 50 mg/kg, for 14 days, promoted a significant reduction in blood pressure in SHRs (Sathasivam and Ki, 2018).

7.3.3 LUTEIN

Lutein is considered a functional nutrient because it has benefits to human health, acting in the prevention of age-related macular degeneration (AMD), cataracts, reduction of oxidative stress in the retina and inhibition of the pathological activity of diabetic retinopathy, improvement of the early stages of atherosclerosis, protection from cardiovascular disease and some types of cancer (Dufossé, 2016; Mehariya et al., 2019).

Hayashi et al. (2014) conducted a study with 40 patients, all with identical degrees of cataracts in both eyes. To analyze changes in antioxidant capacity and oxidative state, they collected aqueous humor before and after taking an oral antioxidant supplement based on lutein. The results showed that the consumption of antioxidant supplements effectively reduces oxidation in aqueous humor, inhibiting the oxidation of surrounding tissues. In another study, the use of *Chlorella* extracts containing lutein reduced the incidence of cancer and prevented macular degeneration. Similarly, carotenoids specifically extracted from *Chlorella ellipsoidea* and *Chlorella vulgaris* inhibited the development of colon cancer (Guedes et al., 2011).

Lutein can also delay the spread of chain reactions such as those initiated by the degradation of polyunsaturated fatty acids (PUFAs) known to contribute to the deterioration of lipid membranes, impairing their integrity. One example is the decline in cognitive ability that accompanies AD, apparently caused by persistent oxidative stress in the brain. Using transgenic mice fed with *Chlorella* sp. containing, the authors found that significant prevention of cognitive impairment occurred (Guedes et al., 2011).

Lutein, in addition to antioxidant, protect diabetic, anticancer, also shows antimicrobial action, being able to prevent gastric infection by *Helicobacter pylori* (Hamidi et al., 2020). Studies of antibacterial products reported with microalgal lutein have shown inhibition of the growth of Gram-positive and Gram-negative bacteria. Some microalgae, such as *Dunaliella*, also produce compounds with antifungal activities (Khan et al., 2018).

7.3.4 ZEAXANTHIN

As oxidative stress induced by ROS is the main reason for inflammatory events that lead to many different medical conditions, such as cancer, diabetes, neurodegenerative, and cardiovascular diseases (CVDs), the quest to understand the mechanisms of oxidative stress and also to find new antioxidant compounds have become great interest to medical scholars. Epidemiological investigations have shown a strong correlation between antioxidants, such as zeaxanthin, and a decreased risk of these chronic diseases such as cardiovascular disease and cancer (Hamidi et al., 2020).

The anti-cancer activity of carotenoids involves a variety of mechanisms, including the induction of cell apoptosis and suppression of cell proliferation. An *in vivo* study showed that β-carotene, astaxanthin, canthaxanthin, and zeaxanthin help to reduce the size and number of liver cancers (Vílchez et al., 2011; Sathasivam and Ki, 2018).

In addition to the antioxidant capacity of carotenoids, other mechanisms are also known, such as the absorption and filter of blue light in eye health. The bioactivity of dietary macular carotenoids (class of xanthophylls: lutein and zeaxanthin) offers a means of strengthening the antioxidant defenses of the macula, as they accumulate, thereby reducing the risk of AMD and/or progression (Dufossé, 2016; Eisenhauer et al., 2017; Sathasivam and Ki, 2018).

Recent findings show that carotenoids such as lycopene, lutein, and zeaxanthin can protect against diabetic retinopathy (Sathasivam and Ki, 2018; Hamidi et al., 2020). Besides, studies report that oral administration of zeaxanthin is effective in managing inflammatory responses induced by UV irradiation (Zhang et al., 2014).

7.3.5 FUCOXANTHIN

Recently, fucoxanthin and its derivatives have been shown to have many beneficial health effects. This carotenoid has been reported to exhibit strong potential as an antioxidant, anti-cancer, anti-obesity, anti-diabetic, anti-inflammatory, anti-hypertensive agent, in addition to the bone protection effect (Xia et al., 2013; Petrushkina et al., 2017; Gille et al., 2018; Khan et al., 2018; Sathasivam and Ki, 2018).

Fucoxanthin also has cardioprotective activity. The administration of this carotenoid in an *in vivo* study showed a reduction in blood triglyceride levels. Rats fed 2 mg/kg of fucoxanthin, showed a significant reduction in the absorption of triglycerides when receiving a diet with 10% soy oil (Sathasivam and Ki, 2018).

Obesity is considered a multifactorial metabolic disorder associated with many complications and diseases such as cancer, CVDs, diabetes mellitus, and aging. As obesity occurs due to excessive deposition of adipose tissue, it can be controlled by reducing adipogenesis. Research has revealed several anti-hyperlipidemic agents from natural sources, such as medicinal plants. Currently, microalgae are being studied as potential sources of these agents. Thus, obesity can be controlled by protecting cells against adipogenesis (Khan et al., 2018). Promising results, in this sense, were obtained in research with obese rats supplemented with fucoxanthin (Maeda et al., 2006).

Fucoxanthin has been used in the treatment of osteoclastic diseases, which suggests that fucoxanthin suppresses osteoclastogenesis, inhibiting osteoclast differentiation and inducing apoptosis in osteoclasts. Also, fucoxanthin food supplements may be useful in preventing bone diseases, such as osteoporosis and rheumatoid arthritis, known to be related to bone resorption (Sathasivam and Ki, 2018).

Another property reported in animal studies is that fucoxanthin in algae is effective in chemoprevention of cancer (Xia et al., 2013). Many reports indicate these natural products' potential to treat cancer and tumors by inhibiting angiogenesis (Khan et al., 2018).

Fucoxanthin also showed therapeutic effects on diabetes and arachidonic acid (AA) synthesis, and the DHA content in the liver of mice, inhibition of skin melanogenesis by the negative regulation of the transcription factors involved. Besides, it showed protection from photooxidation of DNA (Khan et al., 2018).

7.3.6 CANTHAXANTHIN

Carotenoids also can stimulate the immune system, being potentially involved in more than 60 life-threatening diseases, including various forms of cancer, coronary heart disease, premature aging, and arthritis; this is specifically the case for canthaxanthin and astaxanthin, and other non-provitamin carotenoids (Guedes et al., 2011; Hamidi et al., 2020).

According to studies, the effects of canthaxanthin on breast tissue chemically induced carcinogenesis in mice show that ingesting canthaxanthin for three weeks, before cancer induction with dimethylbenzanthracene, can reduce the occurrence of cancer by 65% (Sathasivam and Ki, 2018). Canthaxanthin also has antioxidant, anti-inflammatory, and neuroprotective properties (Sathasivam et al., 2019). Canthaxanthin has therapeutic action as an antitumor agent, suppressing the growth of cancer cells through the induction of apoptosis (Gao et al., 2020).

The data on the bioaccessibility and bioavailability process of microalgae carotenoids are little limited; therefore, a more detailed assessment of the metabolism, biotransformation, and bioactivity of the uptake carotenoids is necessary.

7.4 BIOAVAILABILITY AND BIOACCESSIBILITY OF MICROALGAL CAROTENOIDS

Microalgae are a potential source of bioactive metabolites, including carotenoids that show many bioactive functions. However, in order to mediate such activities, this metabolite must be bioavailable (Gille et al., 2016, 2019). The term bioavailability is defined as the fraction of the ingested carotenoids that are available for utilization in physiological functions (Wood, 2005). A prerequisite for these structures' bioavailability is their bioaccessibility. Bioaccessibility can be set as the portion of the ingested compound released from the matrix that gets incorporated into mixed micelles and is available for absorption (Parada and Aguilera, 2007).

The application of whole microalgal biomass in food systems might have some implications for the bioactive efficacy of carotenoids in the human body (Niccolai et al., 2019). The data generated by previous researchers on the bioaccessibility of carotenoids in microalgae have been compiled in Table 7.2.

The extent to which carotenoids are released upon digestion in the gastrointestinal tract depends on several factors, including the physical entrapment in the matrix, chemical structure (isomeric forms), physicochemical properties of the carotenoid the physiological conditions, and the digestive enzymes (Donhowe and Kong, 2014; Lemmens et al., 2014; Colle et al., 2016).

TABLE 7.2 Bioaccessibility and Bioavailability Assays of Microalgal Carotenoids

Microalgae	**Main Carotenoid**	**Assay**	**Processing Methods**	**References**
D. salina	β-carotene and *cis*-β-carotene	*In vitro* digestion and cells Caco-2	Emulsions	Ferruzzi et al. (2006)
S. almeriensis	Lutein, zeaxanthin, and β-carotene	*In vitro* digestion	Extract	Granado-Lorencio et al. (2009)
N. oculata	β-carotene and lycopene	*In vitro* digestion	Extract	Goh et al. (2009)
C. calcitrans	β-carotene and lycopene	*In vitro* digestion	Extract	Goh et al. (2009)
C. vulgaris	Lutein	*In vitro* digestion and cells Caco-2	Microfluidization	Cha et al. (2011)
C. vulgaris	Lutein and β-carotene	*In vitro* digestion	Sonication	Gille et al. (2016)
C. reinhardtii	Lutein and β-carotene	*In vitro* digestion	Sonication	Gille et al. (2016)
C. vulgaris	Lutein and β-carotene	*In vitro* digestion	Maceration biomass	Gille et al. (2018)
P. tricornutum	Fucoxanthin, zeaxanthin, and β-carotene	*In vitro* digestion	Maceration biomass	Gille et al. (2018)
C. vulgaris	Lutein and β-carotene	*In vivo* (C57BL/6J)	Maceration biomass	Gille et al. (2018)
P. tricornutum	Fucoxanthin, zeaxanthin	*In vivo* (C57BL/6J)	Maceration biomass	Gille et al. (2018)
P. tricornutum	Zeaxanthin, fucoxanthin, and β-carotene	*In vitro* digestion and cells Caco-2	Sonication and cellulase digestion	Gille et al. (2019)
Nannochloropsis sp.	Antheraxanthin, zeaxanthin, and β-carotene	*In vitro* digestion	High pressure	Bernaerts et al. (2020)
Nannochloropsis sp.	Antheraxanthin, zeaxanthin, and β-carotene	*In vitro* digestion	Extract	Bernaerts et al. (2020)

The bioaccessibility of carotenoids of untreated microalgae biomass after *in vitro* digestion seems to be species-dependent. Generally, low bioaccessibility values (0–7%) have been reported for carotenoids in *Chlorella vulgaris and Nannochloropsis* sp. (1–6%) (Gille et al., 2016; Bernaerts et al., 2020). In contrast, higher values were observed for carotenoids in *Phaeodactylum tricornutum* (27–52%) and *Chlamydomonas reinhardtii* (10–20%) (Gille et al., 2016, 2018).

A possible explanation might be the distinct composition of the cell walls of these microalgae species. While *C. vulgaris* contain cellulose polymers in their cell wall, these are absent in *P. tricornutum* and *C. reinhardtii.* However, more research is required to conclude a direct relation between the cell wall composition and carotenoid bioaccessibility. As a matter of fact, while the fragile cell wall of *Isochrysis galbana* would suggest a high bioaccessibility, low values (8–13%) have been reported in this microalga (Bernaerts et al., 2020).

Another factor is the higher efficiency of micellarization of isomers compared to form all-*trans*. Possibly it is a consequence of the acyclic and rigid structure of all-*trans* that limits its incorporation into the micellar fraction during digestion. While, the *cis*-configuration presents a bent to shape to the isomers, which improved their incorporation in micelles. Ferruzzi et al. (2006), in an *in vitro* study with carotenoids from microalgae *Dunaliella salina* showed that micellarization of *cis*-β-carotene exceeded that of all-*trans*-β-carotene.

Micellarization of carotenoids also dependent on lipophilicity. Thus, xanthophylls are generally more bioaccessible than carotenes due to the least hydrophobic character that facilitates transfer to micelles during digestion. Gille et al. (2016), in an *in vitro* study, showed that micellarization of lutein larger that of β-carotene. Gille et al. (2019) have reported that epoxy carotenoids fucoxanthin showed a good bioaccessibility. Further, they presumed that the absorption and metabolism of epoxy carotenoids are different from the other carotenoids.

Among the strategies to modulate the bioavailability and bioaccessibility of carotenoids are the protocols *in vivo* and *in vitro*. *In vivo* studies provide more specific information about the bioavailability of carotenoids, but these studies are more expensive, involve ethical constraints, and until now, there is no optimal animal model to assess carotenoid metabolism (Failla et al., 2008). Therefore, *in vitro* methods may simulate gastrointestinal digestion, and the final absorption process is usually assessed using the Caco-2 cell. Finally, *in vitro* bioaccessibility assay can help with a preliminary assessment of the carotenoid metabolism and estimation of their bioavailability. However, the use of a standardized *in vitro* digestion protocol would aid the production of more comparable data in the future.

7.5 CONCLUSION

Microalgae are excellent sources of bioactive carotenoids applied in various areas of the industry. However, the bioaccessibility and bioavailability of carotenoids require a better understanding and will aid in achieving more desirable high-value compounds and thereby creating noteworthy progress towards the future of microalgae-based biotechnology.

KEYWORDS

- **astaxanthin**
- **bioaccessibility**
- **bioactive compounds**
- **bioavailability**
- **biological properties**
- **canthaxanthin**
- **carotenoids**
- **fucoxanthin**
- **lutein**
- **microalgae**
- **zeaxanthin**
- ***β*-carotene**

REFERENCES

Bernaerts, T. M. M., Verstreken, H., Dejonghe, C., et al., (2020). Cell disruption of *Nannochloropsis* sp. improves *in vitro* bioaccessibility of carotenoids and ω3-LC-PUFA. *Journal of Functional Foods*, *65*, 103770.

Bohn, T., (2018). Metabolic fate of bioaccessible and non-bioaccessible carotenoids. *Food Chemistry, Function and Analysis*, *5*, 165–200.

Borowitzka, M. A., (2013). High-value products from microalgae - their development and commercialization. *Journal of Applied Phycology*, *25*, 743–756.

Cha, K. H., Lee, J. Y., Song, D. G., et al., (2011). Effect of microfluidization on *in vitro* micellization and intestinal cell uptake of lutein from *Chlorella vulgaris*. *Journal of Agricultural and Food Chemistry*, *59*(16), 8670–8674.

Chen, J., Wang, Y., Benemann, J. R., Zhang, X., Hu, H., & Qin, S., (2015). Microalgal industry in China: Challenges and prospects. *J. Appl. Phycol.*, *28*, 715–725.

Colle, I. J., Lemmens, L., Knockaert, G., Van, L. A., & Hendrickx, M., (2016). Carotene degradation and isomerization during thermal processing: A review on the kinetic aspects. *Critical Reviews in Food Science and Nutrition*, *56*(11), 1844–1855.

Cuellar-Bermudez, S. P., Aguilar-Hernandez, I., Cardenas-Chavez, D. L., Ornelas-Soto, N., Romero-Ogawa, M. A., & Parra-Saldivar, R., (2014). Extraction and purification of high-value metabolites from microalgae: essential lipids, astaxanthin and phycobiliproteins. *Microbial Biotechnology*, *8*(2), 190–209.

Donhowe, E. G., & Kong, F., (2014). Beta-carotene: Digestion, microencapsulation, and *in vitro* bioavailability. *Food and Bioprocess Technology*, *7*(2), 33.

Dufossé, L., (2016). Current and potential natural pigments from microorganisms (bacteria, yeasts, fungi, microalgae). In: Carle, R., & Schweigert, R., (eds.), *Handbook on Natural Pigments in Food and Beverages* (pp. 337–354).

Eisenhauer, B., Natoli, S., Liew, G., & Flood, V., (2017). Lutein and zeaxanthin - food sources, bioavailability and dietary variety in age-related macular degeneration protection. *Nutrients*, *9*(2), 120.

El-Akabawy, G., & El-Sherif, N. M., (2019). Zeaxanthin exerts protective effects on acetic acid-induced colitis in rats via modulation of pro-inflammatory cytokines and oxidative stress. *Biomedicine and Pharmacotherapy*, *111*, 841–851.

Failla, M. L., Chitchumroonchokchai, C., & Ishida, B. K., (2008). *In vitro* micellarization and intestinal cell uptake of *cis* isomers of lycopene exceed those of all-*trans*-lycopene. *The Journal of Nutrition*, *138*(3), 482–486.

Fernandes, A. S., Petry, F. C., Mercadante, A. Z., Jacob-Lopes, E., & Zepka, L. Q., (2020). HPLC-PDA-MS/MS as a strategy to characterize and quantify natural pigments from microalgae. *Current Research in Food Science*, *3*, 100–112.

Ferruzzi, M. G., Lumpkin, J. L., Schwartz, S. J., & Failla, M., (2006). Digestive stability, micellarization, and uptake of β-carotene isomers by Caco-2 human intestinal cells. *Journal of Agricultural and Food Chemistry*, *54*(7), 2780–2785.

Gao, X., Xu, H., Zhu, Z., She, Y., & Ye, S., (2020). Improved production of echinenone and canthaxanthin in transgenic *Nostoc* sp. PCC 7120 overexpressing a heterologous crtO gene from *Nostoc flagelliforme*. *Microbiological Research*, *236*, 126455.

Gille, A., Hollenbach, R., Trautmann, A., Posten, C., & Briviba, K., (2019). Effect of sonication on bioaccessibility and cellular uptake of carotenoids from preparations of photoautotrophic *Phaeodactylum tricornutum*. *Food Research International*, *118*, 40–48.

Gille, A., Neumann, U., Louis, S., Bischoff, S. C., & Briviba, K., (2018). Microalgae as a potential source of carotenoids: Comparative results of an *in vitro* digestion method and a feeding experiment with C57BL/6J mice. *Journal of Functional Foods*, *49*, 285–294.

Gille, A., Trautmann, A., Posten, C., & Briviba, K., (2016). Bioaccessibility of carotenoids from *Chlorella vulgaris* and *Chlamydomonas reinhardtii*. *International Journal of Food Sciences and Nutrition*, *67*(5), 507–513.

Goff Le, M., Ferrec Le, E., Mayer, C., et al., (2019). Microalgal carotenoids and phytosterols regulate biochemical mechanisms involved in human health and disease prevention. *Biochimie.*, *167*, 106–118.

Goh, L. P., Loh, S. P., Fatimah, M. Y., & Perumal, K., (2009). Bioaccessibility of carotenoids and tocopherols in marine microalgae *Nannochloropsis* sp. and *Chaetoceros* sp. *Malaysian Journal of Nutrition*, *15*(1), 77–86.

Granado-Lorencio, F., Herrero-Barbudo, C., Acién-Fernández, G., et al., (2009). *In vitro* bioaccesibility of lutein and zeaxanthin from the microalgae *Scenedesmus almeriensis*. *Food Chemistry*, *114*(2), 747–752.

Guedes, A. C., Amaro, H. M., & Malcata, F. X., (2011). Microalgae as sources of carotenoids. *Marine Drugs*, *9*(4), 625–644.

Gul, K., Tak, A., Singh, A. K., Singh, P., Yousuf, B., & Wani, A. A., (2015). Chemistry, encapsulation, and health benefits of β-carotene: A review. *Cogent Food and Agriculture*, *1*, 1018696.

Hamidi, M., Kozani, P. S., Kozani, P. S., Pierre, G., Michaud, P., & Delattre, C., (2020). Marine bacteria versus microalgae: Who is the best for biotechnological production of bioactive compounds with antioxidant properties and other biological applications? *Marine Drugs*, *18*(28), 38.

Harari, A., Harats, D., Marko, D., et al., (2008). A 9-*cis* β-carotene-enriched diet inhibits atherogenesis and fatty liver formation in LDL receptor knockout mice. *The Journal of Nutrition and Disease*, *138*, 1923–1930.

Hayashi, R., Hayashi, S., Arai, K., Sakai, M., Okamoto, H., & Chikuda, M., (2014). The gender-differentiated antioxidant effects of a lutein-containing supplement in the aqueous humor of patients with senile cataracts. *Experimental Eye Research*, *129*, 5–12.

Irazusta, V., Nieto-Peñalver, C. G., Cabral, M. E., Amoroso, M. J., & Figueroa, L. I. C., (2013). Relationship among carotenoid production, copper bioremediation and oxidative stress in *Rhodotorula mucilaginosa* RCL-11. *Process Biochemistry*, *48*(5, 6), 803–809.

Jacob-Lopes, E., Maroneze, M. M., Deprá, M. C., Sartori, R. B., Dias, R. R., & Zepka, L. Q., (2019). Bioactive food compounds from microalgae: An innovative framework on industrial biorefineries. *Current Opinion in Food Science*, *25*, 1–7.

Jomova, K., & Valko, M., (2013). Health protective effects of carotenoids and their interactions with other biological antioxidants. *European Journal of Medicinal Chemistry*, *70*, 102–110.

Khan, M. I., Shin, J. H., & Kim, J. D., (2018). The promising future of microalgae: Current status, challenges, and optimization of a sustainable and renewable industry for biofuels, feed, and other products. *Microbial Cell Factories*, *17*(36), 1–21.

Khoo, K. S., Chew, K. W., Ooi, C. W., Ong, H. C., Ling, T. C., & Show, P. L., (2019). Extraction of natural astaxanthin from *Haematococcus pluvialis* using liquid biphasic flotation system. *Bioresource Technology*, *290*, 121794.

Klassen, J. L., & Foght, J. M., (2011). Characterization of *Hymenobacter* isolates from Victoria upper glacier, Antarctica reveals five new species and substantial non-vertical evolution within this genus. *Extremophiles*, *15*(1), 45–57.

Koller, M., Muhr, A., & Braunegg, G., (2014). Microalgae as versatile cellular factories for valued products. *Algal Research*, *6*, 52–63.

Kopec, R. E., & Failla, M. L., (2018). Recent advances in the bioaccessibility and bioavailability of carotenoids and effects of other dietary lipophiles. *Journal of Food Composition and Analysis*, *68*, 16–30.

Lafarga, T., Clemente, I., & Garcia-Vaquero, M., (2020). Carotenoids from microalgae. In: Galanakis, C. M., (ed.), *Carotenoids: Properties, Processing and Applications* (pp. 149–187). Academic Press.
Lemmens, L., Colle, I., Buggenhout, S. V., Palmero, P., Loey, A. V., & Hendrickx, M., (2014). Carotenoid bioaccessibility in fruit- and vegetable-based food products as affected by product (micro)structural characteristics and the presence of lipids: A review. *Trends in Food Science and Technology*, *38*, 125–135.
Maeda, H., Hosokawa, M., Sashima, T., Takahashi, N., Kawada, T., & Miyashita, K., (2006). Fucoxanthin and its metabolite, fucoxanthinol, suppress adipocyte differentiation in 3T3-L1 cells. *International Journal of Molecular Medicine*, *18*(1), 147–152.
Magosso, M. F., Carvalho, P. C., Shneider, B. U. C., et al., (2016). *Acrocomia aculeata* prevents toxicogenetic damage caused by the antitumor agent cyclophosphamide. *Genetics and Molecular Research*, *15*(2), 15027816.
Maldonade, I. R., Rodriguez-Amaya, D. B., & Scamparini, A. R. P., (2008). Carotenoids of yeasts isolated from the Brazilian ecosystem. *Food Chemistry*, *107*(1), 145–150.
Mehariya, S., Iovine, A., Di Sanzo, G., et al., (2019). Supercritical fluid extraction of lutein from *Scenedesmus almeriensis*. *Molecules*, *24*(7), 1324.
Mulders, K. J., Lamers, P. P., Martens, D. E., & Wijffels, R. H., (2014). Phototrophic pigment production with microalgae: Biological constraints and opportunities. *Journal of Phycology*, *50*(2), 229–242.
Naguib, Y. M. A., (2000). Antioxidant activities of astaxanthin and related carotenoids. *Journal of Agricultural and Food Chemistry*, *48*(4), 1150–1154.
Nascimento, T. C. D., Cazarin, C. B. B., Maróstica, Jr. M. R., Mercadante, A. Z., Jacob-Lopes, E., & Zepka, L. Q., (2020). Microalgae carotenoids intake: Influence on cholesterol levels, lipid peroxidation and antioxidant enzymes. *Food Research International*, *128*, 108770.
Nascimento, T. C., Cazarin, C. B. B., Roberto, M. M., et al., (2019). Microalgae biomass intake positively modulates serum lipid profile and antioxidant status. *Journal of Functional Foods*, *58*, 11–20.
Niccolai, A., Chini, G., Rodol, L., Biondi, N., & Tredici, M. R., (2019). Microalgae of interest as food source: Biochemical composition and digestibility. *Algal Research*, *42*, 101617.
Panis, G., & Carreon, J. R., (2016). Commercial astaxanthin production derived by green alga *Haematococcus pluvialis*: A microalgae process model and a techno-economic assessment all through production line. *Algal Research*, *18*, 175–190.
Parada, J., & Aguilera, J. M., (2007). Food microstructure affects the bioavailability of several nutrients. *Journal Food Science*, *72*(2), R21–R32.
Patias, L. D., Fernandes, A. S., Petry, F. C., Mercadante, A. Z., Jacob-lopes, E., & Zepka, L. Q., (2017). Carotenoid profile of three microalgae/cyanobacteria species with peroxyl radical scavenger capacity. *Food Research International*, *100*, 260–266.
Petrushkina, M., Gusev, E., Sorokin, B., et al., (2017). Fucoxanthin production by heterokont microalgae. *Algal Research*, *24*, 387–393.
Rammuni, M. N., Ariyadasa, T. U., Nimarshana, P. H. V., & Attalage, R. A., (2018). Comparative assessment on the extraction of carotenoids from microalgal sources: Astaxanthin from *H. pluvialis* and β-carotene from *D. salina*. *Food Chemistry*, *277*, 128–134.
Rodriguez-Amaya, D. B., (2001). *A Guide to Carotenoid Analysis in Foods*. Washington, DC: OMNI Research.
Rodriguez-Amaya, D. B., (2015). Status of carotenoid analytical methods and *in vitro* assays for the assessment of food quality and health effects. *Current Opinion in Food Science*, *1*, 56–63.

Saini, R. K., & Keum, Y. S., (2018). Omega-3 and omega-6 polyunsaturated fatty acids: Dietary sources, metabolism, and significance: A review. *Life Sciences, 203*, 255–267.

Sathasivam, R., & Ki, J. S., (2018). A review of the biological activities of microalgal carotenoids and their potential use in healthcare and cosmetic industries. *Marine Drugs, 16*(1), 26.

Sathasivam, R., Radhakrishnan, R., Hashem, A., & Abd-Allah, E. F., (2019). Microalgae metabolites: A rich source for food and medicine. *Saudi Journal of Biological Sciences, 26*(4), 709–722.

Shah, M. M. R., Liang, Y., Cheng, J. J., & Daroch, M., (2016). Astaxanthin-producing green microalga *Haematococcus pluvialis*: From single cell to high value commercial products. *Frontiers in Plant Science, 2016*, 1–7.

Singh, S. K., Kaur, R., Bansal, A., Kapur, S., & Sundaram, S., (2020). Biotechnological exploitation of cyanobacteria and microalgae for bioactive compounds. In: Verma, M. L., & Chandel, A. K., (eds.), *Biotechnological Production of Bioactive Compounds* (pp. 107–137). Elsevier, Oxford.

Trejo-Solís, C., Pedraza-Chaverrí, J., Torres-Ramos, M., et al., (2013). Multiple molecular and cellular mechanisms of action of lycopene in cancer inhibition. *Evidence-Based Complementary and Alternative Medicine, 2013*, 1–17.

Uenojo, M., Maróstica, Jr. M. R., Pastore, G. M., et al., (2007). Carotenoids: Propriedades, aplicações e biotransformação para formação de compostos de aroma. *Química Nova, 30*(3), 616–622.

Vílchez, C., Forján, E., Cuaresma, M., Bédmar, F., Garbayo, I., & Vega, J. M., (2011). Marine carotenoids: Biological functions and commercial applications. *Marine Drugs, 9*(3), 319–333.

Wood, R. J., (2005). Bioavailability: Definition, general aspects and fortificants. In: Caballero, B., Allen, L. H., & Prentice, A., (eds.), *Encyclopedia of Human Nutrition.* Oxford: Elsevier Ltd.

Xavier, A. A. O. X., & Mercadante, A. Z. M., (2019). The bioaccessibility of carotenoids impacts the design of functional foods. *Current Opinion in Food Science, 26*, 1–8.

Xia, S., Wang, K., Wan, L., Li, A., Hu, Q., & Zhang, C., (2013). Production, characterization, and antioxidant activity of fucoxanthin from the marine diatom *Odontella aurita. Marine Drugs, 11*(7), 2667–2681.

Yeh, P. T., Huang, H. W., Yang, C. M., Yang, W. S., & Yang, C. H., (2016). Astaxanthin inhibits expression of retinal oxidative stress and inflammatory mediators in streptozotocin-induced diabetic rats. *PLoS One, 11*(1), e0146438.

Zaghdoudi, K., Pontvianne, S., Framboisier, X., et al., (2015). Accelerated solvent extraction of carotenoids from: Tunisian kaki (*Diospyros kaki* L.), peach (*Prunus persica* L.) and apricot (*Prunus armeniaca* L.). *Food Chemistry, 184*, 131–139.

Zhang, J., Sun, Z., Sun, P., Chen, T., & Chen, F., (2014). Microalgal carotenoids: Beneficial effects and potential in human health. *Food and Function, 5*(3), 413.

CHAPTER 8

MICROALGAL CAROTENOIDS: BIOTECHNOLOGICAL ASPECTS FOR INDUSTRIAL APPLICATIONS

AYŞEGÜL ERDOĞAN[1] and MELTEM CONK DALAY[2]

[1]*Ege University Application and Research Center for Testing and Analysis (EGE MATAL), Bornova, İzmir–35100, Turkey*

[2]*Department of Bioengineering, Faculty of Engineering, Ege University, Bornova, İzmir–35100, Turkey*

ABSTRACT

Microalgae produce a variety of valuable substances for human health. Carotenoids are among these compounds with high antioxidant and colorant properties as well as their health effects. While microalgae have the potential to produce novel and/or high amounts of carotenoids, large-scale production of microalgae and downstream processing of carotenoids present challenges. This chapter provides a broad overview of carotenoid production from microalgae and its extraction, saponification, analysis, identification, and commercialization. The opportunities and challenges facing the carotenoid markets have also been addressed. An emphasis was made on the legal aspects associated with the commercial production of microalgae-based carotenoids for food/feed, nutraceutical, and cosmetic industry worldwide. The microalgal carotenoid market will most likely continue to expand due to the increasing demand for natural products. In terms of various biotechnological aspects, more work needs to be done before carotenoids can be placed on the market.

Algal Metabolites: Biotechnological, Commercial, and Industrial Applications. Jeyabalan Sangeetha, Devarajan Thangadurai, Saher Islam, & Ravichandra Hospet (Eds.)

8.1 INTRODUCTION TO CAROTENOIDS: AN OVERVIEW

Photosynthetic bacteria, some archaea and fungal species, algae, plants, and animals all contain carotenoids, which are tetraterpene pigments. It has been reported to be approximately 850 natural carotenoids so far (Takaichi, 2011).

Carotenoids are generally lipophilic substances and therefore do not dissolve in an aqueous medium. However, carotenoids with highly polar functional groups are excluded. The presence of a long, comprehensive system of conjugated double bonds (or polyene chain) is responsible for light absorption, one of the most prominent features of carotenoids. A chromophore with seven or more double bonds gives the ability to absorb light in the visible range, and as a result, carotenoids display yellow, orange, and red colors.

Most carotenoids have a 40-carbon skeleton and are made up of eight isoprene molecules. A nine conjugated double-linked polyene chain and an end group (R) at both ends of the polyene chain make up their general structure (Britton et al., 1996; Yabuzaki, 2017). The general chemical structure of carotenoids is shown in Figure 8.1.

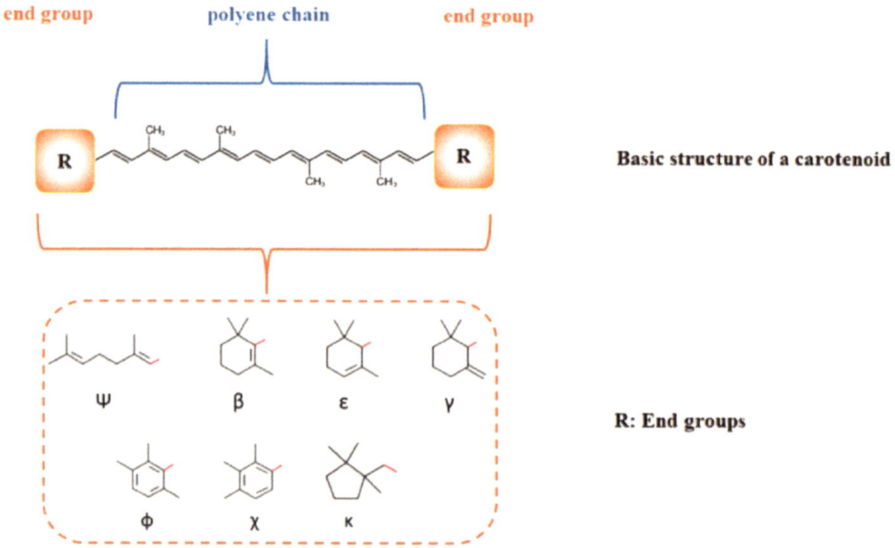

FIGURE 8.1 Chemical structure of carotenoids with different end groups.

Carotenoids are divided into two classes according to the elements they contain: carotenes and xanthophylls. There are about 50 kinds of carotene in nature. Carotenes contain only C and O. On the other hand, xanthophylls are carotenoids that contain additional oxygen atoms and have functional groups like hydroxyl, carbonyl, aldehyde, carboxylic, and epoxide. Some xanthophylls are available as fatty acid esters, glycosides, sulfates, and protein complexes. Xanthophylls come in a variety of shapes and sizes (Maoka, 2009; Britton et al., 2012). Figure 8.2 shows the structures of typical carotenes and xanthophylls. Most carotenoids have a 40-carbon backbone. Higher carotenoids, on the other hand, have a carbon skeleton of 45 or 50. There are roughly 40 distinct higher carotenoids in some species of archaea. Apocarotenoids, on the other hand, are carotenoids made up of carbon skeletons with less than 40 carbons. There are about 120 varieties of apocarotenoids that have been reported to be the degradation products of some plant and animal species (Britton, 1995; Maoka, 2009; Liaaen-Jensen et al., 2012).

Carotenoids are colored pigments with a distinctive visible light absorption spectrum due to the presence of a polyene chain. The location of the maximum light absorption bands (max) for a given carotenoid is a function of the number of conjugated double bonds in the molecule (Figure 8.3). Roman numerals are commonly used to denote absorption maxima (I, II, III). At the absorption maximum (max), the introduction of a new conjugated double bond to the chromophore generates a 20–22 nm bathochromic (higher wavelength) displacement; however this impact varies depending on the presence of different end groups. The chromophore's contribution from the end group-ring is just 9–11 nm. In the case of a ring-ring, the double bond's conjugation is lost, and the chromophore is not formed. Due to the presence of a polyene chain, carotenoids have a characteristic visible light absorption spectrum. The location of a carotenoid's maximum light absorption bands (max) is determined by the number of conjugated double bonds in the molecule (Figure 8.3). Absorption maxima are often represented by Roman numerals (I, II, III). The insertion of a new conjugated double bond to the chromophore causes a 20–22 nm bathochromic (higher wavelength) displacement at the absorption maximum (max), however the impact varies depending on the presence of different end groups. The contribution of the end group-ring to the chromophore is just 9–11 nm. The conjugation of the double bond is lost in the case of a ring-ring, and the chromophore is not generated (Hurst, 2008).

The percent III/II, as well as the maximum values, might be displayed to give an understanding of the spectral fine structure. The percent III/II is the ratio of the heights of the longest-wavelength absorption peak, III,

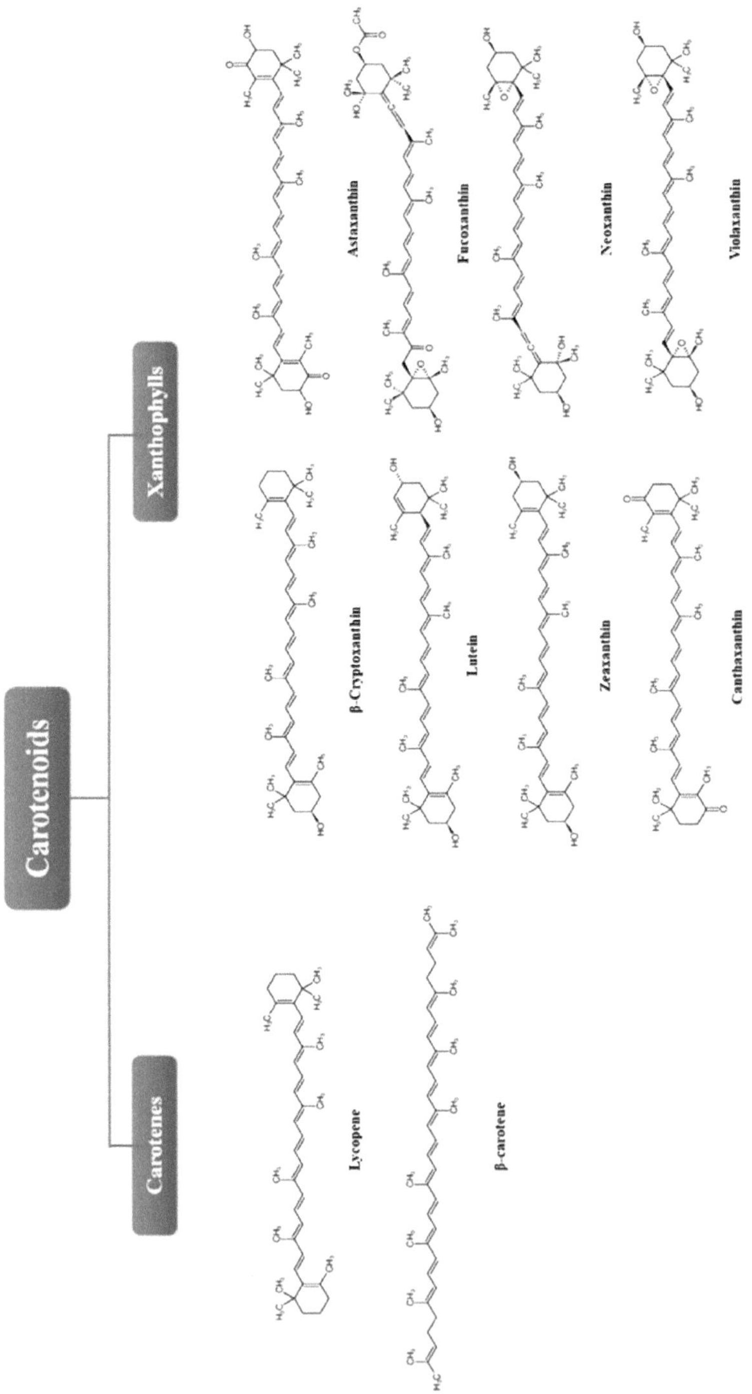

FIGURE 8.2 Classification of carotenoids as carotenes and xanthophylls.

and the middle-wavelength absorption peak, II, multiplied by 100, using the minimum between the two peaks as a baseline. Because the spectrum of conjugated ketocarotenoids like canthaxanthin and echinenone consists of a broad single maximum with no distinct fine structure, the percent III/II is 0 (Britton, 1995).

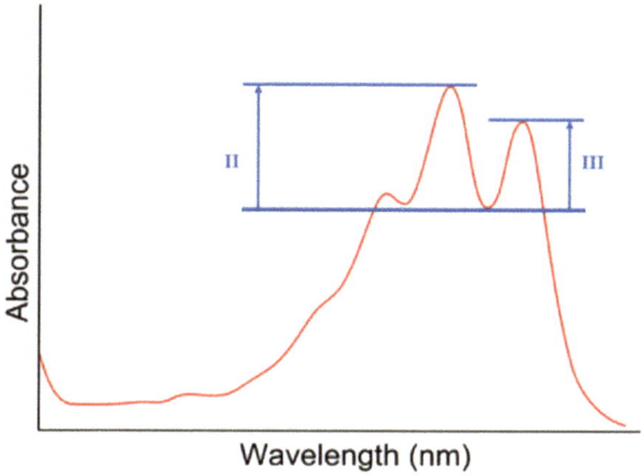

FIGURE 8.3 General ultraviolet (UV)-visible light absorption spectrum of carotenoids (Spectral fine structure: Calculation of %III/II for a carotenoid).

Due to their applications and health-promoting properties, carotenoid demand in the global market is expected to increase from 1.5 billion USD in 2014 to 1.8 billion USD in 2019 (Strati and Oreopoulou, 2014). As a result, carotenoids, and their therapeutic derivatives are the subject of several scientific papers and patent rights around the world. These instances demonstrate that carotenoids are an important vitamin for human health. Carotenoids-rich products can be found in a variety of forms, including food and feed additives, supplements, and natural colorants. Chemically produced carotenoids make up most commercially accessible carotenoids. Some of them, however, come from natural sources (Rodriguez-Amaya, 2016). When compared to natural carotenoids, synthetic carotenoids have several appealing qualities. First, synthetic carotenoids are particularly designed to reduce oxidation and isomerization, making them more stable. To make carotenoid application in foods easier, synthetic carotenoids are created as colloidal suspension, emulsification, and dispersion colloids. They are commonly accessible as water-soluble and stable emulsions on the market. Despite these benefits, chemically manufactured carotenoids have been linked to significant levels of toxicity, carcinogenicity, and teratogenicity, causing concern among health-conscious consumers. As a result, natural sources of carotenoids are in high demand among today's customers (Kirti et al., 2014). This significant change in consumer interest has posed a number of difficulties for manufacturers and consumers when it comes to developing natural carotenoid formulations. Since food matrices contain a wide range of molecular pools, selective recovery of carotenoids necessitates a number of preventive actions and analysis protocols. Selective and efficient extraction procedures, as well as their improvements, are critical to ensuring the recovery of high purity carotenoids.

Research and studies on carotenoids are increasing day by day due to their widespread use in particularly health and food sector. Researchers working in this field targeted specific goals. Some of these are aimed at exploring carotenoids from natural sources. Besides, it is also important to increase the stability of the carotenoids obtained. Thus, their industrial usability is increased and varied. In addition, studies on carotenoid production with high extraction or purification efficiency are also very intense. Researchers are also interested in increasing carotenoid bioavailability. The food-processing and carotenoid-extraction technologies used by the industry result in active, natural formulations commercialized as ready-to-eat carotenoid-rich goods on the market (Nagarajan et al., 2017).

8.2 MICROALGAL CAROTENOIDS

Microalgae are photosynthetic micro-organisms of around 30,000 species. They are classified as Chlorophyta (green algae), Cyanophyta (blue-green algae (BGA)), Rhodophyta (red algae), and Phaeophyta (brown algae), according to their morphological features (Ambati et al., 2019). Table 8.1 shows the major carotenoids produced by microalgae and their benefits.

8.2.1 CAROTENOID BIOSYNTHESIS

Isoprenoids have a subfamily called carotenoids. It's one of the most well-known and well-established classes of all-natural products and biomolecules. Isopentenyl diphosphate (IPP) and isomer dimethylallyl diphosphate are also synthesized by microalgae (DMAPP). Starting from the precursor isopentenyl pyrophosphate (IPP), microalgal carotenoid biosynthesis (carotenegensis) is carried out via the mevalonate (MVA) pathway or the 1-deoxy-D-xylose-5-phosphate (DOXP) pathway (IPP). IPP is enzymatically

TABLE 8.1 Benefits of Carotenoid Pigments from Microalgae (Ambati et al., 2019)

Carotenoid Pigments	Benefits	References
Astaxanthin	Antioxidants, protect from UV rays, bioavailability, food colorant, animal feed, enhance immune functions, improve eye/skin health, anti-aging property, pharmaceutical, nutraceutical, cosmeceutical	Lorenz and Cysewski (2000); Guerin et al. (2003); Higuera-Ciapara et al. (2006); Hussein et al. (2006); Lu et al. (2010); Park et al. (2010); Ranga Rao et al. (2010); Choi et al. (2011); Kidd (2011); Kim et al. (2011); Chan et al. (2012); Maoka et al. (2012); Chew et al. (2013); Dong et al. (2013); Huangfu et al. (2013); Ishiki et al. (2013); Rao et al. (2013); Yamashita (2013); Yang et al. (2013); Ambati et al. (2014)
β-carotene	Vitamin A precursor, antioxidant, food color, malnutrition, prevent macular degeneration, skin burn from UV rays, depression, asthma, infertility, psoriasis, high blood pressure, pharmaceutical, cosmetics, animal feed, fish feed, pet foods	Fujii et al. (1993); Ben-Amotz and Levy (1996); Curtain (2000); Kar (2002); Dufossé et al. (2005); Galaup et al. (2005); Murthy et al. (2005); Shaish et al. (2006); Mogedas et al. (2009); Ranga et al. (2010); Raghunath Reddy et al. (2010); Weber and Grune (2012); Jayappriyan et al. (2013); Donhowe and Kong (2014); Zhang et al. (2014)
Lutein, Zeaxanthin	Antioxidant, food colorant, fish feed, eye disease, health, age-related macular degeneration, cataracts, pharmaceutical, nutraceutical, cosmetics, animal feed	Landrum and Bone (2001); Rao et al. (2006); Ceron et al. (2008); Fernández-Sevilla et al. (2010); Ranga et al. (2010); Raghunath et al. (2010); Maoka (2009); Tokuda et al. (2012); Weber and Grune (2012); Christaki et al. (2013); Bonos et al. (2013)
Fucoxanthin	Anti-oxidant, anti-viral, anti-diabetes, anti-obesity, anti-cancer, anti-inflammatory, anti-allergic, anti-osteoporotic	Maeda et al. (2009); Heo et al. (2010); Peng et al. (2011); Lee et al. (2013); Rokkaku et al. (2013)

Source: Reprinted with permission from Ambati et al., 2019. © Taylor & Francis.

condensed and prolonged to geranylgeranyl-diphosphate (GGDP) farnesyl pyrophosphate via geranylgeranyl diphosphate synthase. The chemical C40, phytoene, is formed when two GGDP molecules are combined head-to-head with phytoene synthase (Takaichi, 2011).

The carotenoid phytoene has three conjugated double bonds and is colorless. Phytoene is desaturated in a stepwise manner by phytoene desaturase to produce lycopene, ζ-carotene, and neurosporene. Lycopene cyclases make α-carotene and β-carotene, which have cyclic terminal end groups. Carotene hydroxylases, ketolases, and epoxidase all produce a number of xanthophylls. Other carotenoids then form with the aid of different enzymes via hydroxylation, re-arrangements, and addition of different functional groups. Enzymatic and genetic investigations have identified these carotenoid biosynthesis pathways in detail (Britton et al., 1998).

Our understanding of carotenoid metabolism and its regulation in microalgae is still limited and has been gathered mainly from information obtained for higher plants. However, studies focusing on carotenoid biosynthesis in algae are few (Takaichi, 2011; Paniagua-Michel et al., 2012). The biosynthetic pathway for carotenoids from microalgae has been given in detail in Figure 8.4 (McGee and Gillespie, 2019).

8.2.2 CULTIVATION AND HARVESTING OF MICROALGAE

To increase the economic viability and environmental impact of microalgae, cost-effective culture and harvesting technologies are necessary. Understanding the combined impact of photobioreactor designs and growth conditions on biomass yield and carotenoid productivity is crucial for overcoming the aforementioned challenges (Okoro et al., 2019). Filtration, centrifugation, flotation, flocculation, gravity sedimentation, electrolytic process, electrophoresis, and magnetic separation are all ways for harvesting microalgae with each method having its advantages and disadvantages (Chen et al., 2011; Safi et al., 2014; Salian and Strezov, 2017). For all microalgae species, a single approach will not achieve great efficiency. Centrifugation is one of the most widely used procedures today, with a harvesting efficiency of over 98%. However, the process's high initial cost and energy consumption make it difficult to deploy. Flocculation is said to be the most cost-effective and preferred method of harvesting (Grima et al., 2003). At a low cost, combining two processes, such as flocculation and sedimentation, can improve biomass recovery (Duan and Gregory, 2003; Barros et al., 2015). This could be a possible alternative of harvesting

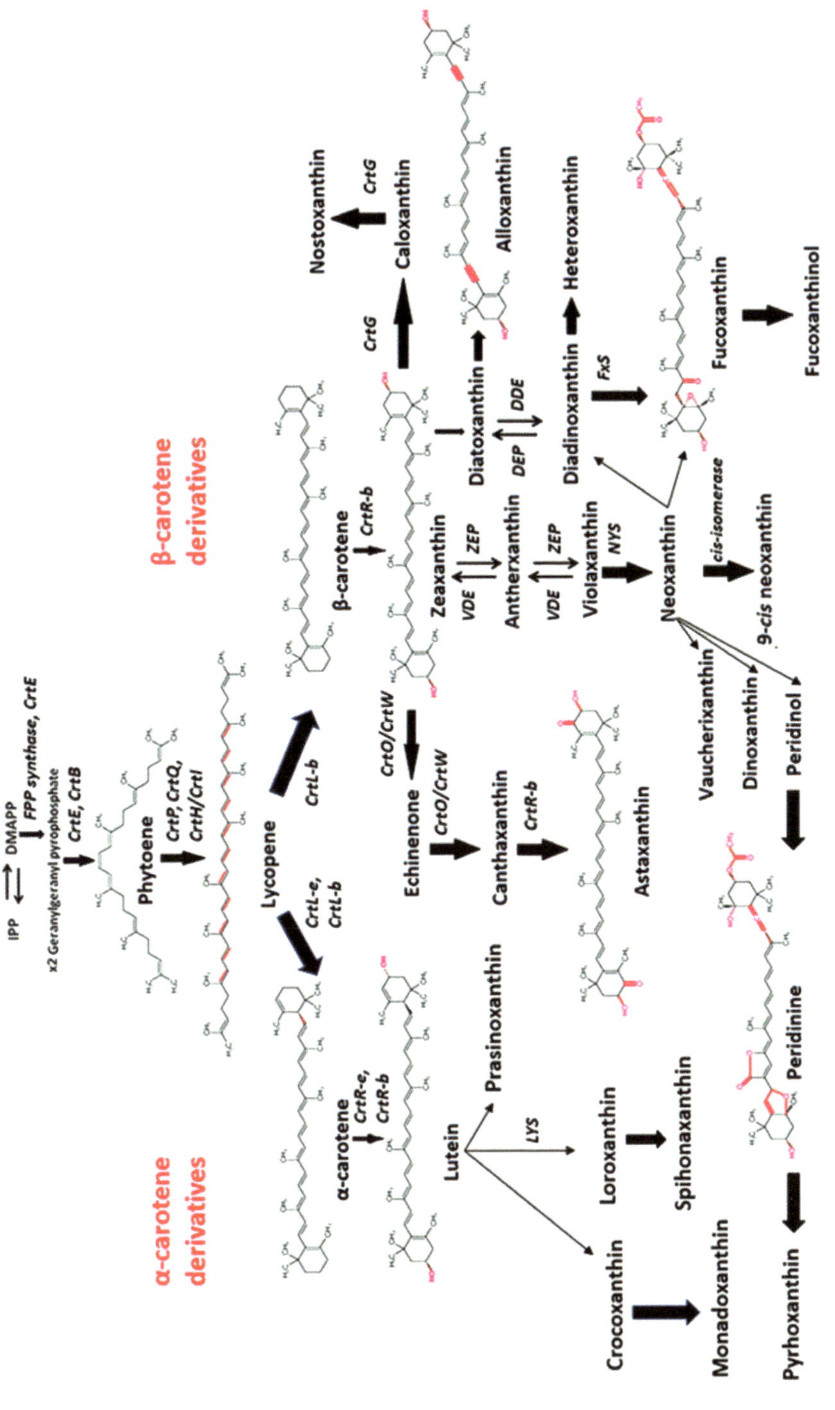

FIGURE 8.4 Biosynthetic pathway involved in carotenoid production in microalgae.

Source: Reprinted with permission from McGee and Gillespie, 2019. © Springer Nature.

microalgae that is low-cost, low-energy, and extremely efficient (Ho et al., 2011; Salim et al., 2011).

8.2.2.1 HAEMATOCOCCUS PLUVIALIS

Haematococcus pluvialis is one of the most commercially cultured microalgae. It is considered as the best astaxanthin source and the main producer organism of this commercial product. It produces astaxanthin with a strong antioxidant effect to protect themselves from unfavorable conditions. Although *Haematococcus* production can be carried out in a single, adjustable photobioreactor, it is generally carried out in two systems in two stages as depicted in Figure 8.5:

1. Inoculum cultures are grown to initiate the production under laboratory conditions.
2. Photobioreactor cultures are done in closed environments because algae can be easily contaminated during this period. In this period, it is aimed to reach the maximum production capacity. Therefore, it is aimed to provide all conditions at an optimum level.
3. In outdoor cultures, it is aimed to triggering the production of astaxanthin with the stress of sunlight, high temperature and nutrient starvation by using ponds or different kinds of photobioreactors:
   - i. **Green Phase:** At this stage, the cells are in the form of mobile with two flagella.
   - ii. **Red Phase:** Stress conditions in outdoor, initiates astaxanthin accumulation in cells. In this phase, *Haematococcus* cells transform their shape, leave their flagella, stop growing, turns into cystic phase and begin to produce Astaxanthin.
4. Water separation and harvesting stage: Centrifugation is the most common method for large-scale harvesting. Separators are used in industrial-scale production.
5. Drying stage: Dehydration is a process that is performed before the metabolite is recovered to extend the shelf life of the product.

   The most known drying techniques used in microalgae are sun drying, spray drying and freeze-drying.
6. Cell disruption and extraction of pure final product is the last step. The solvent extraction and supercritical carbon dioxide (CO_2) extraction are the most effective, and widely used methods for astaxanthin extraction from *H. pluvialis*.

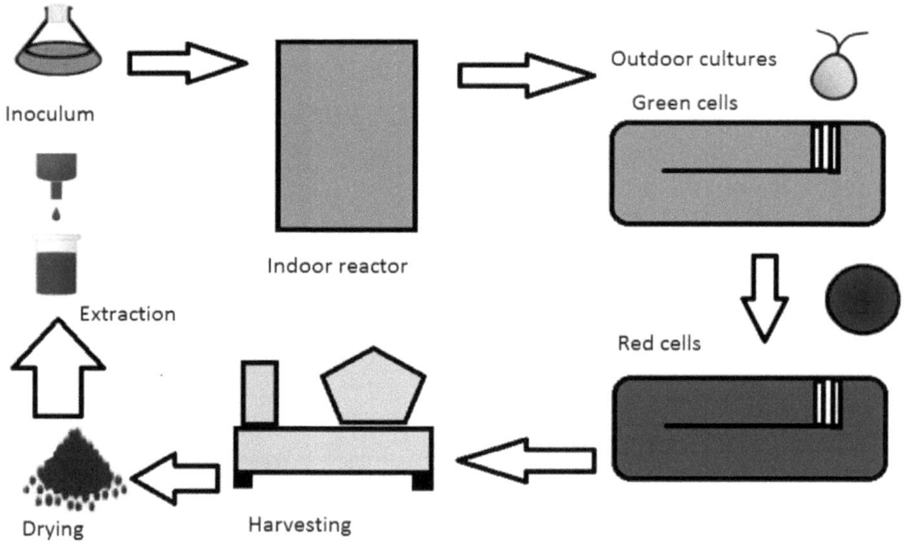

FIGURE 8.5 Astaxanthin production process from *Haematococcus pluvialis*.

8.2.2.2 *DUNALIELLA SALINA*

Dunaliella salina is a green alga that can live in lakes with high salt ratio, its color changes from green to orange-red. In shallow waters, it accumulates a high rate of β-carotene in its content as a result of the decrease in nitrogen amount and high light concentration. Due to this feature, it is produced commercially to obtain β-carotene. At the vegetative production stage, they divide longitudinally mobile form with two flagella. Cysts (aplanospores) occur in extreme conditions. When the media becomes dry, due to too much dilution of salt and high heat and light concentration from the sun, it enables to change its form and turn into cystic phase, accumulate β-carotene in glycerol. Its resistance to high salinity, creates a protective factor against contamination in commercial outdoor cultures. They can be produced in raceway ponds and tubular and panel photobioreactors, plastic bags, as well as shallow soil ponds (Figure 8.6).

8.2.2.3 *CHLORELLA* sp.

Chlorella is a commercially produced green algae because of the valuable chemicals in its content. The ease of its cultivation and rapid reproduction ability provides resistance to contamination. Due to all these features, it is possible to produce *Chlorella* in outdoor ponds as shown in Figure 8.7. Circular ponds are the most frequently used type for the production of *Chlorella*. Their areas are generally around 10 m^2 and their depth is about 50 cm. The circulation of the ponds is provided by a central rotary mixer. Changes in carotenoid composition vary with factors such as light, temperature, nutrient starvation, and growth phase.

FIGURE 8.6 Outdoor cultivation of *Dunaliella*: (a) Open raceway pond*; (b) small shallow experimental culture ponds in Abu Dhabi, UAE (Reprinted with permission from Borowitzka, 2018. © Elsevier); (c) tubular photobioreactor (Reprinted with permission from Milledge, 2013); (d) growth ponds in Australia (Courtesy Cognis).

https://www.ifeu.de/wpcontent/uploads/IFEU_HI_RISE_UoG_2017_Integrated_sustainability_assessment_of_Dunaliella_algae_biorefinery.pdf.

8.2.3 CAROTENOGENESIS RESPONSES IN MICROALGAE

Appropriate metabolic engineering and genetic engineering of microalgae can dramatically affect the quantity of carotenoids present. Random mutagenesis followed by screening has traditionally been employed to enhance product production in microalgal strains used for industrial applications. These methods are, on the whole, slow, and time-consuming. In most cases, mutagenesis, and screening will not provide the target strain. Therefore, it is preferred to use rational engineering, including metabolic and genetic engineering, to change or improve the biochemical capabilities of the microalgal cell. As cell factories, there are some reported advanced technologies that can be used for higher pigment production and development of the microalgal cell (Saini et al., 2020).

Figure 8.8 presents the generalized representation of metabolic engineering processes after the selection of microalgae to produce carotenoids in high amounts or change the type of carotenoid synthesized by microalgae.

FIGURE 8.7 Large scale *Chlorella* cultivation: (a) circular outdoor ponds*; (b) tubular photobioreactors; (c) plastic bags.

Source: Reprinted with permission from Hamed, 2016. © John Wiley and Sons.

Due to the formation of secondary metabolites, microalgae have excellent tolerance to various abiotic stressors. The biosynthesis of carotenoids and other high-value chemicals is triggered by environmental conditions (temperature, light, salinity, and nutrient deficiency alterations) and the usage of additions (such as iron and copper) in the environment (Paliwal et al., 2017; Liang et al., 2018).

It provides crucial evidence about critical genes and metabolic pathways in microalgae physiology to enhance the synthesis of carotenoids and other metabolites from microalgae as a result of recent scientific discoveries in genomes, transcriptomics, and proteomics (Bajhaiya et al., 2017). Carotenoid synthesis is a rather complicated process. Therefore, it contains dozens of genes and enzymes that are very complex to explain.

Wang et al. (2019) reported that mostly, engineering strategies are focused mainly on stoichiometric modulation of pathway components, which are successfully applied to carotenoid production. However, besides the general strategies, a different engineering strategy is required for the production of carotenoids because some enzymes and substrates in the carotenogenic pathway are found in the membrane regions. The production of carotenoids will be restricted to a certain level due to the limited storage space and product toxicity caused by the accumulation of carotenoids in the cytosolic membrane. Since carotenoids accumulate in the cellular membranes or neutral lipid droplets, maximum microbial carotenoid production can be achieved by increasing the formation of cellular membranes or neutral lipid droplets. However, it should be noted that this will vary because the responses of each microalgae species to stress are different.

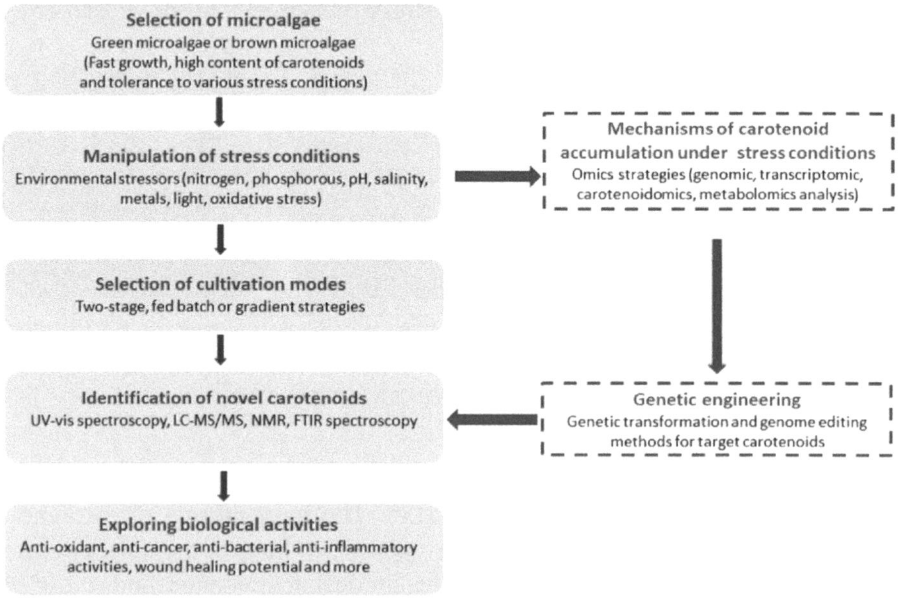

FIGURE 8.8 Schematic representation of a bioprospecting protocol for enhanced production of target carotenoids via manipulating the culture conditions of microalgae.

Source: Adapted from: Chen et al. (2017).

8.3 CAROTENOID ANALYSIS: THE ANALYTICAL METHODOLOGY

Long exposure of heat, oxygen, and light to microalgae adversely affects the stability of carotenoids and lead to structural changes of carotenoids which in turn degrade them. This results in the loss of antioxidant property of carotenoids (Regal et al., 2019).

Carotenoid composition varies among different types of microalgae both qualitatively and quantitatively. For this reason, sample preparation prior to extraction and saponification procedures is the key factor for their analysis. A lot of effort should be spent for accurate and reliable carotenoid analysis. There has been several and important literature summarizing these protocols and precautions that must be taken (Kopec et al., 2012; Rivera Vélez and Canela, 2012; Arvayo-Enríquez et al., 2013; Islam and Schweigert, 2015; Mercadante et al., 2017; Giuffrida et al., 2018).

The lack of a reference standard makes it difficult to analyze the diversity of carotenoid structures in different microalgae. Furthermore, certain other factors reduce the accuracy of analytical methods for identifying and quantifying carotenoids in microalgae. These factors can be summarized as unavailability of commercially available standards, low quantities of some carotenoids, and the presence of interfering substances in biological samples (Oliver and Palou, 2000; Amorim-Carrilho et al., 2014).

In general, a conventional carotenoid analysis procedure in microalgae includes the following steps: Homogenization, carotenoid extraction and/or saponification, and instrumental analysis are all steps in the lyophilization (freeze-drying) or low-temperature drying process (usually, separation by chromatography and detection). Rodriguez-Amaya (2003) has told that the carotenoid analysis is difficult and prone to errors many years ago. To eliminate errors and produce solid quantitative data, the analyst must take advantage of all analytical improvements and emerging technologies accessible.

8.3.1 SAMPLE PREPARATION

Before applying any extraction methods to any biological matrix for carotenoids, the analyst must ensure that the sample has been prepared accurately. In carotenoid analysis from microalgae, the biomass should be dried. It is much better and more accurate to express the carotenoid amount in terms of dry weight. On the other hand, in some cases, some amount of water addition is recommended for better extraction of xanthophylls due to high polarity when compared to other organic solvents (Damergi et al., 2017).

Even at low temperatures, carotenoid loss can be encountered. Therefore, lyophilization is the best choice for drying even though it is more expensive for large scale processes. For this purpose, the harvested microalgae should be thoroughly rinsed to remove all the culture media before freeze-drying in lyophilizator. Finally, a fine powder has been obtained after grinding gently with the aid of a mortar. Figure 8.9 depicts the general procedure to microalgae.

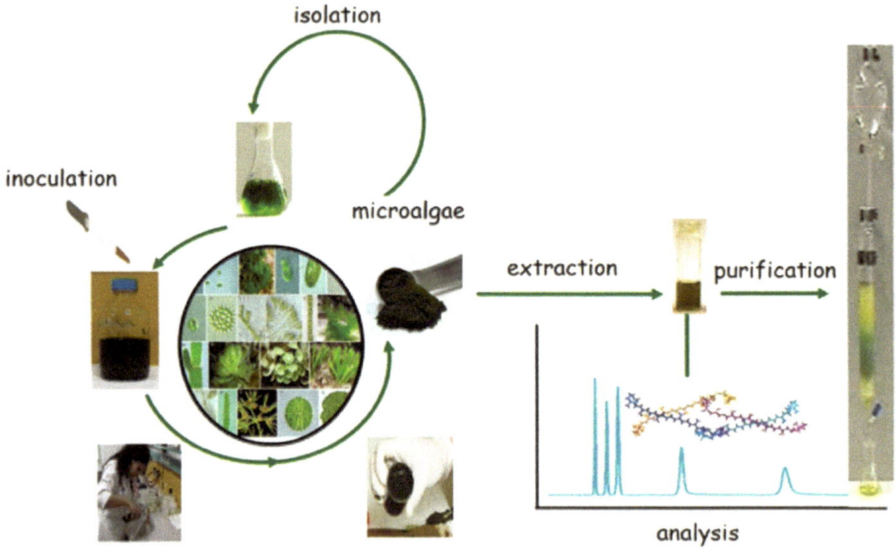

FIGURE 8.9 Procedure for preparation of biomass and analysis of carotenoid profiles from microalgae.

8.3.2 EXTRACTION OF CAROTENOIDS FROM MICROALGAE

Unfortunately, there has been no generalized and standard protocol for the carotenoid extraction in laboratories (Riggi, 2010; Saini and Keum, 2018). In the case of microalgae, homogenization or grinding must be the first step in order to extract or release the analytes from rigid algal cell walls. Figure 8.9 depicts the preparation of algal biomass and general carotenoid analysis applied to microalgae followed by a purification procedure.

After the biomass powder has been prepared, the first step is to employ organic solvents or a combination of these solvents to achieve synergistic effects. As a result, choosing the right solvent is one of the most important aspects of a successful carotenoids extraction (Amorim-Carrilho et al., 2014; Cepeda et al., 2014). In addition, the number of extractions is the second important parameter that should be considered since one-step extraction procedures do not give the maximum amount of carotenoids present in the microalgae. Hence, as in the case of solvent selection number of extractions must be optimized. The extraction procedure should be followed until decoloration has been observed. Moreover, each fraction should be analyzed quantitatively to decide whether carotenoid portions have still been obtained.

There have been so many extraction techniques used in the extraction of carotenoids in literature so far like atmospheric liquid extraction with maceration, Soxhlet extraction, supercritical and subcritical extraction, ultrasound-assisted extraction (UAE), microwave-assisted extraction (MAE), accelerated solvent extraction (ASE), pressurized liquid extraction (PLE), pulsed electric field-assisted extraction (PEF), enzyme-assisted extraction (EAE) and green extraction (Saini and Keum, 2018).

For each extraction process, extraction, and analysis must be performed as rapidly as possible under special conditions of light (dim light, yellow light or dark medium), temperature (low), oxygen (inert atmosphere like N_2 or Ar), and pH (neutral or low basic). Otherwise, carotenoids can be easily oxidized, isomerized, and/or degraded. In order to prevent or minimize the degradation, a neutralizer must also be added during extraction, such as calcium carbonate ($CaCO_3$), sodium bicarbonate ($NaHCO_3$) or magnesium carbonate ($MgCO_3$) to neutralize the organic acids that might liberate from the biomass. Particularly, antioxidant addition to extraction solvents at concentrations of $\approx$0.1% (w/v) is another important factor. Pyrogallol, tert-butylhydroquinone (TBHQ) and butylated hydroxytoluene (BHT) are the most common ones (Cernelic et al., 2013; Erdoğan et al., 2015).

8.3.3 SAPONIFICATION

Carotenes are found in the free form, whereas xanthophyll can be found in either the free or more stable fatty acid esterified forms. After the extraction process, chlorophylls, and lipids are removed by saponification. In this sense, chromatographic separation has been simplified. In addition, saponification allows the esterified xanthophylls to free forms (Olmedilla et al., 2001; Mercadante et al., 2017). However, some alkali sensitive carotenoids can be degraded or structurally change during saponification like astaxanthin and fucoxanthin. Erdoğan et al. (2016) did not perform any saponification procedure after the extraction since fucoxanthin turns into fucoxanthinol in the presence of KOH. In a reported study by Su et al. (2018), a comparison of saponification and enzymolysis for hydrolyzing astaxanthin esters of *H. pluvialis* was made. The astaxanthin esters were hydrolyzed with considerably less degradation and isomerization, according to the findings. Because of the significant degree of oxidation and degradation that occurs during ester hydrolysis, the saponification procedure with even weak alkaline conditions was seen to be unsuitable for hydrolyzing astaxanthin esters. Therefore, this procedure is generally eliminated unless required.

In general carotenoid analysis from microalgae, the recommended procedures involve the addition of methanolic KOH to algal extracts flushed with N_2/Ar gas at room temperature, in the dark by adding synthetic antioxidants. Saponification time depends on the carotenoid diversity and form as well as the chlorophyll *a*nd lipid content of the microalgae of interest. For selective isolation of a free form of carotenoid, this procedure may not be applied. Moreover, if the carotenoid profile does not change after alkali treatment this step can be omitted. It is not necessary for all samples but, despite all the existing drawbacks, saponification is still optimized in certain cases to improve analytical resolution and to reduce analysis time in chromatographic separations (Kimura et al., 1990; Inbaraj et al., 2008).

8.3.4 CHROMATOGRAPHIC SEPARATION: IDENTIFICATION AND DETERMINATION OF CAROTENOIDS

Carotenoids absorb light maximally between 460 nm and 550 nm and appear red, orange, or yellow to us. Their physical and chemical properties thus allow us to determine the carotenoid profiles both qualitatively and quantitatively in any matrix by HPLC coupled to various absorbance detectors (de Quirós and Costa, 2006). Among them diode array detector (DAD) is the most useful one since multiple wavelengths can be selected for screening and the absorbance profiles in the selected region (generally 300–600 nm) could also be recorded during the chromatographic analysis which then allows us the important step for tentative identification of carotenoids. However, many carotenoids might co-elute during the separation as carotenoids have similar structures. Consequently, their determination by absorbance is not possible anymore because their spectra are very similar. For this reason, mass spectrometry (MS) is required as a complementary technique for their identification. Other analytical alternatives have been also reported for the analysis of carotenoids, including infrared (IR), mid-, and near-infrared spectroscopy (MIR and NIR) and Raman spectroscopy. Despite all their interesting applications, the use of these technologies is not very often. In the case of NMR, this technique offers a non-destructive, reliable method for characterizing structural and stereochemical changes in carotenoids, resulting in good carotenoid profiles in real samples with minimal purifying steps, with or without prior LC separations (Aman et al., 2005; Tiziani et al., 2006). The procedure for the identification and determination of carotenoids was schematically represented in Figure 8.10.

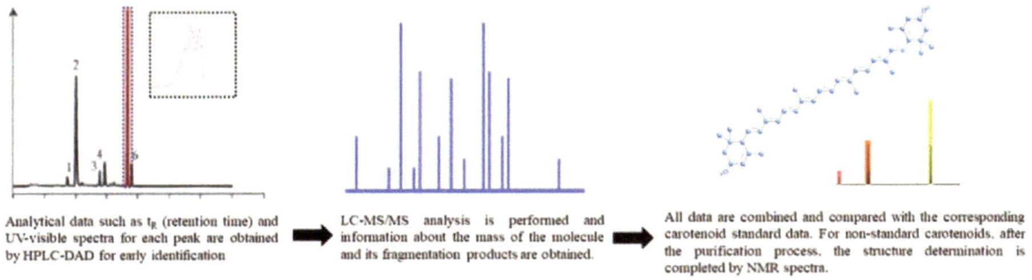

FIGURE 8.10 A brief summary and pathway for the identification and analysis of carotenoids.

Separation of carotenoid pigments can be accomplished using either normal-phase or reverse phase HPLC. However, due to the poor separation of non-polar carotenoids, normal phase HPLC is not suited for carotenoid separation. In comparison, reverse phase HPLC permits a large increase in the contact between the analyte and the non-polar stationary phase, resulting in increased carotenoids resolution (Sander et al., 2000). Among the columns, C_{18} columns with isocratic or gradient mode are preferred for carotenoid separation (Khachik et al., 1997). However, the use of polymeric C_{30} carotenoid columns provides better resolution of carotenoids. It identifies and quantifies geometrical carotenoid isomers with better sensitivity and selectivity compared to C_{18} columns (Bijttebier et al., 2014).

8.4 SWOT ANALYSIS OF CAROTENOIDS: OPPORTUNITIES AND CHALLENGES

Carotenoids use as a natural coloring agent is extensively increasing its application in the food and beverage industry. In addition, carotenoids aid in supporting eye, skin, cardiovascular health, and are being widely used for cancer prevention. Therefore, strong growth of food, pharmaceutical, and cosmetic industry emerges in several countries day by day. Hence, it creates growth opportunities for the market in the future (Ambati et al., 2019; Gogisetty et al., 2019).

Microalgae has a high content of carotenoids, although other natural sources like plants are able to synthesize them (Giuliano et al., 2008). In addition, they can accumulate a certain carotenoid via microalgal biotechnology through directed carotenoid metabolism. By changing/optimizing the culture conditions, both large amounts of biomass and target carotenoids can be obtained. With the development of environmentally friendly extraction technologies, sustainable processing techniques are also involved. This provides a great market opportunity in the field of carotenoids. In addition, biotechnological production of carotenoids from natural sources like microalgae attracts end-users. Thus, interest in the production of natural carotenoids from sustainable sources also increases. Besides, exploration of new microalgae species may lead to the identification of novel and rare carotenoids.

However, cultivation of carotenoid rich microalgae is an expensive process. Therefore, microalgae with high growth rates should be preferred. Besides, microalgae cultivation only for carotenoid production has some disadvantages in terms of producing waste and new biorefinery approaches must be evaluated. Apart from that, when they are grown in open ponds, the species must be selected which is resistant to possible climate changes. Also, the risks of microalgae biomass being contaminated by bacteria and other microbes are quite high. Contaminated biomass may negatively affect the microalgae cultivation through their products or even cause the death of cells.

A recent opinion statement emphasized that microalgae are particularly suitable for the production of high-value products such as food/feed and nutraceuticals, but their widespread applicability in the production of these products still faces some difficulties (Kiesenhofer and Fluch, 2018). A SWOT analysis of carotenoid production from microalgae has been designated in Figure 8.11. Although it has many advantageous aspects, microalgae biotechnology has not yet taken the place it deserves. More efforts should be performed, and collaborative projects with industrial partners should be implemented. There still exists a gap which needs to be filled jointly by academia and industry.

8.5 ROLE OF CAROTENOIDS IN BIOTECHNOLOGY

Recently, there is a great interest in carotenoids due to their bioactive roles and coloring properties, strong antioxidant potential and use of pharmaceuticals. Carotenoids' use in food coloring and dietary supplements has expanded in the global market during the last 10 years, both in terms of industrial and commercial applications. Carotenoids are expected to be worth $ 1.5 billion by 2020, according to experts (Dufossé et al., 2005).

Carotenoids are also used as additives in the food industry to increase the value and attractiveness of products to certain foods (Hsu et al., 2008). Due to the suspected carcinogenic effects of synthetic carotenoids, the natural production of carotenoids has supported the interest in microalgae and high market demand (Abd El-Baky and El-Baroty, 2013). Sustainable industrial production from microalgae in current applications in the feed, food, and beverage market is possible with biotechnology. A safe source of potential food additives and supplements provide sustainable and high quality of life (Plaza et al., 2009).

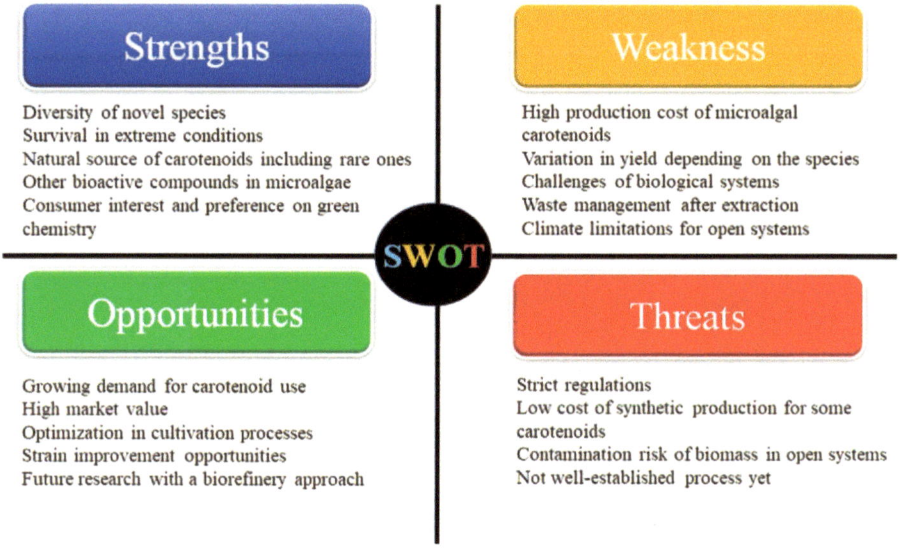

FIGURE 8.11 SWOT analysis of microalgal carotenoid production.

Often marine, and freshwater animals, including fish and crustaceans, owe their bright colors to carotenoids (Matsuno et al., 1985; Bjerkeng, 2008). In terms of animal health, carotenoids are beneficial but not essential. Diets rich in carotenoids can be applied to animal feed to increase their color and value (Christaki et al., 2013). This is something that is desired in the industrial sense, because it increases the nutritional value of seafood and attracts the attention of the customer.

Numerous epidemiological studies show that the consumption of natural carotenoids is associated with the prevention of the risk of developing various degenerative diseases in humans (Schünemann et al., 2002; Catarina Guedes et al., 2013). In this context, microalgal pigments stimulate the immune system and help prevent important diseases. There are reports of the beneficial effects of microalgal carotenoids on humans and animals due to their high antioxidant activity. Carotenoids have been shown to have anti-inflammatory qualities that can aid with arthritis, muscle pain, and carpal tunnel syndrome (Chacón-Lee and González-Mariño, 2010).

In addition, some carotenoids protect the eye macula from any negative photochemical reaction due to its antioxidant functions. Low levels of carotenoids have also been found in people with cataracts and are especially over the age of 65 (Olmedilla et al., 2001; Christaki et al., 2013; Bonos et al., 2013).

Carotenoids are natural pigments that occur widely in nature and shown to be beneficial in human health and play an important role in the prevention of several diseases (Rao and Rao, 2007; Cooperstone and Schwartz, 2016; Wells et al., 2017) as depicted in Figure 8.12.

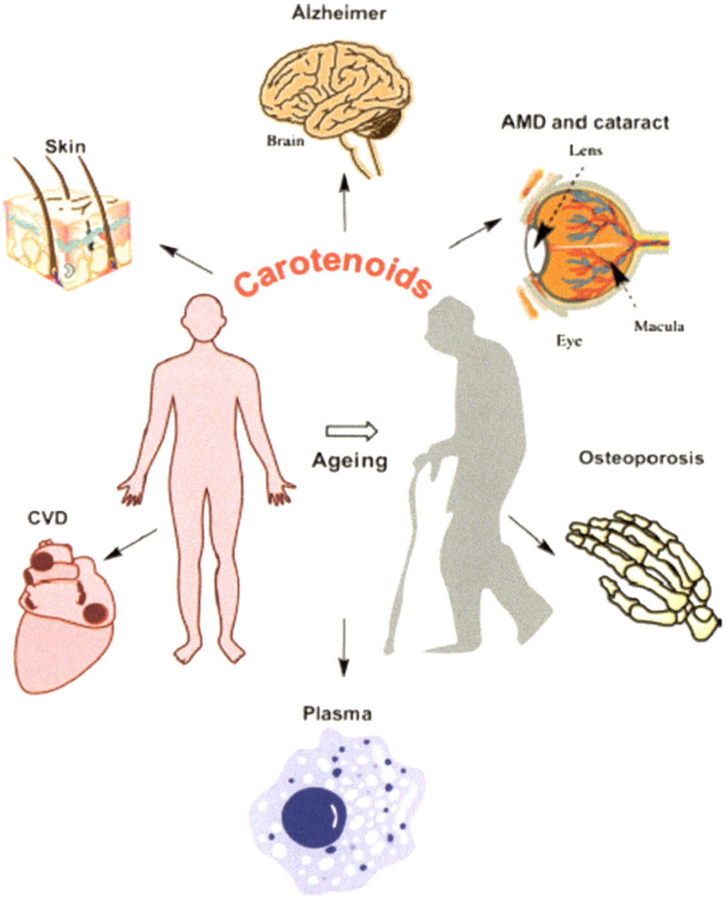

FIGURE 8.12 Suggested effects of carotenoids in the prevention of several diseases.

Source: Reprinted with permission from Rivera-Madrid et al., 2020. © Elsevier.)

In vitro and animal studies provide evidence that carotenoids can protect people against various types of cancer (Figure 8.13). Some authors even reported that specific types of cancer can be treated with specific carotenoids (Rivera-Madrid et al., 2020).

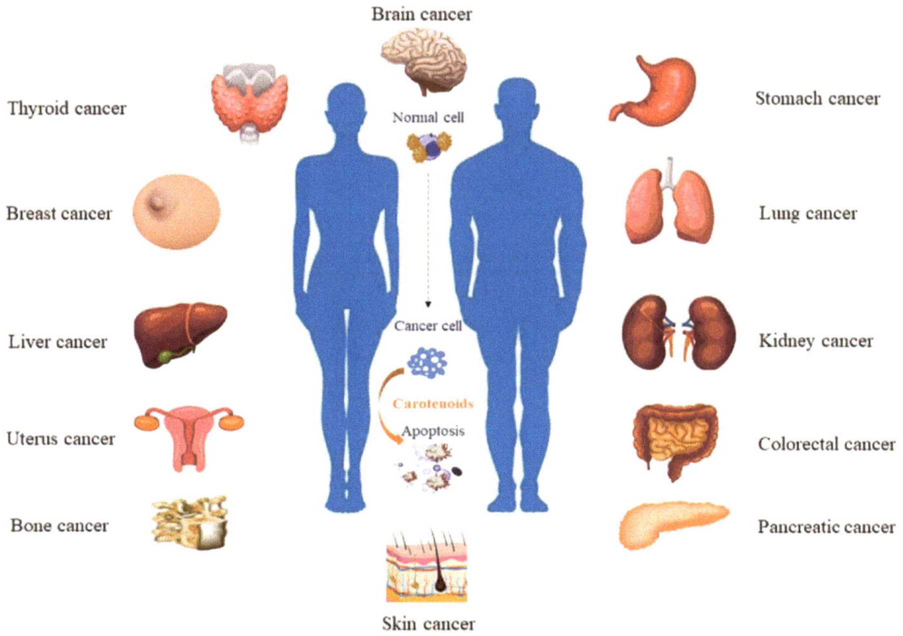

FIGURE 8.13 Role of carotenoids against different types of cancer.
Source: Adapted with permission from: Rivera-Madrid et al. (2020).

Among the algae species, lutein in *Chlorella* spp. and *B. braunii*; astaxanthin and its esters in *H. pluvialis* and *Chlorococcum* spp.; β-carotene in *D. salina*; and fucoxanthin in diatoms have been used for commercial applications. Carotenoid pigments found in algae, such as astaxanthin, β-carotene, lutein, zeaxanthin, and fucoxanthin, have gained increased attention due to their potential applications in food, animal feed, nutraceuticals, medicines, the cosmetics industry, and dietary supplements (Ambati et al., 2014; Bilbao et al., 2016; Lelyana, 2016).

According to global carotenoid market analysis, Europe has a strong and potential market due to the growing demand for animal feed, health supplements and cosmetics. The participation of leading cosmetic industries such as L'Oreal, Unilever, Beiersdorf, and Henkel is predicted to increase the value of the carotenoid market in Europe. In addition, the Asia Pacific region is also expected to increase the use of carotenoids in food supplements and animal feed in the future. Many countries, such as Japan, China, and India, are trying to economically develop carotenoid production technology to improve the carotenoid market in the Asia Pacific region. A number of key vendors play an important role in the production of worldwide carotenoid pigments such as Lycored, Kemin Industries Inc., Divis Laboratories, Allied Biotech Corp., Brenntag BASF Corporation, Naturex SA, FMC Corporation, DSM Nutritional Products, Royal. DSM NVChr. Hansen A/S, Allied Biotechnology, Cognis, Carotech, D.D. Williamson and Co. Inc., Dohler Group and ExcelVite SDN BHD (Gogisetty et al., 2019).

To sum up, the industrial production of carotenoids and their necessity in these areas are better understood if their applications in health and food are considered. This will contribute significantly to the growing economy. For these reasons, microalgae require new integrated emerging technologies to build sustainable solutions for carotenoid production in commercial processes for health and food applications as well as in the field of cosmetics.

8.6 COMMERCIALIZATION ASPECTS OF CAROTENOIDS

Carotenoids are part of a natural and large pigment group, but only a few have had the opportunity to turn into products in an industrial sense. Market shares of carotenoid are limited with -carotene, lutein, zeaxanthin, astaxanthin, and lycopene. They can be used in food, cosmetics, and medicinal applications due to their bioactive and color qualities. The application area of carotenoids is mostly due to its color and protective functions (Bogacz-Radomska et al., 2020).

Capsanthin is widely used in the food and cosmetics industry for the purpose of a natural coloring (Matsufuji et al., 1998; Kim et al., 2009). It is also used in chicken feed to intensify the color of egg yolks (Lokaewmanee et al., 2013). The animal feed industry is the primary user of astaxanthin. Astaxanthin is a good component of dietary supplements and functional foods because of its diverse biological effect (Pogorzelska et al., 2016; Sieradzka and Kołodziejczyk-Czepas, 2016). It is also used in the nutraceutical and pharmaceutical industry and has become very popular in the cosmetics industry. β-Carotene, on the other hand, is an active vitamin A precursor and finds a number of applications in the food, medicine, and cosmetics industry due to its properties (Saini et al., 2015).

Lutein, similar to most carotenoids, has powerful antioxidant properties that contribute to human health. The most unique feature of lutein (and zeaxanthin) is that it can accumulate in the eyeball (macula and retina). Lutein and zeaxanthin lessen the risk of age-related macular degeneration (AMD) due to the characteristics listed above (AMD) (Mozaffarieh et al., 2003). The literature mentions several potential benefits of using lutein as a nutrient or cosmetic agent (Shegokar and Mitri, 2012). In addition, lycopene is also used as a colorant in the food industry. Lycopene is a good active component in cosmetics because of its antioxidant capabilities (Chiu et al., 2007). Moreover, literature emphasized the anticancer properties of lycopene (Mein et al., 2008; Story et al., 2010; Cooperstone et al., 2012). Canthaxanthin, another carotenoid type is used as a colorant in the food industry. It also finds application in the animal feed, pharmaceutical, and cosmetics industry. As Igielska-Kalwat and colleagues emphasize, interest in carotenoids (especially lycopene) as natural active ingredients of cosmetics is growing (Igielska-Kalwat et al., 2014).

Carotenoid applications in the European Union (EU) are underlined in regulation (EC) No. 1333/2008 on food additives of the European Parliament and the Council of 16 December 2008.

> *Food additive is any substance not normally consumed as a food in itself and not normally used as a characteristic ingredient of food, whether or not it has nutritive value, the intentional addition of which to food for a technological purpose in the manufacture, processing, preparation, treatment, packaging, transport, or storage of such food results, or maybe reasonably expected to result, in it or its by-products becoming directly or indirectly a component of such foods.*

The application of carotenoids to foods depends on various and different arrangements or regulations. Carotenoids are mostly employed according to Good Manufacturing Practices in the United States, the EU, Colombia, Canada, Chile, Venezuela, Japan, and South Korea (Bogacz-Radomska et al., 2020).

8.7 CURRENT TRENDS AND FUTURE PERSPECTIVES IN MICROALGAL CAROTENOID INDUSTRY

In order to find, select, and produce new carotenoids from microalgae, considering their cost and applications, they should be considered in many ways. The most important questions to be addressed are given below (Ambati et al., 2019; Gogisetty et al., 2019):

- How are micro-bioreactor technologies developed to lower bioprocess cost and timelines in new biotechnological applications?
- How to make a large-scale carotenoid production micro-photobioreactor using high-power light-emitting diodes (LEDs) and adjustable light intensity?
- How to develop enhanced engineering tools for proteomics, genomics, and metabolomics?
- How to improve long-term algal culture methods that use photosynthetic organisms as a control mechanism for effective biomass and carotenoid production?
- How to identify potential pollutants in open mass algae culture systems?
- How are early detection methods developed to control them in an algae culture?
- How to improve carotenoid production of selected microalgae?
- How to develop sustainable technologies for carotenoid production in the commercial process?
- How to develop high quality carotenoid pigments in the global market for human consumption?
- How can carotenoids be used for drug discovery and other new applications?

Once these procedures are optimized, microalgal carotenoid production will have the potential to be industrialized and compete in the carotenoid market. To date, astaxanthin from *H. pluvialis* and β-carotene from *D. salina* have been two successful examples of commercialized microalgal carotenoid synthesis. While synthetic astaxanthin costs more than US$ 1,000 per kilogram, the retail price is more nearly twice that amount (Li et al., 2011). According to Kleinegris et al. (2011), an estimated cost of US$ 718/kg to produce astaxanthin from microalgae can be considered with a lower microalgae cost. In addition to increasing the production rate and pigment content, several attempts have been made to reduce costs. Tran et al. (2014) tried to reduce the cost by recycling the culture medium. It is currently in the harvest and extraction stages to further reduce production costs (Barba et al., 2015).

Lutein is another significant carotenoid component in the market. Currently, the cost of natural lutein derived from marigold flowers is US$ 500 per kilogram. This is even lower than the number for synthetic products. Economic concerns place a premium on growth rate, nutritional requirements, pigment content, tolerance to environmental variations, and species selection. Additionally, consumer acceptance of novel functional foods may have an effect on the growth of the market for microalgal carotenoids (Freitas et al., 2012). Apart from carotenoids, microalgae may be utilized to produce a variety of important byproducts, including biodiesel, EPA, DHA, vitamins, proteins, and enzymes. Thus, considering additional compounds such as biofuels as a by-product of microalgal carotenoid production may give a solution to reduce production costs. Dineshkumar et al. (2015) obtained lutein and biodiesel using the same algae in a single procedure and achieved satisfactory yield for both products (6 mg/g lutein and 94 mg/g FAME). Several studies have been conducted in this area for integrated single-step biodiesel and lutein production (Araya et al., 2014; Dash et al., 2015; Dineshkumar et al., 2015; Minhas et al., 2016). Fucoxanthin and lipids were also obtained from the hexane-ethanol system (Kim et al., 2012). It is not reported that other products are produced simultaneously with carotenoids, but the biomass itself is an edible product as it is known that microalgae produce sterols, protein, sugar, vitamins, and more. This is common in Korea and Japan. Thus, there is the potential to obtain more than one product from microalgae by developing better separation technologies.

Microalgae require new integrated technologies to create bioactive compounds for health and food purposes, such as carotenoids. In this context, while obtaining carotenoids from microalgae, new technologies are required in order to protect both the biomass and carotenoids from degradation. Apart from this, energy cost must be saved in freeze or oven dryer by developing wet extraction methods for carotenoids derived from algae biomass. Understanding these processes will enable researchers to adapt new microalgal carotenoids to be used in biotechnological applications. Synthetic ones are preferred due to the high cost of technologies currently available for microalgal carotenoid pigment production on an industrial scale. Therefore, to compete with synthetic carotenoids, a sustainable and cost-effective process must be developed. Biotechnologists, biochemists, chemists, chemical engineers, and medical professionals are expected to collaborate to further microalgae research. If carotenoids from algae can be scaled cost-effectively and even produced with a bio-refinery approach, they will have a very important place in the industry and market.

8.8 CONCLUDING REMARKS

Microalgal carotenoid pigments have risen to the top of the bio-based economy's most competitive, enticing, and popular developing fields. Microalgae are an excellent source of carotenoids for mass production due to their high carotenoid content, rapid growth, and a variety of other benefits.

In the cultivation of carotenoid-rich microalgae, biotechnology with high-efficiency photobioreactors and optimum growth conditions are applied. However, a more cost-effective and scalable cultivation approach is currently being researched, and more research should be conducted in this field. There are still significant challenges in downstream processing, particularly in the harvesting and cell disruption operations. While centrifugation is one of the most often used harvesting technologies, it must be combined with appropriate energy reduction techniques. Flocculation is an option, but additional research is needed to make it more effective. MAE and pulsed electric field are the most promising methods in terms of both efficiency and cost among the most recent breakthroughs, when evaluating the scale-up efficiency of various cell disruption methods.

In the future, research should focus on increasing efficiency and lowering costs in microalgal bioprocessing. Advanced metabolic engineering tools can be used to accomplish this. In order to create high-quality carotenoids, storage stability must be considered. It should be emphasized that the link between the physiological mechanisms of carotenogenesis and the cell wall disruption method improves the species-specific growth strategy. New techniques, such as *in situ* extraction and replaceable solvents, may open new opportunities for research. If non-toxic solvents can be eliminated while maintaining bioactivity, exchange solvent extraction has a lot of promise. Wet extraction methods provide another favorable new research element by reducing drying expenses. Furthermore, because microalgal biotechnology creates a wide range of useful goods, research activities should incorporate the simultaneous production of several products in order to develop a more economically attractive and sustainable microalgal business. Microalgal carotenoid synthesis has a lot of potential, but there are a lot of obstacles to overcome. To make processes more competitive, better engineering design and innovation are still required. Microalgal biotechnology and the production of carotenoids from microalgae should be supported by each effort in this area. More efforts should be made to shift to industrial manufacturing by maximizing the benefits of each step while minimizing the drawbacks.

ACKNOWLEDGMENTS

Dr. Ayşegül Erdoğan and Prof. Dr. Meltem Conk Dalay has worked within the COST (European Cooperation in Science and Technology) project linked through a COST Action EUROCAROTEN-CA15136 (European Network to

Advance Carotenoid Research and Applications in Agro-Food and Health) between 2017 and 2021 funded by TÜBİTAK (The Scientific and Technological Research Council of Turkey) with a grant number of 216Z167.

KEYWORDS

- **biosynthesis**
- **carotenogenesis**
- **carotenoids**
- ***Chlorella vulgaris***
- **chromatography**
- **cultivation**
- ***Dunaliella salina***
- **extraction**
- ***Haematococcus pluvialis***
- **microalgae**
- **saponification**

REFERENCES

Abd El-Baky, H. H., & El-Baroty, G. S., (2013). The potential use of microalgal carotenoids as dietary supplements and natural preservative ingredients. *Journal of Aquatic Food Product Technology*, *22*(4), 392–406.

Aman, R., Biehl, J., Carle, R., Conrad, J., Beifuss, U., & Schieber, A., (2005). Application of HPLC coupled with DAD, APcI-MS and NMR to the analysis of lutein and zeaxanthin stereoisomers in thermally processed vegetables. *Food Chemistry*, *92*(4), 753–763.

Ambati, R. R., Gogisetty, D., Aswathanarayana, R. G., Ravi, S., Bikkina, P. N., Bo, L., et al., (2019). Industrial potential of carotenoid pigments from microalgae: Current trends and future prospects. *Critical Reviews in Food Science and Nutrition*, *59*(12), 1880–1902.

Ambati, R. R., Phang, S. M., Ravi, S., & Aswathanarayana, R. G., (2014). Astaxanthin: Sources, extraction, stability, biological activities and its commercial applications: A review. *Marine Drugs*, *12*(1), 128–152.

Amorim-Carrilho, K., Cepeda, A., Fente, C., & Regal, P., (2014). Review of methods for analysis of carotenoids. *TrAC Trends in Analytical Chemistry*, *56*, 49–73.

Araya, B., Gouveia, L., Nobre, B., Reis, A., Chamy, R., & Poirrier, P., (2014). Evaluation of the simultaneous production of lutein and lipids using a vertical alveolar panel bioreactor for three *Chlorella* species. *Algal Research*, *6*, 218–222.

Arvayo-Enríquez, H., Mondaca-Fernández, I., Gortárez-Moroyoqui, P., López-Cervantes, J., & Rodríguez-Ramírez, R., (2013). Carotenoids extraction and quantification: A review. *Analytical Methods, 5*(12), 2916–2924.

Bajhaiya, A. K., Moreira, J. Z., & Pittman, J. K., (2017). Transcriptional engineering of microalgae: Prospects for high-value chemicals. *Trends in Biotechnology, 35*(2), 95–99.

Barba, F. J., Grimi, N., & Vorobiev, E., (2015). New approaches for the use of non-conventional cell disruption technologies to extract potential food additives and nutraceuticals from microalgae. *Food Engineering Reviews, 7*(1), 45–62.

Barros, A. I., Gonçalves, A. L., Simões, M., & Pires, J. C., (2015). Harvesting techniques applied to microalgae: A review. *Renewable and Sustainable Energy Reviews, 41*, 1489–1500.

Ben-Amotz, A., & Levy, Y., (1996). Bioavailability of a natural isomer mixture compared with synthetic all-trans beta-carotene in human serum. *The American Journal of Clinical Nutrition, 63*(5), 729–734.

Bijttebier, S., D'Hondt, E., Noten, B., Hermans, N., Apers, S., & Voorspoels, S., (2014). Ultra high performance liquid chromatography versus high performance liquid chromatography: Stationary phase selectivity for generic carotenoid screening. *Journal of Chromatography A, 1332*, 46–56.

Bilbao, P. G. S., Damiani, C., Salvador, G. A., & Leonardi, P., (2016). *Haematococcus pluvialis* as a source of fatty acids and phytosterols: Potential nutritional and biological implications. *Journal of Applied Phycology, 28*(6), 3283–3294.

Bjerkeng, B., (2008). *Carotenoids in Aquaculture: Fish and Crustaceans* (pp. 237–254). Carotenoids. Springer.

Bogacz-Radomska, L., Harasym, J., & Piwowar, A., (2020). Commercialization aspects of carotenoids. *Carotenoids: Properties, Processing and Applications* (pp. 327–357). Elsevier.

Britton, G., (1995). UV/visible spectroscopy. *Carotenoids: Spectroscopy* (Vol. 1B, pp. 20, 21).

Britton, G., Liaaen-Jensen, S., & Pfander, H., (1998). *Synthesis from a Different Perspective: How Nature Does it* (pp. 1–12). Birkhäuser Verlag: Basel, Switzerland.

Britton, G., Liaaen-Jensen, S., & Pfander, H., (2012). *Carotenoids: Handbook.* Birkhäuser.

Britton, G., Pfander, H., & Liaaen-Jensen, S., (1996). *Carotenoids: Synthesis* (Vol. 2, p. 2) Springer Science and Business Media.

Catarina, G. A., Amaro, H. M., Sousa-Pinto, I., & Xavier, M. F., (2013). Bioactive carotenoids from microalgae. *Bioactive Compounds from Marine Foods: Plant and Animal Sources*, 131–151.

Cernelic, K., Prosek, M., Golc-Wondra, A., Rodic, Z., Simonovska, B., & Puklavec, M., (2013). Influence of synthetic antioxidants on extraction of all-trans-lutein from spinach under air and nitrogen atmosphere. *Food Nutr. Sci., 4*, 195–200. doi: 10.4236/fns.2013.42027.

Ceron, M. C., Campos, I., Sanchez, J. F., Acien, F. G., Molina, E., & Fernandez-Sevilla, J. M., (2008). Recovery of lutein from microalgae biomass: Development of a process for *Scenedesmus almeriensis* biomass. *Journal of Agricultural and Food Chemistry, 56*(24), 11761–11766.

Chacón-Lee, T., & González-Mariño, G. E., (2010). Microalgae for "healthy" foods - Possibilities and challenges. *Comprehensive Reviews in Food Science and Food Safety, 9*(6), 655–675.

Chan, K. C., Pen, P. J., & Yin, M. C., (2012). Anticoagulatory and antiinflammatory effects of astaxanthin in diabetic rats. *Journal of Food Science, 77*(2), H76–H80.

Chen, C. Y., Yeh, K. L., Aisyah, R., Lee, D. J., & Chang, J. S., (2011). Cultivation, photobioreactor design and harvesting of microalgae for biodiesel production: A critical review. *Bioresource Technology, 102*(1), 71–81.

Chew, W., Mathison, B. D., Kimble, L. L., Mixter, P. F., & Chew, B. P., (2013). Astaxanthin decreases inflammatory biomarkers associated with cardiovascular disease in human umbilical vein endothelial cells. *American Journal of Food Science and Technology*, *1*, 1–17.

Chiu, Y., Chiu, C., Chien, J., Ho, G., Yang, J., & Chen, B., (2007). Encapsulation of lycopene extract from tomato pulp waste with gelatin and poly (γ-glutamic acid) as carrier. *Journal of Agricultural and Food Chemistry*, *55*(13), 5123–5130.

Choi, H. D., Kang, H. E., Yang, S. H., Lee, M. G., & Shin, W. G., (2011). Pharmacokinetics and first-pass metabolism of astaxanthin in rats. *British Journal of Nutrition*, *105*(2), 220–227.

Christaki, E., Bonos, E., Giannenas, I., & Florou-Paneri, P., (2013). Functional properties of carotenoids originating from algae. *Journal of the Science of Food and Agriculture*, *93*(1), 5–11.

Cooperstone, J., & Schwartz, S., (2016). Recent insights into health benefits of carotenoids. *Handbook on Natural Pigments in Food and Beverages* (pp. 473–497). Elsevier.

Curtain, C., (2000). Plant biotechnology – the growth of Australia's algal β-carotene industry. *Australasian Biotechnology*, *10*(3), 19–23.

Damergi, E., Schwitzguébel, J. P., Refardt, D., Sharma, S., Holliger, C., & Ludwig, C., (2017). Extraction of carotenoids from *Chlorella vulgaris* using green solvents and syngas production from residual biomass. *Algal Research*, *25*, 488–495.

De Quirós, A. R. B., & Costa, H. S., (2006). Analysis of carotenoids in vegetable and plasma samples: A review. *Journal of Food Composition and Analysis*, *19*(2, 3), 97–111.

Dineshkumar, R., Dash, S. K., & Sen, R., (2015). Process integration for microalgal lutein and biodiesel production with concomitant flue gas CO_2 sequestration: A biorefinery model for healthcare, energy and environment. *RSC Advances*, *5*(90), 73381–73394.

Dong, L. Y., Jin, J., Lu, G., & Kang, X. L., (2013). Astaxanthin attenuates the apoptosis of retinal ganglion cells in db/db mice by inhibition of oxidative stress. *Marine Drugs*, *11*(3), 960–974.

Donhowe, E. G., & Kong, F., (2014). Beta-carotene: Digestion, microencapsulation, and *in vitro* bioavailability. *Food and Bioprocess Technology*, *7*(2), 338–354.

Duan, J., & Gregory, J., (2003). Coagulation by hydrolyzing metal salts. *Advances in Colloid and Interface Science*, *100*, 475–502.

Dufossé, L., Galaup, P., Yaron, A., Arad, S. M., Blanc, P., Murthy, K. N. C., et al., (2005). Microorganisms and microalgae as sources of pigments for food use: A scientific oddity or an industrial reality? *Trends in Food Science and Technology*, *16*(9), 389–406.

Erdoğan, A., Çağır, A., Dalay, M. C., & Eroğlu, A. E., (2015). Composition of carotenoids in *Scenedesmus protuberans*: Application of chromatographic and spectroscopic methods. *Food Analytical Methods*, *8*(8), 1970–1978.

Fernández-Sevilla, J. M., Fernández, F. A., & Grima, E. M., (2010). Biotechnological production of lutein and its applications. *Applied Microbiology and Biotechnology*, *86*(1), 27–40.

Freitas, A. C., Rodrigues, D., Rocha-Santos, T. A., Gomes, A. M., & Duarte, A. C., (2012). Marine biotechnology advances towards applications in new functional foods. *Biotechnology Advances*, *30*(6), 1506–1515.

Fujii, Y., Sakamoto, S., Ben-Amotz, A., & Nagasawa, H., (1993). Effects of beta-carotene-rich algae *Dunaliella bardawil* on the dynamic changes of normal and neoplastic mammary cells and general metabolism in mice. *Anticancer Research*, *13*(2), 389–393.

Giuffrida, D., Donato, P., Dugo, P., & Mondello, L., (2018). Recent analytical techniques advances in the carotenoids and their derivatives determination in various matrixes. *Journal of Agricultural and Food Chemistry*, *66*(13), 3302–3307.

Giuliano, G., Tavazza, R., Diretto, G., Beyer, P., & Taylor, M. A., (2008). Metabolic engineering of carotenoid biosynthesis in plants. *Trends in Biotechnology*, *26*(3), 139–145.

Grima, E. M., Belarbi, E. H., Fernández, F. A., Medina, A. R., & Chisti, Y., (2003). Recovery of microalgal biomass and metabolites: Process options and economics. *Biotechnology Advances*, *20*(7, 8), 491–515.

Guerin, M., Huntley, M. E., & Olaizola, M., (2003). *Haematococcus* astaxanthin: Applications for human health and nutrition. *Trends in Biotechnology*, *21*(5), 210–216.

Heo, S. J., Yoon, W. J., Kim, K. N., Ahn, G. N., Kang, S. M., Kang, D. H., et al., (2010). Evaluation of anti-inflammatory effect of fucoxanthin isolated from brown algae in lipopolysaccharide-stimulated RAW 264.7 macrophages. *Food and Chemical Toxicology*, *48*(8, 9), 2045–2051.

Higuera-Ciapara, I., Felix-Valenzuela, L., & Goycoolea, F., (2006). Astaxanthin: A review of its chemistry and applications. *Critical Reviews in Food Science and Nutrition*, *46*(2), 185–196.

Ho, S. H., Chen, C. Y., Lee, D. J., & Chang, J. S., (2011). Perspectives on microalgal CO_2-emission mitigation systems: A review. *Biotechnology Advances*, *29*(2), 189–198.

Hsu, Y. W., Tsai, C. F., Chang, W. H., Ho, Y. C., Chen, W. K., & Lu, F. J., (2008). Protective effects of *Dunaliella salina* – A carotenoid-rich alga, against carbon tetrachloride-induced hepatotoxicity in mice. *Food and Chemical Toxicology*, *46*(10), 3311–3317.

Huangfu, J., Liu, J., Sun, Z., Wang, M., Jiang, Y., Chen, Z. Y., et al., (2013). Antiaging effects of astaxanthin-rich alga *Haematococcus pluvialis* on fruit flies under oxidative stress. *Journal of Agricultural and Food Chemistry*, *61*(32), 7800–7804.

Hurst, W. J., (2008). *Methods of Analysis for Functional Foods and Nutraceuticals*. CRC Press.

Hussein, G., Sankawa, U., Goto, H., Matsumoto, K., & Watanabe, H., (2006). Astaxanthin, a carotenoid with potential in human health and nutrition. *Journal of Natural Products*, *69*(3), 443–449.

Igielska-Kalwat, J., Gościańska, J., & Nowak, I., (2014). Zastosowanie likopenu w dermokosmetykach. *Przemysł Chemiczny*, *93*(7), 1110–1113.

Inbaraj, B. S., Lu, H., Hung, C., Wu, W., Lin, C., & Chen, B., (2008). Determination of carotenoids and their esters in fruits of *Lycium barbarum* Linnaeus by HPLC–DAD–APCI–MS. *Journal of Pharmaceutical and Biomedical Analysis*, *47*(4, 5), 812–818.

Ishiki, M., Nishida, Y., Ishibashi, H., Wada, T., Fujisaka, S., Takikawa, A., et al., (2013). Impact of divergent effects of astaxanthin on insulin signaling in L6 cells. *Endocrinology*, *154*(8), 2600–2612.

Islam, K., & Schweigert, F. J., (2015). Comparison of three spectrophotometric methods for analysis of egg yolk carotenoids. *Food Chemistry*, *172*, 233–237.

Jayappriyan, K., Rajkumar, R., Venkatakrishnan, V., Nagaraj, S., & Rengasamy, R., (2013). *In vitro* anticancer activity of natural β-carotene from *Dunaliella salina* EU5891199 in PC-3 cells. *Biomedicine and Preventive Nutrition*, *3*(2), 99–105.

Kar, H. K., (2002). Efficacy of beta-carotene topical application in melasma: An open clinical trial. *Indian Journal of Dermatology, Venereology, and Leprology*, *68*(6), 320.

Khachik, F., Spangler, C. J., Smith, J. C., Canfield, L. M., Steck, A., & Pfander, H., (1997). Identification, quantification, and relative concentrations of carotenoids and their metabolites in human milk and serum. *Analytical Chemistry*, *69*(10), 1873–1881.

Kidd, P., (2011). Astaxanthin, cell membrane nutrient with diverse clinical benefits and anti-aging potential. *Alternative Medicine Review*, *16*(4), 355–364.

Kiesenhofer, D. P., & Fluch, S., (2018). The promises of microalgae — Still a long way to go. *FEMS Microbiology Letters*, *365*(1), fnx257.

Kim, J. H., Chang, M. J., Choi, H. D., Youn, Y. K., Kim, J. T., Oh, J. M., et al., (2011). Protective effects of *Haematococcus* astaxanthin on oxidative stress in healthy smokers. *Journal of Medicinal Food, 14*(11), 1469–1475.
Kim, S. M., Kang, S. W., Kwon, O. N., Chung, D., & Pan, C. H., (2012). Fucoxanthin as a major carotenoid in *Isochrysis* aff. *galbana*: Characterization of extraction for commercial application. *Journal of the Korean Society for Applied Biological Chemistry, 55*(4), 477–483.
Kim, S., Ha, T. Y., & Hwang, I. K., (2009). Analysis, bioavailability, and potential healthy effects of capsanthin, natural red pigment from *Capsicum* spp. *Food Reviews International, 25*(3), 198–213.
Kimura, M., Rodriguez-Amaya, D. B., & Godoy, H. T., (1990). Assessment of the saponification step in the quantitative determination of carotenoids and provitamins A. *Food Chemistry, 35*(3), 187–195.
Kirti, K., Amita, S., Priti, S., Mukesh, K. A., & Jyoti, S., (2014). Colorful world of microbes: Carotenoids and their applications. *Advances in Biology, 2014*, 837891. https://doi.org/10.1155/2014/837891.
Kleinegris, D. M., Janssen, M., Brandenburg, W. A., & Wijffels, R. H., (2011). Two-phase systems: Potential for in situ extraction of microalgal products. *Biotechnology Advances, 29*(5), 502–507.
Kopec, R. E., Cooperstone, J. L., Cichon, M. J., & Schwartz, S. J., (2012). Analysis methods of carotenoids. *Analysis of Antioxidant-Rich Phytochemicals, 4*, 105–148.
Landrum, J. T., & Bone, R. A., (2001). Lutein, zeaxanthin, and the macular pigment. *Archives of Biochemistry and Biophysics, 385*(1), 28–40.
Lee, J. C., Hou, M. F., Huang, H. W., Chang, F. R., Yeh, C. C., Tang, J. Y., et al., (2013). Marine algal natural products with anti-oxidative, anti-inflammatory, and anti-cancer properties. *Cancer Cell International, 13*(1), 1–7.
Lelyana, R., (2016). Role of marine-natural ingredient fucoxanthin on body's immune response of obesity. *Journal of Nanomedicine and Nanotechnology, 7*(2).
Li, J., Zhu, D., Niu, J., Shen, S., & Wang, G., (2011). An economic assessment of astaxanthin production by large scale cultivation of *Haematococcus pluvialis*. *Biotechnology Advances, 29*(6), 568–574.
Liang, M. H., Zhu, J., & Jiang, J. G., (2018). Carotenoids biosynthesis and cleavage related genes from bacteria to plants. *Critical Reviews in Food Science and Nutrition, 58*(14), 2314–2333.
Lokaewmanee, K., Yamauchi, K., & Okuda, N., (2013). Effects of dietary red pepper on egg yolk color and histological intestinal morphology in laying hens. *Journal of Animal Physiology and Animal Nutrition, 97*(5), 986–995.
Lorenz, R. T., & Cysewski, G. R., (2000). Commercial potential for *Haematococcus* microalgae as a natural source of astaxanthin. *Trends in Biotechnology, 18*(4), 160–167.
Lu, Y. P., Liu, S. Y., Sun, H., Wu, X. M., Li, J. J., & Zhu, L., (2010). Neuroprotective effect of astaxanthin on H_2O_2-induced neurotoxicity *in vitro* and on focal cerebral ischemia *in vivo*. *Brain Research, 1360*, 40–48.
Maeda, H., Hosokawa, M., Sashima, T., Murakami-Funayama, K., & Miyashita, K., (2009). Anti-obesity and anti-diabetic effects of fucoxanthin on diet-induced obesity conditions in a murine model. *Molecular Medicine Reports, 2*(6), 897–902.
Maoka, T., (2009). Recent progress in structural studies of carotenoids in animals and plants. *Archives of Biochemistry and Biophysics, 483*(2), 191–195.
Maoka, T., Tokuda, H., Suzuki, N., Kato, H., & Etoh, H., (2012). Anti-oxidative, anti-tumor-promoting, and anti-carcinogensis activities of nitroastaxanthin and nitrolutein, the reaction products of astaxanthin and lutein with peroxynitrite. *Marine Drugs, 10*(6), 1391–1399.

Matsufuji, H., Nakamura, H., Chino, M., & Takeda, M., (1998). Antioxidant activity of capsanthin and the fatty acid esters in paprika (*Capsicum annuum*). *Journal of Agricultural and Food Chemistry, 46*(9), 3468–3472.

Matsuno, T., Ookubo, M., & Komori, T., (1985). Carotenoids of tunicates, III. The structural elucidation of two new marine carotenoids, amarouciaxanthin A and B. *Journal of Natural Products, 48*(4), 606–613.

Mc Gee, D., & Gillespie, E., (2019). The bioactivity and chemotaxonomy of microalgal carotenoids. *Biodiversity and Chemotaxonomy* (pp. 215–237). Springer.

Mein, J. R., Lian, F., & Wang, X. D., (2008). Biological activity of lycopene metabolites: Implications for cancer prevention. *Nutrition Reviews, 66*(12), 667–683.

Mercadante, A. Z., Rodrigues, D. B., Petry, F. C., & Mariutti, L. R. B., (2017). Carotenoid esters in foods - A review and practical directions on analysis and occurrence. *Food Research International, 99*, 830–850.

Minhas, A. K., Hodgson, P., Barrow, C. J., & Adholeya, A., (2016). A review on the assessment of stress conditions for simultaneous production of microalgal lipids and carotenoids. *Frontiers in Microbiology, 7*, 546.

Mogedas, B., Casal, C., Forján, E., & Vílchez, C., (2009). β-Carotene production enhancement by UV-A radiation in *Dunaliella bardawil* cultivated in laboratory reactors. *Journal of Bioscience and Bioengineering., 108*(1), 47–51.

Mozaffarieh, M., Sacu, S., & Wedrich, A., (2003). The role of the carotenoids, lutein and zeaxanthin, in protecting against age-related macular degeneration: A review based on controversial evidence. *Nutrition Journal, 2*(1), 1–8.

Murthy, K. C., Vanitha, A., Rajesha, J., Swamy, M. M., Sowmya, P., & Ravishankar, G. A., (2005). *In vivo* antioxidant activity of carotenoids from *Dunaliella salina* — A green microalga. *Life Sciences, 76*(12), 1381–1390.

Nagarajan, J., Ramanan, R., Raghunandan, M., Galanakis, C., & Krishnamurthy, N., (2017). Carotenoids. *Nutraceutical and Functional Food Components*. Amsterdam: Elsevier Inc.

Okoro, V., Azimov, U., Munoz, J., Hernandez, H. H., & Phan, A. N., (2019). Microalgae cultivation and harvesting: Growth performance and use of flocculants - A review. *Renewable and Sustainable Energy Reviews, 115*, 109364.

Oliver, J., & Palou, A., (2000). Chromatographic determination of carotenoids in foods. *Journal of Chromatography A, 881*(1, 2), 543–555.

Olmedilla, B., Granado, F., Blanco, I., Vaquero, M., & Cajigal, C., (2001). Lutein in patients with cataracts and age-related macular degeneration: A long-term supplementation study. *Journal of the Science of Food and Agriculture, 81*(9), 904–909.

Paliwal, C., Mitra, M., Bhayani, K., Bharadwaj, S. V., Ghosh, T., Dubey, S., et al., (2017). Abiotic stresses as tools for metabolites in microalgae. *Bioresource Technology, 244*, 1216–1226.

Paniagua-Michel, J., Olmos-Soto, J., & Ruiz, M. A., (2012). Pathways of carotenoid biosynthesis in bacteria and microalgae. *Microbial Carotenoids from Bacteria and Microalgae* (pp. 1–12). Springer.

Park, J. S., Chyun, J. H., Kim, Y. K., Line, L. L., & Chew, B. P., (2010). Astaxanthin decreased oxidative stress and inflammation and enhanced immune response in humans. *Nutrition and Metabolism, 7*(1), 18.

Peng, J., Yuan, J. P., Wu, C. F., & Wang, J. H., (2011). Fucoxanthin, a marine carotenoid present in brown seaweeds and diatoms: Metabolism and bioactivities relevant to human health. *Marine Drugs, 9*(10), 1806–1828.

Plaza, M., Herrero, M., Cifuentes, A., & Ibanez, E., (2009). Innovative natural functional ingredients from microalgae. *Journal of Agricultural and Food Chemistry, 57*(16), 7159–7170.

Pogorzelska, E., Hamulka, J., & Wawrzyniak, A., (2016). Astaksantyna–budowa, właściwości i możliwości zastosowania w żywności funkcjonalnej. *Żywność Nauka Technologia Jakość, 23*(1).

Ranga, R. A., Raghunath, R. R., Baskaran, V., Sarada, R., & Ravishankar, G., (2010). Characterization of microalgal carotenoids by mass spectrometry and their bioavailability and antioxidant properties elucidated in rat model. *Journal of Agricultural and Food Chemistry, 58*(15), 8553–8559.

Rao, A. R., Sarada, R., Baskaran, V., & Ravishankar, G. A., (2006). Antioxidant activity of *Botryococcus braunii* extract elucidated *in vitro* models. *Journal of Agricultural and Food Chemistry, 54*(13), 4593–4599.

Rao, A. R., Sindhuja, H., Dharmesh, S. M., Sankar, K. U., Sarada, R., & Ravishankar, G. A., (2013). Effective inhibition of skin cancer, tyrosinase, and antioxidative properties by astaxanthin and astaxanthin esters from the green alga *Haematococcus pluvialis*. *Journal of Agricultural and Food Chemistry, 61*(16), 3842–3851.

Rao, A. V., & Rao, L. G., (2007). Carotenoids and human health. *Pharmacological Research, 55*(3), 207–216.

Regal, P., Lamas, A., Fente, C. A., Franco, C. M., & Cepeda, A., (2019). Analysis and metabolomics of carotenoids. *Carotenoids: Properties, Processing and Applications*, 189.

Riggi, E., (2010). Recent patents on the extraction of carotenoids. *Recent Patents on Food, Nutrition and Agriculture, 2*(1), 75–82.

Rivera, V. S. M., & Canela, I. G. R., (2012). Analytical tools for the analysis of carotenoids in diverse materials. *Journal of Chromatography A, 1224*, 1–10.

Rivera-Madrid, R., Carballo-Uicab, V. M., Cárdenas-Conejo, Y., Aguilar-Espinosa, M., & Siva, R., (2020). Overview of carotenoids and beneficial effects on human health. *Carotenoids: Properties, Processing and Applications* (pp. 1–40). Elsevier.

Rodriguez-Amaya, D. B., (2016). Natural food pigments and colorants. *Current Opinion in Food Science, 7*, 20–26.

Rodriguez-Amaya, D., (2003). Food carotenoids: Analysis, composition and alterations during storage and processing of foods. *Forum of Nutrition.*

Rokkaku, T., Kimura, R., Ishikawa, C., Yasumoto, T., Senba, M., Kanaya, F., et al., (2013). Anticancer effects of marine carotenoids, fucoxanthin and its deacetylated product, fucoxanthinol, on osteosarcoma. *International Journal of Oncology, 43*(4), 1176–1186.

Safi, C., Zebib, B., Merah, O., Pontalier, P. Y., & Vaca-Garcia, C., (2014). Morphology, composition, production, processing and applications of *Chlorella vulgaris*: A review. *Renewable and Sustainable Energy Reviews, 35*, 265–278.

Saini, D. K., Chakdar, H., Pabbi, S., & Shukla, P., (2020). Enhancing production of microalgal biopigments through metabolic and genetic engineering. *Critical Reviews in Food Science and Nutrition, 60*(3), 391–405.

Saini, R. K., & Keum, Y. S., (2018). Carotenoid extraction methods: A review of recent developments. *Food Chemistry, 240*, 90–103.

Saini, R. K., Nile, S. H., & Park, S. W., (2015). Carotenoids from fruits and vegetables: Chemistry, analysis, occurrence, bioavailability and biological activities. *Food Research International, 76*, 735–750.

Salian, K., & Strezov, V., (2017). Biofuels from microalgae. *Encyclopedia of Sustainable Technologies* (pp. 107–120). Elsevier.

Salim, S., Bosma, R., Vermuë, M. H., & Wijffels, R. H., (2011). Harvesting of microalgae by bio-flocculation. *Journal of Applied Phycology, 23*(5), 849–855.

Sander, L. C., Sharpless, K. E., & Pursch, M., (2000). C30 stationary phases for the analysis of food by liquid chromatography. *Journal of Chromatography A, 880*(1, 2), 189–202.

Schünemann, H. J., McCann, S., Grant, B. J., Trevisan, M., Muti, P., & Freudenheim, J. L., (2002). Lung function in relation to intake of carotenoids and other antioxidant vitamins in a population-based study. *American Journal of Epidemiology, 155*(5), 463–471.

Shaish, A., Harari, A., Hananshvili, L., Cohen, H., Bitzur, R., Luvish, T., et al., (2006). 9-cis β-carotene-rich powder of the alga *Dunaliella bardawil* increases plasma HDL-cholesterol in fibrate-treated patients. *Atherosclerosis, 189*(1), 215–221.

Shegokar, R., & Mitri, K., (2012). Carotenoid lutein: A promising candidate for pharmaceutical and nutraceutical applications. *Journal of Dietary Supplements, 9*(3), 183–210.

Sieradzka, M., & Kołodziejczyk-Czepas, J., (2016). Astaxanthin–carotenoid antioxidant with cardioprotective properties. *Problems of Hygiene and Epidemiology, 97*(3), 197–206.

Story, E. N., Kopec, R. E., Schwartz, S. J., & Harris, G. K., (2010). An update on the health effects of tomato lycopene. *Annual Review of Food Science and Technology, 1*, 189–210.

Strati, I., & Oreopoulou, V., (2014). Recovery of carotenoids from tomato processing by-products: a review. *Food Research International, 65*, 311–321.

Su, F., Xu, H., Yang, N., Liu, W., & Liu, J., (2018). Hydrolytic efficiency and isomerization during de-esterification of natural astaxanthin esters by saponification and enzymolysis. *Electronic Journal of Biotechnology, 34*, 37–42.

Takaichi, S., (2011). Carotenoids in algae: Distributions, biosyntheses and functions. *Marine Drugs, 9*(6), 1101–1118.

Tiziani, S., Schwartz, S. J., & Vodovotz, Y., (2006). Profiling of carotenoids in tomato juice by one- and two-dimensional NMR. *Journal of Agricultural and Food Chemistry, 54*(16), 6094–6100.

Tran, D., Doan, N., Louime, C., Giordano, M., & Portilla, S., (2014). Growth, antioxidant capacity and total carotene of *Dunaliella salina* DCCBC15 in a low cost enriched natural seawater medium. *World Journal of Microbiology and Biotechnology, 30*(1), 317–322.

Wang, C., Zhao, S., Shao, X., Park, J. B., Jeong, S. H., Park, H. J., et al., (2019). Challenges and tackles in metabolic engineering for microbial production of carotenoids. *Microbial Cell Factories, 18*(1), 1–8.

Weber, D., & Grune, T., (2012). The contribution of β-carotene to vitamin A supply of humans. *Molecular Nutrition and Food Research, 56*(2), 251–258.

Wells, M. L., Potin, P., Craigie, J. S., Raven, J. A., Merchant, S. S., Helliwell, K. E., et al., (2017). Algae as nutritional and functional food sources: Revisiting our understanding. *Journal of Applied Phycology, 29*(2), 949–982.

Yabuzaki, J., (2017). *Carotenoids Database: Structures, Chemical Fingerprints and Distribution among Organisms*. Database.

Yamashita, E., (2013). Astaxanthin as a medical food. *Functional Foods in Health and Disease, 3*(7), 254–258.

Yang, Y., Kim, B., & Lee, J. Y., (2013). Astaxanthin structure, metabolism, and health benefits. *Journal of Human Nutrition and Food Science, 1*(1003), 1–1003.

Zhang, J., Sun, Z., Sun, P., Chen, T., & Chen, F., (2014). Microalgal carotenoids: Beneficial effects and potential in human health. *Food and Function, 5*(3), 413–425.

PART III

MACROALGAL METABOLITES: PHARMACEUTICAL, THERAPEUTIC, NUTRACEUTICAL, AND INDUSTRIAL POTENTIAL

CHAPTER 9

MACROALGAE: CURRENT APPLICATIONS IN DAILY LIFE AND PROSPECTS FOR THE FUTURE

ADRIANA LEANDRO,[1] DIANA PACHECO,[1] SARA GARCÍA-POZA,[1] JOÃO COTAS,[1] JOÃO CARLOS MARQUES,[1] LEONEL PEREIRA,[1] and ANA MARTA MENDES GONÇALVES[1,2]

[1]*University of Coimbra, MARE-Marine, and Environmental Sciences Center, Department of Life Sciences, Coimbra–3001-456, Portugal*

[2]*Department of Biology and CESAM, University of Aveiro, Aveiro–3810-193, Portugal*

ABSTRACT

During the last centuries, marine macroalgae have been traditionally used by populations from worldwide coastal areas, especially in Asian countries, as food, feed, medicinal, or components of compost and soil fertilizer in agricultural practices. Marine macroalgae play a crucial role in the aquatic trophic chain with ecological relevance in oxygen production, carbon dioxide (CO_2) sequestration, nutrient recycling, and are used as food source habitat for many other organisms, while enhancing water quality, providing several ecosystem services. They are capable of inhabiting in shallow waters at the edge of the world's oceans, under severe conditions. For this reason, they are able to produce bioactive substances (secondary metabolites) for defense against different stressors (e.g., predators, epiphytes, UV radiation, temperature, and salinity fluctuations). Thus, the interest in macroalgae and their by-products is increasing, as they provide valuable raw material

Algal Metabolites: Biotechnological, Commercial, and Industrial Applications. Jeyabalan Sangeetha, Devarajan Thangadurai, Saher Islam, & Ravichandra Hospet (Eds.)

and compounds that are already applied in many products and industries (e.g., pharmaceutical, biomedical, food, feed nutraceutical, cosmeceutical, and agriculture), being present in daily products. However, the scientific community is committed to tackle current issues with the development of new products based on marine macroalgae with industrial applications.

9.1 INTRODUCTION

Marine macroalgae or seaweeds refer to macroscopic, benthic, and autotrophic organisms ubiquitously distributed along with the coastal systems (Gibson et al., 1990; Stefels and Van Boekel, 1993). Seaweeds belong to Plantae Kingdom, being organized in three large groups, according to the thallus pigmentation: brown Ochrophyta—Phaeophyceae (brown seaweed), Rhodophyta (red seaweed) and Chlorophyta (green seaweed) (Chan et al., 2006).

Seaweeds are one of the primary producers with the highest biomass production in the Oceans, being in the basis of marine food webs (Häder and Figueroa, 1997; Christie et al., 2009; Harley et al., 2012). They are responsible for the O_2 produced in aquatic environments (Häder and Figueroa, 1997).

Seaweeds are a useful tool for environmental quality assessment since they are sensitive and influenced by external (biotic and abiotic) factors (Van Der Strate et al., 2002). They also perform bioremediation of contaminated waters, through bio-absorption and bioaccumulation of organic substances and metals (Castine et al., 2013; Wu et al., 2015; Yu et al., 2016; Neveux et al., 2018).

Marine macroalgae produce a range of structural molecules (primary metabolites), such as the proteins, lipids, and carbohydrates, and, in response to the exposure to certain external stimuli (protection against herbivory, UV radiation, for example) they can synthesize other interesting compounds (secondary metabolites). Polysaccharides are their main component, but seaweeds are also composed of other structural, storage, and bioactive compounds, such as peptides, omega 3 fatty acids, carotenoids, phenolics, vitamins, and minerals (Kumar et al., 2008; Hamed et al., 2015).

Indeed, seaweeds are originally an important source of food, supplements, medicines, and fertilizers in numerous coastal communities, not only in Pacific and Asian cultures, but also in some regions of Europe, including Ireland, Brittany, and Wales (Mouritsen et al., 2013). Seaweeds are an underestimated and adequate source of food and food ingredients. Their high content of bioactive compounds makes them a potential source of food preservatives and other useful molecules (Martelli et al., 2020). Their metabolites, and associated biological activities, are highly relevant for multiple nutraceutical, cosmetic, and pharmaceutical applications, which is why they are considered to have high nutritional value (Smit, 2004; MacArtain et al., 2008; Stengel et al., 2011). Seaweed-derived compounds have been studied and revealed extraordinary properties to be applied in various products of several industries (Figure 9.1), such as pharmaceutical, biomedical, food and nutraceutical, cosmetic and cosmeceutical, feed, and agriculture, among other even more emergent applications (Leandro et al., 2020). Although seaweeds are an excellent raw material for biorefineries and industries, their content of valuable compounds is dependent on species, season of harvest, and environmental seaweed growth conditions (Peng et al., 2015; Sharma et al., 2018; Zóllmann et al., 2019). Thus, it makes its research and characterization indispensable for the development of future commercially accepted products.

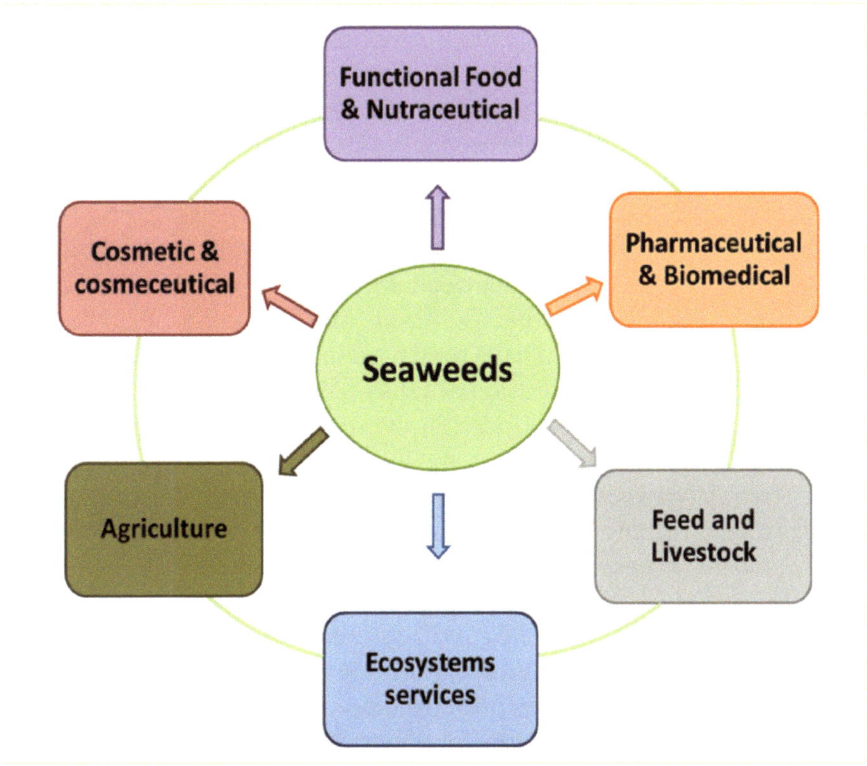

FIGURE 9.1 Main sectors of seaweed and seaweed-derived extracts applications.

This chapter provides an overview of how macroalgae and their derivative by-products are widespread introduced into the market and, consequently, in our daily lives.

9.2 FUNCTIONAL FOOD AND NUTRACEUTICAL

Seaweeds have been part of traditional diets and consumed by coastal communities worldwide since the beginning of human civilization (Dillehay et al., 2008) and have been employed by humans as medicine and food for at least 13,000 years (Teas et al., 2004).

Since ancient times, seaweeds have been employed as food in Asian countries like Japan, Korea, and China (Darcy-Vrillon, 1993; McDowell, 2003). Nowadays, Western countries have noticed the benefits of seaweeds and so they are also being consumed in Europe, North America, and South America (McHugh et al., 2003; Mouritsen et al., 2013). In the last years, the unfavorable connotation of seaweeds has been disappearing and people have accepted seaweeds not only as a healthy but also tasty food (Holdt and Kraan, 2011; Edwards et al., 2012; Mouritsen et al., 2013). The growing knowledge of their nutritional and gastronomic values has generated a permanent increase in their culinary uses and consumption. Because fresh seaweed is a highly perishable product, it can also be found as dehydrated, salted, freeze-dried, frozen, and in canned products, as well as in products such as cookies and pates (Herrero et al., 2010).

In Japan, 21 species of seaweed are regularly involved in the diet and in Korea more than 40 kinds of seaweed are usually applied as food (Arasaki and Arasaki, 1983). In the Pacific basin, in Hawaii and other Polynesian islands, 29 types of seaweed have been used in food, medicine, and as part of religious celebrations in precolonial times (Abbott, 1978; Schönfeld-Leber, 1979). Seaweeds are still included in the diets of a large number of indigenous people living there (Teas et al., 2004).

In general, seaweed used for human consumption belongs to the genera *Porphyra* spp. (commercial denomination – Nori), *Palmaria palmata* (Dulse), and *Chondrus crispus* (Irish moss) (Rhodophyta), *Laminaria* (Kombu), *Undaria pinnatifida* (Wakame) (Phaeophyceae) (McHugh et al., 2003). In addition, other genera such as *Ulva* (Chlorophyta) are making strong inroads into European markets. The growing offer in products generates the need to provide detailed information to the consumer, not only about the commercial name, but also about the scientific name of each species included in the products marketed (European Parliament, 2000).

Obesity and malnutrition are the main problems associated with the daily diet; urbanization, economic development and changes in lifestyle strongly affect food habits. Numerous reports confirm a direct relationship between diet and health, and people believe that food has more impact on health than exercise and heredity (Cockbill, 1994; Espín et al., 2007).

Marine algae species possess all the nutrients such as minerals, vitamins, trace elements, lipids, especially the long-chain polyunsaturated fatty acids (PUFAs), carbohydrates, and even proteins. In addition, in terms of carbohydrates, seaweeds are a particularly rich source of dietary fiber, mainly due to indigestible sulfated polysaccharides (Jimenez-Escrig and Sanchez-Muniz, 2000; Norziah and Ching, 2000; Wong and Cheung, 2001; Gómez-Ordóñez et al., 2010; Pereira, 2011). Many people use seaweed products without even realizing it, as they are part of the daily-use products like processed food items (e.g., creams, canned or meat products) (Ito and Hori, 1989; Mabeau and Fleurence, 1993; Kraan, 2012).

Nutraceutical is defined as any nutritional food product with increased health benefits; in other words, it is a combination of nutrients and pharmaceuticals (Tanna and Mishra, 2018). These can be essential and non-essential natural compounds with proven health benefits, and they can be taken either as part of a food product or as a supplement (Biesalski et al., 2009). Seaweeds have recently emerged as a wide source of metabolites with unique structures and nutritional and therapeutic activities, and this field is becoming one of the most relevant challenges in food science and technology (Plaza et al., 2008–2010). Overall, seaweeds are considered a promising functional food. Nevertheless, much research, on their role in nutrition and disease prevention, remains to be done before science-based dietary recommendations can be given for edible seaweeds (Smit, 2004). In this sense, systematic research on some nutritional and health-promoting attributes of specific seaweeds normally consumed in several European (Denis et al., 2010) countries and worldwide (Gressler et al., 2010) has been in the recent past recognized as a prelude to future rational economic exploitation (Jiménez-Escrig et al., 2012).

Red seaweeds (Rhodophyta), such as *Palmaria palmata*, *Porphyra* spp., *Gelidiella acerosa*, *Gracilaria cornea*, are widely used in traditional dishes and food products and revealed nutraceutical effects. For example, *P. palmata* protein hydrolysate has inhibitory effects on renin and is also employed to formulate bread (Fitzgerald et al., 2014). Similarly, antioxidant, cardioprotective, and antidiabetic activities have been reported for protein hydrolysates of the red seaweed *P. palmata* (Harnedy and Fitzgerald, 2011). The British

Nutrition Foundation recommends 8.7 mg iron/day for adult males (British Nutrition Foundation, 2016) and the seaweed *P. palmata* contains on average 6.4 mg of iron per 8 g serving, in comparison with only 1.2–3.1 mg in a 100 g portion of lean beef (Branscheid and Judas, 2011). Vitamins A, C, and E are found in high amounts in the red seaweed *Gelidiella acerosa* (Syad et al., 2013). Other red seaweeds (Rhodophyta), such as *Solieria filiformis* and *Gracilaria cornea*, have the ability to produce pigments, polysaccharides, and other nutraceuticals with antioxidant activity (Tanna and Mishra, 2018).

Previous comparative studies of fatty acid (FA) profiles in species from different phyla reported that red macroalgae species have high levels of eicosapentaenoic acid (EPA, C20:5(n-3)), arachidonic acid (ARA, C20:4(n-6)), oleic acid (C18:1(n-9)), palmitic acid (C16:0), and stearic acid (C18:0), which possess healthy properties (Galloway et al., 2012; Kelly and Scheibling, 2012).

"Hana Tsunomata" is one of the many commercial items available that contains seaweed as its main ingredient. "Hana Tsunomata" is gluten-free, low in calories, and an excellent source of iron, magnesium, and fiber; it is used in vegetarian and vegan dishes and is produced from different strains of red seaweed *Chondrus crispus* Stackhouse (Tanna and Mishra, 2018).

In addition, supplementation with the red seaweed *C. crispus* and the brown seaweed *Undaria pinnatifida* and *Saccharina japonica* reported to have prebiotic effects in rats and mice, respectively. Also, the results included increased beneficial gut bacteria and reduced pathogenic bacteria (Liu et al., 2015; Kim et al., 2016). An extract of sulfated polysaccharide from red algae *Palisada perforata* (as *Laurencia papillosa*) (composed of kappa and lambda carrageenans) has been indicated as a possible inexpensive dietary supplement for daily food intake. The extract showed dose-dependent antiproliferative activity towards the MCF-7 human breast cancer (Ghannam et al., 2018). Carrageenans from red seaweed *Stenogramma interruptum* have anticoagulant (Wijesekara et al., 2011) and antihyperlipidemic properties (Panlasigui et al., 2003). In the 57th meeting of the JECFA—Joint Food and Agriculture Organization (FAO) of the United Nations/World Health Organization (WHO) Expert Committee on Food Additives the safety of carrageenans in food was validated (FAO, 2006). Nevertheless, JECFA declared that for use in infant formulas carrageenans are not recommended (FAO/WHO, 2007).

Brown seaweeds are completely integrated in the ordinary diet in East Asia, with particular consumption in Japan, China, and the Republic of Korea. The main compound present in brown seaweeds is fucoidan, a sulfated polysaccharide with a pivotal role in nutritional and health human values. For example, fucoidan extracted from *U. pinnatifida* has been discovered really powerful to be integrated in food and used its biological activity to reduce diseases or used as therapy (Ruocco et al., 2016). Researchers have shown the use of fucoidan as base therapy to cure arthritis, blood homeostasis, influenza, and malaria. Fucoidans also exhibit anti-inflammatory activity (Phull and Kim, 2017; Tanna and Mishra, 2019). Another experiment conducted by Tanna and Mishra (2019) extracted fucoidan from *S. japonica*. Its molecular weight is very low, allowing the use of fucoidan from *S. japonica* in food supplements. During the study, no toxicity or mutagens were observed in 5,000 μg/mL of brown seaweed (Hwang et al., 2016; Tanna and Mishra, 2019).

Fucus vesiculosus (Phaeophyceae) is among the edible seaweeds with the highest amount of dietary fiber and antioxidants (Jiménez-Escrig et al., 2001; Rupérez and Toledano, 2003). Moreover, monitoring the consumption of *F. vesiculosus* on 122 people with osteoarthritis (Fitton et al., 2016), it was noticed a reduction of the disease, showing no toxic compounds in the blood samples. The experiment was carried out for 12 weeks using 85% fucoidan extract, the equivalent of 300 mg (Fitton et al., 2016; Tanna and Mishra, 2018). The compounds extracted from *F. vesiculosus* are various, recognizable in protein, minerals, iodine, vitamins, and monounsaturated and PUFAs (Herbreteau et al., 1997; Rupérez and Saura-Calixto, 2001; Rupérez et al., 2002). Therefore, the positive health effects are given by dietary fibers (non-digestible polysaccharides) and polyphenols. These fibers have proven to have a powerful consequence on blood pressure and cholesterol (Jimenez-Escrig and Sanchez-Muniz, 2000) and, thanks to the several advantages that brown seaweed extracts reflect on human health, a lot of ingredients for functional foods and dietary supplements were used, such as fucoidan powders *F. vesiculosus* capsules or antioxidant extracts (Díaz-Rubio et al., 2009). Additionally, polyphenols of seaweeds can inhibit hepatic gluconeogenesis and reduce the activity of digestive enzymes such as α-amylase, α-glucosidase, lipase, and aldose reductase (Lee et al., 2012; Lopes et al., 2017).

InSea2™ released in the market *A. nodosum* and *F. vesiculosus* phlorotannins; consumption of these algae occurs in a positive way, by the promotion of 90% reduction of the postprandial blood glucose and 40% reduction of insulin secretion (Roy et al., 2011). The advantage of using brown seaweed extracts could be the avoidance of resorting to the existing anti-diabetic drugs, which sometimes may have strong side-effects (Lopes et al., 2017). The effect of the commercial algae extracts available from InSea2™ were studied on 12 women and 11 men (Paradis et al., 2011). *F. vesiculosus* and *Ascophyllum nodosum* extracts were used and it was discovered that these algae contained natural α-amylase and α-glucosidase inhibitors. The anti-diabetic properties were proven through the inhibition of digestive enzymes that convert polysaccharides into simple sugars in the intestine and increase blood glucose levels (Roy et al., 2011).

Tanemura and colleagues (2014) examined the postprandial blood glucose profiles of 12 healthy adults after a meal with and without the inclusion of fresh, cooked, whole *U. pinnatifida* (Wakame), or its sporophylls (Mekabu) (cooked and fresh). Results showed that glucose levels were extremely lower in people who consumed the meal with "Mekabu." The presence of fucoxanthin fraction and the polysaccharide content of the cooked "Mekabu" sporophylls made possible the anti-glycemic effect (Tanemura et al., 2014). It was concluded that the inclusion of fresh brown seaweeds in meals could be powerful even in monitoring blood glucose levels in people affected by type 2 diabetes.

Phlorotannins (oligomeric polyphenols) are frequent in brown algae families. Other polyphenol extracted from brown algae is eckol. It is isolated from *Ecklonia* sp. of Laminariaceae family, and it has shown scavenging activity on lung fibroblast cells (Kang et al., 2005). The commercial activity of SEANOL®, registered trade name of Botamedi Inc (new corporate name of "Livechem Inc" since 2012) and Simply Healthy, represent the sales of dietary ingredients based of phlorotannins extracted from *Ecklonia cava* brown algae (Douglas et al., 2016; "Seanol®" 2020). Botamedi Inc and Simply Healthy are the only ones with authorization that allows the extraction of compounds for diet from *E. cava*. SEANOL®-brand includes: SEANOL®-P (a.k.a. SEAPOLYNOL®) and SEANOL®-F, constituted by a particular class of health-supporting phlorotannins called "eckols." SEAPOLYNOL® (a.k.a. SEANOL®-P) was acknowledged in 2008 by US FDA as a new fundamental ingredient for functional food. In 2018, European Food Safety Authority (EFSA) reconfirmed its good properties and EU authorized it as Novel Food Ingredient (NFI) (Turck et al., 2017; "Seanol®" 2020).

The company Diana Food Inc. not only offers to customers a healthy solution for the food and beverage industry, but it is also specialized in products based on macroalgae. The production of food is carefully considered by experts since the harvest of fruits, vegetables, and the treatment of meats and seafood. Counting on strong experience and knowledge, the quality of the food is guaranteed. HealSea™ is based on phytotannins extracted from *F. vesiculosus* and it has been on the market since it was discovered very helpful for cardiovascular health (Freile-Pelegrín and Robledo, 2013).

Macroalgae are considered as an important element for diets and health, because they are also a good source of phytosterols. It has been discovered the potential of phytosterols in reducing blood cholesterol and to block the development of colon cancer (Lopes et al., 2013). Among the health properties exerted by brown seaweeds, anti-obesity, and antidiabetic activities (Miyashita et al., 2013) must be taken into account. Throughout the presence of PUFAs, seaweeds may be used to produce food with low-fat content. The most consumed food products are the ones that include ω-3 LC-PUFAs and replacement elements could be high-value lipids widely present in seaweeds and considered as a valid substitute of PUFAs (Mišurcová et al., 2011).

Kim et al. (2010) conducted research using *Sargassum thunbergii* (Mertens ex Roth) Kuntze. The study confirmed the use of PUFAs from this alga as a requirement for functional food for the prevention of ROS-mediated oxidative damage in the human system (Kim et al., 2010). Moreover, the genus *Cystoseira* also showed nutraceutical value specified by PUFA composition (Vizetto-Duarte et al., 2015). The extraction of these compounds is treated by the nutraceutical industry (Vizetto-Duarte et al., 2015).

A key molecule present in brown seaweeds is fucoxanthin, a lipid that exerts anti-obesity properties (Miyashita et al., 2013; Gammone and D'Orazio, 2015). Fucoxanthin extracted from *Gongolaria barbata* (as *Cystoseira barbata*) is even used as an integrated ingredient of turkey meat sausages, improving the nutritional value of the meat (Sellimi et al., 2017; Shannon and Abu-Ghannam, 2019).

Shannon and Abu-Ghannam (2017) found anti-obesity properties due to fucoxanthin of brown algae. A study conducted on 50 women and men demonstrated that, by taking fucoxanthin capsules at a dosage of 1 mg per day for four weeks, the total fat mass, subcutaneous fat area, waist circumference, and right thigh circumference significantly decreased. Also, for the group submitted to the consumption of 3 mg per day, significant decreases in visceral fat, basal metabolic rate, abdominal fat and circumferences of the neck, arm, and thigh were verified (Hitoe and Shimoda, 2017).

It is possible to confirm that a seaweed diet implemented with a healthy lifestyle could be a good obesity treatment (Shannon and Abu-Ghannam, 2019). The brown algae *F. vesiculosus* and *A. nodosum* are famous for their property of reduction of blood glucose level. Tests were conducted in animals (Apostolidis and Lee, 2011), but it is clear that fucoxanthin has a good effect on insulin sensitiveness and reduction of glucose levels in the blood. The study was also conducted in diabetic mice. These results are plausible due to the production of cells that limit the secretion of glucagon (Maeda, 2015).

Among the other elements extracted from seaweeds, iron, copper, and magnesium (as in the case of *Laminaria* and *Undaria*) are taken into account

(Rajapakse and Kim, 2011). The amount of iodine in *L. digitata* showed an important biological activity for the treatment of myxoedema and goiter, while *U. pinnatifida* is good to prevent tumors (Holdt and Kraan, 2011). Generally, iodine content in brown seaweeds could reach up to 1.2% of DW (Holdt and Kraan, 2011).

Heart diseases and obesity are usually consequences of regular consumption of a huge amount of saturated fat, salt, and artificial additives. Through elimination or substitution of these compounds with seaweed extracts, the possibility to reduce diseases increases. For this reason, seaweed components have been used as healthy and functional ingredients in meat (Shannon and Abu-Ghannam, 2019). The advantage of substituting seaweed compounds in meat products is the incidence of lower levels of fat and salt, less not-natural additives and the increase of fibers and polyunsaturated fat amount (Cofrades et al., 2017). Additionally, polyphenolic compounds, selenium, and vitamins A, C, and E in seaweed play the role of natural antioxidants and preservatives inside the meat matrix. As an example, introducing *Undaria pinnatifida*, *Porphyra umbilicalis*, and *Himanthalia elongata* into pork frankfurters, beef burgers, and restructured poultry steaks, the thrombogenic index and the fatty acids have a positive decrease (López-López et al., 2009).

The enrichment of seaweed products within meats was tested on animals and they showed an improved lipoprotein metabolism. It means that these elements may be developed as an intervention for hyper-cholesterolaemic in human diet (Olivero-David et al., 2011). Another experiment was carried out by Astorga-España et al. (2017). The authors incorporated dried seaweeds in meat-based foods, and the results were positive; the PUFAs content decreased in a relevant way. The algae used for this experiment were *Durvillaea antarctica* (present at 3% in fettuccine; 5% in bread; and 5% in hamburger), *Macrocystis pyrifera* (Phaeophyceae), added at 8% in breadsticks and 7% in fritters, while 28% *Pyropia columbina* (Rhodophyta) was used in parsley pesto (Astorga-España et al., 2015).

Nevertheless, high polyphenolic content of seaweeds can reduce the digestibility of algal proteins (Wong and Cheung, 2001). Algae still represent a good alternative to proteins derived from animals (Shannon and Abu-Ghannam, 2019). Just to mention different cases, *U. pinnatifida* has an amino acid score of 1.0, like the amount within egg and soy; *Pyropia*/*Porphyra* spp. 0.91, and *Saccharina latissima* (as *Laminaria saccharina*) 0.82 (Murata and Nakazoe, 2001).

One of the major risks for health is the obesity. Obesity combined with lack of a proper quantity of minerals and vitamins can lead to the development of other diseases such as type 2 diabetes, hypertension, dyslipidemia, and coronary heart disease (Medina-Remón et al., 2016). The Organization for Economic Cooperation and Development reported that, in 2015, 38.2% of the adult population in the United States of America (USA) was obese, opposed to only 3.7% in Japan, and 5.3% in Korea (OECD, 2017). This data suggests that the consumption of seaweed fibers within the food in Asia also improves life quality. Fucoxanthin of brown seaweeds has anti-obesity properties (Shannon and Abu-Ghannam, 2017), while carotenoids and alginate have been discovered to have positive effects on appetite, satiety, blood glucose and cholesterol (Kim et al., 2008; Brown et al., 2014).

U. pinnatifida powder was used on 36 Japanese patients with hypertension, consuming 3.3 g/day. After 4 weeks of treatment, the diastolic pressure fell by 9 mmHg and by 8 mmHg after 8 weeks. Moreover, after the first control (4 weeks) hypercholesterolemia decreased as well by 8% (Hata et al., 2001). A similar study was conducted by Teas et al. (2009).

Moreover, *Saccharina japonica* (as *Laminaria japonica*) was widely studied for its antioxidant properties, already used for centuries in Korea. Consuming fermented *S. japonica* for a month significantly increased levels of catalase (CAT) and superoxide dismutase (Kang et al., 2012).

Daily portion of *U. pinnatifida* and *Sacchoriza polyschides* (as *Gigantea bulbosa*) when mixed with food was able to balance blood glucose levels and to decrease serum triglyceride concentrations; additionally, it was found a major occurrence of high-density lipoprotein (HDL) cholesterol in subjects with type 2 diabetes (Kim et al., 2008).

As conclusion, it is possible to confirm that brown seaweeds contain a huge amount of aspartic acid, responsible for a particular taste, and glutamic acid, a key element for a taste of umami. However, the most original source of flavor-enhancing is considered to be *S. japonica*.

Many studies have reported pharmaceutical activities of polysaccharides obtained from several seaweeds (Tanna and Mishra, 2019). Sulfated acidic polysaccharides from green seaweeds are ulvans and have gelling and rheological properties almost the same as gum Arabic due to solubility in water. Ulvans were created as vegan alternatives to beef-derived gelatin due to consumer concerns about bovine BSE contamination (Kraan, 2012). In addition, ulvan from *Ulva australis* (as *Ulva pertusa*) exhibited to be more active against superoxide radicals than vitamin C, although its decreasing capacity was lower than that of vitamin C (Xing et al., 2005).

Crude polysaccharide from the green alga *Ulva lactuca* exhibited hepatoprotective effects in rats with galactosamine-induced hepatitis through its antihyperlipidemic and antiperoxidative actions (Sathivel et al., 2008). Sulfated polysaccharide obtained from *U. lactuca* showed antiviral activity against Japanese encephalitis virus (JEV) in infected C3H/HeN mice (Chiu et al., 2012) and has proven to cause improvements in lipid profiles and to exhibit antioxidant activity in hypercholesterolemic rats (Hassan et al., 2011).

Sulfated polysaccharide from the green alga *Ulva rigida* (as *Ulva armoricana*) has antimicrobial and immunomodulatory activities and exhibits the ability to stimulate cytokine expression in intestinal epithelial cells. These properties may be employed to improve the resistance of farm animals to various types of infections (Berri et al., 2016). Overall, these studies propose that supplementation of seaweeds and their polysaccharides as part of regular food consumption will improve human health and reduce several symptoms of disease conditions (Tanna and Mishra, 2019).

Peptides can link to the active site of the angiotensin-I converting enzyme (ACE), reducing blood pressure (Shannon and Abu-Ghannam, 2019). For instance, peptides obtained from the red alga *Gracilariopsis lemaneiformis* have potent ACE – inhibitory activity *in vitro* (Cao et al., 2017). Sulfated polysaccharide compounds obtained from *Ulva prolifera* (as *Enteromorpha prolifera*) lessen hyperlipidemia in animal models – one of the main causes of heart disease (Ren et al., 2018).

To decrease any possible harmful effects, seaweeds are consumed with fast food or other food in some countries. The quality and texture of noodles has been researched employing different quantities of seaweed supplements, and it was discovered that 6% supplementation with a polysaccharide from the green seaweed *Monostroma nitidum* generated higher cooking yield, better water absorption of noodles, and richness of dietary fiber (Chang et al., 2011). The tropical green seaweed *Caulerpa racemosa* contains a high amount of many beneficial nutrients (Tanna and Mishra, 2018). Semisweet biscuits were prepared employing the natural antioxidant polysaccharides of *C. racemosa* in an innovative approach (Kumar et al., 2018).

9.3 PHARMACEUTICAL AND BIOMEDICAL

Seaweeds have been used for centuries as a natural drug for diverse health diseases in traditional herbal medicine (Chengkui and Junfu, 1984; Jarald et al., 2008). Asian, Celtic, British, and Roman people, namely coastal zones inhabitants, took advantage of these natural resources, for 1,000 years (Arnason et al., 1981; Smit, 2004; Freile-Pelegrín and Tasdemir, 2019). According to ancient literature, some seaweeds were used for healing wounds, as vermifuges or anthelmintic, and even for curing goiter, cough, and asthma, among others (Chengkui and Junfu, 1984; Freile-Pelegrín and Tasdemir, 2019). In China, this folk medicine using seaweeds remains in vogue and it is possible to purchase in local herbal drug stores, for instance, 'Haicao' and 'Kunpu,' respectively, *Sargassum* spp., especially *S. fusiforme*, and *Ulva lactuca* (Chengkui and Junfu, 1984).

Since elderly times the empiric knowledge about seaweed health benefits instigates continuous research in order to discover new drugs from species of the three groups of macroalgae. The ultimate goal in this field is to have medicines as much natural as it is possible, that are effective against the problems/diseases and do not cause (or minimum possible) secondary effects or other damages.

Several studies have demonstrated the value of macroalgae-derived compounds for pharmaceutical and biomedical applications, in the treatment of various diseases like cancer, diabetes, and cardiovascular diseases (CVDs) (Pangestuti and Kim, 2011b; Lopes et al., 2013; Alves et al., 2018), taking advantage of different biological activities, including antioxidant, antimicrobial, anticancer, anti-diabetic, and anti-thrombotic properties (Zhang et al., 2007; Rengasamy et al., 2014; Okimura et al., 2019).

Still, there are a few factors that can lead a seaweed-derived compound to act differently (sometimes oppositely), depending on the species of origin, and the methodology of extraction used (Shanab, 2007; Oliveira et al., 2017). The external conditions during the seaweed growth also interfere in its molecular constitution and, consequently, on its bioactivities and effectiveness (Ravikumar et al., 2002; Oliveira et al., 2017). Thus, these are the main issues that difficult the implementation of these compounds in drugs or biomedical commercialized applications, once it must ensure simultaneously the efficacy and security of the molecules/extracts.

Seaweeds of the red group (Rhodophyta) have been applied in diverse medical and therapeutical practices since ancient times. Registers from the 17th Century revealed folk medicine applications of an unspecified mixture of co-occurring red seaweeds, in Ireland, *Chondrus crispus* and *Mastocarpus stellatus*. Currently, these two similar species are still used to make traditional infusions and cough medicines to prevent or treat colds, bronchitis, and chronic coughs (Pereira, 2018b).

Nowadays, these, and other red seaweed species are being characterized and their components are still being further studied and revealed several medicinal proprieties, such as anticoagulant, anti-inflammatory, immunomodulatory, and antitumoral activities (Cotas et al., 2020). Polysaccharides like agar, carrageenan, and porphyran are the most studied biomolecules with pharmaceutical interest.

Studies revealed that agar obtained from cold-water extraction of *Gracilaria* spp. has antitumoral activity. Moreover, hydrolysates of agar (agar-oligosaccharides) present activity against glycosidase, also being antioxidant (Fernández et al., 1989).

Carrageenan could have an important role in the pharmaceutical sector, in drug delivery systems (Li et al., 2014). Several studies have demonstrated the anticoagulant and antithrombotic properties of carrageenan, with λ-carrageenan showing higher anticoagulant potential than κ-carrageenan (Shanmugam and Mody, 2000).

Carrageenan also showed antioxidant properties (De Souza et al., 2007), antitumoral potential (Cotas et al., 2020), and antiviral activity (Wang, 2012). Indeed, the iota-carrageenan, due to its proven antiviral properties and safety, is already applied in a nasal spray, Bisolviral®, that is commercialized by the pharmaceutical company Boehringer Ingelheim (Ron Eccles et al., 2010; Koenighofer et al., 2014; Eccles et al., 2015).

Besides that, red seaweeds generally have high protein content, up to 45% of dry weight, and their bioactive peptides have diverse actions, such as immunomodulatory, antibacterial, antithrombotic, and antihypertensive effects (Cian et al., 2012).

These seaweeds also synthesize photo-protective molecules against UV radiation, such as the pigmented compounds (e.g., chlorophylls, phycobiliproteins (PBP), and carotenoids) and mycosporine-like amino acids. These molecules have an antioxidant bioactivity that could result in cancer-preventive compounds (Nunes et al., 2018; Lalegerie et al., 2019).

Red seaweeds also synthesize phenolic compounds (PCs) that exhibit free-radical scavenging properties that could act as suppressors in some pathologies (Namvar et al., 2012; Torres et al., 2019).

Brown species, such as *Sargassum* spp., have also been applied in traditional Chinese medicine, for more than 2,000 years, to treat several diseases (e.g., goiter) (Liu et al., 2012). Seaweeds are also producers of compounds that have been researched and are in a good position to be used in these sectors.

Currently, alginate, the main brown seaweed polysaccharide, is widely applied in a variety of pharmaceutical and biomedical products (Lee and Mooney, 2012). As a biocompatible and non-harmful (non-toxic) polymer, alginate is used as an excipient and rheological agent (e.g., emulsifying or thickener agent) in drugs or in drug-delivery systems, and it also can be used to develop hydrogels for regenerative medicine (Aderibigbe and Buyana,

2018; Kaczmarek-Pawelska, 2020). Thus, alginate is useful as biomaterial/biopolymer since it has the ability to be turned into a hydrogel, binding large amounts of water, drugs, and bioactive substances, which provides two- or three-dimensional support to the cells, in tissue bioengineering (Pielesz et al., 2011; Kaczmarek-Pawelska, 2020). Alginate is not degraded in mammals, but the ionically cross-linked alginate hydrogels can be dissolved by the release of divalent ions (e.g., calcium, sodium, or potassium) (Zhao et al., 2010). However, there are new techniques to dissolve the alginate hydrogels in aqueous media and to be removed from the human body, through the oxidation of alginate. These hydrogels have potential as a vehicle for drug delivery for various compounds and applications (Al-Shamkhani and Duncan, 1995; Bouhadir et al., 2001).

Alginate gel is significant for cell cultures, in biomedical research and development studies. It is an excellent cell culture substrate, as it can provide two- and three-dimensional support to the cells/tissue (Lee and Mooney, 2012).

Then, as alginate is biocompatible, it could form biomaterials/bioartificial implants through microencapsulation of living cells, tissues, that could easily be transferred from *in vitro* dishes to *in vivo*, as it could be introduced into the body (Zimmermann et al., 2007). Therefore, alginate nanocomposites are used in bone restructuration (Christenson et al., 2007), and for cardiac regeneration (Liberski et al., 2016).

Alginate is a natural chelating agent against heavy metals poisoning, and it is demonstrated to be an operational coadjuvant in food poisoning medical cases (Eliaz et al., 2007). Furthermore, sodium alginate is the active ingredient in a commercial formula of Reckitt Benckiser (an Irish Company), the Gaviscon® Advance, used for the treatment of gastro-oesophageal, creating a mechanical anti-reflux barrier within the fundus of the stomach (Washington, 1990). This brown polysaccharide is also biomaterial used for dental impressions (Nallamuthu et al., 2006, 2012; Sinurat et al., 2020).

Another polysaccharide extracted from brown seaweeds species, such as *Saccharina japonica*, *U. pinnatifida* and *F. vesiculosus*, is fucoidan, which contains considerable percentages of L-fucose and sulfate ester groups (Yang et al., 2013; Luthuli et al., 2019). Several studies reported the biological activities of this polymer, including antibacterial, antioxidant, antiviral, and antitumor (Magalhaes et al., 2011; Zhang et al., 2015; Oliveira et al., 2017). Similarly, to alginate, fucoidan is a biopolymer also applied in wound dressings (Aderibigbe and Buyana, 2018). Also, fucoidan extracted from the red alga *Cladosiphon okamuranus* showed a significant inhibition of the dengue virus (DENV-2) in baby hamster kidney (BHK-21) cells in a dose-dependent manner (Hidari et al., 2008).

Besides that, fucoidan showed synergistic effect with current anticancer agents (Zhang et al., 2013; Atashrazm et al., 2015). However, despite the promising medical actions, fucoidan has not yet been developed as a regulated therapeutic agent, because of the effect's variability, depending, for example, on the sulfation degree of the molecule (Oliveira et al., 2017).

Another example of the huge importance of poly- and oligosaccharides of the seaweeds for pharmaceutical and biomedical sectors is the sodium oligomannate, which is a mixture of oligosaccharides with the degree of polymerization ranging from 2 to 10, that was approved in China, for phase 2 clinical trials and that constitute a hope for the neurodegenerative Alzheimer's disease (AD) (Wang et al., 2019; Ferrari and Sorbi, 2020). Sodium oligomannate remodels gut microbiota and suppresses gut bacterial amino acids that are contributing to favor the inflammation in the neurons, and, in that way, it ends up inhibiting AD progression (Wang et al., 2019).

This specific extract of brown seaweeds is being developed by the drug's manufacturer Shanghai Green, promoting the cognitive function in people with mild-to-moderate AD (Wang et al., 2019; Ferrari and Sorbi, 2020).

PCs of the brown seaweeds, such as the phlorotannins, also have demonstrated outstanding antioxidant action and other therapeutical properties. In fact, their bioactivities have been exploited in different areas, from cardiovascular to neuroprotective (Freile-Pelegrín and Robledo, 2013). They have appliance in CVDs, since they contribute to the arteriosclerosis prevention and to the increase of protective high-density lipoprotein cholesterol (HDL-C). Furthermore, phlorotannins have neuroprotective action and maybe the key to the treatment of neurodegenerative diseases (Pangestuti and Kim, 2011b).

By the way, *Eisenia bicyclis* (Arame) extracts inhibited β-amyloid cleavage enzyme activity (Jung et al., 2010). Similar effects were verified when using phlorotannins of *Ishige foliacea*, *Ecklonia maxima* and *E. cava*, and giving that these tannins could be applied in some treatment for AD (Pangestuti and Kim, 2011; Um et al., 2018; Lee et al., 2018; Wang et al., 2018). A phloroglucinol polymer 6,6′-bieckol obtained from *E. cava* is an active ingredient in a formulation that inhibits HIV-1 (Kim et al., 2013).

A light-harvesting pigment, fucoxanthin, a xanthophyll found in brown macroalgae (e.g., *Fucus* spp.) revealed anti-diabetic, anti-obesity, and antitumoral properties (Maeda et al., 2005; Rokkaku et al., 2013; Maeda, 2015; Martin, 2015; Satomi, 2017). This molecule is under investigation in the phase II clinical trial to be a drug commercially available for metabolic disorders (Ortiz, 2020).

Equally to what happens in the other groups of seaweeds, some seaweed compounds have valuable and promising bioactivities to be applied in the pharmaceutical industry, from antioxidant and anti-inflammatory to cardiovascular prevention and antitumoral properties (Pengzhan, 2003; Moo-Puc et al., 2011; Toskas et al., 2011; Yu-Qing et al., 2016).

An example of a compound with great pharmacological potential is a specific xanthophyll of the genus *Caleurpa* (Chlorophyta), named caleurpin, that has an antiviral effect against HSV-1, stopping the viral replication; however, pre-clinical studies are still needed (Macedo et al., 2012).

Another investigation indicated the crude extract of *Cladophora patentiramea* as vasorelaxant effect (Lim and Mok, 2010). This could be a possible new therapy for cardiovascular pathologies, ameliorating the endothelium function, in order to improve the health status and increase the lifespan of the patients (Sena et al., 2018).

Green seaweed compounds also demonstrate having antitumoral activity (Moo-Puc et al., 2011; Abirami and Kowsalya, 2012). For instance, *Udotea flabellum* extracts showed inhibitory effect of cancer cell lines (Moo-Puc et al., 2011). In another study, methanolic and aqueous extracts of *Ulva lactuca* (as *Ulva fasciata*) were tested in mice with lymphoma, and the treatments promoted a reduction of cancer cells (Abirami and Kowsalya, 2012).

Despite the several studies that demonstrate the potentialities, further *in vivo* and clinical studies are still needed to guarantee the reproducibility of the effect and the safety of the substance before it can be in the market (Wang et al., 2015).

9.4 COSMETIC AND COSMECEUTICAL

The regulation EC No 1223/2009 defines that a cosmetic product is a substance to be used topically or in the oral cavity (European Parliament, 2009). From a global perspective, the cosmetic market was evaluated in 425 billion € in 2016 (Couteau, 2016), assuming its relevance in the European market with a value of 77 billion € ("Cosmetics Europe," 2020). Nevertheless, the cosmeceutical concept is attracting industries' interest. This concept arises from the alliance between pharmaceutical and cosmetic properties (Vermeer, 1996; Kim, 2014; Zappelli et al., 2016). The consumer is increasingly concerned about the natural source and sustainability of cosmetic products (Couteau, 2016; Zappelli et al., 2016). In this context, there are several studies that highlight the bioactivities of seaweeds' secondary and primary metabolites as health promotor agents with application in pharmaceutical and food industries (Holdt and Kraan, 2011; Pangestuti and Kim, 2011a; Bhatia et al., 2015).

Thus, there are already in the market cosmetic products with seaweeds' compounds in their constitution. For instance, Chlorofiltrat® Ulva HG is a skincare product richly based on a hydroglycolic extract from *Ulva lactuca* (Chlorophyta) ("Chlorofiltrat® Ulva," 2020). Moreover, "Sealgae" is a line of thalassotherapy products based on green and brown seaweed extracts with moisturizing properties (Pereira, 2018a; Sealgae, 2020).

Gelcarin® PC 379 and Gelcarin® PC 812 are available commercial products based on the red seaweed *Chondrus crispus* extract rich in sulfated polysaccharides with gelling, thickener, and stabilizer properties used in shampoos, foaming products or exfoliants and lipstick and deodorants ("Gelcarin PC 379" n.d.). Carrageenan from the same mentioned seaweed is therefore also a base for Gelalg®, a product with skin moisturizing and lifting effect ("GELALG™," 2020). Another red macroalga's, *Gracilaria conferta*, powder is the resource to the formulation of *Gracilaria* hydrogel, providing anti-ageing, anti-wrinkle, anti-cellulite, firming, and brightening benefits through its vitamin, mineral, and amino acid content ("Gracilaria Hydrogel," 2020).

Seaweeds possess antioxidant and photoprotective compounds in order to prevent DNA damage (Mercurio et al., 2015). By this, *Neopyropia tenera* (formerly known as *Porphyra tenera*) and *Corallina officinalis* are suppliers for sunscreen products as commercially available products, namely At UV Protector P and Phycocorail® ("At UV PROTECTOR P" 2020; Yasham, 2020). Besides all the commercially available products based on red seaweed, hydrolyzed Rhodophyta extract rich in sulfated polysaccharides is applied in shampoos formulations, aiming for hair protection, and repairing through their content in Calcium (Ca), Zinc (Zn) and Manganese (Mn) (SEPPIC, 2020). Brown seaweeds, such as *U. pinnatifida* and *S. fusiforme* (formerly known as *Hizikia fusiformis*) (Phaeophyceae), are also a source of many compounds with cosmeceutical applications. For example, their peptide extract enables the development of Wakamine 1% (peptidic extract), Wakamine™ XP and Morechem™ Hijiki extract which are products with a tyrosinase inhibitory activity and acting as a whitening agent ("Wakamine™," 2020; "Morechem™ Hijiki Extract," 2020).

Polysaccharides extracted from species of the genus *Fucus*, namely *F. vesiculosus*, are feedstock for cosmeceutical products, as the "Akomarine® Fucus" is used in cosmetic formulations with the aim of promoting skin softness, slimming, and elasticity ("Akomarine® Fucus," 2020). Phlorotannins extracted from the brown seaweed *F. serratus* are important dental care agents in Dentactive® (Yasham, 2020).

Seaweeds from Laminariaceae are used by Presperse Inc. who developed the product Phycolanine™ based on aqueous extracts for skincare and color cosmetic applications ("Phycolanine," 2020). For instance, Phycoboreane® is based on *Laminaria Hyperborea* extract with lipolytic activity applied in anti-cellulite products (SEPPIC, 2020).

Here are highlighted some of the products already established in the market, in which bioactive compounds from seaweed are incorporated. Thus, scientific evidence of health promotion of these bioactive compounds is emphasized by researchers (Zhang et al., 2010; Kendel et al., 2015; Pérez et al., 2016).

9.5 AGRICULTURE

Seaweeds have been used as soil fertilizer since ancient times, with evidence before the Roman Empire in Europe. This usage was only in coastal zones and the tradition was very similar among coastal communities. Seaweeds were collected or harvest and stockpiled in depilated fields to dry and be washed off by the rain, before the mixture was put in cultivation soil or transformed into compost with other organic wastes, straw, or peat to be scattered in soil, turning infertile soil into a fertile one near the shore. There were transportation systems to take seaweeds to inland field, far from the shore but there were limitations of 10 km from the location, which happened in Ireland and Portugal (McHugh et al., 2003; Craigie, 2011; Chatzissavvidis and Therios, 2014; Kim and Chojnacka, 2015; Pereira and Cotas, 2019).

Until the end of World War II, the seaweed usage by coastal communities was a tradition. The liquid seaweed fertilizers were investigated initially by the biochemist R.F. Milton after WWII until he found best-liquefied extract, based on his experience in the alginate industry. This extract was patented and commercialized after 1949, by the first company specialized in seaweeds' extracts for agriculture, the Maxicrop (Craigie, 2011). This permitted the overall accessibility of seaweeds' extracts for agriculture independent of the coastal proximity.

Nowadays, there are a lot of seaweed extracts commercially available for agriculture, such as the extracts of the brown seaweeds (kelps) *Ecklonia maxima* and *Ascophyllum nodosum*, designated, respectively, Kelpak® and PROFERTIL® (Basak, 2008; ADP, 2020; Slack et al., 2020).

The extracts have a wide range of characteristics, such as ranging from colorless to dark color and varying in odor strength, viscosity, and particulate matter content. The methods of extraction production are rare to be published, thus they are maintained as industrial secret. However, the extracts are commonly made by aqueous, alkali or acidic extraction methods, or even by physical methods, such as low-temperature cell disruption method (Craigie, 2011).

This field is in a constant evolution to obtain the best product, and usually is based on brown seaweeds extracts, because of the high knowledge and bioavailability of brown seaweeds in nature and in aquaculture, such as *A. nodosum*. Some extracts produced by the Maxicrop, Acadian Seaplants and Algae have high content of humic-like polyphenols or polyphenols, derivate to main use of brown seaweeds in this type of industry (Craigie, 2011). One example is the *A. nodosum* analyzed from Norway and Nova Scotia that has 15–25% of extractable polyphenols with high molecular weight (Glombitza, 1977).

The commercial fertilizer extracts are not one seaweed compound isolated but a mixture of various seaweeds' compounds, such as PCs, polysaccharides (one of the major components), and minerals (Craigie, 2011; Battacharyya et al., 2015).

There is the commercialization of a solution with seaweed PCs to be a biostimulant for mycorrhizal and rhizobial symbiosis that can happen as fertilization or treatment based mainly on *Fucus* and *Ascophyllum* genera (Briand and Salamagne, 2016). Also, the seaweed phenols can have a protection factor to diminish the plants diseases by anti-microbial and antioxidant activities (Chojnacka, 2012; Wijesinghe and Jeon, 2012).

Seaweed polysaccharides, such as alginate and laminarin, are soil conditioners, having a positive influence on soil structure (by facilitating to bring soil crumbs and retaining water) (Kraan, 2012). The alginate can suppress various types of pathogens (Sultana et al., 2011). Actually, alginate pellets were developed as transporter material for biocontrol agents that have been described to diminish the reproduction of *Rhizoctonia* disease in potatoes (Jacobs et al., 2003), whereas in integration of *A. nodosum* extract into the planting medium triggered a deferral and reduced frequency of *Verticillium* wilt in pepper plants (Jayaraj et al., 2008). Consequently, it is demonstrated that alginates are involved in host defense mechanisms (Craigie, 2011). It is of particular relevance in agriculture to activate self-protective mechanisms in plants, resulting in protection against diseases or insect damage (Chatzissavvidis and Therios, 2014).

Laminarin is an elicitor that causes strong defense responses against different pathogens (Romanazzi et al., 2016; Garde-Cerdán et al., 2017) and is implicated in the induction of genes that encode various pathogenesis-related

(PR) proteins with antimicrobial properties, acting as biostimulant in plants (Van Loon and Van Strien, 1999; Khan et al., 2009). It was also shown that an efficient inducer of defense responses in vine cells and crop plants and capable of controlling the development of the fungi *Botrytis cinerea* and *Plasmopara viticola* in infected plants is the β-1,3-glucan laminarin derived from brown algae *Laminaria digitata* (Aziz et al., 2003). The foliar application of laminarin revealed good effectiveness in strawberry protection against pathogens as an alternative to the number of fungicide sprays (Meszka and Bielenin, 2011). The demand for biocontrol preparations has increased in the last year because pre-harvest synthetic fungicides retain residues in fruits and vegetables. An extract from the brown seaweed *Laminaria digitata*, Vacciplant® (Goëmar), which contains laminarin, is registered as a phytopharmaceutical (Reg. 1007/2009; Kim and Chojnacka, 2015). Mercier et al. (2001) showed that laminarin and carrageenan polysaccharides from macroalgae induce efficient signaling and defense gene expression in tobacco leaves.

The polysaccharides can have other functions in the fertilizer formulas. For instance, as a product emulsifier or even as a slow-release delivery system of active ingredient in fertilizer products preventing the fast-release of the active ingredient, which is the leading downside related with traditional fertilizer formulations.

The mineral content is also very important in the seaweed extract because seaweeds are very rich in minerals and trace elements, such as iodine, iron, magnesium, calcium, phosphorous, sodium, and potassium that are very important, not only for seaweeds' health, but also for the crops' health and growth (Chatzissavvidis and Therios, 2014).

Nevertheless, the lack of information of the compounds present in the seaweed's fertilizer is a key factor that comprises the full understanding of fertilizer compounds effects in the plant mechanism. Nevertheless, there is a wide range of scientific work exploring this thematic, in the number of reviews (e.g., Khan et al., 2009; Craigie, 2011; Battacharyya et al., 2015). However, even within scientific researchers, there is not a specific explanation of how the seaweed extracts actuate in the plant (Chatzissavvidis and Therios, 2014). There is a wide range of questions in this field still under study, to clarify this mechanism of action. On the other hand, the best explanation can be related to various molecules present in seaweeds, such as macro and micro-nutrients, polysaccharides, vitamins, amino acids, and phytohormones; these molecules can have effects separated or have a synergic effect (Craigie, 2011; Chatzissavvidis and Therios, 2014). The future of seaweed fertilizers extracts passes through underlying and understanding the mechanism of action.

The main advantages in agriculture of the seaweed extracts are higher yields, enhance of the vigor, better assimilation of the soil nutrients, higher resistance against diseases and pathogens, longer shelf life of the vegetables and fruit, improved seed germination and better defense and resistance against abiotic stresses (Chatzissavvidis and Therios, 2014).

9.6 FEED AND LIVESTOCK

The seaweed feeds usages in animal husbandry dates from at least the Roman Empire, even been mentioned by the ancient Greeks, in the coastal areas, the seaweed usage for agriculture and animal feed is interconnected from the beginning (Makkar et al., 2016).

The traditional methods of application are similar to the agriculture. The animal graze or eat directly in the beach or the seaweeds were collected in the coastal zone, dried, and stockpiled in the barns or silages to be used as feed to sheep, horses, or cattle during the fodder scarcity, likewise during the winter. This was documented in the early 20th Century (Evans and Critchley, 2014).

These seaweed-based animal feeds are commonly described in France (Brittany and Normandy), Ireland, and Great Britain and Scandinavia region, due to the Second World War. Seaweed was one of the frequent animals' feeds used in Europe (Chapman and Chapman, 1980). However, the nutritional science reported that seaweeds had low nutritive value to the livestock (Evans and Critchley, 2014), lowering the interest in animal feed. However, in the 60's of the last century the seaweeds had gained again interest with the first attempts for seaweed cultivation and the seaweed fertilizer industry gaining more attention (McHugh and FAO, 2002).

Seaweed is mainly used as extract or milled to the animal feed, mainly as a supplement to the daily normal use of feed. Seaweed composition reports a great difference in the final contents of the components. However, they have proteins, minerals, lipids (mainly, PUFAs), polysaccharides, PCs, and vitamins that are very relevant for seaweed nutritional value for human and animal feed (Gaillard et al., 2018). It is recognized that green and red seaweeds have higher nutritional value than brown algae, owing it to the lower protein and high mineral content. Still, brown seaweed has a greater and varied content on bioactive molecules with more commercial interest. Depending on the composition, seaweeds can be used and act in distinct ways in the animal organism, such as prebiotics, protein, and/or energy source (Makkar et al., 2016).

The seaweeds' compounds have been reported as bioavailable for animals from the colon, specifically phenols, or interact with the intestine microbiome (Chojnacka, 2012).

When compared with others feed additives, the seaweeds have better protein quality and are more different than the soy flour. Still, the protein content is a limiting factor to choose the seaweed species to explore for animal feed (Angell et al., 2016).

The best use that is now gaining more attention, is the use of seaweeds as nutraceutical for animals, with various seaweeds species presenting favorable bioactivities for inclusion in alimentation of rats, chickens, laying hens, pigs, and other animals (Makkar et al., 2016; Li et al., 2018). Being relevant for reducing the antibiotics in the animal industry and also enhancing the quality of the food, likewise, eggs, and meat (Craigie, 2011). Furthermore, the content of minerals is important for good work of animal organism and the seaweeds can have a high mineral content (Craigie, 2011; Cotas et al., 2020).

The normal seaweed dosage for animal feed is little doses under 10%, because the over-dosage will have negative effects on animal health. Therefore, it is a systematic analysis of the seaweed or extract quality to extrapolate the dosages. The seaweed supplement origins an improvement in productivity (animal growth, milk, and eggs production), quality (meat and other animal-origin products) and safety (reduction of pathogen microorganisms in the animals and derivative products for consumption). Overall, the seaweeds ameliorated the immune system of the animals (Saker et al., 2001; Nagayama et al., 2002; Shibata et al., 2002; Spiers et al., 2004; Bach et al., 2008; Makkar et al., 2016).

Nowadays, there are already some companies exploring seaweeds as an animal feed supplement. In addition, new research appears, also was the use of the red alga *Asparagopsis armata* as tool for reduction of the ruminant methane production by inhibition of the methanogenic archaea proliferation in the ruminant intestine (Regal et al., 2020).

9.7 SEAWEEDS' ECOLOGICAL RELEVANCE AND ECOSYSTEMS SERVICES

Seaweeds are part of an ecosystem where they play a pivotal role as primary producers and as a shelter or nursery habitat for an enormous biodiversity (Satheesh et al., 2017). They assume their ecological relevance as oxygen producers, carbon dioxide (CO_2) and nutrient recycling uptake, providing important ecosystem services (Werner et al., 2016).

However, estuarine and coastal areas have been targeted with several stressors from natural (e.g., global warming; extreme climatic events) or anthropogenic origin (e.g., pollution; resources overexploitation) (Ryan et al., 2017). Contaminants could arise from domestic, agricultural, and industrial effluents and their physicochemical constitution differs according to their source (Muñoz et al., 2009). In this context, estuarine and coastal areas are endangered by metals, pesticides, and nutrients, such as nitrate and phosphate (Chowdhury et al., 2016; Menció et al., 2016; Emparan et al., 2018). However, the pollutant quantification only in soil and sediment is not representative of the toxicity for biota (Aly et al., 2013). Hence, seaweed assume their relevance as bioindicators, to monitor pollution that threats these ecosystems, due to their sensitivity to some pollutants and their ability to tolerate and bioaccumulate them (Doshi et al., 2008). For instance, seaweed species from the genus *Ulva* (Chlorophyta) are seen as a good bioindicator since they have a widespread global distribution and their growth capability under high concentrations of trace elements (Villares et al., 2002; Ustunada et al., 2011; Allam et al., 2016). Researchers have shown that brown seaweeds, such as *F. serratus*, *F. vesiculosus* and *Ascophyllum nodosum*, are biological indicators of Fe, Cu, Zn, and As pollution from acid mine drainage (Chalkley et al., 2019). Therefore, the presence and biomass of these seaweed species allows the detection of these elements (Bonanno et al., 2020). Moreover, this strategy is used all over the world to assess pollution outbreaks since 1990. For example, researchers used 12 seaweed species, namely *Dictyota kunthii* (previously, *Glossophora kunthii*), *Lessonia nigrescens*, *Petalonia fascia* (Phaeophyceae), *Asterfilopsis furcellata* (as *Gymnogongrus furcellatus*), *Chondracanthus chamissoi* (formerly known as *Gigartina chamissoi*), *Chondrus canaliculatus*, *Gelidium* sp., *Halopteris hordacea*, *Pyropia columbina* (as *Porphyra columbina*) (Rhodophyta), *Colpomenia sinuosa*, *Ulva* sp. (as *Enteromorpha* sp.) (Chlorophyta), in order to assess metal pollution from anthropogenic origin (Vasquez and Guerra, 1996). Also, in Italy, *Ulva lactuca* is monitored along the coast to assess metal pollution (Bonanno et al., 2020).

Nevertheless, not only seaweeds' presence and biomass quantification indicate the pollution levels, but also their composition. For instance, the fatty acid profile of *Pterocladiella capillacea* (Rhodophyta), *U. lactuca* (Chlorophyta) and *Sargassum hornschuchii* (Phaeophyceae) revealed that, under stress conditions, their lipidic profile suffered variations, indicating the pollution levels of bisphenol (Mohy El-Din, 2018).

Because seaweeds are a multirole organism in marine ecosystems, they are key biological elements of the quality of a certain water body. That is why, along the years, several tools to assess ecological quality status were developed under the scope of the European Water Framework Directive (European Parliament, 2000), such as Marine Macroalgae Assessment Tool (MarMAT) (Neto et al., 2012), ecological evaluation index (EEI) (Orfanidis et al., 2003), Macroalgae Species Richness (Wells et al., 2003) and cartography of littoral and upper-sublittoral rocky-shore communities (CARLIT) (Ballesteros et al., 2007).

Seaweeds' presence in ecosystems is not only important due to their ecological relevance, but also because of their potential as bioremediation agents (Neveux et al., 2018). They are key elements to environmental detoxification (Deniz and Ersanli, 2018). For instance, when large amounts of nutrient-rich effluents are discharged in water bodies, seaweeds metabolize those nutrients and therefore contribute to tackling the risk of eutrophication (Saldarriaga-Hernandez et al., 2020). A study demonstrated that the brown macroalga *Saccharina japonica*'s cultivation contributed to the improvement of water quality, relatively to its physicochemical parameters, as also increased the ecosystem biodiversity (Jiang et al., 2020).

For this reason, seaweeds are included in the concept of integrated multi-trophic aquaculture as a biofilter, in order to clean the effluents produced by this industry (Chávez-Crooker and Obreque-Contreras, 2010; Roleda and Hurd, 2019). Their ability to accumulate other types of pollutants, such as antibiotics (Rosa et al., 2020) and metals (Deniz and Ersanli, 2018; Maurya et al., 2019; Costa et al., 2020), is highlighted by the scientific community. Moreover, brown macroalgae are a source of by-products, such as alginate, that possesses the capability of binding with metals, which is a useful tool to bioremediation of different types of effluents (Davis et al., 2003; Khashei et al., 2018; Shi et al., 2018; El Bestawy, 2019; Zhang et al., 2019).

However, to achieve success in bioremediation using macroalgae at an industrial scale, it is necessary to consider differences in morphological, physiological, and economic aspects in order to make this process feasible (Hadley et al., 2015).

9.8 OTHER APPLICATIONS

Macroalgae and their derived compounds have a wide range of potential applications. In addition to those mentioned above, investigation in this field of research is evolving in order to develop novel products in order to tackle current challenges.

One of these applications is in avoidance of biofouling, which is a process that leads to economic losses, due to vessels damage and with harmful ecological impacts derived from the chemicals used in paints (e.g., tri-butyl-tin, and copper) (Armstrong et al., 2000; Andersson et al., 2020).

The search for novel compounds with antifouling activity is raising interest among researchers, which support marine seaweeds as a source of natural compounds with this bioactivity. In this context, it was investigated if crude extracts from brown and red seaweeds had anti-fouling effect. Among the macroalgae tested, it was the crude extract from the red seaweed *Laurencia obtusa* that showed the inhibition of other organisms' settlement (Da Gama et al., 2002).

A novel approach was also proposed by Pinteus et al. (2020), based on two invasive Portuguese seaweeds (*Asparagopsis armata* and *Sargassum muticum*), in order to prevent medusa polyp's settlement. While other researchers found the opportunist green seaweed *Ulva lactuca* synthesizes biogenic silver-nano particles, which presented remarkable antifouling properties (Chellammal et al., 2020).

Moreover, marine macroalgae could be applied in the textile industry, in the first instance, as dyes. Natural dyes were the primary color source for textiles since pre-historic times (Cristea, 2003). These non-synthetically obtained dyes increase the confidence and acceptability of a product by the customers and also respond to sustainable environmental issues (Hill, 1997). Seaweeds can be exploited and used as a natural dye source and produce colors palettes on textiles. For example, shades obtained from the species *Caulerpa lentillifera* (Chlorophyta) range between greenish-yellow and reddish-brown, depending on the mordant used, which gave acceptable fastness properties even without mordant (Kadir et al., 2014).

Seaweed could potentially be used for the yarn and dye, as it is applied in projects as the "SEA ME" of the Nienke Hoogvliet studio, which designed products like a rug made of sea algae yarn, knotted by hand in an old fishing net (Hoogvliet and Jongerius, 2020). This project was outlined in order to draw attention to this raw material, adding to the re-utilization of plastic waste. This project demonstrates that a yarn made of marine macroalgae could offer a sustainable option for the textile industry (Hoogvliet and Jongerius, 2020).

There is a patented process that embeds the seaweed firmly within a natural cellulose fiber, resulting in the SeaCell®, which is produced using an innovative and eco-friendly production method that does not release chemicals as waste (Tschopp et al., 2020). This textile benefits from the positive properties of the seaweed that are permanently preserved within the fiber, even after multiple washing cycles. The seaweed is smashed and joined to the fiber, forming the fabric, where bioactive compounds of the seaweed contribute to activate cell regeneration, relieve skin diseases, reduce inflammation, and soothe itchiness. Also due to its high level of antioxidants, it protects the skin against harmful free radicals (Tschopp et al., 2020). These potentials of incorporate bioactive compounds, that provide beneficial effects, in fabrics, magnify their uses for skin medicinal purposes (Janarthanan and Kumar, 2018).

Within these potentials, there is the antibacterial action (Durán et al., 2007). For a double interest, being the reducer agent of silver ions and the stabilizer agent (El-Rafie et al., 2013). For example, incorporations of silver nanoparticles (nano-Ag) in fabrics, enables develop antiseptic dressings or bandages, for biomedical uses (Durán et al., 2007; El-Rafie et al., 2013). Seaweed derived-polysaccharides could be used in this nano-Ag, having the ability to immobilize silver on the cotton fibers bringing an antimicrobial property to the final textile product (El-Rafie et al., 2013). This is an emerging field that could be further exploited, enhancing a low-cost, green approach to apply in large-scale textile productions.

Additionally, marine macroalgae may be a solution to the plastic issue. Marine, costal, and terrestrial ecosystems are threatened by the amount of non-biodegradable plastics pressure, causing harmful effects throughout the food chain (European Chemicals Agency, 2020). Thus, biopolymers derived from brown seaweed, such as alginate, are potential candidates to develop biodegradable plastic at an industrial scale (Pereira and Cotas, 2020).

The green macroalga *U. lactuca* biosynthesizes lactic acid, which is a precursor of polylactic acid. For this reason, Helmes et al. (2018) have shown that this opportunist species can be an important resource for bioplastic manufacture. Among the same taxonomic group, *Cladophora rupestris* (Chlorophyta) biosynthesizes polyhydroxybutyrate (PHB), a natural biopolymer (Stabili et al., 2014).

Brown seaweeds (kelp) are also resources that provide a natural solution to replace plastic in disposable ware. A US-based company produces straws that are plastic-free, durable, vegan, and with the possibility of being edible (Briganti et al., 2020). These environmental-friendly products are made with natural ingredients, are biodegradable and edible, although the straws were not invented to be a snack, but to be a less-impacting alternative to the plastic straws. Some companies have already demonstrated the interest to use these straws, which are scheduled to be commercially available (Briganti et al., 2020).

Brown seaweeds could either be used for completely distinct products or industries. An emerging example of that is the application of alginate as a supportive agent for silicon nano-powder in a battery. Alginate improves the performance of lithium batteries stabilizing the battery anode (Kovalenko et al., 2011).

9.9 CONCLUSION

This chapter comprises a general overview and insights regarding the uses of the seaweeds, from the most traditional practices to the current and future applications. In fact, seaweeds have been employed for millenniums as food and medicines, and still be an interesting natural raw ingredient to be applied in many sectors, such as the nutraceutical and pharmaceutical industries.

Therefore, on one hand, some traditional uses of seaweeds remain rooted in some civilizations, on the other, with scientific progress, investigations, and discoveries about seaweeds and their extracts' properties increase the potential of their application in products that we use daily.

Besides that, there is more awareness of sustainable use of natural resources, rather than synthetic and processed products with eventual harmful side effects to the consumer and the environment. In addition to the ecosystem services, seaweeds provide oxygen production, CO_2 sequestration, food source, and shelter for many other organisms, among other ecological roles. Seaweeds are of low-cost and sustainable cultivation, do not compete for arable land, and do not produce greenhouse gases. Therefore, the interest on these resources is rising, as well as in green methodologies and blue and circular economy. Thus, research and development on macroalgae-derived products are areas in expansion, and we are continuously gathering information and creating new opportunities for future applications.

ACKNOWLEDGMENTS

This work is financed by national funds through FCT – Foundation for Science and Technology, IP, within the scope of the projects UIDB/04292/2020 – MARE – Marine and Environmental Sciences Center and UIDP/50017/2020+UIDB/50017/2020 (by FCT/MTCES) granted to CESAM – Center for Environmental and Marine Studies. Adriana Leandro thanks FCT for the financial support provided through the doctoral grant SFRH/BD/143649/2019 funded by National Funds and Community Funds through FSE. Diana Pacheco thanks to the PTDC/BIA-CBI/31144/2017-POCI-01 project – 0145-FEDER-031144 – MARINE INVADERS, co-financed by the ERDF through POCI (Operational Program Competitiveness and Internationalization) and by the FCT (IP). Sara García-Poza thanks to the project MENU – Marine Macroalgae: Alternative recipes for a daily nutritional diet (FA_05_2017_011) which co-financed this research, funded by the Blue Fund under Public Notice No. 5 – Blue Biotechnology. João Cotas thanks MARE, UC, and Universidade Nova de Lisboa, through its Chemistry Department of the Faculty of Sciences and Technology within the scope of the research project "Algae in Gastronomy – Development of innovative techniques for conservation and use – Alga4Food" (MAR-01.03.01-FEAMP-0016). Ana M. M. Gonçalves acknowledges the University of Coimbra for the contract IT057-18-7253.

KEYWORDS

- **agriculture**
- **cosmetic and cosmeceutical**
- **ecosystem services**
- **food and feed**
- **macroalgae**
- **nutraceutical**
- **pharmaceutical**

REFERENCES

Ab Kadir, M. I., Wan, A. W. Y., Ahmad, M. R., et al., (2014). Utilization of eco-colorant from green seaweed on textile dyeing. In: Ahmad, M. R., & Yahya, M. F., (eds.), *Proceedings of the International Colloquium in Textile Engineering, Fashion, Apparel and Design 2014 (ICTEFAD 2014)* (pp. 79–83). Singapore: Springer.

Abbott, I. A., (1978). The uses of seaweed as food in Hawaii. *Economic Botany, 32*(4), 409–412.

Abirami, R. G., & Kowsalya, S., (2012). Anticancer activity of methanolic and aqueous extract of *Ulva fasciata* in albino mice. *International Journal of Pharmacy and Pharmaceutical Sciences, 4*(2), 681–684.

Aderibigbe, B. A., & Buyana, B., (2018). Alginate in wound dressings. *Pharmaceutics, 10*(2), 42.

ADP, (2020). *Profertil – Adubos de Portugal.* https://www.adp-fertilizantes.pt/pt/agricultura/produtos/categorias/foliares/profertil/ (accessed on 7 April 2022).

Akomarine® Fucus-Akott-Datasheet, (2020). https://cosmetics.specialchem.com/product/i-akott-akomarine-fucus (accessed on 7 April 2022).

Allam, H., Aouar, A., Benguedda, W., & Bettioui, R., (2016). Use of sediment and algae for biomonitoring the coast of Honaïne (Far West Algerian). *Open Journal of Ecology, 6*(4), 159–166.

Al-Shamkhani, A., & Duncan, R., (1995). Radioiodination of alginate via covalently-bound tyrosinamide allows monitoring of its fate *in vivo*. *Journal of Bioactive and Compatible Polymers*, *10*(1), 4–13.

Alves, C., Silva, J., Pinteus, S., et al., (2018). From marine origin to therapeutics: The antitumor potential of marine algae-derived compounds. *Frontiers in Pharmacology*, *9*, 777.

Aly, W., Williams, I. D., & Hudson, M. D., (2013). Metal contamination in water, sediment and biota from a semi-enclosed coastal area. *Environmental Monitoring and Assessment*, *185*(5), 3879–3895.

Andersson, J., Oliveira, D. R., Yeginbayeva, I., Leer-Andersen, M., & Bensow, R. E., (2020). Review and comparison of methods to model ship hull roughness. *Applied Ocean Research*, *99*, 102119.

Angell, A. R., Angell, S. F., Nys, R., & Paul, N. A., (2016). Seaweed as a protein source for mono-gastric livestock. *Trends in Food Science and Technology*, *54*, 74–84.

Apostolidis, E., & Lee, C. M., (2011). Brown seaweed-derived phenolic phytochemicals and their biological activities for functional food ingredients with focus on *Ascophyllum nodosum*. In: *Handbook of Marine Macroalgae* (pp. 356–370). Chichester, UK: John Wiley & Sons, Ltd.

Arasaki, S., & Arasaki, T., (1983). *Low Calorie, High Nutrition Vegetables from the Sea to Help You Look and Feel Better* (Vol. 99). Tokyo: Japan Publications.

Armstrong, E., Boyd, K. G., & Burgess, J. G., (2000). Prevention of marine biofouling using natural compounds from marine organisms. *Biotechnology Annual Review*, *6*, 221–241.

Arnason, T., Hebda, R. J., & Johns, T., (1981). Use of plants for food and medicine by native peoples of Eastern Canada. *Canadian Journal of Botany*, *59*(11), 2189–2325.

Astorga-España, M. S., Galdón, B. R., Rodríguez, E. M., & Romero, C. D., (2015). Mineral and trace element concentrations in seaweeds from the sub-Antarctic ecoregion of Magallanes (Chile). *Journal of Food Composition and Analysis*, *39*, 69–76.

AT UV Protector P-Athena-Datasheet, (2020). https://cosmetics.specialchem.com/product/i-athena-at-uv-protector-p (accessed on 7 April 2022).

Atashrazm, F., Lowenthal, R. M., Woods, G. M., Holloway, A. F., & Dickinson, J. L., (2015). Fucoidan and cancer: A multifunctional molecule with anti-tumor potential. *Marine Drugs*, *13*(4), 2327–2346.

Aziz, A., Poinssot, B., Daire, X., et al., (2003). Laminarin elicits defense responses in grapevine and induces protection against *Botrytis cinerea* and *Plasmopara viticola*. *Molecular Plant-Microbe Interactions*, *16*(12), 1118–1128.

Bach, S. J., Wang, Y., & McAllister, T. A., (2008). Effect of feeding sun-dried seaweed (*Ascophyllum nodosum*) on fecal shedding of *Escherichia coli* O157:H7 by feedlot cattle and on growth performance of lambs. *Animal Feed Science and Technology*, *42*(1, 2), 17–32.

Ballesteros, E., Torras, X., Pinedo, S., García, M., Mangialajo, L., & Torres, M., (2007). A new methodology based on littoral community cartography dominated by macroalgae for the implementation of the European water framework directive. *Marine Pollution Bulletin*, *55*(1–6), 172–180.

Basak, A., (2008). Effect of preharvest treatment with seaweed products, Kelpak® and Goëmar BM 86®, on fruit quality in apple. *International Journal of Fruit Science*, *8*(1, 2), 1–14.

Battacharyya, D., Babgohari, M. Z., Rathor, P., & Prithiviraj, B., (2015). Seaweed extracts as biostimulants in horticulture. *Scientia Horticulturae*, *196*, 39–48.

Berri, M., Slugocki, C., Olivier, M., et al., (2016). Marine-sulfated polysaccharides extract of *Ulva armoricana* green algae exhibits an antimicrobial activity and stimulates cytokine expression by intestinal epithelial cells. *Journal of Applied Phycology*, *28*, 2999–3008.

Bestawy, E. E., (2019). Efficiency of immobilized cyanobacteria in heavy metals removal from industrial effluents. *Desalination and Water Treatment, 159*, 66–78.
Bhatia, S., Sharma, K., & Bera, T., (2015). Structural characterization and pharmaceutical properties of porphyran. *Asian Journal of Pharmaceutics, 9*(2), 93.
Biesalski, H. K., Dragsted, L. O., Elmadfa, I., et al., (2009). Bioactive compounds: Safety and efficacy. *Nutrition, 25*(11, 12), 1206–1211.
Bonanno, G., Veneziano, V., & Piccione, V., (2020). The alga *Ulva lactuca* (Ulvaceae, Chlorophyta) as a bioindicator of trace element contamination along the coast of Sicily, Italy. *Science of the Total Environment, 699*, 134329.
Bouhadir, K. H., Alsberg, E., & Mooney, D. J., (2001). Hydrogels for combination delivery of antineoplastic agents. *Biomaterials, 22*(19), 2625–2633.
Branscheid, W., & Judas, M., (2011). Detection of bone in meat. In: Nollet, L. M. L., & Toldra, F., (eds.), *Handbook of Analysis of Edible Animal By-Products* (pp. 247–285). Boca Raton; CRC Press.
Briand, X., & Salamagne, S. C., (2016). *Use of Phlorotannins as a Stimulant for Mychorrhizal and Rhizobial Symbioses*. WO2017032954A1. https://patents.google.com/patent/WO2017032954A1/en (accessed on 7 April 2022).
Briganti, C., Tucker, L. A., Zweifel, I., Bhatia, M., & Stanton, K., (2020). *Loliware*. https://www.loliware.com/ (accessed on 7 April 2022).
British Nutrition Foundation (BNF), (2016). *Nutrition Requirements: Reference Nutrient Intakes for Minerals*. https://www.nutrition.org.uk/healthy-sustainable-diets/vitamins-and-minerals/?level=Health%20professional (accessed on 7 April 2022).
Brown, E. M., Allsopp, P. J., Magee, P. J., et al., (2014). Seaweed and human health. *Nutrition Reviews, 72*(3), 205–216.
Cao, D., Lv, X., Xu, X., Yu, H., Sun, X., & Xu, N., (2017). Purification and identification of a novel ACE inhibitory peptide from marine alga *Gracilariopsis lemaneiformis* protein hydrolysate. *European Food Research and Technology, 243*(10), 1829–1837.
Castine, S. A., McKinnon, D. A., Paul, N. A., Trott, L. A., & De Nys, R., (2013). Wastewater treatment for land-based aquaculture: Improvements and value-adding alternatives in model systems from Australia. *Aquaculture Environment Interactions, 4*(3), 285–300.
Chalkley, R., Child, F., Al-Thaqafi, K., Dean, A. P., White, K. N., & Pittman, J. K., (2019). Macroalgae as spatial and temporal bioindicators of coastal metal pollution following remediation and diversion of acid mine drainage. *Ecotoxicology and Environmental Safety, 182*, 109458.
Chan, C., Ho, C., & Phang, S., (2006). Trends in seaweed research. *Trends in Plant Science, 11*(4), 165, 166.
Chang, H. C., Chen, H. H., & Hu, H. H., (2011). Textural changes in fresh egg noodles formulated with seaweed powder and full or partial replacement of cuttlefish paste. *Journal of Texture Studies, 42*(1), 61–71.
Chapman, V. J., & Chapman, D. J., (1980). *Seaweeds and Their Uses*. Netherlands: Springer.
Chatzissavvidis, C., & Therios, I., (2014). Role of algae in agriculture. In: Pomin, V. H., (ed.), *Seaweeds: Agricultural Uses, Biological and Antioxidant Agents* (pp. 1–37). New York: Nova Science Publishers, Inc.
Chávez-Crooker, P., & Obreque-Contreras, J., (2010). Bioremediation of aquaculture wastes. *Current Opinion in Biotechnology, 21*(3), 313–317.
Chellammal, S., Kalaiselvi, P., & Subramanian, G., (2020). *Antifouling Effects of Silver Nano Particles Synthesized from Tropical Seaweeds*. https://www.researchgate.net/publication/284247172 (accessed on 7 April 2022).

Chengkui, Z., & Junfu, Z., (1984). Chinese seaweeds in herbal medicine. In: Bird, C. J., & Ragan, M. A., (eds.), *Eleventh International Seaweed Symposium* (pp. 152–154). Dordrecht: Springer Netherlands.

Chiu, Y. H., Chan, Y. L., Li, T. L., & Wu, C. J., (2012). Inhibition of Japanese encephalitis virus infection by the sulfated polysaccharide extracts from *Ulva lactuca*. *Marine Biotechnology*, *14*(4), 468–478.

Chlorofiltrat® Ulva HG-Codif–Datasheet, (2020). https://cosmetics.specialchem.com/product/i-codif-chlorofiltrat-ulva-hg (accessed on 7 April 2022).

Chojnacka, K., (2012). Biologically active compounds in seaweed extracts-the prospects for the application. *The Open Conference Proceedings Journal*, *3*(1), 20–28.

Chowdhury, S., Mazumder, M. A. J., Al-Attas, O., & Husain, T., (2016). Heavy metals in drinking water: Occurrences, implications, and future needs in developing countries. *Science of the Total Environment*, *569*, *570*, 476–488.

Christenson, E. M., Anseth, K. S., Van, D. B. J. J. P., et al., (2007). Nanobiomaterial applications in orthopedics. *Journal of Orthopaedic Research*, *25*(1), 11–22.

Christie, H., Norderhaug, K. M., & Fredriksen, S., (2009). Macrophytes as habitat for fauna. *Marine Ecology Progress Series*, *396*, 231–243.

Cian, R. E., Martínez-Augustin, O., & Drago, S. R., (2012). Bioactive properties of peptides obtained by enzymatic hydrolysis from protein byproducts of *Porphyra columbina*. *Food Research International*, *49*(1), 364–372.

Cockbill, C. A., (1994). Food law and functional foods. *British Food Journal*, *96*(3), 3, 4.

Cofrades, S., Benedí, J., Garcimartin, A., Sánchez-Muniz, F. J., & Jimenez-Colmenero, F., (2017). A comprehensive approach to formulation of seaweed-enriched meat products: From technological development to assessment of healthy properties. *Food Research International*, *99*, 1084–1094.

Cosmetics Europe, (2020). *Cosmetics Europe: The Personal Care Association: Cosmetics Industry*. https://cosmeticseurope.eu/cosmetics-industry/ (accessed on 7 April 2022).

Costa, M., Henriques, B., Pinto, J., et al., (2020). Influence of toxic elements on the simultaneous uptake of rare earth elements from contaminated waters by estuarine macroalgae. *Chemosphere*, *252*, 126562.

Cotas, J., Leandro, A., Pacheco, D., Gonçalves, A. M. M., & Pereira, L., (2020). A comprehensive review of the nutraceutical and therapeutic applications of red seaweeds (Rhodophyta). *Life*, *10*(3), 19.

Cotas, J., Marques, V., Afonso, M. B., Rodrigues, C. M. P., & Pereira, L., (2020). Antitumor potential of *Gigartina pistillata* carrageenans against colorectal cancer stem cell-enriched tumor spheres. *Marine Drugs*, *18*(1), 50.

Couteau, C., & Coiffard, L., (2016). In: Levine, J., & Fleurence, I., (eds.), *Seaweed in Health and Disease Prevention* (1st edn.) Netherlands, Amsterdam: Elsevier.

Craigie, J. S., (2011). Seaweed extract stimuli in plant science and agriculture. *Journal of Applied Phycology*, *23*(3), 371–393.

Cristea, D., (2003). Identification and quantitative HPLC analysis of the main flavonoids present in weld (*Reseda luteola* L.). *Dyes and Pigments*, *57*(3), 267–272.

Darcy-Vrillon, B., (1993). Nutritional aspects of the developing use of marine macroalgae for the human food industry. *International Journal of Food Sciences and Nutrition*, *44*, 23–35.

Davis, T. A., Llanes, F., Volesky, B., Diaz-Pulido, G., McCook, L., & Mucci, A., (2003). ^{1}H-NMR study of Na alginates extracted from *Sargassum* spp. in relation to metal biosorption. *Applied Biochemistry and Biotechnology*, *110*(2), 75–90.

De Souza, M. C. R., Marques, C. T., Dore, C. M. G., Da Silva, F. R. F., Rocha, H. A. O., & Leite, E. L., (2007). Antioxidant activities of sulfated polysaccharides from brown and red seaweeds. *Journal of Applied Phycology, 19*(2), 153–160.

Denis, C., Morançais, M., Li, M., et al., (2010). Study of the chemical composition of edible red macroalgae *Grateloupia turuturu* from Brittany (France). *Food Chemistry, 119*(3), 913–917.

Deniz, F., & Ersanli, E. T., (2018). An ecofriendly approach for bioremediation of contaminated water environment: Potential contribution of a coastal seaweed community to environmental improvement. *International Journal of Phytoremediation, 20*(3), 256–263.

Díaz-Rubio, M. E., Pérez-Jiménez, J., & Saura-Calixto, F., (2009). Dietary fiber and antioxidant capacity in *Fucus vesiculosus* products. *International Journal of Food Sciences and Nutrition, 60*(2), 23–34.

Dillehay, T. D., Ramirez, C., Pino, M., Collins, M. B., Rossen, J., & Pino-Navarro, J. D., (2008). Monte Verde: Seaweed, food, medicine, and the peopling of South America. *Science, 320*(5877), 784–786.

Doshi, H., Seth, C., Ray, A., & Kothari, I. L., (2008). Bioaccumulation of heavy metals by green algae. *Current Microbiology, 56*(3), 246–255.

Douglas, T. E. L., Dokupil, A., Reczyńska, K., et al., (2016). Enrichment of enzymatically mineralized gellan gum hydrogels with phlorotannin-rich *Ecklonia cava* extract Seanol® to endow antibacterial properties and promote mineralization. *Biomedical Materials, 11*(4), 045015.

Durán, N., Marcato, P. D., De Souza, G. I. H., Alves, O. L., & Esposito, E., (2007). Antibacterial effect of silver nanoparticles produced by fungal process on textile fabrics and their effluent treatment. *Journal of Biomedical Nanotechnology, 3*(2), 203–208.

Eccles, R., Meier, C., Jawad, M., Weinmüllner, R., Grassauer, A., & Prieschl-Grassauer, E., (2010). Efficacy and safety of an antiviral iota-carrageenan nasal spray: A randomized, double-blind, placebo-controlled exploratory study in volunteers with early symptoms of the common cold. *Respiratory Research, 16*(1), 1–10.

Eccles, R., Winther, B., Johnston, S. L., Robinson, P., Trampisch, M., & Koelsch, S., (2015). Efficacy and safety of iota-carrageenan nasal spray versus placebo in early treatment of the common cold in adults: The ICICC trial. *Respiratory Research, 16*(1), 1–11.

Edwards, M. D., Løvstad-Holdt, S., & Hynes, S., (2012). Algal eating habits of phycologists attending the ISAP Halifax conference and members of the general public. *Journal of Applied Phycology, 24*(3), 627–633.

El-Din, S. M. M., (2018). Fatty acid profiling as bioindicator of chemical stress in marine *Pterocladia capillacea, Sargassum hornschuchii* and *Ulva lactuca. International Journal of Environmental Science and Technology, 15*(4), 791–800.

Eliaz, I., Weil, E., & Wilk, B., (2007). Integrative medicine and the role of modified citrus pectin/alginates in heavy metal chelation and detoxification-five case reports. *Forschende Komplementarmedizin, 14*(6), 358–364.

El-Rafie, H. M., El-Rafie, M. H., & Zahran, M. K., (2013). Green synthesis of silver nanoparticles using polysaccharides extracted from marine macro algae. *Carbohydrate Polymers, 96*(2), 403–410.

Emparan, Q., Harun, R., & Danquah, M. K., (2018). Role of phycoremediation for nutrient removal from wastewaters: A review. *Applied Ecology and Environmental Research, 17*(1), 889–915.

Espín, J. C., García-Conesa, M. T., & Tomás-Barberán, F. A., (2007). Nutraceuticals: Facts and fiction. *Phytochemistry, 68*(22–24), 2986–3008.

European Chemicals Agency, (2020). *Position Paper on EU Restriction of Microplastic – EEB: The European Environmental Bureau.* https://eeb.org/library/ngo-position-on-eu-restriction-of-microplastic/ (accessed on 7 April 2022).

European Parliament, (2000). Directive 2000/60/EC of the European parliament and of the council of 23 October 2000 establishing a framework for community action in the field of water policy. *Official Journal of the European Communities.*

European Parliament, (2009). Regulation (EC) No 1223/2009 of the European parliament and of the council of 30 November 2009 on cosmetic products. *Official Journal of the European Union.*

Evans, F. D., & Critchley, A. T., (2014). Seaweeds for animal production use. *Journal of Applied Phycology, 26*(2), 891–899.

FAO, (2006). *Compendium of Food Additives Specifications. FAO Food and Nutrition Paper* (Vol. 4). Rome, Italy: Food and Agriculture Organization of the United Nations. http://www.fao.org/3/a-a0691e.pdf (accessed on 7 April 2022).

FAO/WHO, (2007). Summary and Conclusions of the Sixty-Eighth Meeting of the Joint FAO/WHO Expert Committee on Food Additives (JECFA). WHO Technical Report Series, no. 12. http://apps.who.int/iris/bitstream/10665/44514/1/WHO_TRS_959_eng.pdf (accessed on 7 April 2022).

Fernández, L. E., Valiente, O. G., Mainardi, V., Bello, J. L., Vélez, H., & Rosado, A., (1989). Isolation and characterization of an antitumor active agar-type polysaccharide of *Gracilaria dominguensis. Carbohydrate Research, 190*(1), 77–83.

Ferrari, C., & Sorbi, S., (2020). An efficacy and safety study of sodium oligo-mannurarate (GV-971) capsule for the treatment of Alzheimer's disease. *Clinical Trials.* https://clinicaltrials.gov/ct2/show/NCT02293915 (accessed on 7 April 2022).

Fitton, H., Meyers, S., Brooks, L., et al., (2016). Effects of fucoidan from *Fucus vesiculosus* in reducing symptoms of osteoarthritis: A randomized placebo-controlled trial. *Biologics: Targets and Therapy, 2016*(10), 81–88.

Fitzgerald, C., Gallagher, E., Doran, L., Auty, M., Prieto, J., & Hayes, M., (2014). Increasing the health benefits of bread: Assessment of the physical and sensory qualities of bread formulated using a renin inhibitory *Palmaria palmata* protein hydrolysate. *LWT-Food Science and Technology, 56*(2), 398–405.

Freile-Pelegrín, Y., & Robledo, D., (2013). Bioactive phenolic compounds from algae. In: Hernández-Ledesma, B., & Herreron, (eds.), *Bioactive Compounds from Marine Foods: Plant and Animal Sources* (pp. 113–129). New Jersey: Wiley.

Freile-Pelegrín, Y., & Tasdemir, D., (2019). Seaweeds to the rescue of forgotten diseases: A review. *Botanica Marina, 62*(3), 211–226.

Friting, B., Heitz, T., & Legrand, M., (1998). Antimicrobial proteins in induced plant defense. *Current Opinion in Immunology, 10*(1), 16–22.

Gaillard, C., Bhatti, H. S., Novoa-Garrido, M., Lind, V., Roleda, M. Y., & Weisbjerg, M. R., (2018). Amino acid profiles of nine seaweed species and their in-situ degradability in dairy cows. *Animal Feed Science and Technology, 241*, 210–222.

Galloway, A. W. E., Britton-Simmons, K. H., Duggins, D. O., Gabrielson, P. W., & Brett, M. T., (2012). Fatty acid signatures differentiate marine macrophytes at ordinal and family ranks. *Journal of Phycology, 48*(4), 956–965.

Gama, B. A. P., Pereira, R. C., Carvalho, A. G. V., & Coutinho, R., (2002). The effects of seaweed secondary metabolites on biofouling. *Biofouling, 18*(1), 13–20.
Gammone, M., & D'Orazio, N., (2015). Anti-obesity activity of the marine carotenoid fucoxanthin. *Marine Drugs, 13*(4), 2196–2214.
Garde-Cerdán, T., Mancini, V., Carrasco-Quiroz, M., et al., (2017). Chitosan and laminarin as alternatives to copper for *Plasmopara viticola* control: Effect on grape amino acid. *Journal of Agricultural and Food Chemistry, 65*(34), 7379–7386.
Gelcarin PC 379 – Givaudan Active Beauty–Datasheet, (2020). https://cosmetics.specialchem.com/product/i-givaudan-active-beauty-gelcarin-pc-379 (accessed on 7 April 2022).
Ghannam, A., Murad, H., Jazzara, M., Odeh, A., & Allaf, A. W., (2018). Isolation, structural characterization, and antiproliferative activity of phycocolloids from the red seaweed *Laurencia papillosa* on MCF-7 human breast cancer cells. *International Journal of Biological Macromolecules, 108*, 916–926.
Gibson, J. A. E., Garrick, R. C., Burton, H. R., & McTaggart, A. R., (1990). Dimethylsulfide and the alga *Phaeocystis pouchetii* in Antarctic coastal waters. *Marine Biology, 104*(2), 339–346.
Glombitza, K. W., (1977). Highly hydroxylated phenols of the phaeophyceae. In: Faulkner, D. J., & Fenical, W. H., (eds.), *Marine Natural Products Chemistry* (pp. 191–204). Boston: Springer.
Gómez-Ordóñez, E., Jiménez-Escrig, A., & Rupérez, P., (2010). Dietary fiber and physicochemical properties of several edible seaweeds from the northwestern Spanish coast. *Food Research International, 43*(9), 2289–2294.
Gracilaria Hydrogel–Sealaria–Datasheet, (2020). https://cosmetics.specialchem.com/product/i-sealaria-gracilaria-hydrogel (accessed on 7 April 2022).
Gressler, V., Sumie, N., Toyota, M., Colepicolo, P., Mancini, J., Pavan, R., & Pinto, E., (2010). Lipid, fatty acid, protein, amino acid and ash contents in four Brazilian red algae species. *Food Chemistry, 120*, 585–590.
Häder, D. P., & Figueroa, F. L., (1997). Photoecophysiology of marine macroalgae. *Photochemistry and Photobiology, 66*(1), 1–14.
Hadley, S., Wild-Allen, K., Johnson, C., & Macleod, C., (2015). Modelling macroalgae growth and nutrient dynamics for integrated multi-trophic aquaculture. *Journal of Applied Phycology, 27*(2), 901–916.
Hamed, I., Özogul, F., Özogul, Y., & Regenstein, J. M., (2015). Marine bioactive compounds and their health benefits: A review. *Comprehensive Reviews in Food Science and Food Safety, 14*(4), 446–465.
Harley, C. D. G., Anderson, K. M., Demes, K. W., et al., (2012). Effects of climate change on global seaweed communities. *Journal of Phycology, 48*(5), 1064–1078.
Harnedy, P. A., & Fitzgerald, R. J., (2011). Bioactive proteins, peptides, and amino acids from macroalgae. *Journal of Phycology, 47*(2), 218–232.
Hassan, S., El-Twab, S. A., Hetta, M., & Mahmoud, B., (2011). Improvement of lipid profile and antioxidant of hypercholesterolemic albino rats by polysaccharides extracted from the green alga *Ulva lactuca* Linnaeus. *Saudi Journal of Biological Sciences, 18*(4), 333–340.
Hata, Y., Nakajima, K., Uchida, J., Hidaka, H., & Nakano, T., (2001). Clinical effects of brown seaweed, *Undaria pinnatifida* (Wakame), on blood pressure in hypertensive subjects. *Journal of Clinical Biochemistry and Nutrition, 30*, 43–53.
Helmes, R. J. K., López-Contreras, A. M., Benoit, M., et al., (2018). Environmental impacts of experimental production of lactic acid for bioplastics from *Ulva* spp. *Sustainability, 10*(7), 2462.

Herbreteau, F., Coiffard, L. J. M., Derrien, A., & De Roeck-Holtzhauer, Y., (1997). The fatty acid composition of five species of macroalgae. *Botanica Marina*, *40*(1–6), 25–28.

Herrero, B., Madriñán, M., Vieites, J. M., & Espiñeira, M., (2010). Rapid identification of seaweeds in food products by PCR combined with ALF-RFLP and FINS methodologies. *Journal of Agricultural and Food Chemistry*, *58*(22), 11586–11592.

Hidari, K. I. P. J., Takahashi, N., Arihara, M., Nagaoka, M., Morita, K., & Suzuki, T., (2008). Structure and anti-dengue virus activity of sulfated polysaccharide from a marine alga. *Biochemical and Biophysical Research Communications*, *376*(1), 91–95.

Hill, D. J., (2008). Is there a future for natural dyes? *Review of Progress in Coloration and Related Topics*, *27*(1), 18–25.

Hitoe, S., & Shimoda, H., (2017). Seaweed fucoxanthin supplementation improves obesity parameters in mild obese Japanese subjects. *Functional Foods in Health and Disease*, *7*(4), 246–262.

Holdt, S. L., & Kraan, S., (2011). Bioactive compounds in seaweed: Functional food applications and legislation. *Journal of Applied Phycology*, *23*, 543–597.

Hoogvliet, N., & Jongerius, T., (2020). *Studio Nienke Hoogvliet*. https://www.nienkehoogvliet.nl/portfolio/seame/ (accessed on 7 April 2022).

Hwang, P. A., Yan, M. D., Lin, H. T., Li, K. L., & Lin, Y. C., (2016). Toxicological evaluation of low molecular weight fucoidan *in vitro* and *in vivo*. *Marine Drugs*, *14*(7), 121.

Igor, K., Zdyrko, B., Magasinski, A., et al., (2011). A major constituent of brown algae for use in high-capacity Li-ion batteries. *Science*, *334*(6052), 75–79.

Ito, K., & Hori, K., (1989). Seaweed: Chemical composition and potential food uses. *Food Reviews International*, *5*(1), 101–144.

Jacobs, H., Gray, S. N., & Crump, D. H., (2003). Interactions between nematophagous fungi and consequences for their potential as biological agents for the control of potato cyst nematodes. *Mycological Research*, *107*(1), 47–56.

Janarthanan, M., & Kumar, M. S., (2018). The properties of bioactive substances obtained from seaweeds and their applications in textile industries. *Journal of Industrial Textiles*, *48*(1), 361–401.

Jarald, E., Joshi, S. B., & Jain, D. C., (2008). Diabetes and herbal medicines. *Iranian Journal of Pharmacology and Therapeutics*, *7*(1), 97–100.

Jayaraj, J. A., Rahman, W. M., & Punja, Z. K., (2008). Seaweed extract reduces foliar fungal diseases on carrot. *Crop Protection*, *27*(10), 1360–1366.

Jiang, Z., Liu, J., Li, S., et al., (2020). Kelp cultivation effectively improves water quality and regulates phytoplankton community in a turbid, highly eutrophic bay. *Science of the Total Environment*, *707*, 135561.

Jimenez-Escrig, A., & Sanchez-Muniz, F. J., (2000). Dietary fiber from edible seaweeds: Chemical structure, physicochemical properties and effects on cholesterol metabolism. *Science*, *20*(4), 585–598.

Jiménez-Escrig, A., Gómez-Ordóñez, E., & Rupérez, P., (2012). Brown and red seaweeds as potential sources of antioxidant nutraceuticals. *Journal of Applied Phycology*, *24*(5), 1123–1132.

Jiménez-Escrig, A., Jiménez-Jiménez, I., Pulido, R., & Saura-Calixto, F., (2001). Antioxidant activity of fresh and processed edible seaweeds. *Journal of the Science of Food and Agriculture*, *81*, 530–534.

Jung, H. A., Oh, S. H., & Choi, J. S., (2010). Molecular docking studies of phlorotannins from *Eisenia bicyclis* with BACE1 inhibitory activity. *Bioorganic and Medicinal Chemistry Letters*, *20*(11), 3211–3215.

Kaczmarek-Pawelska, A., (2020). Alginate-based hydrogels in regenerative medicine. In: *Alginates - Recent Uses of This Natural Polymer* (pp. 81–94). London: IntechOpen.

Kang, E., Choi, Y. Y., Chae, S. K., Moon, J. H., Chang, J. Y., & Lee, S. H., (2012). Microfluidic spinning of flat alginate fibers with grooves for cell-aligning scaffolds. *Advanced Materials, 24*(31), 4271–4277.

Kang, K. A., Lee, K. H., Chae, S., et al., (2005). Eckol isolated from *Ecklonia cava* attenuates oxidative stress induced cell damage in lung fibroblast cells. *FEBS Letters, 579*(28), 6295–6304.

Kelly, J. R., & Scheibling, R. E., (2012). Fatty acids as dietary tracers in benthic food webs. *Marine Ecology Progress Series, 446*, 1–22.

Kendel, M., Wielgosz-Collin, G., Bertrand, S., Roussakis, C., Bourgougnon, N., & Bedoux, G., (2015). Lipid composition, fatty acids and sterols in the seaweeds *Ulva armoricana*, and *Solieria chordalis* from Brittany (France): An analysis from nutritional, chemotaxonomic, and antiproliferative activity perspectives. *Marine Drugs, 13*(9), 5606–5628.

Khan, W., Rayirath, U. P., Subramanian, S., et al., (2009). Seaweed extracts as biostimulants of plant growth and development. *Journal of Plant Growth Regulation, 28*, 386–399.

Khashei, S., Etemadifar, Z., & Rahmani, H. R., (2018). Immobilization of *Pseudomonas putida* PT in resistant matrices to environmental stresses: A strategy for continuous removal of heavy metals under extreme conditions. *Annals of Microbiology, 68*(12), 931–942.

Kim, J. A., Kong, C. S., & Kim, S. K., (2010). Effect of *Sargassum thunbergii* on ROS mediated oxidative damage and identification of polyunsaturated fatty acid components. *Food and Chemical Toxicology, 48*(5), 1243–1249.

Kim, J. Y., Yu, D. Y., Kim, J. A., Choi, E. Y., Lee, C. Y., & Hong, Y. H., (2016). Effects of *Undaria pinnatifida* and *Laminaria japonica* on rat's intestinal microbiota and metabolite. *Journal of Nutrition and Food Sciences, 6*(3–10), 4172.

Kim, M. M., Kim, S. W., & Lee, S. L., (2013). *HIV-1 Inhibiting Pharmaceutical Composition Containing an Ecklonia Cava: Derived Phloroglucinol Polymer Compound.* US20130338218A1. https://patents.google.com/patent/US20130338218A1/en (accessed on 7 April 2022).

Kim, M. S., Kim, J. Y., Choi, W. H., & Lee, S. S., (2008). Effects of seaweed supplementation on blood glucose concentration, lipid profile, and antioxidant enzyme activities in patients with type 2 diabetes mellitus. *Nutrition Research and Practice, 2*(2), 62.

Kim, S. K., & Chojnacka, K., (2015). *Marine Algae Extracts: Processes, Products, and Applications*, (Vol. 2, p. 345). London: Wiley.

Kim, S. K., (2014). Marine cosmeceuticals. *Journal of Cosmetic Dermatology, 13*(1), 56–67.

Koenighofer, M., Lion, T., Bodenteich, A., et al., (2014). Carrageenan nasal spray in virus confirmed common cold: Individual patient data analysis of two randomized controlled trials. *Multidisciplinary Respiratory Medicine, 9*(1), 1–12.

Kraan, S., (2012). Algal polysaccharides, novel applications and outlook. In: *Carbohydrates – Comprehensive Studies on Glycobiology and Glycotechnology* (pp. 399–522). London: InTechOpen.

Kumar, A., Krishnamoorthy, E., Devi, H. M., et al., (2018). Influence of sea grapes (*Caulerpa racemosa*) supplementation on physical, functional, and anti-oxidant properties of semi-sweet biscuits. *Journal of Applied Phycology, 30*, 1393–1403.

Kumar, C. S., Gasesan, P., Suresh, P. V., & Bhaskar, N., (2008). Seaweeds as a source of nutritionally beneficial compounds – A review. *Journal of Food Science and Technology, 45*(1), 1–13.

Lalegerie, F., Lajili, S., Bedoux, G., Taupin, L., Stiger-Pouvreau, V., & Connan, S., (2019). Photo-protective compounds in red macroalgae from Brittany: Considerable diversity in mycosporine-like amino acids (MAAs). *Marine Environmental Research, 147*, 37–48.

Leandro, A., Pereira, L., & Gonçalves, A. M. M., (2020). Diverse applications of marine macroalgae. *Marine Drugs, 18*(1), 17.

Lee, K. Y., & Mooney, D. J., (2012). Alginate: Properties and biomedical applications. *Progress in Polymer Science, 37*(1), 106–126.

Lee, S. H., Park, M. H., Han, J. S., Jeong, Y., Kim, M., & Jeon, Y. J., (2012). Bioactive compounds extracted from gamtae (*Ecklonia cava*) by using enzymatic hydrolysis, a potent α-glucosidase and α-amylase inhibitor, alleviates postprandial hyperglycemia in diabetic mice. *Food Science and Biotechnology, 21*(4), 1149–1155.

Lee, S., Youn, K., Kim, D., Ahn, M. R., Yoon, E., Kim, O. Y., & Jun, M., (2018). Anti-neuroinflammatory property of phlorotannins from *Ecklonia cava* on Aβ25-35-induced damage in PC12 cells. *Marine Drugs, 17*(1), 7.

Li, L., Ni, R., Shao, Y., & Mao, S., (2014). Carrageenan and its applications in drug delivery. *Carbohydrate Polymers, 103*(15), 1–11.

Li, Q., Luo, J., Wang, C., et al., (2018). Ulvan extracted from green seaweeds as new natural additives in diets for laying hens. *Journal of Applied Phycology, 30*(3), 2017–2027.

Liberski, A., Latif, N., Raynaud, C., Bollensdorff, C., & Yacoub, M., (2016). Alginate for cardiac regeneration: From seaweed to clinical trials. *Global Cardiology Science and Practice, 2016*(1), e201604.

Lim, Y. L., & Mok, S. L., (2010). *In vitro* vascular effects produced by crude aqueous extract of green marine algae, *Cladophora patentiramea* (Mont.) Kützing, in aorta from normotensive rats. *Medical Principles and Practice, 19*(4), 260–268.

Liu, J., Kandasamy, S., Zhang, J., et al., (2015). Prebiotic effects of diet supplemented with the cultivated red seaweed *Chondrus crispus* or with fructo-oligo-saccharide on host immunity, colonic microbiota and gut microbial metabolites. *Complementary and Alternative Medicine, 15*(1), 1–12.

Liu, L., Heinrich, M., Myers, S., & Dworjanyn, S. A., (2012). Towards a better understanding of medicinal uses of the brown seaweed *Sargassum* in traditional Chinese medicine: A phytochemical and pharmacological review. *Journal of Ethnopharmacology, 142*(3), 591–619.

Loon, L., Van, C., & Van, S. E. A., (1999). The families of pathogenesis-related proteins, their activities, and comparative analysis of PR-1 type proteins. *Physiological and Molecular Plant Pathology, 55*(2), 85–97.

Lopes, G., Andrade, P. B., & Valentão, P., (2017). Phlorotannins: Towards new pharmacological interventions for diabetes mellitus type 2. *Molecules, 22*(1), 1–21.

Lopes, G., Sousa, C., Valentão, P., & Andrade, P. B., (2013). Sterols in algae and health. In: Hernández-Ledesma, B., & Herrero, M., (eds.), *Bioactive Compounds from Marine Foods: Plant and Animal Sources* (pp. 173–191). London: Wiley.

López-López, I., Bastida, S., Ruiz-Capillas, C., et al., (2009). Composition and antioxidant capacity of low-salt meat emulsion model systems containing edible seaweeds. *Meat Science, 83*(3), 492–498.

Luthuli, S., Wu, S., Cheng, Y., Zheng, X., Wu, M., & Tong, H., (2019). Therapeutic effects of fucoidan: A review on recent studies. *Marine Drugs, 17*(9), 487.

Mabeau, S., & Fleurence, J., (1993). Seaweed in food products: Biochemical and nutritional aspects. *Trends in Food Science and Technology, 4*(4), 103–107.

MacArtain, P., Gill, C. I. R., Brooks, M., Campbell, R., & Rowland, I. R., (2008). Nutritional value of edible seaweeds. *Nutrition Reviews, 65*(12), 535–543.

Macedo, N. R. P. V., Ribeiro, M. S., Villaça, R. C., et al., (2012). Caulerpin as a potential antiviral drug against herpes simplex virus type 1. *Brazilian Journal of Pharmacognosy, 22*(4), 861–867.
Maeda, H., (2015). Nutraceutical effects of fucoxanthin for obesity and diabetes therapy: A review. *Journal of Oleo Science, 64*(2), 125–132.
Maeda, H., Hosokawa, M., Sashima, T., Funayama, K., & Miyashita, K., (2005). Fucoxanthin from edible seaweed, *Undaria pinnatifida*, shows antiobesity effect through UCP1 expression in white adipose tissues. *Biochemical and Biophysical Research Communications., 332*(2), 392–397.
Magalhaes, K. D., Costa, L. S., Fidelis, G. P., et al., (2011). Anticoagulant, antioxidant and antitumor activities of heterofucans from the seaweed *Dictyopteris delicatula*. *International Journal of Molecular Sciences, 12*(5), 3352–3365.
Makkar, H. P. S., Tran, G., Heuzé, V., et al., (2016). Seaweeds for livestock diets: A review. *Animal Feed Science and Technology, 212*, 1–17.
Martelli, F., Favari, C., Mena, P., et al., (2020). Antimicrobial and fermentation potential of *Himanthalia elongata* in food applications. *Microorganisms, 8*(2), 248.
Martin, L., (2015). Fucoxanthin and its metabolite fucoxanthinol in cancer prevention and treatment. *Marine Drugs, 13*(8), 4784–4798.
Maurya, P. K., Malik, D. S., Yadav, K. K., Kumar, A., Kumar, S., & Kamyab, H., (2019). Bioaccumulation and potential sources of heavy metal contamination in fish species in River Ganga basin: Possible human health risks evaluation. *Toxicology Reports, 6*, 472–481.
McDowell, L. R., (2003). *Minerals in Animal and Human Nutrition*. Netherlands: Elsevier.
McHugh, D. J., (2002). *Prospects for Seaweed Production in Developing Countries*. Rome (Italy): FAO.
McHugh, D. J., Food and Agriculture Organization of the United Nations (FAO), (2003). *A Guide to the Seaweed Industry*. Italy: Food and Agriculture Organization of the United Nations. *FAO Fisheries Technical Paper*. Rome. http://www.fao.org/docrep/006/y4765e/y4765e00.htm (accessed on 7 April 2022).
Medina-Remón, A., Kirwan, R., Lamuela-Raventós, R. M., & Estruch, R., (2016). Dietary patterns and the risk of obesity, type 2 diabetes mellitus, cardiovascular diseases, asthma, and neurodegenerative diseases. *Critical Reviews in Food Science and Nutrition, 58*(2), 262–296.
Menció, A., Mas-Pla, J., Otero, N., et al., (2016). Nitrate pollution of groundwater; all right..., but nothing else?. *Science of the Total Environment, 539*, 241–251.
Mercier, L., Lafitte, C., Borderies, G., Briand, X., Esquerré-Tugayé, M. T., & Fournier, J., (2001). The algal polysaccharide Carrageenans can act as an elicitor of plant defense. *New Phytologist, 149*(1), 43–51.
Mercurio, D. G., Wagemaker, T. A. L., Alves, V. M., Benevenuto, C. G., Gaspar, L. R., & Maia, C. P. M. B. G., (2015). *In vivo* photoprotective effects of cosmetic formulations containing UV filters, vitamins, *Ginkgo biloba* and red algae extracts. *Journal of Photochemistry and Photobiology B: Biology, 153*, 121–126.
Meszka, B., & Bielenin, A., (2011). Activity of laminarin in control of strawberry diseases. *Phytopathologia, 62*, 15–23.
Mišurcová, L., Ambrožová, J., & Samek, D., (2011). Seaweed lipids as nutraceuticals. *Advances in Food and Nutrition Research, 64*, 339–355.
Miyashita, K., Mikami, N., & Hosokawa, M., (2013). Chemical and nutritional characteristics of brown seaweed lipids: A review. *Journal of Functional Foods, 5*(4), 1507–1517.

Moo-Puc, R., Robledo, D., & Freile-Pelegrin, Y., (2011). Enhanced antitumoral activity of extracts derived from cultured *Udotea flabellum* (Chlorophyta). *Evidence-Based Complementary and Alternative Medicine*, *20*11, 1–7.

Morechem™ Hijiki Extract, (2020). https://www.knowde.com/products/morechem-hijiki-extract (accessed on 7 April 2022).

Mouritsen, O. G., Dawczynski, C., Duelund, L., Jahreis, G., Vetter, W., & Schröder, M., (2013). On the human consumption of the red seaweed dulse (*Palmaria palmata* L.). *Journal of Applied Phycology*, *25*(6), 1777–1791.

Muñoz, I., Gómez-Ramos, M. J., Agüera, A., Fernández-Alba, A. R., García-Reyes, J. F., & Molina-Díaz, A., (2009). Chemical evaluation of contaminants in wastewater effluents and the environmental risk of reusing effluents in agriculture. *TrAC – Trends in Analytical Chemistry*, *28*(6), 676–694.

Murata, M., & Nakazoe, J. I., (2001). Production and use of marine algae in Japan. *Japan Agricultural Research Quarterly*, *35*(4), 281–290.

Nagayama, K., Iwamura, Y., Shibata, T., Hirayama, I., & Nakamura, T., (2002). Bactericidal activity of phlorotannins from the brown alga *Ecklonia kurome*. *Journal of Antimicrobial Chemotherapy*, *50*(6), 889–893.

Nallamuthu, N. A., Braden, M., & Patel, M. P., (2012). Some aspects of the formulation of alginate dental impression materials-setting characteristics and mechanical properties. *Dental Materials*, *28*(7), 756–762.

Nallamuthu, N., Braden, M., & Patel, M. P., (2006). Dimensional changes of alginate dental impression materials. *Journal of Materials Science: Materials in Medicine*, *17*(12), 1205–1210.

Namvar, F., Mohamed, S., Fard, S. G., et al., (2012). Polyphenol-rich seaweed (*Eucheuma cottonii*) extract suppresses breast tumor via hormone modulation and apoptosis induction. *Food Chemistry*, *130*(2), 376–382.

Neto, J. M., Gaspar, R., Pereira, L., & Marques, J. C., (2012). Marine macroalgae assessment tool (MarMAT) for intertidal rocky shores. Quality assessment under the scope of the European water framework directive. *Ecological Indicators*, *19*, 39–47.

Neveux, N., Bolton, J. J., Bruhn, A., Roberts, D. A., & Ras, M., (2018). The bioremediation potential of seaweeds: Recycling nitrogen, phosphorus, and other waste products. In: Barre, S., & Bates, S. S., (eds.), *Blue Biotechnology: Production and Use of Marine Molecules* (pp. 217–239). New Jersey: Wiley.

Norziah, M. H., & Ching, C. Y., (2000). Nutritional composition of edible seaweed *Gracilaria changgi*. *Food Chemistry*, *68*(1), 69–76.

Nunes, N., Valente, S., Ferraz, S., Barreto, M. C., & Pinheiro De, C. M. A. A., (2018). Nutraceutical potential of *Asparagopsis taxiformis* (Delile) trevisan extracts and assessment of a downstream purification strategy. *Heliyon*, *4*(11), e00957.

Okimura, T., Jiang, Z., Liang, Y., Yamaguchi, K., & Oda, T., (2019). Suppressive effect of ascophyllan HS on postprandial blood sugar level through the inhibition of α-glucosidase and stimulation of glucagon-like peptide-1 (GLP-1) secretion. *International Journal of Biological Macromolecules*, *125*, 453–458.

Oliveira, C., Ferreira, A. S., Novoa-Carballal, R., et al., (2017). The key role of sulfation and branching on fucoidan antitumor activity. *Macromolecular Bioscience*, *15*(5), 1–13.

Olivero-David, R., Schultz-Moreira, A., Vázquez-Velasco, M., et al., (2011). Effects of Nori- and Wakame-enriched meats with or without supplementary cholesterol on arylesterase activity, lipaemia and lipoproteinaemia in growing Wistar rats. *British Journal of Nutrition*, *106*(10), 1476–1486.

Orfanidis, S., Panayotidis, P., & Stamatis, N., (2003). An insight to the ecological evaluation index (EEI). *Ecological Indicators*, *3*, 27–33.

Ortiz, M. G., (2020). Effect of fucoxanthin on the components of the metabolic syndrome, insulin sensitivity and insulin secretion (NCT03613740). *Clinical Trials*. https://clinicaltrials.gov/ct2/show/results/NCT03613740 (accessed on 7 April 2022).

Pangestuti, R., & Kim, S. K., (2011a). Biological activities and health benefit effects of natural pigments derived from marine algae. *Journal of Functional Foods*, *3*(4), 255–266.

Pangestuti, R., & Kim, S. K., (2011b). Neuroprotective effects of marine algae. *Marine Drugs*, *9*(5), 803–818.

Panlasigui, L. N., Baello, O. Q., Dimatangal, J. M., & Dumelod, B. D., (2003). Blood cholesterol and lipid-lowering effects of carrageenan on human volunteers. *Asia Pacific Journal of Clinical Nutrition*, *12*(2), 209–214.

Paradis, M. E., Couture, P., & Lamarche, B., (2011). A randomized crossover placebo-controlled trial investigating the effect of brown seaweed (*Ascophyllum nodosum* and *Fucus vesiculosus*) on post challenge plasma glucose and insulin levels in men and women. *Applied Physiology, Nutrition, and Metabolism*, *36*(6), 913–919.

Parliament, Council of European, (2000). DIRECTIVE 2000/13/EC – on the approximation of the laws of the member states relating to the labelling, presentation and advertising of foodstuffs. *Official Journal of the European Union*.

Peng, Y., Hu, J., Yang, B., et al., (2015). Chemical composition of seaweeds. In: Tiwari, B. K., & Troy, D. J., (eds.), *Seaweed Sustainability: Food and Non-Food Applications* (pp. 79–124). San Diego: Academic Press.

Pengzhan, Y., (2003). Antihyperlipidemic effects of different molecular weight sulfated polysaccharides from *Ulva pertusa* (Chlorophyta). *Pharmacological Research*, *48*(6), 543–549.

Pereira, L., & Cotas, J., (2019). Historical use of seaweed as an agricultural fertilizer in the European Atlantic area. In: Pereira, L., Bahcevandziev, K., & Joshi, N. H., (eds.), *Seaweeds as Plant Fertilizer, Agricultural Biostimulants and Animal Fodder* (pp. 1–22). Boca Raton: CRC Press.

Pereira, L., & Cotas, J., (2020). Alginates - a general overview. In: Pereira, L., & Cotas, J., (eds.), *Alginates – Recent Uses of This Natural Polymer* (pp. 3–18). London: IntechOpen.

Pereira, L., (2011). A review of the nutrient composition of selected edible seaweeds. In: Pomin, V. H., (eds.), *Seaweed: Ecology, Nutrient Composition and Medicinal Uses* (pp. 15–47). New York: Nova Science Publishers.

Pereira, L., (2018a). Seaweeds as source of bioactive substances and skin care therapy — Cosmeceuticals, algotheraphy, and thalassotherapy. *Cosmetics*, *5*(4), 68.

Pereira, L., (2018b). *Therapeutic and Nutritional Uses of Algae* (p. 672). Boca Raton, FL, USA: CRC Press.

Pérez, M., Falqué, E., & Domínguez, H., (2016). Antimicrobial action of compounds from marine seaweed. *Marine Drugs*, *14*(3), 52.

Phull, A. R., & Kim, S. J., (2017). Fucoidan from *Undaria pinnatifida* regulates type II collagen and COX-2 expression via MAPK and PI3K pathways in rabbit articular chondrocytes. *Biologia*, *72*, 1362–1369.

Phycolanine – Presperse (Sumitomo Corporation) – Datasheet, (2020). https://cosmetics.specialchem.com/product/i-presperse-phycolanine (accessed on 7 April 2022).

Pielesz, A., MacHnicka, A., & Sarna, E., (2011). Antibacterial activity and scanning electron microscopy (SEM) examination of alginate-based films and wound dressings. *Ecological Chemistry and Engineering*, *18*(2), 197–210.

Plaza, M., Cifuentes, A., & Ibáñez, E., (2008). In the search of new functional food ingredients from algae. *Trends in Food Science and Technology*, *19*(1), 31–39.
Plaza, M., Herrero, M., Cifuentes, A., & Ibáñez, E., (2009). Innovative natural functional ingredients from microalgae. *Journal of Agricultural and Food Chemistry*, *57*(16), 7159–7170.
Plaza, M., Santoyo, S., Jaime, L., et al., (2010). Screening for bioactive compounds from algae. *Journal of Pharmaceutical and Biomedical Analysis*, *51*(2), 450–455.
Rajapakse, N., & Kwon, K. S. K., (2011). Nutritional and digestive health benefits of seaweed. *Advances in Food and Nutrition Research*, *64*, 17–28.
Ravikumar, S., Anburajan, L., Ramanathan, G., & Kaliaperumal, N., (2002). Screening of seaweed extracts against antibiotic resistant post operative infectious pathogens. *Seaweed Research and Utilization*, *24*, 95–99.
Regal, A. L., Alves, V., Gomes, R., et al., (2020). Drying process, storage conditions, and time alter the biochemical composition and bioactivity of the anti-greenhouse seaweed *Asparagopsis taxiformis*. *European Food Research and Technology.*, *246*(4), 781–793.
Ren, R., Gong, J., Zhao, Y., et al., (2018). Sulfated polysaccharide from *Enteromorpha prolifera* suppresses SREBP-1c and ACC expression to lower serum triglycerides in high-fat diet-induced hyperlipidaemic rats. *Journal of Functional Foods*, *40*, 722–728.
Rengasamy, K. R. R., Kulkarni, M. G., Stirk, W. A., & Staden, J. V., (2014). Advances in algal drug research with emphasis on enzyme inhibitors. *Biotechnology Advances*, *32*(8), 1364–1381.
Rokkaku, T., Kimura, R., Ishikawa, C., et al., (2013). Anticancer effects of marine carotenoids, fucoxanthin and its deacetylated product, fucoxanthinol, on osteosarcoma. *International Journal of Oncology*, *43*(4), 1176–1186.
Roleda, M. Y., & Hurd, C. L., (2019). Seaweed nutrient physiology: Application of concepts to aquaculture and bioremediation. *Phycologia*, *58*(5), 552–562.
Romanazzi, G., Sanzani, S. M., Yang, B. Y., Tian, S., Martínez, P. G., & Alkan, N., (2016). Induced resistance to control postharvest decay of fruit and vegetables. *Postharvest Biology and Technology*, *122*, 82–94.
Rosa, J., Leston, S., Crespo, D., et al., (2020). Uptake of enrofloxacin from seawater to the macroalgae *Ulva* and its use in IMTA systems. *Aquaculture*, *516*, 734609.
Roy, M. C., Anguenot, R., Fillion, C., Beaulieu, M., Bérubé, J., & Richard, D., (2011). Effect of a commercially-available algal phlorotannins extract on digestive enzymes and carbohydrate absorption *in vivo*. *Food Research International*, *44*(9), 3026–3029.
Ruocco, N., Costantini, S., Guariniello, S., & Costantini, M., (2016). Polysaccharides from the marine environment with pharmacological, cosmeceutical and nutraceutical potential. *Molecules*, *21*(5), 1–16.
Rupérez, P., & Saura-Calixto, F., (2001). Dietary fiber and physicochemical properties of edible Spanish seaweeds. *European Food Research and Technology*, *212*(3), 349–354.
Rupérez, P., & Toledano, G., (2003). Indigestible fraction of edible marine seaweeds. *Journal of the Science of Food and Agriculture*, *83*(12), 1267–1272.
Rupérez, P., Ahrazem, O., & Leal, J. A., (2002). Potential antioxidant capacity of sulfated polysaccharides from the edible marine brown seaweed *Fucus vesiculosus*. *Journal of Agricultural and Food Chemistry*, *50*(4), 840–845.
Ryan, M. H., Tibbett, M., Lambers, H., et al., (2017). Pronounced surface stratification of soil phosphorus, potassium and sulfur under pastures upstream of a eutrophic wetland and estuarine system. *Soil Research*, *55*(7), 657–669.
Saker, K. E., Allen, V. G., Fontenot, J. P., et al., (2001). Tasco-forage: II. Monocyte immune cell response and performance of beef steers grazing tall fescue treated with a seaweed extract. *Journal of Animal Science*, *79*(4), 1022.

Saldarriaga-Hernandez, S., Hernandez-Vargas, G., Iqbal, H. M. N., Barceló, D., & Parra-Saldívar, R., (2020). Bioremediation potential of *Sargassum* sp. biomass to tackle pollution in coastal ecosystems: Circular economy approach. *Science of the Total Environment, 1*(715), 136978.
Satheesh, S., Siddik, A., Baakdah, M., & Al-Sofyani, A., (2017). An introduction to the ecological significance of seaweeds on coastal ecosystems. In: Nabdit, E., (ed.), *Biotechnological Applications of Seaweeds* (p. 130). New York: Nova Science Publishers, Inc.
Sathivel, A., Raghavendran, H. R. B., Srinivasan, P., & Devaki, T., (2008). Anti-peroxidative and anti-hyperlipidemic nature of *Ulva lactuca* crude polysaccharide on D-galactosamine induced hepatitis in rats. *Food and Chemical Toxicology, 46*(10), 3262–3267.
Satomi, Y., (2017). Antitumor and cancer-preventative function of fucoxanthin: A marine carotenoid. *Anticancer Research, 37*(4), 1557–1562.
Schönfeld-Leber, B., (1979). Marine algae as human food in Hawaii, with notes on other Polynesian islands. *Ecology of Food and Nutrition, 8*(1), 47–59.
Sealgae, (2020). http://www.sealgae.pt/ (accessed on 7 April 2022).
Seanol®. (2020). http://seanolinstitute.org/ssc/ (accessed on 7 April 2022).
Sellimi, S., Benslima, A., Ksouda, G., Montero, V. B., Hajji, M., & Nasri, M., (2017). Safer and healthier reduced nitrites Turkey meat sausages using lyophilized *Cystoseira barbata* seaweed extract. *Journal of Complementary and Integrative Medicine, 15*(1), 1–14.
Sena, C. M., Leandro, A., Azul, L., Seiça, R., & Perry, G., (2018). Vascular oxidative stress: Impact and therapeutic approaches. *Frontiers in Physiology, 9*, 1668.
SEPPIC, (2020). *Raw Materials from SEPPIC.* https://www.seppic.com/en/beauty-care (accessed on 7 April 2022).
Shanab, S. M. M., (2007). Antioxidant and antibiotic activities of some seaweeds (Egyptian isolates). *International Journal of Agriculture and Biology, 9*, 220–225.
Shanmugam, M., & Mody, K. H., (2000). Heparinoid-active sulphated polysaccharides from marine algae as potential blood anticoagulant agents. *Current Science, 79*(12), 1672–1683.
Shannon, E., & Abu-Ghannam, N., (2017). Optimisation of fucoxanthin extraction from Irish seaweeds by response surface methodology. *Journal of Applied Phycology, 29*(2), 1027–1036.
Shannon, E., & Abu-Ghannam, N., (2019). Seaweeds as nutraceuticals for health and nutrition. *Phycologia, 58*(5), 563–577.
Sharma, S., Neves, L., Funderud, J., Mydland, L. T., Øverland, M., & Horn, S. J., (2018). Seasonal and depth variations in the chemical composition of cultivated *Saccharina latissima*. *Algal Research, 32*, 107–112.
Shi, X., Zhou, G., Liao, S., Shan, S., Wang, G., & Guo, Z., (2018). Immobilization of cadmium by immobilized *Alishewanella* sp. WH16-1 with alginate-lotus seed pods in pot experiments of Cd-contaminated paddy soil. *Journal of Hazardous Materials, 357*, 431–439.
Shibata, T., Yamaguchi, K., Nagayama, K., Kawaguchi, S., & Nakamura, T., (2002). Inhibitory activity of brown algal phlorotannins against glycosidases from the viscera of the turban shell *Turbo cornutus*. *European Journal of Phycology, 37*(4), 493–500.
Sinurat, E., Subaryono, S., & Rambe, A. O., (2020). Making of alginate dental impression material from *Sargassum duplicatum*. In: *E3S Web of Conferences*, EDP Sciences 147, 03015.
Slack, R., Friedmann, P., & Schewes, R., (2020). Kelpack®. https://www.kelpak.com/home.html (accessed on 7 April 2022).
Smit, A. J., (2004). Medicinal and pharmaceutical uses of seaweed natural products: A review. *Journal of Applied Phycology, 16*, 245–262.
Spiers, D. E., Eichen, P. A., Leonard, M. J., et al., (2004). Benefit of dietary seaweed (*Ascophyllum nodosum*) extract in reducing heat strain and fescue toxicosis: A comparative evaluation. *Journal of Thermal Biology, 29*(7, 8), 753–757.

Stabili, L., Acquaviva, M. I., Biandolino, F., et al., (2014). Biotechnological potential of the seaweed *Cladophora rupestris* (Chlorophyta, Cladophorales) lipidic extract. *New Biotechnology*, *31*(5), 436–444.

Stefels, J., & Van, B. W. H. M., (1993). Production of DMS from dissolved DMSP in axenic cultures of the marine phytoplankton species *Phaeocystis* sp. *Marine Ecology Progress Series*, *97*(1), 11–18.

Stengel, D. B., Connan, S., & Popper, Z. A., (2011). Algal chemodiversity and bioactivity: Sources of natural variability and implications for commercial application. *Biotechnology Advances*, *29*, 483–501.

Strate, H. J. V. D., Boele-Bos, S. A., Olsen, J. L., Zande, L. V. D., & Stam, W. T., (2002). Phylogeographic studies in the tropical seaweed *Cladophoropsis membranacea* (Chlorophyta, Ulvophyceae) reveal a cryptic species complex. *Journal of Phycology*, *38*(3), 572–582.

Sultana, V., Baloch, G. N., Ara, J. A., Tariq, M. R., & Ehteshamul-Haque, S., (2011). Comparative efficacy of a red alga *Solieria robusta*, chemical fertilizers and pesticides in managing the root diseases and growth of soybean. *Pakistan Journal of Botany*, *43*(1), 1–6.

Syad, A. N., Shunmugiah, K. P., & Devi, K. P. D., (2013). Seaweeds as nutritional supplements: Analysis of nutritional profile, physicochemical properties and proximate composition of *G. acerosa* and *S. wightii*. *Biomedicine and Preventive Nutrition*, *3*(2), 139–144.

Tanemura, Y., Yamanaka-Okumura, H., Sakuma, M., Nii, Y., Taketani, Y., & Takeda, E., (2014). Effects of the intake of *Undaria pinnatifida* (Wakame) and its sporophylls (Mekabu) on postprandial glucose and insulin metabolism. *The Journal of Medical Investigation*, *61*(3, 4), 291–297.

Tanna, B., & Mishra, A., (2018). Metabolites unravel nutraceutical potential of edible seaweeds: An emerging source of functional food. *Comprehensive Reviews in Food Science and Food Safety*, *17*(6), 1613–1624.

Tanna, B., & Mishra, A., (2019). Nutraceutical potential of seaweed polysaccharides: Structure, bioactivity, safety, and toxicity. *Comprehensive Reviews in Food Science and Food Safety*, *18*(3), 817–831.

Teas, J., Baldeón, M. E., Chiriboga, D. E., Davis, J. R., Sarriés, A. J., & Braverman, L. E., (2009). Could dietary seaweed reverse the metabolic syndrome?. *Asia Pacific Journal of Clinical Nutrition*, *18*(2), 145–154.

Teas, J., Pino, S., Critchley, A., & Braverman, L. E., (2004). Variability of iodine content in common commercially available edible seaweeds. *Thyroid*, *14*(10), 36–41.

Torres, P., Santos, J. P., Chow, F., & Santos, D. Y. A. C., (2019). A comprehensive review of traditional uses, bioactivity potential, and chemical diversity of the genus *Gracilaria* (*Gracilariales, Rhodophyta*). *Algal Research*, *37*, 288–306.

Toskas, G., Hund, R. D., Laourine, E., Cherif, C., Smyrniotopoulos, V., & Roussis, V., (2011). Nanofibers based on polysaccharides from the green seaweed *Ulva rigida*. *Carbohydrate Polymers*, *84*(3), 1093–1102.

Turck, D., Bresson, J. L., Burlingame, B., et al., (2017). Safety of *Ecklonia cava* phlorotannins as a novel food pursuant to Regulation (EC) No 258/97. *EFSA Journal*, *15*(10), e05003.

Um, M. Y., Lim, D. W., Son, H. J., Cho, S., & Lee, C., (2018). Phlorotannin-rich fraction from *Ishige foliacea* brown seaweed prevents the scopolamine-induced memory impairment via regulation of ERK-CREB-BDNF pathway. *Journal of Functional Foods*, *40*, 110–116.

Ustunada, M., Erduğan, H., Yilmaz, S., Akgul, R., & Aysel, V., (2011). Seasonal concentrations of some heavy metals (Cd, Pb, Zn, and Cu) in *Ulva rigida* J. Agardh (Chlorophyta) from Dardanelles (Canakkale, Turkey). *Environmental Monitoring and Assessment*, *177*(1), 337–342.

Vasquez, J. A., & Guerra, N., (1996). The use of seaweeds as bioindicators of natural and anthropogenic contaminants in Northern Chile. *Hydrobiologia, 326*, 327–333.

Vermeer, B., (1996). Cosmeceuticals: A proposal for rational definition, evaluation, and regulation. *Archives of Dermatology, 132*(3), 337–340.

Villares, R., Puente, X., & Carballeira, A., (2002). Seasonal variation and background levels of heavy metals in two green seaweeds. *Environmental Pollution, 119*(1), 79–90.

Vizetto-Duarte, C., Pereira, H., Sousa, C. B., et al., (2015). Fatty acid profile of different species of algae of the *Cystoseira* genus: A nutraceutical perspective. *Natural Product Research, 29*(13), 1264–1270.

Wakamine™ 1%-Givaudan Active Beauty–Datasheet, (2020). https://cosmetics.specialchem.com/product/i-givaudan-active-beauty-wakamine-1 (accessed on 7 April 2022).

Wang, H. M. D., Chen, C. C., Huynh, P., & Chang, J. S., (2015). Exploring the potential of using algae in cosmetics. *Bioresource Technology, 184*, 355–362.

Wang, J., Zheng, J., Huang, C., et al., (2018). Eckmaxol, a phlorotannin extracted from *Ecklonia maxima*, produces anti-β-amyloid oligomer neuroprotective effects possibly via directly acting on glycogen synthase kinase 3β. *ACS Chemical Neuroscience, 9*(6), 1349–1356.

Wang, W., Wang, S. X., & Guan, H. S., (2012). The antiviral activities and mechanisms of marine polysaccharides: An overview. *Marine Drugs, 10*(12), 2795–2816.

Wang, X., Sun, G., Feng, T., et al., (2019). Sodium oligomannate therapeutically remodels gut microbiota and suppresses gut bacterial amino acids-shaped neuroinflammation to inhibit Alzheimer's disease progression. *Cell Research, 29*(10), 787–803.

Washington, N., (1990). Investigation into the barrier action of an alginate gastric reflux suppressant, liquid Gaviscon®. *Drug Investigation, 20*(1), 23–30.

Wells, E., Wilkinson, M., Wood, P., & Scanlan, C. M., (2003). *Water Framework Directive Development of Classification Tools for Ecological Assessment: Macroalgae Species Richness* UK TAG Report – Macroalgae on Intertidal Rocky Shores.

Werner, F. J., Graiff, A., & Matthiessen, B., (2016). Even moderate nutrient enrichment negatively adds up to global climate change effects on a habitat-forming seaweed system. *Limnology and Oceanography, 61*, 1891–1899.

Wijesekara, I., Pangestuti, R., & Kim, S. K., (2011). Biological activities and potential health benefits of sulfated polysaccharides derived from marine Algae. *Carbohydrate Polymers, 84*(1), 14–21.

Wijesinghe, W. A. J. P., & Jeon, Y., (2012). Biological activities and potential industrial applications of fucose rich sulfated polysaccharides and fucoidans isolated from brown seaweeds: A review. *Carbohydrate Polymers, 88*(1), 13–20.

Wong, K. H., & Cheung, P. C. K., (2001). Nutritional evaluation of some subtropical red and green seaweeds part II. *In vitro* protein digestibility and amino acid profiles of protein concentrates. *Food Chemistry, 72*(1), 11–17.

Wu, H., Huo, Y., Han, F., Liu, Y., & He, P., (2015). Bioremediation using *Gracilaria chouae* co-cultured with *Sparus macrocephalus* to manage the nitrogen and phosphorous balance in an IMTA system in Xiangshan Bay, China. *Marine Pollution Bulletin, 91*(1), 272–279.

Xing, R., Liu, S., Yu, H., Guo, Z., Li, Z., & Li, P., (2005). Preparation of high-molecular weight and high-sulfate content chitosans and their potential antioxidant activity *in vitro*. *Carbohydrate Polymers, 61*(2), 148–154.

Yang, L., Wang, P., Wang, H., et al., (2013). Fucoidan derived from *Undaria pinnatifida* induces apoptosis in human hepatocellular carcinoma SMMC-7721 cells via the ROS-mediated mitochondrial pathway. *Marine Drugs, 11*(6), 1961–1976.

Yasham, (2020). *Hair Care Actives from Seaweeds.* https://www.yasham.in/hair-care-seaweeds.html (accessed on 7 April 2022).

Yu, Z., Robinson, S. M. C., Xia, J., Sun, H., & Hu, C., (2016). Growth, bioaccumulation and fodder potentials of the seaweed *Sargassum hemiphyllum* grown in oyster and fish farms of south China. *Aquaculture, 464*, 459–468.

Yu-Qing, T., Mahmood, K., Shehzadi, R., & Ashraf, M. F., (2016). *Ulva lactuca* and its polysaccharides: Food and biomedical aspects. *Journal of Biology, 6*(1), 140–151.

Zappelli, C., Barbulova, A., Apone, F., & Colucci, G., (2016). Effective active ingredients obtained through biotechnology. *Cosmetics, 3*(4), 39.

Zhang, H., Su, M., Li, M., et al., (2019). Efficient removal of Pb(II) ions from aqueous solution by novel PVA-sodium alginate immobilized sulfate reducing bacteria. *Desalination and Water Treatment, 138*, 280–290.

Zhang, J., Tiller, C., Shen, J., et al., (2007). Antidiabetic properties of polysaccharide- and polyphenolic-enriched fractions from the brown seaweed *Ascophyllum nodosum. Canadian Journal of Physiology and Pharmacology, 85*(11), 1116–1123.

Zhang, W., Oda, T., Yu, Q., & Jin, J., (2015). Fucoidan from *Macrocystis pyrifera* has powerful immune-modulatory effects compared to three other fucoidans. *Marine Drugs, 13*(3), 1084–1104.

Zhang, Z., Teruya, K., Yoshida, T., Eto, H., & Shirahata, S., (2013). Fucoidan extract enhances the anti-cancer activity of chemotherapeutic agents in MDA-MB-231 and MCF-7 breast cancer cells. *Marine Drugs, 11*(1), 81–98.

Zhang, Z., Wang, F., Wang, X., Liu, X., Hou, Y., & Zhang, Q., (2010). Extraction of the polysaccharides from five algae and their potential antioxidant activity *in vitro. Carbohydrate Polymers, 82*(1), 118–121.

Zhao, X., Huebsch, N., Mooney, D. J., & Suo, Z., (2010). Stress-relaxation behavior in gels with ionic and covalent crosslinks. *Journal of Applied Physics, 107*(6), 63509.

Zimmermann, H., Shirley, S. G., & Zimmermann, U., (2007). Alginate-based encapsulation of cells: Past, present, and future. *Current Diabetes Reports, 7*(4), 314–320.

Zollmann, M., Robin, A., Prabhu, M., et al., (2019). Green technology in green macroalgal biorefineries. *Phycologia, 58*(5), 516–534.

CHAPTER 10

THERAPEUTIC POTENTIAL OF MACROALGAE AND ITS BENEFICIAL HEALTH IMPACTS IN AQUACULTURE

PRABHAKAR MISHRA, NEELAKANTA SARVASHIVA KIRAN, and DAVID ELISHA HENRY

Department of Biotechnology, School of Applied Sciences, REVA University, Bengaluru–560064, Karnataka, India

ABSTRACT

On account of macroalgae, *Chlorophyta* accounts for just a small portion of worldwide seaweed biomass landings. *Rhodophyta* and *Phaeophyta*, on the other hand, account for the majority of the remaining biomass production, totaling more than 32 million tons each year. In comparison to conventional terrestrial resources, marine macroalgae are considered to be alternative promising fish feed resources with numerous advantages, including high growth rate, exceptional tolerance to be potentially cultivated in marine water, and non-competition of arable lands for cultivation. Furthermore, macroalgae are a rich source of various bioactive components, such as complex carbohydrates, proteins, lipids, and pigments which are gaining popularity due to their potential to improve aquaculture and fish health. Nutritional treatment with seaweed and its extracts has been shown to boost fish immunological health and growth performance. The use of diverse macroalgae in fish diets has resulted in significant advances in productive indices and overall health. Macroalgae and their compounds have also lately been offered as a safer, alternative, preventive, and therapeutic treatment for controlling pathogenic illnesses in farmed fish. Macroalgae have been found

Algal Metabolites: Biotechnological, Commercial, and Industrial Applications. Jeyabalan Sangeetha, Devarajan Thangadurai, Saher Islam, & Ravichandra Hospet (Eds.)

to be a potential source of antimicrobial and immunomodulatory compounds against a variety of pathogenic species in several investigations. This chapter focuses on the health advantages of macroalgae and its extracts, as well as its prospective application in aquaculture for medicinal characteristics.

10.1 INTRODUCTION

Seaweeds and other benthic marine algae that are visible to the naked eye are categorized under a single term known as Macroalgae. Large-sized Macroalgae are commonly called seaweeds, though they are typically not "weeds." The ecological roles of Macroalgae and plants are quite indistinguishable, and they are treated as marine aquatic plants because of their photosynthetic property. Macroalgae are completely devoid of structures like roots, leaves, flowers, and other vascular tissues, which makes them differ from the other marine plant species. From the ancient periods, seaweeds have been used diversely in various aspects of marine biology, medicine, and nutritious diet for the well-being of both marine life and humans.

Seaweeds have gained a lot of value and attention in recent decades due to its distinctive biological importance (Figure 10.1). In the current scenario, these macroalgae are being extensively applied in the aquaculture field. Aquaculture, typically known as aquafarming, is the method for cultivation of various marine species, which includes fishes, mollusks, marine plants, and algae. Interest in Aquaculture has spontaneously increased from the late 1950s. According to FAO, in 2018 the global fish production had peaked up to 179 million tons from which 47% has been utilized for direct human consumption (FAO, 2020) (Figure 10.2). The remaining percent (53%) for non-food uses such as for production of fish oils and reduction to fishmeal. The capture fishery production has obtained a static in the late 1980s; since then, aquaculture has been responsible for the spontaneous growth of fish supply for human consumption. Aquaculture products are a rich source for protein supplements and other nutraceutical aspects, simultaneously the aquaculture is highly vulnerable to diverse microbial pathogens resulting in a higher economical loss. Inadequate water treatment, lack of hygiene facilities, change in temperatures, and poor nutritional levels lead to the spread of infections, stress, and immune suppression in aquatic plants and animals which cause an elevation in the mortality rates to a great extent. Several veterinary drugs and primeval control techniques which practiced to terminate these economic losses.

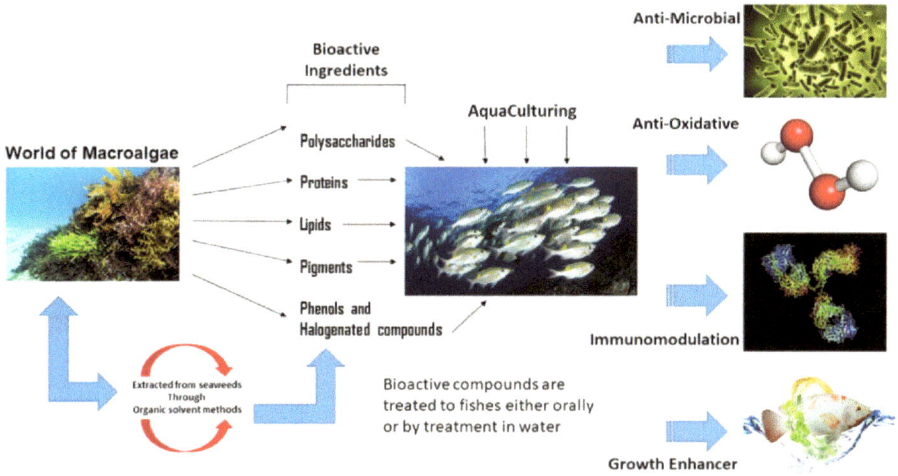

FIGURE 10.1 Pictorial representation of the applicative therapeutic potential of macroalgae in aquaculture.

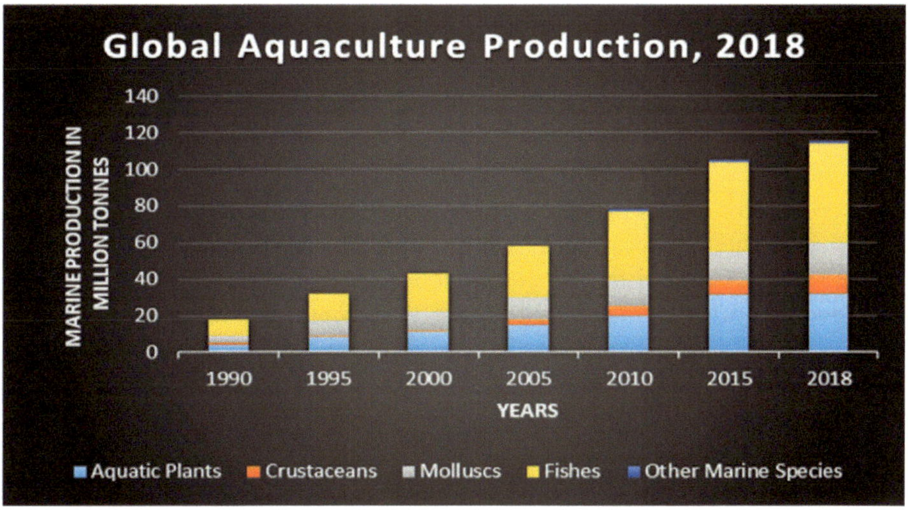

FIGURE 10.2 Graphical representation of aquaculture production around the globe.

Commercial use of these synthetically synthesized veterinary drugs such as chloramphenicol in aquaculture proved to be very much fruitful in controlling many disease-causing microorganisms, but the use of these drugs was restricted as it was found to cause certain immunopathological effects in humans (Reverter et al., 2014). Due to these upcoming negative circumstances, the use of eco-friendly control measures gained a lot of attention worldwide. Seaweeds are naturally incorporated with different bioactive compounds which prove to be highly beneficial as a therapeutic agent. Seaweed extracts provided a wide range of advantageous properties which included anti-stress, antioxidant, growth promotion, immune booster, and disease resistance property in marine animals. The presence of these bioactive compounds in the aquaculture proved to enhance appetite and anti-pathogenic properties in fishes. Use of these seaweed extracts and essential oils was found to be completely eco-friendly and cost-effective and also highly biodegradable compared to the synthetically synthesized veterinary drugs.

10.2 MACROALGAE: DIVERSITY, DISTRIBUTION, AND ECOLOGICAL SIGNIFICANCE

Macroalgae have been distributed worldwide in various aquatic habitats, both in freshwater and marine ecosystems. It is estimated that the marine environment is home to around 80% of the total macroalgal population, and the other 20% found in the freshwaters. Macroalgae are diversified into four main phyla based on the pigments which they accommodate: *Rhodophyta* [red algae], *Chlorophyta* [green algae], *Phaeophyta* [brown algae] (Wan et al., 2019; Vilar et al., 2020) and *Cyanophyta* [blue-green algae (BGA)]. Over 12,000 species of macroalgae have been prescribed worldwide till date from which over 7,000 species within over 700 genera belong to *Rhodophyta*, more than 2,000 species from 265 genera to *Phaeophyta* and over 1,500 species categorized under *Chlorophyta* and *Cyanophyta* respectively (Macreadie et al., 2017). Green seaweeds that are found in the freshwater habitats are classified into a special category known as *Charophyta* (Williams and Smith, 2007), which consists of only 36 prescribed species till date. Macroalgae usually thrive in temperatures between 10°C and 23°C, but some macroalgae were also found in the regions of Antarctica, however, with a relatively low diversity of approximately 120 species (Pellizzari et al., 2017). The taxonomy of macroalgae is found to be highly vexed, with many recent changes being made due to the gathered molecular data. Although red algae are more diverse, brown algae are the ones which are the largest extending from over 30 to 50 meters in length.

Seaweeds are distributed horizontally in various zonations which are supralittoral, littoral, and sublittoral regions of the seas and oceans. Green macroalgae are mostly predominant in the littoral zone and some commonly found green algae include species of *Ulva*, *Caulerpa*, and *Enteromorpha*. Brown and red macroalgae inhabit the regions of the sublittoral zones and mostly include the species of *Sargassum*, *Gracilaria* and *Ceramium* (Figure 10.3). The blue-green (BG) macroalgae are usually found in the supralittoral zones which mainly include the species of *Lyngbya*, *Spirulina*, and *Oscillatoria*.

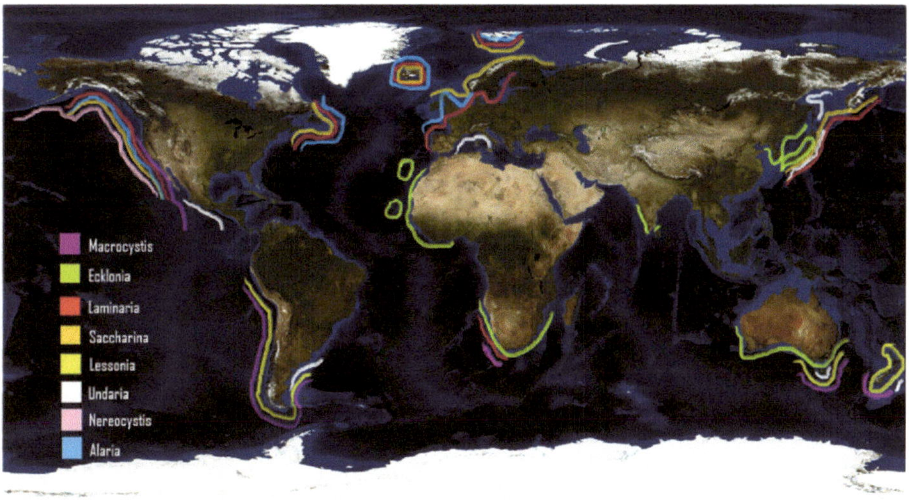

FIGURE 10.3 Worldwide distribution of dominant genera of macroalgal species.

Macroalgae play a very crucial role in maintaining the ecological balance of the aquatic ecosystems. They have a high level of contributions as primary producers of organic matter which is consumed by various herbivorous fishes, crabs, sea urchins and zooplanktons. Macroalgae frequently leak organic carbon into the waters, which is consumed by a diverse variety of bacteria. Organic nitrogen is a very crucial element in the ecosystem, macroalgae mainly the species belonging to the *Cyanophyta* help in the fixation of organic nitrogen in the ocean waters, mainly at the reefs. Macroalgae can also be titled as the "Constructors of the Aquatic Ecosystems" for their important contributions in depositing Calcium Carbonate [$CaCO_3$] for the framework of coral reefs (Eyre et al., 2014). Some macroalgae live within the skeleton of the dead corals for the progress of reef erosion which help in the destruction and removal of the carcass in the ocean floor. Macroalgae play a vital role in the degradation of the reefs, especially during the phase shifts (Cheal et al., 2010) when the ample corals are restored by macroalgae.

10.3 SEAWEED FARMING: CULTIVATION, TECHNIQUES, AND CHALLENGES

Since the late 17th Century, Seaweed farming has observed a lot of attention from aquatic farmers worldwide due to its various biological beneficiaries and economic demand. Seaweeds were first propagated in 1670 at Tokyo Bay, Japan; thenceforth, the cultivation of seaweeds metastasized to various parts of the world. Seaweed farming has emerged as an interestingly contentious biomass production suitor for food and other related uses. However, 99% of the complete global seaweed production is contributed by only 8 Asian countries and the rest 1% from the other countries. There are approximately 150 countries worldwide with coastal linings who are yet to start seaweed farming (Radulovich et al., 2015). Worldwide production of seaweeds is primarily obtained by two means: harvesting from wild stocks and through aquaculture. According to the FAO, the global seaweed production in 2016 was estimated to be over 30.1 million tons gathering an economical value of over $11.7 billion (FAO, 2018). Presently about 80% of the total seaweed production is used for direct human consumption and flavoring additives and the rest 20% for extraction of hydrocolloids, which has various diversified uses in cosmetics, medicines, and industrial production.

Even though there are thousands of species of seaweeds distributed in diverse habitats of oceans, their growth is restricted only to the shallow ocean waters. The reason behind this is seaweeds require both an attachment to the ocean floor substrate for growth and adequate quantity of sunlight passing through the ocean surface to photosynthesize. The main principle to be followed for a successful seaweed farming would be the pre-cultivation choices (Figure 10.4). The selection of an appropriate site for cultivation which has optimum environmental conditions is important, which include: sufficient amount of sunlight for optimized photosynthetic rate, water volumetric rate and capacity of nutrient uptake by the water. This evokes the necessity of making sure that the seaweeds grow in place and the desired spatial arrangements while overcoming the possible perilous outbreaks. These hazardous counters include storms, salinity, ocean currents, herbivore attacks, temperature, and other biological changes. Unlike other marine plants, seaweeds, due to the lack of typical plant structures, uptake nutrients and gases from the water through their whole-body surface, hence, the dispense of nutrients to the seaweeds through the motion of water is very essential. Most importantly, unlike fish aquaculture, there is no utilization of pesticides, antibiotics, hormones, and other chemical components in seaweed farming. This marks as a major difference between seaweed farming, fish aquaculture and agriculture proving that seaweed farming is very much an eco-friendly aspect of aquaculture (Kim et al., 2017).

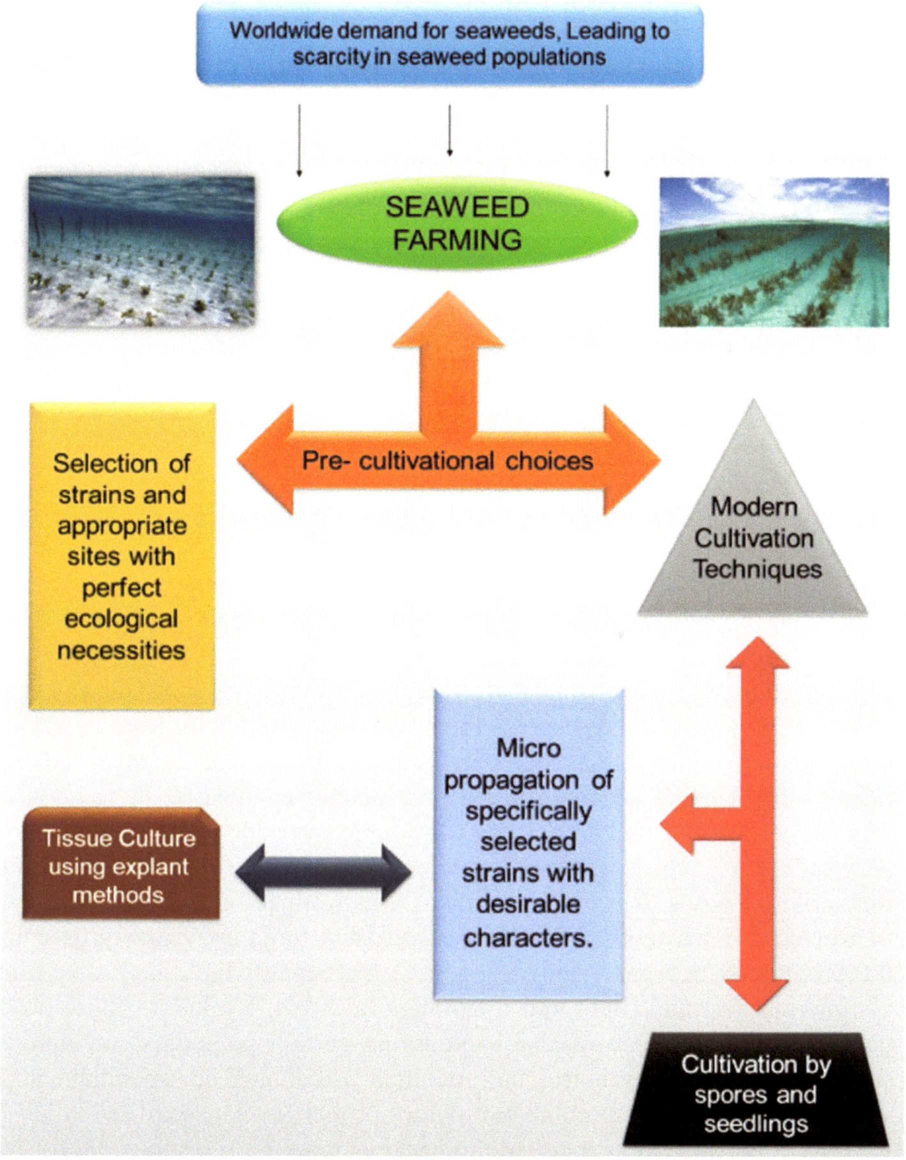

FIGURE 10.4 Various methodology of seaweed farming.

Despite control and maintenance is an important aspect of seaweed farming, complete control of these biological factors is not viable. Following these measures that are to be taken into consideration, many cultivation techniques have emerged since the recent years which proved to be both advantageous and disadvantageous in its ways. The most likely and modern technique of seaweed farming practiced in recent times is the propagule [seedlings] production farming. Seaweed populates through spore formation by sexual, asexual, or vegetative propagation. Many nursery methods have been developed for the production of healthy propagules, including sexual and asexual procedures. Over 90 species of seaweeds are recorded for propagule production by tip culture technique which includes species of *Enteromorpha*, *Saccharina*, *Undaria*, *Pyropia*, *Gracilaria*, and *Sargassum*. After the production of propagules, the planting is done using the main cultivation techniques basically by the use of nets and ropes aligned in rows and columns consisting of an adequate number of seaweed propagules per area. The number of propagules per area must be appropriately maintained to avoid overcrowding of the biomass.

10.3.1 A SCENARIO OF SEAWEED FARMING IN INDIA

With a vast coastal line extending up to 17,000 km, commercial interest of seaweed farming in India had emerged primarily in the regions of Mandapam, Tamil Nadu in 2006. Since then, multiple seaweed cultivation industries manifested in all major coastal towns of India. Abundant growth of seaweeds in India has been observed across the coasts of Tamil Nadu, Kerala, Karnataka, Gujarat, Odisha, Lakshadweep, and Andaman and Nicobar Islands (Figure 10.5). Rich seaweed beds have also been located in regions of Mumbai, Ratnagiri, and Goa. India is home to over 700 species of different seaweeds from which about 60 species have been cultivated for various industrial purposes. It is estimated that India produces around 600,000 tons of wet weight seaweeds annually. Seaweeds in India are majorly used for the production of Agar, Alginates, and Carrageenan which have very high commercial demand (Seth and Shanmugam, 2016). Excluding these products, seaweeds are also used for extracting certain secondary metabolites such as kainic acid, laminarin, and fucoidan which have commendable antiviral, anti-fungus, and immunostimulatory properties. The seaweed industry in India is designated as a cottage industry and are only responsible for the production of natural stock of agar-yielding red seaweeds such as *Gelidiella acerosa* and *Gracilaria edulis* and brown seaweeds such as *Sargassum* and

Turbinaria for the production of algin. There are over 40 seaweed industries found in and around Madurai, Tamil Nadu which are accountable for the production of 120 million tons of agar and 500 million tons of algin annually in India (Mantri et al., 2017).

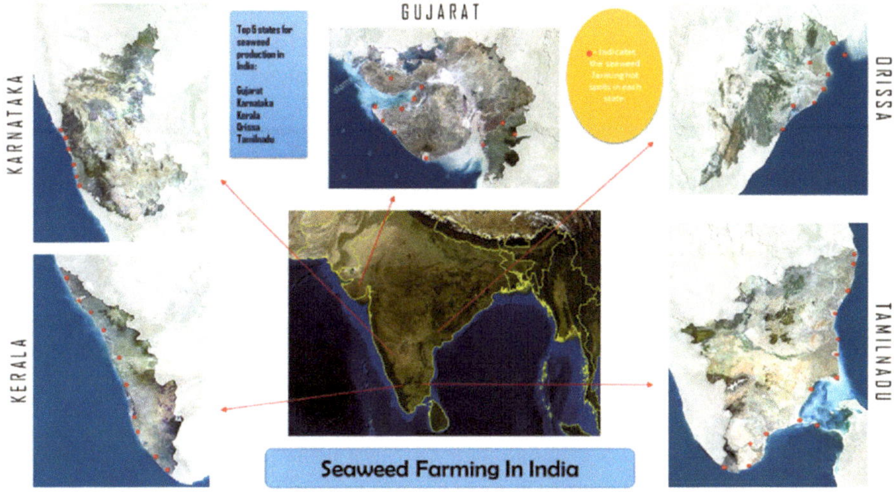

FIGURE 10.5 Pictorial representation of seaweed farming in India.

10.3.2 CHALLENGES FACED IN SEAWEED FARMING

Despite the countless advantages of seaweed farming, there are certain minor challenges that are faced in the aspect of seaweed farming. Thousands of laborers worldwide work for a living taking up seaweed farming as an occupation. Recent researchers conclude that seaweed farmers have significantly poor health than non-seaweed farmers, especially in women. They are reported issues related to back pains, respiratory problems, eye problems, allergies, and injuries caused by animals and sharp shells. These problems were a result of poor working conditions, hunger, overexposure to sunlight and salinity of the seawater. Many of the women farmers also complain about the low income they generate, which is most of the time below the level of poverty. Over 70% of women, seaweed farmers illustrate their wish of getting their children educated to avoid them getting into the same farming occupation and the tragic continues that the majority of the women seaweed farmers still go on with their occupation of seaweed farming to aid the education for their children.

10.4 ACTIVE INGREDIENTS OF MACROALGAE

Seaweeds have been diversely used for various purposes such as food, cosmetics (Bedoux et al., 2014), medicine, and other uses since the ancient periods. The scientific research on the active ingredients of macroalgae began in the late 1970s, even though the therapeutic properties of seaweeds were known to human knowledge from ages. Humans have successfully established various analytical methods for extracting chemical compounds from macroalgae which have miscellaneous beneficial biological advantages. The active ingredients extracted from macroalgae can be indexed into two types based on their mechanism of absorption: the high molecular weight materials such as dietary fibers and the lower molecular weight materials which include trace beneficial elements and certain immunostimulatory hormones. The high molecular active ingredients are found to be having a higher absorption rate compared to the lower molecular weight materials. The phytochemicals such as agar, alginates, and carrageenan extracted from various macroalgae have proven to be extremely favorable for human and fish consumption due to their distinctive properties such as antimicrobial, antioxidant, anti-cancer (Antunes et al., 2011), anti-inflammatory, immunostimulatory, anti-diabetic (Sharifuddin et al., 2015), growth enhancement and also mosquitocidal (Yu et al., 2014). Seaweed extracts are also rich in growth hormones such as auxins, cytokines, and gibberellins. The yield of active ingredients in macroalgae greatly depends on the time of harvest of seaweeds. In the modern biotechnological world, the demand for bioactive compounds synthesized by heavy organic techniques is significantly high, due to which the production of seaweed extract with specific desirable properties remains as a challenge. Seaweed extract is comprised of a wide range of biological molecules which greatly differ with different species of seaweeds (Table 10.1).

10.4.1 POLYSACCHARIDES, STRUCTURAL, AND DIETARY FIBERS

Polysaccharides are complex polymeric sugars consisting of a large number of monomers linked by glycosidic bonds. Marine seaweeds comprise of many polysaccharides based on which they are classified into different groups. They play a vital role in various forms such as mycopolysaccharides, cell wall composition and structural polysaccharides. The extracted polysaccharides are diversely used in the manufacturing of cosmetics, emulsifiers, stabilizers, thickeners, food, feed and beverages. The concentration of polysaccharides found in macroalgae could vary from the range of 4% to 75% of dry weight. The highest concentration of polysaccharides is found in species of brown and red macroalgae (Pádua et al., 2015) such as *Ascophyllum*, *Porphyra*, and *Palmaria*, however, certain species of green macroalgae such as *Ulva* is known to have a high content of polysaccharides of approximately 65% of the total dry weight.

TABLE 10.1 Various Bioactive Ingredients Extracted from Different Macroalgal Species

Seaweed Species	Extracted Bioactive Compounds	References
	Chlorophyta	
Caulerpa spp.	Polysaccharides, proteins, and lipids	Chinnadurai and Kalyanasundaram (2013); Peñalver et al. (2020); Zhong et al. (2020)
	Chlorophyll and carotenoids	
	Phenols	
Codium spp.	Phenols	
Ulva spp.	Polysaccharides, proteins, and lipids	
	Chlorophyll and carotenoids	
	Phenols	
Chaetomorpha spp.	Chlorophyll and carotenoids	
	Rhodophyta	
Porphyra spp.	Polysaccharides, proteins, and lipids	Chinnadurai and Kalyanasundaram (2013); Peñalver et al. (2020); Zhong et al. (2020)
Centroceras spp.	Phenols	
Dasya spp.	Phenols	
Grateloupia spp.	Polysaccharides, proteins, and lipids	
	Chlorophyll and carotenoids	
	Phenols	
Hypnea spp.	Chlorophyll and carotenoids	
Chondrus spp.	Polysaccharides, proteins, and lipids	
Gracilaria spp.	Phycobiliproteins	Sudhakar et al. (2015); Saluri et al. (2019)
Furcellaria spp.		
	Phaeophyta	
Sargassum spp.	Polysaccharides, proteins, and lipids	Chinnadurai and Kalyanasundaram (2013); Peñalver et al. (2020); Zhong et al. (2020)
	Phenols	
Ecklonia spp.	Phenols	
Padina spp.	Chlorophyll and carotenoids	
Saccharina spp.	Polysaccharides, proteins, and lipids	
	Ochrophyta	
Undaria spp.	Polysaccharides, proteins, and lipids	Peñalver et al. (2020)
Durvillaea spp.	Polysaccharides, proteins, and lipids	

The composition, structure, and chemical properties of dietary and structural fibers found in macroalgae vary extensively from species to species. The total fiber content in seaweeds varies from 33% to 62% of the dry weight which is found to be very much greater than that of the higher plants. The undigested parts of polysaccharides form an important part of the total fiber content in seaweeds. The fibers present in macroalgae are classified into two types: water-soluble fibers and water-insoluble fibers. Water-insoluble fibers include cellulose, mannans, and xylans, which fall under the category of structural fibers, and water-soluble fibers such as agars, alginic acid, furonan, and laminarin which are termed as dietary fibers. The fibers present in seaweeds have a wide range of properties, namely antimicrobial, antitumor, anti-herpetic, anti-coagulant, and it is also found to reduce the risk of obesity and diabetes. Some seaweed species which have a high content of fibers include *Sargassum*, *Undaria*, *Laminaria*, and *Porphyra.*

10.4.2 PROTEINS, PEPTIDES, AND AMINO ACIDS

The biological, chemical, and structural properties of proteins, peptides, and amino acids extracted from seaweeds have not been intensely studied to that extent of polysaccharides but yet they have been proven to be used to extract various essential amino acids (EAAs) such as glutamic and aspartic acids. The total concentration of proteins present in macroalgae is found to be less than 6% of the total dry weight. The concentration and composition of amino acids depend on the species, site, and the time of harvest of seaweeds. The protein content in brown algae is found to be comparatively less than green and red macroalgae. Species belonging to *Palmaria* and *Porphyra* are proven to have high amounts of acidic amino acids such as glycine and arginine. Brown macroalgae are specifically found to have high amounts of threonine, valine, lysine, leucine, and alanine. Studies have proven that macroalgae have certain specialized proteins known as lectins that bind to carbohydrates to form complexes which have anti-bacterial, antiviral, and anti-inflammatory properties (Dumay and Morançais, 2016).

10.4.3 LIPIDS: FATTY ACIDS, PHOSPHOLIPIDS (PLS), GLYCOLIPIDS (GL), AND STEROLS

Macroalgae comprise of some really interesting lipid compounds which have very high profitable and commercial values. One of the most important lipids extracted from the seaweeds is the PUFA's (Kendel et al., 2015) which primarily consists of omega-3 fatty acids such as eicosapentaenoic acid (EPA) and docosahexaenoic acid (DHA) and omega-6 fatty acids namely γ-linolenic acid and arachidonic acid (AA). The total concentration of ω-3 and ω-6 fatty acids vary differently in individual seaweed species based on the genetic and environmental variations. Phospholipid compounds are a major and the most crucial classification of lipids which consists of very high concentrations of fatty acids mainly EPA and DHA; they together constitute approximately 10% to 20% of the total lipid content present in macroalgae. PLs play an extremely predominant role in transportation and maintenance of macroalgal structures. Certain lipid molecules found in macroalgae are found to be bind with carbohydrate compounds known as glycolipids (GL) which provide energy and serve as biomarkers of cell recognition, these compounds contribute to around 20% to 52% of the total lipid content. GL are primarily embraced with compounds such as monoglycosyl-diacylglycerol (MGDG), diglycosyldiacylglycerol (DGDG), trimethyl-beta-alaninediacylglycerol, and sulfaquinovosyldiacyl-glycerol which have high commercial and economical values, despite their high commercial demands these compounds are known for their magnificent antimicrobial, anti-tumor, and anti-depressive activities. Despite these compounds, macroalgae consist of certain hydroxyl groups containing lipids helping in the stabilization of the cell membranes known as sterols. Sterols extracted from seaweeds generally consist of compounds such as cholesterol, fucosterol, desmosterol, ergosterol, and de-hydroxy ergosterol. Sterols constitute around 2% to 76% of the total lipid content in macroalgae. These compounds play a vital role for their presence in the fluid mosaic model for the durability, stability, and strength of the algal cell membranes.

10.4.4 PIGMENTS

In the diverse world of macroalgae, pigments present in them play an extremely vital role in the classification of seaweeds and also for the production of color. In simple words, pigments can be elucidated as compounds that produce color based on selective-color absorption. The pigments present in seaweeds are extracted mainly by organic solvent methods using chloroform, acetone, methanol, and hexane (Zhu et al., 2017). These bio-pigments not only persist bio-active properties but also have certain sensorial properties, that is, they are used as natural food colorants which in the modern world has gained immense economical demands. It is estimated that by the end of 2020, the food coloring industries will have an increased net worth of approximately 3.75 billion US Dollars (Aryee et al., 2018). Despite their role as a food additive, they also possess properties such as antioxidant, anti-inflammatory, anti-angiogenic, anti-diabetic, and immune-modulatory activities. Some primarily important pigments found in seaweeds are chlorophyll, carotenoids, and phycobiliproteins (PBP) (Chojnacka et al., 2012).

10.4.4.1 CHLOROPHYLL

Chlorophyll is green cyclic tetrapyrollic lipid-soluble pigments found in all higher plants, cyanobacteria, and macroalgae. The most abundant chlorophyll present in the seaweeds (Chen and Roca, 2018) is the chlorophyll α, primarily found in green macroalgae, while chlorophyll β and chlorophyll χ are found in red and brown macroalgae, respectively. Majority of the chlorophylls in seaweeds are analogous to light-harvesting chlorophyll-binding polypeptides. The total chlorophyll χontent present in seaweeds would range between 550 and 3,000 mg/kg of dry weight. Chlorophyll pigments in seaweeds are transformed into secondary derivatives mainly pheophytin, pyropheophytin, and pheophorbide, these derivatives are known to have anti-mutagenic effects and play a crucial role in preventing cancer and tumor induction. In specific, pheophorbide which has been extracted from brown macroalgae has also been studied to show positive effects for the treatment of neurodegenerative disorders such as Alzheimer's diseases (ADs) (Chen et al., 2017).

10.4.4.2 CAROTENOIDS

Carotenoids are pigments responsible for vibrant colors such as bright red, brilliant yellow, and shades of orange and are primarily composed of tetraterpenes backbone. Carotenoids are widely abundant and predominantly found in plants, algae, and other micro-organisms. They are generally diversified into two classes depending on the pigment content and color production: carotenes and xanthophylls (Poojary et al., 2016). Carotenes are red color containing pigments, namely α-carotene, β-carotene and lycopene, these compounds are strictly hydrocarbons. Xanthophylls are pigments producing orange and yellow colors due to compounds such as lutein, zeaxanthin, neoxanthin, violaxanthin, flavoxanthin, and fucoxanthin. These compounds are structurally similar to carotenes except for the presence of oxygen molecule in xanthophylls. Till date, over 600 types of carotenoids have been reported in different species worldwide. Despite the fact that carotenoids are present in many organisms, marine macroalgae serve as an interesting source for the extraction of carotenoids. Carotenoids extracted from seaweeds have a market value that is said to rise above 1.53 billion US dollars until the end of 2021, this necessity is because of various therapeutic activities that carotenoids can be used for, such as removal of free radicals, antioxidant, and antiseptic properties.

10.4.4.3 PHYCOBILIPROTEINS (PBP)

Unlike the carotenoids and chlorophyll pigments, PBP are light-harvesting pigmented proteins (Sudhakar et al., 2015) specifically present in red macroalgae, though they contribute only up to 2% to 6% of the total pigment content in seaweeds, they are extremely important for aiding out photosynthesis in seaweeds which lack higher amounts of chlorophyll pigments. PBP are the compounds that are primarily responsible for the fluorescence of seaweeds belonging to the class of Cyanophyta. These pigments are classified into two types based on the fluorescence they produce: phycoerythrocyanin and phycocyanin (PC). Phycoerythrocyanin produces a soft pink and orange fluorescence and phycocyanin produce a blue and red fluorescence. PBP extracted from seaweeds are predominantly used for producing natural dyes and colored creams which have lots of economical values.

10.4.5 PHENOLS AND HALOGENATED COMPOUNDS

Phenols are secondary derivatives present in seaweeds which generally works as an anti-oxidative agent and takes part in redox reactions (Audibert et al., 2010). Primarily hydroxycinnamic acids, hydroxybenzoic acids, stilbenes, anthocyanins, and flavonols are the compounds present in seaweeds that are classified under the group of phenols. Phenols play a vital role in the protection of macroalgae from pathogens, epiphytes, and marine gazers. These compounds also act as a photoprotective agent which helps in sheltering seaweeds against cytotoxic damage caused by UV radiation. These bioactive PCs serve as an important source in several fields like medicine, cosmetics, agriculture, and food.

Aquatic habitats are rich in halogens, these elements are utilized by macroalgae to produce halogenated compounds. The most pre-dominant halogenated compounds found in seaweeds are bromo and chloro substituted terpenes and phenols. Halogenated compounds are primarily extracted from red and brown macroalgae and widely used for producing anti-bacterial, anti-fungal, antiviral, and anti-larval agents (Bianco et al., 2013).

10.5 ROLE OF MACROALGAE IN AQUACULTURE

Fish aquaculture is one of globally most important economic and nutritive culturing processes that take place worldwide. As estimated by FAO, fish aquaculture has reached an economic value of about 263.6 billion US Dollars by the end of 2018. With the commercial demand of fish aquaculture surging to peaks spontaneously year after year, the challenges for production of higher biomass and healthy fishes also increases due to predatory fish, harmful algal blooms and other fish pathogenic microbes which lead to heavy losses of edible fish in high masses. The utilization of seaweeds or seaweed extracts in fish aquaculture has gained up interest in modern times due to their immense bioactive qualities. Despite their high nutritional value, seaweeds contain many numbers of pigments, active ingredients, defensive, and storage compounds and other secondary metabolites that show beneficial effects on farmed fish. These compounds were studied to show increased fish growth, developed immune system, stress resistance, higher physiology and increase in muscle quality and mass. The presence of seaweed or seaweed extracts in the fields of fish aquaculture also reduces the growth and habituation of deathly fish pathogens that cause loss of fishes in immense biomass (Figure 10.6).

Green macroalgae such as species belonging to *Ulva* and *Enteromorpha* help in removing harmful nitrogen compounds such as inorganic nitrogen (Yokoyama and Ishihi, 2010), ammonia, and nitrous oxides which are excreted from fish and upon accumulation are potent chemicals known to poison edible fishes from the waters by absorbing them and storing them in the form of non-hazardous compounds such as nitrites. Due to this nitrogen absorbing property of seaweeds, they can be used as bioremediators and biofilters in fish aquaculture (Aníbal et al., 2014). Species belonging to *Laminaria*, *Gracilaria*, and *Porphyra* are also well known for decreasing the risk of eutrophication if used in the sites of fish aquaculture (Xu et al., 2011). In recent studies, it is proven that in the presence of *Artemia*, the bioremediation property of seaweeds is enhanced, which shows more increase in biomass of fishes and a higher yield content while compared to the untreated waters where farming of fishes was conducted (Marinho-Soriano et al., 2011).

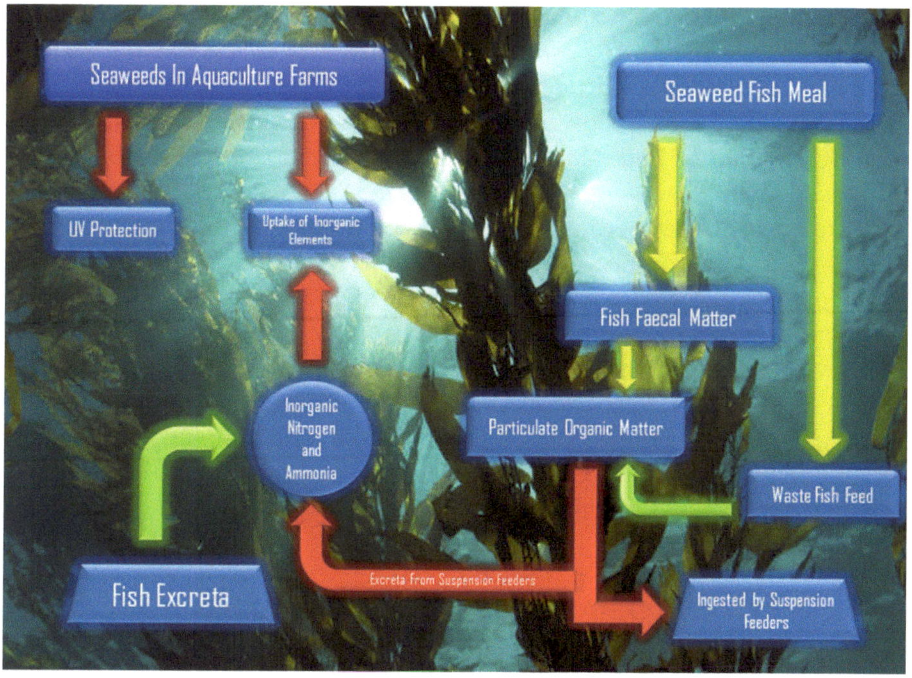

FIGURE 10.6 Mechanistic approach of seaweeds with its distinctive application in aquaculture.

Until the recent times, chemotherapy was used to prevent disease outbreaks among fishes which were found to show extreme negative effects directly to fishes and indirectly to the humans and environment due to the use of hazardous chemicals and synthetic drugs. Hence, the usage of seaweed extract in fish and shellfish aquaculture was practiced, which was found to improve appetite and stimulate weight gain due to the immunostimulatory and anti-parasitic activities of the bioactive ingredients found in macroalgae.

Macroalgae is used as a potential feed for fish and shellfish worldwide due to the presence of micronutrients and small molecules such as carotenoids and phlorotannins, which have properties such as antibiotic, anti-tumor, anti-nociceptive, neuroprotective, anti-adhesion, and anti-HIV activities. Seaweed species such as *Codium fragile*, *Hypnea cervicornis* and *Solieria robusta* are known to have anti-inflammatory and mitogenic activities due to the high PBP content such as phycocyanin, allophycocyanin, phycoerythrin, and phycoerythrocyanin present in them (Dumay et al., 2014). Certain seaweeds are found to inhibit the growth of harmful microalgal blooms that cause the death of edible fishes in a large scale, green macroalgal species such as *Ulva fasciata*, *Ulva pertusa*, and *Ulva linza* cause rapid lysis in the cells of harmful algal blooms of *Prorocentrum micans*, *Prorocentrum donghaiense* and *Heterosigma akashiwo* (Tang and Gobler, 2011). Red macroalgae such as *Corallina pilulifera* show algicidal effect on *Cochlodinium polykrikoides*, *Karenia mikimotoi* and *Akashiwo sanguinea*, which are responsible for fish deaths at high rates.

Presence of seaweed extracts in the fields of fish, shrimps, and shellfish aquacultures observed an increase in the yield at a large scale when compared to the fields with the absence of these extracts (Cunha et al., 2019). For example, the presence of *Gracilaria lemaneiformis*, showed an observed increase in the yield of black sea breams with high muscle growth (Xuan et al., 2013). Seaweed extracts being completely natural and showing absolutely no side effects and negative consequences directly or indirectly on the health of edible aquatic species, has gained plenty of interest and commercial demand worldwide.

10.6 ANTI-MICROBIAL ACTIVITY OF MACROALGAE

Microbial diseases associated with fishes has gained a lot of attention and is a worldwide problem in the field of aquaculture. Use of antibiotics primarily quinolones, tiamulines, sulfonamides, oxytetracycline, and erythromycin as a treatment for microbial diseases (Cortés et al., 2014) in fishes has been used since the past few decades. Although the antibiotics gave a positive result in controlling the diseases, the drawback was that it was temporary due to the development of resistance in the microbes against the specific antibiotic (Priyadharshini et al., 2011). Since then, the interest in naturally produced antimicrobial agents increased especially the ones extracted from the marine organisms. Since the last two decades, the use of macroalgae as a therapeutic agent in aquaculture has increased spontaneously, in specific, species belonging to *Sargassum* [Brown seaweed] and *Asparagopsis* [Red seaweed] exhibit a wider range of antimicrobial activity (Vatsos and Rebours, 2015). The antimicrobial activity of seaweeds varies from species to species and also the method of extraction of the bioactive compounds (Table 10.2). Despite, these numerous benefits of seaweed extract the use of these bioactive compounds as antimicrobial agents in commercial scale is very limited except for a few parts in Asia, where India stands out as the highest consumer of seaweeds for their antimicrobial property.

TABLE 10.2 Antimicrobial Activity of Seaweed Extracts Against Various Fish Pathogens

Seaweed Species	Activity	Effecting Organism	References
Chaetomorpha aerea	Anti-bacterial	*Aeromonas hydrophila*	Rizzo et al. (2017)
Ulva fasciata	Anti-bacterial	*Aeromonas hydrophila*	
		Aeromonas salmonicida	
		Photobacterium damselae	
Cystoseira barbata	Anti-bacterial	*Salmonella typhi*	
		Vibrio cholerae	
Dictyopteris polypodioides	Anti-bacterial	*Staphylococcus aureus*	
		Salmonella typhi	
		Vibrio anguillarum	
Fucus virsoides	Anti-bacterial	*Aeromonas salmonicida*	
		Photobacterium damselae	
		Vibrio parahaemolyticus	
Sargassum uticum	Anti-bacterial	*Vibrio harveyi*	
Undaria pinnatifida	Anti-bacterial	*Pseudomonas aeruginosa*	
		Vibrio harveyi	
		Vibrio cholerae	
Gracilaria gracilis	Anti-bacterial	*Vibrio harveyi*	
Hypnea cornuta	Anti-bacterial	*Vibrio anguillarum*	
Chaetomorpha linum	Anti-fungal	*Rhizopus* spp.	Shajahan et al. (2016)
		Fusarium spp.	
		Aspergillus spp.	
Turbinaria conoides	Anti-fungal	*Fusarium* spp.	
		Aspergillus spp.	
Sargassum wightii	Anti-fungal	*Fusarium* spp.	
		Aspergillus spp.	
Halimeda opuntia	Anti-fungal	*Rhizopus* spp.	
		Aspergillus spp.	
Gracilaria verrucosa	Anti-fungal	*Aspergillus* spp.	
Polysiphonia morrowii	Antiviral	Infectious hematopoietic necrosis virus (IHNV)	Kim et al. (2011)
		Infectious pancreatic necrosis virus (IPNV)	

10.6.1 ANTI-BACTERIAL ACTIVITY

Diseases caused due to bacterial origin is responsible for high mortality rates among fishes. Fishes being the most consumed food by humans worldwide, bacteria-infected fishes also make up a source for indirect human infections. Bacteria mainly belonging to the genus of *Vibrio* such as *V. harveyi*, *V. vulnificus*, *V. parahaemolyticus*, *V. alginolyticus* and *V. salmonicida* (Cavallo et al., 2013) are the primary cause for the heavy loss of fishes in the aquaculture sites, despite these species' other certain bacteria namely *Streptococcus agalactiae*, *Flavobacterium columnare* and *Aeromonas hydrophila* are also pathogenic to fishes (Kurniawati and Adam, 2016). Marine macroalgae being the most accessible and the only source for agar, alginates, carrageenan, and other secondary metabolites such as alkaloids, flavonoids, tannins, and PCs in the ocean waters from which the anti-bacterial compounds can be extracted, are widely taken into consideration because of being completely natural and providing a permanent bacterial resistance and development of non-specific immunity in fishes. Seaweed extracts that are widely being used for anti-bacterial activity are extracted from the species primarily *Ceramium rubrum*, *Cladophora rupestris*, *Padina gymnospora* (Thanigaivel et al., 2015), *Gracilaria dura*, *Laminaria digitata*, *Gracilariopsis longissima* and *Ulva prolifera*, which have excellent anti-bacterial properties against a wide range of fish pathogenic bacteria. Extracts from these are directly been used into the aquaculture waters or even given as oral intake for farmed fishes based on the convenience of the culturing methods used.

10.6.2 ANTI-FUNGAL ACTIVITY

In the world of macroalgae, presence of natural bioactive components with diverse biologically beneficial activities is immense. The anti-fungal property of seaweeds is complimented as one the most essential biological benefits which macroalgae processes. Fungal diseases in fishes are responsible for over 50%–60% loss in aquaculture annually. Tremendous economic loss in fish and shrimp aquaculture due to fungus, made researchers gain attention towards the usage of natural anti-fungal agents extracted from seaweeds. A fungal infection that causes the primary loss in fishes is the Saprolegniasis [cotton wool disease] caused by *Saprolegnia parasitica* is responsible for over 50% loss of fish due to fungal infections. Certain other species of fish pathogenic fungus include *Candida albicans* and *Aspergillus fumigatus* (Ertürk and Taş, 2011) and *Fusarium* species which are also responsible for the mass death of fishes. Extracts from seaweeds such as species belonging to *Sargassum* (Peres et al., 2012), *Laminaria*, *Ulva*, *Ascophyllum*, and *Gracilaria* produce excellent bioactive ingredients especially, the naturally occurring halogenated compounds, terpenes, and polyphenols which are well known for their antifungal properties. Direct ingestion into fish feed or diffusing the seaweed extracts into the aquaculture waters has been done worldwide, but in recent studies, it is found that metal nanoparticles such as silver nanoparticles (AgNPs) synthesized using these seaweed extracts also show remarkable antifungal properties against a wide range of pathogenic fungus, since the process of synthesis of metal nanoparticles in highly expensive and not be afforded in all means, the use of nanoparticles for the treatment of fishes against fungal infection in significantly low (Kumar et al., 2013).

10.6.3 ANTI-VIRAL ACTIVITY

Unlike the other microbial infections, viral infections in fishes in the sites of aquaculture pose a serious threat to the economic crisis of aquaculture worldwide. Several highly contagious viral diseases have been identified which are very much prone in the fields of aquaculture. Infectious hematopoietic necrosis virus (IHNV) and infectious pancreatic necrosis virus (IPNV) (Kim et al., 2011) are the most harmful and hazardous fish pathogenic virus that cause a high amount of fish deaths annually worldwide, 80% methanolic extract of the red macroalga *Polysiphonia morrowii* is known to work excellently against these viruses by inhibiting the metabolic processes of the virus when treated to fishes either by oral ingestion or treated water. The most reactive antiviral bioactive ingredients present in macroalgae is the sulfated polysaccharides and fucoidans, which have an immense effect in inhibiting the metabolic processes of the viruses, thus, in turn decreasing the multiplication and growth rate of the viral cells (Gomaa and Elshoubaky, 2016) (Figure 10.7). Extracts from macroalgae belonging to the class of *Phaeophyta* work best against viral effects compared to the classes of *Rhodophyta* and *Chlorophyta* as the content of polysaccharides and fucoidans are extremely high and their functions and alternatively powerful than the chemotherapeutic antiviral drugs. Interestingly, these seaweed extracts also give viral resistance against herpes simplex virus (HSV) and Reft valley fever virus (RVFV) that, very well known to cause serious diseases and illness not only in fishes but also humans and other animal species.

10.7 ANTIOXIDANT ACTIVITY OF MACROALGAE

Antioxidants in simple words are molecules that prevent the oxidation of any substrate compounds present in the body. Antioxidants have a diverse function in the biological systems which includes protection against oxidative damage and for aiding several pathways of the cell (Souza et al., 2011). One of the most concerning problems in metabolic cycles is the release of reactive oxygen species (ROS), which lead to ageing and decreases immune resistance towards several diseases in fishes (Vega et al., 2020). Until recent times, several synthetic antioxidants such as butylated hydroxytoluene (BHT), butylated hydroxy-anisole (BHA) (Cox et al., 2010), t-butylhydroxyquinone (TBHQ) and propyl gallate (PG) have been used in fish aquacultures, however, it was observed to show a range of toxic and hazardous effects. Therefore, the demand for naturally occurring antioxidants peaked up due to the awareness of food quality and safety. Seaweeds being rich in bioactive ingredients, is one of the best sources for naturally occurring antioxidants in the form of polysaccharides, proteins, lipids, or even PCs (Sotoudeh and Mardani, 2017).

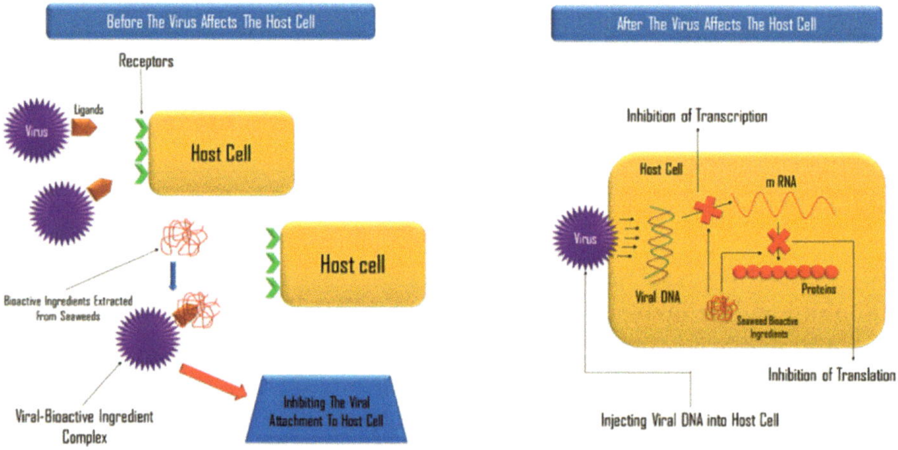

FIGURE 10.7 Representation of antiviral efficacy of macroalgal species.

Antioxidants found in seaweeds not only work on preventing the process of oxidation but also have a very crucial role to play as radical scavenging agents (Figure 10.8). The bioactive molecules in macroalgae scavenge out the free reactive radicals and reduce them to unharmful species which in turn prove to be beneficial for the metabolism of aquatic organisms. Compounds such as 2,2-azino-bis-3-ethylbenzothiaz-oline-6-sulfonic acid (ABTS) and 2,2-diphenyl-1-picrylhydrazyl (DPPH) present in seaweeds act as radical scavenging agents and are primary molecules to determine prevention of oxidation process (Sathya et al., 2017). The extraction of antioxidants from macroalgae is carried out by organic solvent methods mainly by the usage of methanol and ethanol, however, ethanol extraction provided a 50% higher yield of antioxidants than that of extraction by methanol. Certain antioxidants which are proven to be completely natural that are extracted from these extraction methods like gallocatechin, epicatechin, and catechin gallate. Brown and red macroalgae have higher contents and stronger molecules of antioxidants when compared to that of green macroalgae because brown macroalgae have molecules known as phlorotannins, which are the polymers of Phloroglucinol and also carotenoid pigments which include astaxanthin and fucoxanthin and are known to have excellent anti-oxidative properties. Methanol extracts of selected species of seaweeds such as *Padina tetrastromatica*, *Ulva lactuca*, *Acanthophora spicifera*, *Gracilaria foliifera*, *Caulerpa scalpelliformis* and *Sargassum wightii* (Linga Prabu et al., 2013) showed excellent radical scavenging activity (Sachindra et al., 2010) against reactive species of nitrous oxide (NO) and peroxide ions (H_2O_2) (Athiperumalsami et al., 2010).

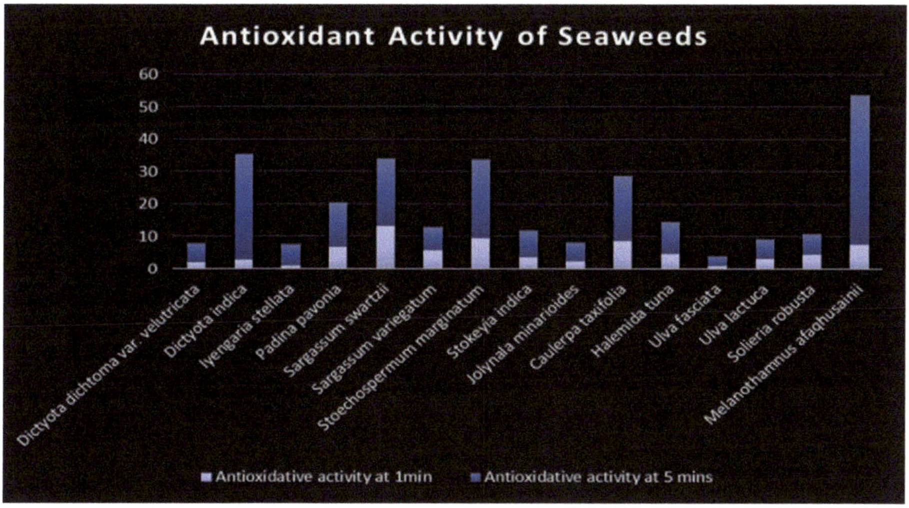

FIGURE 10.8 Graphical representation of the antioxidant activity of seaweeds.

Studies have shown that if species of seaweeds belonging to *Gracilaria*, *Fucus*, and *Ulva* are given as dietary supplements for farmed fishes, they observed increased metabolic rates and developed innate immunity due to the presence of strong antioxidants. Seaweed extracts are widely used in fish storage industries especially for the frozen storage of fish mince, due to the presence of naturally occurring compounds such as polyphenols, α-tocopherol, and ascorbic acid the oxidation of protein is gradually reduced during the process of frozen storage for a long period of time (Wang et al., 2017). Seaweed extracts due to their immense anti-oxidative activity are also helpful for controlling or developing resistance against diseases causing pathogens in fishes (Vinosha et al., 2020). For example, the use of *Chaetomorpha antennina* extract in the sites of shrimp aquaculture witnessed an increase in resistance of shrimps against the pathogen *Vibrio parahaemolyticus* (Thanigaivel et al., 2014).

10.8 IMMUNOSTIMULATORY PROPERTY OF SEAWEEDS IN FISHES

Since the past three decades, fish aquaculture has expanded day by day to become the fastest-growing food-producing sectors worldwide. Immunomodulation has been the most demanding process for developing resistance in fishes against a wide range of diseases which farmed fish are highly prone to be affected with. In order to acquire resistance to diseases that can spread through aquaculture non-specific defense mechanisms must be induced in the metabolism and studies showed that this is undoubtedly possible by treatment with seaweed extracts or including seaweed as a part of dietary supplements for the aquatic organisms because of their vast range of immunostimulatory bioactive ingredients. Usage of immunostimulants as an alternative to drugs, antibiotics, and chemicals is widely practiced in the fields of modern aquaculture, in comparison to vaccines immunostimulants extracted from seaweeds is proven to enhance the innate immunity to a great extent (Figure 10.9).

Immunostimulants extracted from seaweeds are mainly lipopolysaccharides and compounds of glycan. Especially compounds such as β-1,3-glycan and β-1,6-glycan show immense work as immunostimulants (Bobadilla et al., 2013) enacting in diverse ways in the metabolic processes of fishes and providing resistance against a wide range of infectious diseases. Sulfated polysaccharides prominently found in the red macroalgae *Gracilaria fisheri* is known to develop resistance in shrimps against the white spot syndrome virus (WSSV) (Wongprasert et al., 2014) and *Vibrio harveyi* infections which are primarily observed in black shrimps, *Penaeus monodon* (Kanjana et al., 2011). *Pangasianodon hypophthalmus*, commonly known as sutchi catfish (Prabu et al., 2016) is a highly cultivated fish in the parts of southeast Asia, besides its high exploitation in cuisine, it is also highly prone to be infected by the bacteria *Aeromonas hydrophila*. Fucoidan rich seaweed extracts are used in these aquaculture farms to develop immunity against this infection.

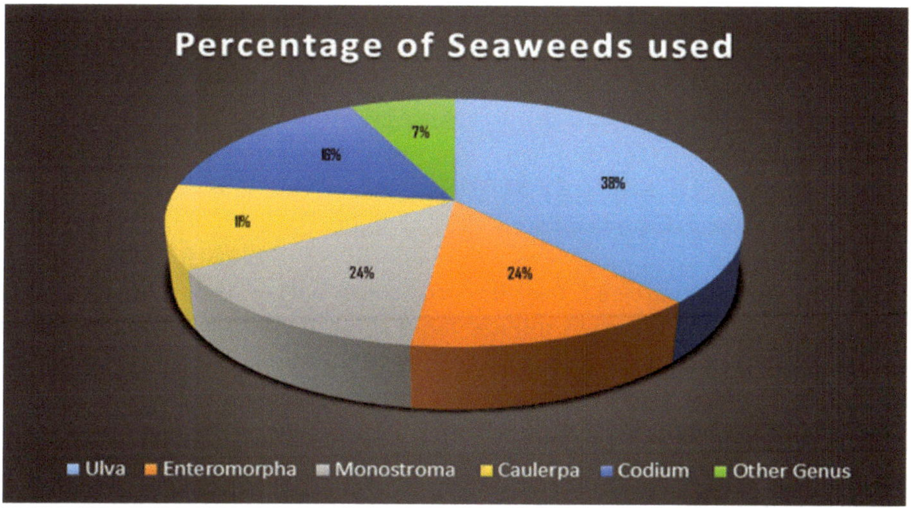

FIGURE 10.9 Graphical representation of seaweeds applicated as immunostimulants.

Polysaccharides in seaweeds have a vital role to play as immunostimulants when used as a dietary supplement in aquaculture. Brown macroalgae polysaccharides primarily alginates which include polymers of β-D-mannuronic acid and α-L-guluronic acid and complexes of laminarin are widely used as immunostimulants for developing resistance against a wide range of bacterial and viral diseases (Okolie et al., 2017). For example, sulfated polysaccharides found in seaweed species such as *Kappaphycus alvarezii*, *Chondrococcus hornemanni* (Karuppiah, 2014), *Chondrus crispus* (Rudtanatip et al., 2018) and *Ulva pertusa* (Tabarsa et al., 2012) are used heavily as dietary food supplements in aquaculture of mussels, shellfish, and certain species of fishes typically *Oreochromis niloticus*, *Lates calcarifer* and *Cyprinus carpio* (Radhika and Mohaideen, 2016) for developing resistance against infections and diseases caused by bacteria such as *Vibrio parahaemolyticus* (Sakthivel et al., 2015) and *Streptococcus agalactiae* (Lee et al., 2020). Polysaccharides extracted from seaweeds are also utilized as probiotic agents in fishes for early development of innate immunity.

Despite these pure biomolecules, certain inorganic substituted molecules present in macroalgae and metal nanoparticles synthesized using seaweed extracts also accommodate immunostimulatory properties. Low molecular weight sodium alginate extracted from macroalgae is known to boost the immune system in Nile Tilapia fishes and provide growth-stimulating effects in fish probiotic bacteria such as *Lactobacillus plantarum* (Van Doan et al., 2016).

ZnO (zinc oxide) nanoparticles synthesized using extracts from *Sargassum wightii* are known to immunomodulate and enhance the disease-resistance properties in green tiger shrimps *Penaeus semisulcatus* (Ishwarya et al., 2018).

10.9 EFFECT OF DIETARY SEAWEEDS ON THE GROWTH PERFORMANCE OF FISHES

Food supplements are the most important necessity in the process of aquaculture, for producing high muscle mass, high quality, and healthy fishes. Dietary food and fish meal makeup to be the most expensive basic needs and take up to 30% to 60% of the total expenditure value utilized over the farming of fishes and shrimps. Its high expense is due to the formulation of fish meal with an adequate amount of all nutritive components, while proteins and EAAs being the most important nutritive component which has to be supplemented to fishes. Increase in the production of edible fishes worldwide, the need of fish meal also continues to increase simultaneously each year. The economic crisis being the outmost problem in aquaculture, food supplements using plant proteins which are primarily extracted from marine macroalgae pose a good turn for production of high-quality fishes at a very low expense (Figure 10.10).

Species of macroalgae belonging primarily to *Ulva*, *Sargassum*, *Laminaria*, *Porphyra*, and *Gracilaria* are used as food supplements to feed farmed fishes worldwide, not only providing anti-oxidative activity and resistance to a wide range of microbial diseases, the bioactive compounds in macroalgae mainly the proteins, polysaccharides, and lipids benefit the fishes for gaining more muscle mass and have a better quality than the untreated fishes. As a matter of fact, food supplements made from seaweeds are completely natural and show no negative side effects on the metabolism of fishes. For example, polysaccharides extracted from the species of *Enteromorpha* show immense positive growth performance on the *Fenneropenaeus merguiensis* [banana shrimps] by observable high muscle mass and higher quality (Liu et al., 2020). Dietary supplements acquired from species such as *Gracilaria arcuata* and *Sargassum wightii* also show excellent growth performance in fish species such as *Oreochromis niloticus* [Nile Tilapia] (Younis et al., 2018) and *Labeo rohita* (Gora et al., 2018) respectively. Red tilapia fishes also respond tremendously when green macroalgal species belonging to *Ulva* was given as dietary supplements which resulted in observed weight gain of fishes with a healthy metabolic rate (El-Tawil, 2010).

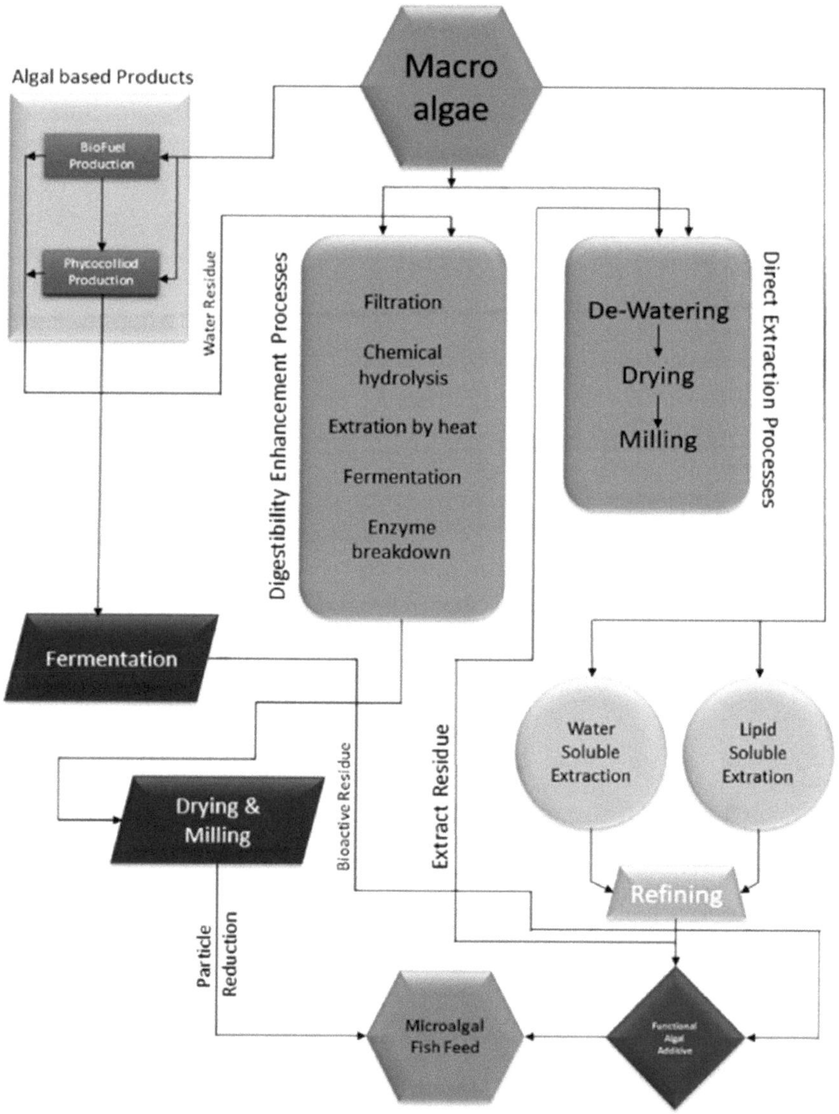

FIGURE 10.10 Application of the macroalgal species towards the production of fish feed.

10.10 CONCLUSION

Marine macroalgae are potentially the best source for production and extraction of naturally acquired bioactive ingredients in the ocean's waters. The compounds extracted from seaweeds can very well be utilized as an alternative to the synthetically synthesized chemotherapeutic drugs, with its wide range of biologically important characteristics which can be excellent mechanisms dealing with anti-oxidative, antimicrobial, anti-tumor, anti-coagulant, and immunomodulation activities, without disturbing the normal metabolism of the organism, in fact, enhancing the metabolic rates and providing a healthy development in the innate immunity of the fishes.

Despite the fact, that seaweeds have numerous beneficial properties not only in the biological and metabolic factors but also the ecological importance, the use of seaweeds in aquaculture processes in the modern world is quite less, it may be to the fact that there is no actual awareness regarding the immense beneficial properties of macroalgae. Not to forget the fact that seaweeds not only enhance the biological and physical properties of fishes but also, can be used for treating humans and other higher organisms of the animal world.

Though not much is known about the complete significance of the macroalgae in recent years, it sure seems to be pacing up rapidly in great rates. Furthermore, the great diverse world of macroalgae with beneficial bio-metabolic properties and remarkable capabilities associated with the extracted bioactive ingredients does possess the potential to expand its health and economic beneficial values in the fields of aquaculture, turning out be a potent applicant not only in aquaculture but also in pharmaceuticals, food, agriculture, and other diverse bio-productive fields.

KEYWORDS

- **antimicrobial**
- **Chlorophyta**
- **immunomodulatory**
- **macroalgae**
- **Phaeophyta**
- **Rhodophyta**

REFERENCES

Aníbal, J., Madeira, H. T., Carvalho, L. F., Esteves, E., Veiga-Pires, C., & Rocha, C., (2014). Macroalgae mitigation potential for fish aquaculture effluents: An approach coupling nitrogen uptake and metabolic pathways using *Ulva rigida* and *Enteromorpha clathrata*. *Environmental Science and Pollution Research, 21*(23), 13324–13334.

Antunes, E. M., Afolayan, A. F., Chiwakata, M. T., et al., (2011). Identification and *in vitro* anti-esophageal cancer activity of a series of halogenated monoterpenes isolated from the South African seaweeds *Plocamium suhrii* and *Plocamium cornutum*. *Phytochemistry, 72*(8), 769–772.

Aryee, A. N., Agyei, D., & Akanbi, T. O., (2018). Recovery and utilization of seaweed pigments in food processing. *Current Opinion in Food Science, 19*, 113–119.

Athiperumalsami, T., Rajeswari, V. D., Poorna, S. H., Kumar, V., & Jesudass, L. L., (2010). Antioxidant activity of seagrasses and seaweeds. *Botanica Marina, 53*(3), 251–257.

Audibert, L., Fauchon, M., Blanc, N., Hauchard, D., & Ar Gall, E., (2010). Phenolic compounds in the brown seaweed *Ascophyllum nodosum*: Distribution and radical-scavenging activities. *Phytochemical Analysis, 21*(5), 399–405.

Bedoux, G., Hardouin, K., Burlot, A. S., & Bourgougnon, N., (2014). Bioactive components from seaweeds: Cosmetic applications and future development. *Advances in Botanical Research, 71*, 345–378.

Bennett, S., & Bellwood, D. R., (2011). Latitudinal variation in macroalgal consumption by fishes on the great barrier reef. *Marine Ecology Progress Series, 426*, 241–252.

Bianco, E. M., Pires, L., Santos, G. K., et al., (2013). Larvicidal activity of seaweeds from northeastern Brazil and of a halogenated sesquiterpene against the dengue mosquito (*Aedes aegypti*). *Industrial Crops and Products, 43*, 270–275.

Bobadilla, F., Rodriguez-Tirado, C., Imarai, M., Galotto, M. J., & Andersson, R., (2013). Soluble β-1,3/1,6-glucan in seaweed from the southern hemisphere and its immunomodulatory effect. *Carbohydrate Polymers, 92*(1), 241–248.

Cavallo, R., Acquaviva, M., Stabili, L., Cecere, E., Petrocelli, A., & Narracci, M., (2013). Antibacterial activity of marine macroalgae against fish pathogenic *Vibrio* species. *Open Life Sciences, 8*(7), 646–653.

Cheal, A. J., MacNeil, M. A., Cripps, E., et al., (2010). Coral–macroalgal phase shifts or reef resilience: Links with diversity and functional roles of herbivorous fishes on the Great Barrier Reef. *Coral Reefs, 29*(4), 1005–1015.

Chen, K., & Roca, M., (2018). *In vitro* bioavailability of chlorophyll pigments from edible seaweeds. *Journal of Functional Foods, 41*, 25–33.

Chen, K., & Roca, M., (2018). *In vitro* digestion of chlorophyll pigments from edible seaweeds. *Journal of Functional Foods, 40*, 400–407.

Chen, K., Ríos, J. J., Pérez-Gálvez, A., & Roca, M., (2017). Comprehensive chlorophyll χomposition in the main edible seaweeds. *Food Chemistry, 228*, 625–633.

Chinnadurai, S., & Kalyanasundaram, G., (2013). Estimation of major pigment content in seaweeds collected from Pondicherry coast. *International Journal of Science and Technology, 9*(1), 522–525.

Chojnacka, K., Saeid, A., Witkowska, Z., & Tuhy, L., (2012). *Biologically Active Compounds in Seaweed Extracts - The Prospects for the Application*. Bentham Open, Sharjah, United Arab Emirates.

Cian, R. E., Salgado, P. R., Drago, S. R., González, R. J., & Mauri, A. N., (2014). Development of naturally activated edible films with antioxidant properties prepared from red seaweed *Porphyra columbina* biopolymers. *Food Chemistry, 146*, 6–14.

Cortés, Y., Hormazábal, E., Leal, H., et al., (2014). Novel antimicrobial activity of a dichloromethane extract obtained from red seaweed *Ceramium rubrum* (Hudson) (*Rhodophyta:*

Florideophyceae) against *Yersinia ruckeri* and *Saprolegnia parasitica*, agents that cause diseases in salmonids. *Electronic Journal of Biotechnology*, *17*(3), 126–131.
Cox, S., Abu-Ghannam, N., & Gupta, S., (2010). An assessment of the antioxidant and antimicrobial activity of six species of edible Irish seaweeds. *International Food Research Journal*, *17*(1), 205–220.
Cunha, M. E., Quental-Ferreira, H., Parejo, A., et al., (2019). Understanding the individual role of fish, oyster, phytoplankton and macroalgae in the ecology of integrated production in earthen ponds. *Aquaculture*, *512*, 734297.
De Corato, U., Salimbeni, R., De Pretis, A., Avella, N., & Patruno, G., (2017). Antifungal activity of crude extracts from brown and red seaweeds by a supercritical carbon dioxide technique against fruit postharvest fungal diseases. *Postharvest Biology and Technology*, *131*, 16–30.
De'ath, G., & Fabricius, K., (2010). Water quality as a regional driver of coral biodiversity and macroalgae on the Great Barrier Reef. *Ecological Applications*, *20*(3), 840–850.
Diaz, P. G., & McCook, L., (2008). *Macroalgae (seaweeds) in China.* Townsville (AU): Great Barrier Reef Marine Park Authority.
Dumay, J., & Morançais, M., (2016). Proteins and pigments. In: *Seaweed in Health and Disease Prevention* (pp. 275–318). Academic Press, Cambridge.
Dumay, J., Morancais, M., Munier, M., Le Guillard, C., & Fleurence, J., (2014). Phycoerythrins: Valuable proteinic pigments in red seaweeds. *Advances in Botanical Research*, *71*, 321–343.
El-Tawil, N. E., (2010). Effects of green seaweeds (*Ulva* sp.) as feed supplements in red tilapia (*Oreochromis* sp.) diet on growth performance, feed utilization and body composition. *J Arabian Aquac. Soc.*, *5*, 179–193.
Ertürk, Ö., & Taş, B., (2011). Antibacterial and antifungal effects of some marine algae. *Kafkas Univ. Vet. Fak. Derg.*, S121–S124.
Eyre, B. D., Andersson, A. J., & Cyronak, T., (2014). Benthic coral reef calcium carbonate dissolution in an acidifying ocean. *Nature Climate Change*, *4*(11), 969–976.
Farvin, K. S., & Jacobsen, C., (2013). Phenolic compounds and antioxidant activities of selected species of seaweeds from the Danish coast. *Food Chemistry*, *138*(2, 3), 1670–1681.
Farvin, K. S., & Jacobsen, C., (2015). Antioxidant activity of seaweed extracts: *In vitro* assays, evaluation in 5% fish oil-in-water emulsions and characterization. *Journal of the American Oil Chemists' Society*, *92*(4), 571–587.
Fernando, I. S., Kim, M., Son, K. T., Jeong, Y., & Jeon, Y. J., (2016). Antioxidant activity of marine algal polyphenolic compounds: A mechanistic approach. *Journal of Medicinal Food*, *19*(7), 615–628.
Ferraces-Casais, P., Lage-Yusty, M. A., De Quirós, A. R. B., & López-Hernández, J., (2012). Evaluation of bioactive compounds in fresh edible seaweeds. *Food Analytical Methods*, *5*(4), 828–834.
Ferraces-Casais, P., Lage-Yusty, M. A., De Quirós, A. R. B., & López-Hernández, J., (2013). Rapid identification of volatile compounds in fresh seaweed. *Talanta*, *115*, 798–800.
Fröcklin, S., De La Torre-Castro, M., Lindström, L., Jiddawi, N. S., & Msuya, F. E., (2012). Seaweed mariculture as a development project in Zanzibar, East Africa: A price too high to pay? *Aquaculture*, *356*, *357*, 30–39.
Ganesan, M., Trivedi, N., Gupta, V., Madhav, S. V., Reddy, C. R., & Levine, I. A., (2019). Seaweed resources in India – current status of diversity and cultivation: Prospects and challenges. *Botanica Marina*, *62*(5), 463–482.
Garcia-Vaquero, M., & Hayes, M., (2016). Red and green macroalgae for fish and animal feed and human functional food development. *Food Reviews International*, *32*(1), 15–45.

Gomaa, H. H., & Elshoubaky, G. A., (2016). Antiviral activity of sulfated polysaccharides carrageenan from some marine seaweeds. *International Journal of Current Pharmaceutical Review and Research*, *7*(1), 34–42.

Gora, A. H., Sahu, N. P., Sahoo, S., et al., (2018). Effect of dietary *Sargassum wightii* and its fucoidan-rich extract on growth, immunity, disease resistance and antimicrobial peptide gene expression in *Labeo rohita. International Aquatic Research*, *10*(2), 115–131.

Gubelit, Y. I., & Kovalchuk, N. A., (2010). Macroalgal blooms and species diversity in the transition zone of the eastern Gulf of Finland. *Hydrobiologia*, *656*(1), 83–86.

Guihéneuf, F., Gietl, A., & Stengel, D. B., (2018). Temporal and spatial variability of mycosporine-like amino acids and pigments in three edible red seaweeds from western Ireland. *Journal of Applied Phycology*, *30*(4), 2573–2586.

Gutiérrez-Gamboa, G., Garde-Cerdán, T., Martínez-Lapuente, L., Costa, B. S. D., Rubio-Bretón, P., & Pérez-Álvarez, E. P., (2020). Phenolic composition of Tempranillo Blanco (*Vitis vinifera* L.) grapes and wines after biostimulation via a foliar seaweed application. *Journal of the Science of Food and Agriculture*, *100*(2), 825–835.

Hamid, N., Ma, Q., Boulom, S., et al., (2015). Seaweed minor constituents. In: *Seaweed Sustainability* (pp. 193–242). Academic Press, Cambridge.

Hernández, A. J., Romero, A., Gonzalez-Stegmaier, R., & Dantagnan, P., (2016). The effects of supplemented diets with a phytopharmaceutical preparation from herbal and macroalgal origin on disease resistance in rainbow trout against *Piscirickettsia salmonis*. *Aquaculture*, *454*, 109–117.

Holdt, S. L., & Kraan, S., (2011). Bioactive compounds in seaweed: Functional food applications and legislation. *Journal of Applied Phycology*, *23*(3), 543–597.

Hughes, A. D., Kelly, M. S., Black, K. D., & Stanley, M. S., (2012). Biogas from Macroalgae: Is it time to revisit the idea?. *Biotechnology for Biofuels*, *5*(1), 86.

Ishwarya, R., Vaseeharan, B., Subbaiah, S., et al., (2018). *Sargassum wightii*-synthesized ZnO nanoparticles–from antibacterial and insecticidal activity to immunostimulatory effects on the green tiger shrimp *Penaeus semisulcatus*. *Journal of Photochemistry and Photobiology B: Biology*, *183*, 318–330.

Ismail, M. M., Gheda, S. F., & Pereira, L., (2016). Variation in bioactive compounds in some seaweeds from Abo Qir Bay, Alexandria, Egypt. *Rendiconti Lincei*, *27*(2), 269–279.

Jebasingh, S. E. J., Rajesh, R. P., & Lakshmikandan, M., (2011). Antibacterial activity of seaweed *Ulva lactuca* against fish pathogens isolated from marine fish *Katsuwonus pelamis*. *International Journal of Pharmacy and Technology*, *3*(2), 2306–2314.

Kanjana, K., Radtanatip, T., Asuvapongpatana, S., Withyachumnarnkul, B., & Wongprasert, K., (2011). Solvent extracts of the red seaweed *Gracilaria fisheri* prevent *Vibrio harveyi* infections in the black tiger shrimp Penaeus monodon. *Fish and Shellfish Immunology*, *30*(1), 389–396.

Kariyawasam, I., (2016). Seaweed mariculture: A potential multi-million dollar industry. *Mar. Environ. Prot. Auth. Sri Lanka*, *2016*, 44–49.

Karuppiah, V., (2014). Evaluation of antibacterial activity and immunostimulant of red seaweed *Chondrococcus hornemanni* (Kuetzing, 1847) against marine ornamental fish pathogens. *J. Coast. Life Med.*, *2*, 61–69.

Kendel, M., Couzinet-Mossion, A., Viau, M., Fleurence, J., Barnathan, G., & Wielgosz-Collin, G., (2013). Seasonal composition of lipids, fatty acids, and sterols in the edible red alga *Grateloupia turuturu*. *Journal of Applied Phycology*, *25*(2), 425–432.

Kendel, M., Wielgosz-Collin, G., Bertrand, S., Roussakis, C., Bourgougnon, N., & Bedoux, G., (2015). Lipid composition, fatty acids and sterols in the seaweeds *Ulva armoricana*, and

Solieria chordalis from Brittany (France): An analysis from nutritional, chemotaxonomic, and antiproliferative activity perspectives. *Marine Drugs*, *13*(9), 5606–5628.
Keyrouz, R., Abasq, M. L., Le Bourvellec, C., et al., (2011). Total phenolic contents, radical scavenging and cyclic voltammetry of seaweeds from Brittany. *Food Chemistry*, *126*(3), 831–836.
Khalid, S., Abbas, M., Saeed, F., Bader-Ul-Ain, H., & Suleria, H. A. R., (2018). *Therapeutic Potential of Seaweed Bioactive Compounds*. IntechOpen, London.
Kim, J. K., Yarish, C., Hwang, E. K., et al., (2017). Seaweed aquaculture: Cultivation technologies, challenges and its ecosystem services. *Algae*, *32*(1), 1–13.
Kim, S. Y., Kim, S. R., Oh, M. J., Jung, S. J., & Kang, S. Y., (2011). *In vitro* antiviral activity of red alga, *Polysiphonia morrowii* extract and its bromophenols against fish pathogenic infectious hematopoietic necrosis virus and infectious pancreatic necrosis virus. *The Journal of Microbiology*, *49*(1), 102–106.
Kindleysides, S., Quek, S. Y., & Miller, M. R., (2012). Inhibition of fish oil oxidation and the radical scavenging activity of New Zealand seaweed extracts. *Food Chemistry*, *133*(4), 1624–1631.
Kraan, S., (2013). Pigments and minor compounds in algae. In: *Functional Ingredients from Algae for Foods and Nutraceuticals* (pp. 205–251). Woodhead Publishing, Cambridge.
Kumar, P., Selvi, S. S., & Govindaraju, M., (2013). Seaweed-mediated biosynthesis of silver nanoparticles using *Gracilaria corticata* for its antifungal activity against *Candida* spp. *Applied Nanoscience*, *3*(6), 495–500.
Kurniawati, I., & Adam, A., (2016). Antibacterial effect of *Gracilaria verrucosa* bioactive on fish pathogenic bacteria. *The Egyptian Journal of Aquatic Research*, *42*(4), 405–410.
Lee, P. T., Wen, C. M., Nan, F. H., Yeh, H. Y., & Lee, M. C., (2020). Immunostimulatory effects of *Sarcodia suiae* water extracts on Nile tilapia *Oreochromis niloticus* and its resistance against *Streptococcus agalactiae*. *Fish and Shellfish Immunology*, *103*, 159–168.
Li, Y. X., & Kim, S. K., (2011). Utilization of seaweed-derived ingredients as potential antioxidants and functional ingredients in the food industry: An overview. *Food Science and Biotechnology*, *20*(6), 1461–1466.
Linga, P. D., Sahu, N. P., Pal, A. K., & Narendra, A., (2013). Isolation and evaluation of antioxidant and antibacterial activities of fucoidan rich extract (FRE) from Indian brown seaweed, *Sargassum wightii*. *Continental J. Pharmaceutical Sciences*, *7*(1), 9–16.
Liu, W. C., Zhou, S. H., Balasubramanian, B., Zeng, F. Y., Sun, C. B., & Pang, H. Y., (2020). Dietary seaweed (*Enteromorpha*) polysaccharides improve growth performance involved in regulation of immune responses, intestinal morphology and microbial community in banana shrimp *Fenneropenaeus merguiensis*. *Fish and Shellfish Immunology*, *104*, 202–212.
Macreadie, P. I., Jarvis, J., Trevathan-Tackett, S. M., & Bellgrove, A., (2017). Seagrasses and macroalgae: Importance, vulnerability and impacts. *Climate Change Impacts on Fisheries and Aquaculture: A Global Analysis* (pp. 729–770). Wiley-Blackwell, Hoboken, New Jersey.
Mantri, V. A., Eswaran, K., Shanmugam, M., et al., (2017). An appraisal on commercial farming of *Kappaphycus alvarezii* in India: Success in diversification of livelihood and prospects. *Journal of Applied Phycology*, *29*(1), 335–357.
Marinho-Soriano, E., Azevedo, C. A. A., Trigueiro, T. G., Pereira, D. C., Carneiro, M. A. A., & Camara, M. R., (2011). Bioremediation of aquaculture wastewater using macroalgae and Artemia. *International Biodeterioration and Biodegradation*, *65*(1), 253–257.
Martins, C. D. L., Ramlov, F., Carneiro, N. P. N., et al., (2013). Antioxidant properties and total phenolic contents of some tropical seaweeds of the Brazilian coast. *Journal of Applied Phycology*, *25*(4), 1179–1187.

Messyasz, B., Pikosz, M., & Treska, E., (2018). Biology of freshwater macroalgae and their distribution. In: *Algae Biomass: Characteristics and Applications* (pp. 17–31). Springer, Cham.

Michalak, I., & Chojnacka, K., (2015). Algae as production systems of bioactive compounds. *Engineering in Life Sciences*, *15*(2), 160–176.

Milhazes-Cunha, H., & Otero, A., (2017). Valorisation of aquaculture effluents with microalgae: The integrated multi-trophic aquaculture concept. *Algal Research*, *24*, 416–424.

Mouritsen, O. G., Bagatolli, L. A., Duelund, L., Garvik, O., Ipsen, J. H., & Simonsen, A. C., (2017). Effects of seaweed sterols fucosterol and desmosterol on lipid membranes. *Chemistry and Physics of Lipids*, *205*, 1–10.

Mystikou, A., Peters, A. F., Asensi, A. O., et al., (2014). Seaweed biodiversity in the south-western Antarctic peninsula: Surveying macroalgal community composition in the Adelaide Island/marguerite bay region over a 35-year time span. *Polar Biology*, *37*(11), 1607–1619.

Natrah, F. M. I., Defoirdt, T., Sorgeloos, P., & Bossier, P., (2011). Disruption of bacterial cell-to-cell communication by marine organisms and its relevance to aquaculture. *Marine Biotechnology*, *13*(2), 109–126.

Nauer, F., Amorim, A. M., Santos, J. P., Chow, F., & Oliveira, M. C., (2018). Physiological plasticity in morphological variations of red seaweed *Hypnea pseudomusciformis* (Gigartinales, *Rhodophyta*) uncovered by molecular, antioxidant capacity and pigments content data. *Brazilian Journal of Botany*, *41*(3), 567–577.

Nurilmala, M., Anwar, E., Luthfiyana, N., & Hidayat, T., (2017). Identification of bioactive compounds of seaweed *Sargassum* sp. and *Eucheuma cottonii* Doty as a raw sunscreen cream: Bioactive compounds of seaweed as a raw sunscreen cream. *Proceedings of the Pakistan Academy of Sciences: B. Life and Environmental Sciences*, *54*(4), 311–318.

Okolie, C. L., Rajendran, S. R., Udenigwe, C. C., Aryee, A. N., & Mason, B., (2017). Prospects of brown seaweed polysaccharides (BSP) as prebiotics and potential immunomodulators. *Journal of Food Biochemistry*, *41*(5), e12392.

Pádua, D., Rocha, E., Gargiulo, D., & Ramos, A. A., (2015). Bioactive compounds from brown seaweeds: Phloroglucinol, fucoxanthin and fucoidan as promising therapeutic agents against breast cancer. *Phytochemistry Letters*, *14*, 91–98.

Pal, A., Kamthania, M. C., & Kumar, A., (2014). Bioactive compounds and properties of seaweeds — A review. *Open Access Library Journal*, *1*(4), 1–17.

Peixoto, M. J., Svendsen, J. C., Malte, H., et al., (2016). Diets supplemented with seaweed affect metabolic rate, innate immune, and antioxidant responses, but not individual growth rate in European seabass (*Dicentrarchus labrax*). *Journal of Applied Phycology*, *28*(3), 2061–2071.

Pellizzari, F., Silva, M. C., Silva, E. M., et al., (2017). Diversity and spatial distribution of seaweeds in the South Shetland Islands, Antarctica: An updated database for environmental monitoring under climate change scenarios. *Polar Biology*, *40*(8), 1671–1685.

Peña, V., Bárbara, I., Grall, J., Maggs, C. A., & Hall-Spencer, J. M., (2014). The diversity of seaweeds on maerl in the NE Atlantic. *Marine Biodiversity*, *44*(4), 533–551.

Peñalver, R., Lorenzo, J. M., Ros, G., Amarowicz, R., Pateiro, M., & Nieto, G., (2020). Seaweeds as a functional ingredient for a healthy diet. *Marine Drugs*, *18*(6), 301.

Peng, Y., Xie, E., Zheng, K., et al., (2013). Nutritional and chemical composition and antiviral activity of cultivated seaweed *Sargassum naozhouense* Tseng et Lu. *Marine Drugs*, *11*(1), 20–32.

Pereira, D. C., Trigueiro, T. G., Colepicolo, P., & Marinho-Soriano, E., (2012). Seasonal changes in the pigment composition of natural population of *Gracilaria domingensis* (Gracilariales, *Rhodophyta*). *Revista Brasileira de Farmacognosia*, *22*(4), 874–880.

Peres, J. C. F., Carvalho, L. R. D., Gonçalez, E., Berian, L. O. S., & Felicio, J. D., (2012). Evaluation of antifungal activity of seaweed extracts. *Ciência e Agrotecnologia*, *36*(3), 294–299.

Poojary, M. M., Barba, F. J., Aliakbarian, B., et al., (2016). Innovative alternative technologies to extract carotenoids from microalgae and seaweeds. *Marine Drugs*, *14*(11), 214.

Prabu, D. L., Sahu, N. P., Pal, A. K., Dasgupta, S., & Narendra, A., (2016). Immunomodulation and interferon-gamma gene expression in sutchi catfish, *Pangasianodon hypophthalmus*: Effect of dietary fucoidan rich seaweed extract (FRSE) on pre and post-challenge period. *Aquaculture Research*, *47*(1), 199–218.

Priyadharshini, S., Bragadeeswaran, S., Prabhu, K., & Ran, S. S., (2011). Antimicrobial and hemolytic activity of seaweed extracts *Ulva fasciata* (Delile 1813) from Mandapam, Southeast coast of India. *Asian Pacific Journal of Tropical Biomedicine*, *1*(1), S38, S39.

Radhika, D. D., & Mohaideen, A., (2016). Effect of utilization of seaweeds as an immunostimulant in *Aeromonas hydrophila* infected *Cyprinus carpio*. *World J. Pharm. Pharm. Sci.*, *5*, 851–857.

Radulovich, R., Neori, A., Valderrama, D., Reddy, C. R. K., Cronin, H., & Forster, J., (2015). *Farming of Seaweeds: Seaweed Sustainability*, 27–59.

Reverter, M., Bontemps, N., Lecchini, D., Banaigs, B., & Sasal, P., (2014). Use of plant extracts in fish aquaculture as an alternative to chemotherapy: Current status and future perspectives. *Aquaculture*, *433*, 50–61.

Rizzo, C., Genovese, G., Morabito, M., et al., (2017). Potential antibacterial activity of marine macroalgae against pathogens relevant for aquaculture and human health. *J. Pure Appl. Microbiol.*, *11*, 1695–1706.

Rudtanatip, T., Lynch, S. A., Wongprasert, K., & Culloty, S. C., (2018). Assessment of the effects of sulfated polysaccharides extracted from the red seaweed Irish moss *Chondrus crispus* on the immune-stimulant activity in mussels *Mytilus* spp. *Fish and Shellfish Immunology*, *75*, 284–290.

Sachindra, N. M., Airanthi, M. K. W. A., Hosokawa, M., & Miyashita, K., (2010). Radical scavenging and singlet oxygen quenching activity of extracts from Indian seaweeds. *Journal of Food Science and Technology*, *47*(1), 94–99.

Sakthivel, M., Deivasigamani, B., Rajasekar, T., Kumaran, S., & Alagappan, K. M., (2015). Immunostimulatory effects of polysaccharide compound from seaweed *Kappaphycus alvarezii* on Asian seabass (*Lates calcarifer*) and it's resistance against *Vibrio parahaemolyticus*. *J. Mar. Biol. Oceanogr.*, *4*, 2.

Saluri, M., Kaldmäe, M., & Tuvikene, R., (2019). Extraction and quantification of phycobiliproteins from the red alga *Furcellaria lumbricalis*. *Algal Research*, *37*, 115–123.

Samocha, T. M., Fricker, J., Ali, A. M., Shpigel, M., & Neori, A., (2015). Growth and nutrient uptake of the macroalga *Gracilaria tikvahiae* cultured with the shrimp *Litopenaeus vannamei* in an integrated multi-trophic aquaculture (IMTA) system. *Aquaculture*, *446*, 263–271.

Satheesh, S., & Wesley, S. G., (2012). Diversity and distribution of seaweeds in the Kudankulam coastal waters, south-eastern coast of India. *Biodiversity Journal*, *3*(1), 79–84.

Sathya, R., Kanaga, N., Sankar, P., & Jeeva, S., (2017). Antioxidant properties of phlorotannins from brown seaweed *Cystoseira trinodis* (Forsskål) C. Agardh. *Arabian Journal of Chemistry*, *10*, S2608–S2614.

Seth, A., & Shanmugam, M., (2016). Seaweeds as agricultural crops in India: New vistas. In: *Innovative Saline Agriculture* (pp. 441–473). Springer, New Delhi.

Sharifuddin, Y., Chin, Y. X., Lim, P. E., & Phang, S. M., (2015). Potential bioactive compounds from seaweed for diabetes management. *Marine Drugs*, *13*(8), 5447–5491.

Shi, Q., Wang, A., Lu, Z., Qin, C., Hu, J., & Yin, J., (2017). Overview on the antiviral activities and mechanisms of marine polysaccharides from seaweeds. *Carbohydrate Research*, *453*, 1–9.

Sotoudeh, E., & Mardani, F., (2018). Antioxidant-related parameters, digestive enzyme activity and intestinal morphology in rainbow trout (*Oncorhynchus mykiss*) fry fed graded levels of red seaweed, *Gracilaria pygmaea*. *Aquaculture Nutrition*, *24*(2), 777–785.

Souza, B. W., Cerqueira, M. A., Martins, J. T., et al., (2011). Antioxidant potential of two red seaweeds from the Brazilian coasts. *Journal of Agricultural and Food Chemistry*, *59*(10), 5589–5594.

Sudhakar, M. P., Jagatheesan, A., Perumal, K., & Arunkumar, K., (2015). Methods of phycobiliprotein extraction from *Gracilaria crassa* and its applications in food colorants. *Algal Research*, *8*, 115–120.

Tabarsa, M., Han, J. H., Kim, C. Y., & You, S. G., (2012). Molecular characteristics and immunomodulatory activities of water-soluble sulfated polysaccharides from *Ulva pertusa*. *Journal of Medicinal Food*, *15*(2), 135–144.

Tang, Y. Z., & Gobler, C. J., (2011). The green macroalga, *Ulva lactuca*, inhibits the growth of seven common harmful algal bloom species via allelopathy. *Harmful Algae*, *10*(5), 480–488.

Tariq, A., Athar, M., Ara, J., Sultana, V., Ehteshamul-Haque, S., & Ahmad, M., (2015). Biochemical evaluation of antioxidant activity and polysaccharides fractions in seaweeds. *Global Journal of Environmental Science and Management*, *1*(1), 47–62.

Teagle, H., Hawkins, S. J., Moore, P. J., & Smale, D. A., (2017). The role of Kelp species as biogenic habitat formers in coastal marine ecosystems. *Journal of Experimental Marine Biology and Ecology*, *492*, 81–98.

Terasaki, M., Narayan, B., Kamogawa, H., et al., (2012). Carotenoid profile of edible Japanese seaweeds: An improved HPLC method for separation of major carotenoids. *Journal of Aquatic Food Product Technology*, *21*(5), 468–479.

Thanigaivel, S., Chandrasekaran, N., Mukherjee, A., & Thomas, J., (2015). Investigation of seaweed extracts as a source of treatment against bacterial fish pathogen. *Aquaculture*, *448*, 82–86.

Thanigaivel, S., Vijayakumar, S., Mukherjee, A., Chandrasekaran, N., & Thomas, J., (2014). Antioxidant and antibacterial activity of *Chaetomorpha antennina* against shrimp pathogen *Vibrio parahaemolyticus*. *Aquaculture*, *433*, 467–475.

Valente, L. M., Rema, P., Ferraro, V., et al., (2015). Iodine enrichment of rainbow trout flesh by dietary supplementation with the red seaweed *Gracilaria vermiculophylla*. *Aquaculture*, *446*, 132–139.

Van, D. H., Hoseinifar, S. H., Tapingkae, W., Tongsiri, S., & Khamtavee, P., (2016). Combined administration of low molecular weight sodium alginate boosted immunomodulatory, disease resistance and growth-enhancing effects of *Lactobacillus plantarum* in Nile tilapia (*Oreochromis niloticus*). *Fish and Shellfish Immunology*, *58*, 678–685.

Vatsos, I. N., & Rebours, C., (2015). Seaweed extracts as antimicrobial agents in aquaculture. *Journal of Applied Phycology*, *27*(5), 2017–2035.

Vega, J., Álvarez-Gómez, F., Güenaga, L., Figueroa, F. L., & Gómez-Pinchetti, J. L., (2020). Antioxidant activity of extracts from marine macroalgae, wild-collected and cultivated, in an integrated multi-trophic aquaculture system. *Aquaculture*, *522*, 735088.

Vinosha, M., Palanisamy, S., Anjali, R., et al., (2020). Sulfated galactan from *Halymenia dilatata* enhance the antioxidant properties and prevents *Aeromonas hydrophila* infection in tilapia fish: *In vitro* and *in vivo* study. *International Journal of Biological Macromolecules, 158*(1), 569–579.

Wan, A. H., Davies, S. J., Soler-Vila, A., Fitzgerald, R., & Johnson, M. P., (2019). Macroalgae as a sustainable aquafeed ingredient. *Reviews in Aquaculture, 11*(3), 458–492.

Wang, L., Wang, X., Wu, H., & Liu, R., (2014). Overview on biological activities and molecular characteristics of sulfated polysaccharides from marine green algae in recent years. *Marine Drugs, 12*(9), 4984–5020.

Wang, T., Li, Z., Yuan, F., Lin, H., & Pavase, T. R., (2017). Effects of brown seaweed polyphenols, α-tocopherol, and ascorbic acid on protein oxidation and textural properties of fish mince (*Pagrosomus major*) during frozen storage. *Journal of the Science of Food and Agriculture, 97*(4), 1102–1107.

Williams, S. L., & Smith, J. E., (2007). A global review of the distribution, taxonomy, and impacts of introduced seaweeds. *Annu. Rev. Ecol. Evol. Syst., 38*, 327–359.

Wongprasert, K., Rudtanatip, T., & Praiboon, J., (2014). Immunostimulatory activity of sulfated galactans isolated from the red seaweed *Gracilaria fisheri* and development of resistance against white spot syndrome virus (WSSV) in shrimp. *Fish and Shellfish Immunology, 36*(1), 52–60.

Wyatt, K. H., Rober, A. R., Schmidt, N., & Davison, I. R., (2014). Effects of desiccation and rewetting on the release and decomposition of dissolved organic carbon from benthic macroalgae. *Freshwater Biology, 59*(2), 407–416.

Xu, D., Gao, Z., Zhang, X., et al., (2011). Evaluation of the potential role of the macroalga *Laminaria japonica* for alleviating coastal eutrophication. *Bioresource Technology, 102*(21), 9912–9918.

Xuan, X., Wen, X., Li, S., Zhu, D., & Li, Y., (2013). Potential use of macro-algae *Gracilaria lemaneiformis* in diets for the black sea bream, *Acanthopagrus schlegelii*, juvenile. *Aquaculture, 412*, 167–172.

Yadav, S. K., Palanisamy, M., & Murthy, G. V. S., (2015). Economically important seaweeds of Kerala coast, India: A review. *Elixir Biosciences, 82*, 32147–32153.

Yokoyama, H., & Ishihi, Y., (2010). Bioindicator and biofilter function of *Ulva* spp. (*Chlorophyta*) for dissolved inorganic nitrogen discharged from a coastal fish farm — Potential role in integrated multi-trophic aquaculture. *Aquaculture, 310*(1, 2), 74–83.

Younis, E. S. M., Al-Quffail, A. S., Al-Asgah, N. A., Abdel-Warith, A. W. A., & Al-Hafedh, Y. S., (2018). Effect of dietary fish meal replacement by red algae, *Gracilaria arcuata*, on growth performance and body composition of Nile tilapia *Oreochromis niloticus*. *Saudi Journal of Biological Sciences, 25*(2), 198–203.

Yu, K. X., Jantan, I., Ahmad, R., & Wong, C. L., (2014). The major bioactive components of seaweeds and their mosquitocidal potential. *Parasitology Research, 113*(9), 3121–3141.

Zhong, B., Robinson, N. A., Warner, R. D., Barrow, C. J., Dunshea, F. R., & Suleria, H. A., (2020). LC-ESI-QTOF-MS/MS characterization of seaweed phenolics and their antioxidant potential. *Marine Drugs, 18*(6), 331.

Zhu, Z., Wu, Q., Di, X., et al., (2017). Multistage recovery process of seaweed pigments: Investigation of ultrasound-assisted extraction and ultra-filtration performances. *Food and Bioproducts Processing, 104*, 40–47.

CHAPTER 11

PHARMACEUTICAL AND NUTRACEUTICAL POTENTIAL OF SEAWEED SUSTAINABLE EXPLOITATION

SARA GARCÍA-POZA,[1] ADRIANA LEANDRO,[1] JOÃO COTAS,[1] DIANA PACHECO,[1] JOÃO CARLOS MARQUES,[1] LEONEL PEREIRA,[1] and ANA MARTA MENDES GONÇALVES[1,2]

[1]*University of Coimbra, MARE-Marine, and Environmental Sciences Center, Department of Life Sciences, Coimbra–3001-456, Portugal*

[2]*Department of Biology and CESAM, University of Aveiro, Aveiro–3810-193, Portugal*

ABSTRACT

Seaweeds are autotrophic marine organisms able to produce multiple molecules, being primary and secondary metabolites. In recent years, several studies investigated the bioactivities of seaweed-derived molecules. Many proprieties are being described, for example, antioxidant, anti-inflammatory, antiviral, antitumoral, antidiabetic, and antiobesity, in order to potentially apply in pharmaceutical and nutraceutical sectors. In the last decades, consumer's demand has increased so much that it has exceeded the capacity of the market supply based on wild nature reserves, generating a need to cultivate the main species to be consumed. To make this a sustainable process, and avoid the limitation of the wild stocks, it is needed to resort to aquaculture systems. Since ancient times, there has been harvesting and

Algal Metabolites: Biotechnological, Commercial, and Industrial Applications. Jeyabalan Sangeetha, Devarajan Thangadurai, Saher Islam, & Ravichandra Hospet (Eds.)

cultivation of algae, but due to the increase in research and the recognition of the properties of its compounds, aquaculture has been growing quickly in the last years. The cultivation method ranges from offshore to land-based. Integrated multitrophic aquaculture (IMTA) is increasingly used due to its great advantages. Algae aquaculture is a promising area with the possibility of improving methods and techniques with great potential in the pharmaceutical and nutraceutical areas.

11.1 INTRODUCTION

By 2050, it is estimated a demographic growth of one-third of the current global population (Godfray et al., 2010). This, associated with the increment of limited availability of cultivable area and access to freshwater, leads to the research of new and sustainable food sources (Bleakley and Hayes, 2017).

Marine habitats are becoming a source of a wide range of biomolecules, being these the target of the research for a demand of novel pharmaceutical and nutraceutical products. Still, it is important to take into account wildlife is important to the homeostasis of the world and overexploitation can be very harmful to the environment and to the humankind. So, there are new policies in the scope of blue economy relatively to sustainability, eco-friendly methods and to recover the environmental systems (Babb, 2015; Wenhai et al., 2019; Rajak et al., 2020). The policies are driven to find new solutions that are ecologically neutral or positive, in detriment of negative. This is happening in the European Union (EU) nowadays with the restriction on synthetic products, such as plastic and chemical additives in food (FSAI, 2015; Anadón et al., 2018; Council of the EU, 2020; European Chemicals Agency, 2020).

Zero hunger, marine life protection, responsible consumption and production are central issues included in the 2030 Agenda and on UN Sustainable Development Goals, recognizing the importance of ensuring food access, specially to people in vulnerable conditions, sustainable use of ocean resources and to avoid over-extraction of environmental resources, respectively (United Nations, 2016).

Historically, seaweeds are described as sea vegetables, because they are the basis of the marine food chain and are one of the most explored marine resources in the world, mainly in Asia as food and pharmaceutical resources (Pereira, 2016). From a commercial point of view, seaweeds are in the front line of the development of new pharmaceutical and nutraceutical components products, due to the myriad of compounds they produce (Kim, 2011; Villanueva et al., 2014). Currently, the awareness in nutraceutical food products are growing due to their well-being beneficial properties. Consumer awareness is increasing regarding questions about food safety, namely food processing, storage, and additives source (O'Sullivan et al., 2010; Lordan et al., 2011). On a research and development level, their potential for the treatment of chronic diseases are explored on a large-scale basis (Choudhary and Grover, 2012; Brown et al., 2014; Ding et al., 2019).

The Ocean represents a reservoir of bioactive compounds, where seaweeds are very rich in some health-promoting molecules, such as ω-3 fatty acids, essential amino acids (EAAs), vitamins A, B (B1, B2, B9, and B12), C, D, E, K, essential minerals (calcium, iron, iodine, magnesium, phosphorus, potassium, zinc, copper, manganese, selenium, and fluoride) and also in dietary fibers (Rajapakse and Kim, 2011; Pereira, 2016; Shannon and Abu-Ghannam, 2019). These molecules are elements responsible for maintaining cell homeostasis, thus promoting human health through its dietary consumption (Schwager et al., 2008).

Compounds synthesized by seaweeds are already widely applied in the food industry, because of the countless properties to increase food security and quality. For instance, seaweeds are applied as a functional ingredient in commercial applications as a stabilizer, emulsifier, thickening agent, texture modifier, and phytochemicals improved with vitamins and as a dietary fiber (Leandro et al., 2020).

In the pharmaceutical area, seaweeds' compounds have multi-role actions as anti-diabetic, anti-inflammatory, antimicrobial, antiviral, antiallergic, antidiabetic, antioxidant, antiphotoaging, antipruritic, hepatoprotective, hypotension, neuroprotective, tissue regeneration, and anticancer agents (Magalhães et al., 2011; Nwosu et al., 2011; Pangestuti and Kim, 2011; Lopes, 2014; Im et al., 2017; Khalil et al., 2017; Murray et al., 2018; Mekinić et al., 2019; Cotas et al., 2020b; Valado et al., 2020).

From an industrial point of view, the challenge is that, in their wild habitats, seaweeds' compounds quantity and diversity are very sensitive to abiotic and biotic factors, such as species, life stage, size, age, and reproductive status, locations, depth, nutrient enrichment, salinity, light intensity exposure, ultraviolet (UV) radiation, intensity of herbivore and harvesting season (Domínguez, 2013).

For example, a study points that at different stages of the life cycle of the red seaweed *Gigartina pistillata*, the female gametophyte and tetrasporophyte phase, the carrageenan synthesized has different chemical conformations. Despite that, both demonstrated antitumor potential against colorectal cancer stem-like cells (Cotas et al., 2020b). Another question is wild seaweed overexploitation related with the seaweeds' consumption. There is the necessary quantity to prevent environmental pressures of its growing demand (Jung et al., 2013). These are challenging tasks, because it is necessary to perform ecological assessment studies of each species to minimize harvesting harmful impacts. Besides that, each species holds a complex biochemical and biological variability, that is not fully understood by the scientific community yet (Domínguez, 2013; Singh and Sidana, 2013). In this context, seaweed farming is a sustainable option to tackle environmental pressures on macroalgae's wild stocks.

Seaweed global market is a sector in enlargement, with a yearly growing rate of 8.9%. Presently, it is estimated at 11.7 billion US$ and the previsions point out that by 2024 it will scope the market worth of 22.13 billion US$ (Ferdouse et al., 2018). With FAO reporting four species/genus being intensively cultivated (above 2,000,000 tons), *Eucheuma* spp., *Saccharina japonica* (formerly known as *Laminaria japonica*), *Gracilaria* spp. and *Undaria pinnatifida* with a net joint over near 25,000,000 tons produced in 2016, and 85% of cultivated seaweed was produced by China and Indonesia. These seaweeds are mainly cultivated to supply the hydrocolloid and food industries (Ferdouse et al., 2018).

Seaweed farming is the opportunity to deliver a trustworthy, harmless, and sustainable seaweed biomass with guaranteed quality. Consequently, with the effort on the characterization of the natural variability of the seaweeds compounds and aquaculture systems optimization, seaweed farming has been done with several species, such as *Kappaphycus alvarezii* (Rhodophyta) and *Saccharina latissima* (Phaeophyceae) (Barry et al., 2015; Michalak and Chojnacka, 2015). On the other hand, there is also the possibility of exploring the potentialities of wild seaweeds in seaweeds that have rapid growth rate or seaweeds that have a great and reliable biomass in the wild, which happens with *Ascophyllum nodosum* (Phaeophyceae), *Gelidium* sp. and *Gracilaria gracilis* (Rhodophyta).

In the future, in farming seaweeds will be performed optimized or new cultivation methods (to the improvement of seaweed overall quality), and not merely to the phycocolloid industry that currently take place. Pigments and other industries, such as the substitute of animal-based gelatin with seaweed polymers, or textile industry for natural pigments (Ferdouse et al., 2018), or the PUFAs enrichment industry (nowadays, based in fish discards) will be some of the performed. Moreover, the up-to-date techniques to extract and purify some of the seaweed's compounds characterized are overpriced and economically unfeasible for scale-up, but there is an exception to this fact, seaweed polysaccharides, that are already used in several industries, such as pharmaceutical, food, and biomedical industries. The investigation in this area is evolving to achieve more proficient and ecological extraction methods (Michalak and Chojnacka, 2014, 2015).

Seaweed polysaccharides are the most researched and commercially used compounds extracted from seaweeds, and derived from a high commercial burden, the extraction and purification techniques are already cutting-edge in the standings of cost efficiency and economic viability. For example, alginate, agar, and carrageenan are mostly used in the food and pharmaceutical industries as multipurpose components, such as stabilizers, emulsifiers, and homogenizers and this particularity already promoted the growth in the seaweed aquaculture, mainly in Asia and South America (Michalak and Chojnacka, 2015; FAO, 2018). The highest percentage of seaweed provided by aquaculture is to provide biomass to phycocolloid industry.

Seaweeds are arousing interest as one of the most attractive natural sources, for their wide range of compounds, having a high potential for the improvement of novel therapeutic and nutraceutical products. Seaweeds present benefits over other natural sources of bioactive compounds due to their low-cost, feasibility, and environmentally friendly production methods, a high extraction yield, and, most important, to the availability of biomass in natural ecosystem to assure the aquaculture steady (Cotas et al., 2020a, b).

At this time, one of the main goals is to ensure that wild seaweed species and seaweeds produced in aquaculture have similar or higher concentrations of targeted biomolecules, in demand to ensure feedstock for extraction and ecological sustainability. Seaweed aquaculture will be highlighted in the next years to answer this growing market and its high demand (Cotas et al., 2020a). The path of seaweed aquaculture to the future must pass by the biomolecule's identification and an effort to reproduce in an aquaculture system (Cotas et al., 2020a, b).

The request for seaweeds and their by-products has been rising globally, and so has the attentiveness in their production and the attraction of stakeholders to capitalize more extensively the production of various algal species that will fill diverse economic segment's needs (Ashkenazi et al., 2019). In the last years, the world production of seaweed, has more than tripled, up from 10.6 million tons in 2000 to 32.4 million tons in 2018 (FAO, 2020), exhibiting the potential of this market segment.

This chapter is a compendium of seaweeds' main compounds that have nutraceutical and pharmaceutical potential, focusing on the potential already described in literature and explaining the importance of aquaculture in detriment of wild harvesting.

11.2 BIOACTIVE COMPOUNDS FROM SEAWEEDS

11.2.1 FATTY ACIDS

Typically, seaweeds have low levels of lipids, in which the largest parcel are polyunsaturated fatty acids (PUFAs) (MacArtain et al., 2007; van Ginneken et al., 2011; Makkar et al., 2016). Fatty acid profiles vary not only between taxonomic groups, species, and genetic status, but also depend on environmental conditions, such as light, salinity, and pollution levels, and on location and seasonality (Fleury et al., 2011; Galloway et al., 2012; Pereira et al., 2012).

11.2.1.1 PUFAs

There is an increasing commercial interest in long-chain fatty acids, specifically in essential PUFA, which are significantly present in macroalgae (Pereira et al., 2012). In fact, seaweeds contain significantly higher levels of PUFAs than land vegetables (Darcy-Vrillon, 1993). Seaweed accumulates PUFAs when there is a reduction of the environmental temperature (Kayama et al., 1985; Khotimchenko, 1991). Seaweeds from the same genus, from different parts of the world normally have almost the same fatty acid profile even though the actual content of PUFAs is affected by the site of collection. Thus, cold water species have a higher level of PUFAs, demonstrating a potential use in nutraceuticals or in food formulations (Kumar et al., 2008). Whereas, warm water species lipidic profile have a higher content of saturated fatty acids (Narayan et al., 2004).

Green seaweeds (Chlorophyta) are characterized for having high amounts of Palmitidonic (16:4ω-3) PUFAs in most of the species (Kumar et al., 2008) and α-linolenic acid (ALA, C18:3 ω-3) is typical of order Ulvales (Khotimchenko, 1993; Thompson, 1996; Khotimchenko et al., 2002; Li et al., 2002; Kumari et al., 2010; Pereira et al., 2012). Green algae contain larger amounts of palmitolinolenic (16:3ω3) and palmitidonic (16:4ω-3) PUFAs compared with brown and red algae (Sato, 1975). Brown macroalgae (Phaeophyceae) present mainly linoleic acid (LA; C18:2 (n-6), arachidonic acid (AA,

20:4ω-6) and eicosapentaenoic acid (EPA, 20:5ω-3) (Johns et al., 1979; Khotimchenko et al., 1991; Khotimchenko, 1998; Li et al., 2002; Narayan et al., 2004). For instance, red macroalgae (Rhodophyta) are characterized by having a high level of C20:4 arachidonic (AA, 20:4ω-6), and C20:5 eicosapentaenoic (EPA, 20:5ω-3) fatty acids and a lower concentration of palmitoleic (C16:1) and oleic (C18:1) acids (Kumar et al., 2008). Palmitic acid (16:0) is predominant in all groups of seaweeds, since it is a structural saturated fatty acid (SFA) (Gressler et al., 2010).

PUFA are the major components of cell membrane phospholipids (PLs) and are extremely important in human metabolism (Calder, 2010). They have a role in the regulation of membrane fluidity, electron, and oxygen transport, as well as thermal adaptation (Funk, 2001). Additionally, they are precursors of hormone-like signaling molecules, which include thromboxanes, prostaglandins, and leukotrienes (Radwan, 1991; Poudyal et al., 2013). Many beneficial properties have been attributed to PUFA, such as antibacterial, anti-inflammatory, antioxidant, antitumoral, prevention of cardiac diseases, among others (Field and Schley, 2004; Ward and Singh, 2005; Schmitz and Ecker, 2008; Das et al., 2009; Huang and Ebersole, 2010; Mozaffarian and Wu, 2011). Due to these great bioactivities, PUFAs are indicated for nutraceutical and pharmaceutical industries.

Seaweeds could even be employed as ingredients in hatchery/nursery diets in shrimp/fish culture due to their requirement of PUFA enrichment for growth and development (Narayan et al., 2004).

11.2.1.2 PHYTOSTEROLS

The interest in these compounds emerged as a consequence of their nutraceutical properties, as they are not endogenously produced by humans but have beneficial effects on human health (Lopes et al., 2013). Phytosterols are a group of cholesterol-like molecules found exclusively in plants and algae, composing their cellular membranes. In these organisms, they play important functions: control membrane fluidity, permeability (stabilizing phospholipid bilayers), and in signal transduction, acting as hormones or are hormonal precursors (Piironen et al., 2000; Mirjana et al., 2009; Lopes et al., 2013). They are structurally similar and functionally analogous to cholesterol in vertebrates (Piironen et al., 2000). As other compounds produced by seaweeds, external abiotic factors differences lead to different phytosterols profiles, as well as variability between species and development stage of the seaweed (Piironen et al., 2000; Lopes et al., 2013).

Depending on the species, different algae have a specific favorable season for vegetative growth and reproduction, and the accumulation of sterols probably corresponds to the rapid phase of cell growth (Piironen et al., 2000; Lopes et al., 2013). Thus, Rhodophyta achieves its maximum biomass production in spring, from March to June (Guiry and Dawes, 1992; Schoschina, 1996). Whereas in Phaeophyta and Chlorophyta it is in summer, and that can be an explanation for the higher amounts of sterols in those periods (Turna et al., 2002). Moreover, an investigation demonstrated the influence of the salinity and temperature on phytosterol profiles of Ulvaceae species (Kapetanović et al., 2005).

Cholesterol is one of the major sterols present in all groups of seaweed (Kamenarska et al., 2003; Lopes et al., 2011). Brown and green algae contain high levels of other C29 sterols, especially fucosterol and isofucosterol, respectively (Patterson, 1971; Fattorusso et al., 1980; Lopes et al., 2013; Torres et al., 2019). Chlorophyceae also present high quantity of ergosterol and 24-ethylcholesterol (Sánchez-Machado et al., 2004).

Phytosterols exhibit some health-promoting effects, such as anticancer and hypocholesterolemic, antiobesity, and anti-inflammatory properties (Suzuki et al., 1998; Platt et al., 2004; Patch et al., 2006; Lopes et al., 2013).

In this case, seaweeds are an excellent source of esterified phytosterols and can be used as a food (nutraceutical). Furthermore, there is a positive correlation between phytosterol consumption and a reduction in the risk of some chronic diseases (Moreau et al., 2002; EFSA, 2008). Phytosterols may be of great relevance to the development of new drugs and functional foods (Lee et al., 2008).

11.2.2 PROTEINS

Proteins are polymers of amino acids and are one of the main nutrients of which there will be a high demand in the future. Some macroalgae are known to have great protein content, similar to meat, egg, soybean, and milk (Fleurence, 1999; Bleakley and Hayes, 2017). In addition, the fact that marine macroalgae do not require fresh water or arable land to be cultivated (van Krimpen et al., 2013) makes them a suitable and effective alternative protein source, which would be mandatory to fulfil the demand of consumers (Holdt and Kraan, 2011; Bleakley and Hayes, 2017).

Many studies demonstrated that all seaweeds are rich in proteins and they contain every EAA (Vieira et al., 2018). Comparing all the three taxonomic seaweed groups, in general, the red and green species contain relatively higher protein levels, with an average value of 4–50% (w/w) dry weight, than the brown species, which contain between 1% and 29% (w/w) dry weight (Rodrigues et al., 2015; Kadam et al., 2017). For example, *Palmaria palmata* and *Neopyropia tenera* (formerly *Pyropia tenera*) (Rhodophyta) have been reported to contain 33% and 47% d.w., respectively (Fleurence, 1999). According to the literature, some small variances in protein content correspond to geographical, environmental, and seasonal factors. In general, the highest protein contents in seaweeds are found during winter-early spring and the lowest in summer-early autumn. This could be explained by the fact that during the autumn and winter, seaweeds possess higher ability to store nitrogen due to lower nitrogen metabolism (and degradation) (Martínez and Rico, 2008; Patarra et al., 2011; Peinado et al., 2014; Vieira et al., 2018). In seaweed aquaculture, it is also important to know that the nitrogen levels in the water affect the growth of seaweed, and consequently, the protein profile (Diamahesa et al., 2017).

Tryptophan and lysine are the limiting amino acids found in macroalgae (Kolb et al., 1999; Dawczynski et al., 2007; Volkmann et al., 2008). Then, in red seaweed, there are also in low concentrations leucine and isoleucine, while in brown species there are methionine, cysteine, and lysine (Kakinuma et al., 2001; Dawczynski et al., 2007; Mišurcová et al., 2014). On the other hand, aspartic acid and glutamic acid constitute a relatively large proportion of the total amino acids in many seaweed species (Macartain et al., 2007).

With their good nutritional values and protein composition, seaweeds are a valuable resource for athletes requiring high levels of protein, especially for vegan athletes for whom eggs and dairy whey protein may not be suitable (Du, 2015).

PBP (phycoerythrin, phycocyanin, allophycocyanin, phycoerythrocyanin) are large light energy-capturing protein complexes anchored to thylakoid membranes present on the algae (Tandeau De Marsac, 2003; Sekar and Chandramohan, 2008). For example, in *P. palmata*, phycoerythrin constitutes a significant proportion of the overall protein content in red algae, with levels of 1.2% (total dry weight (d.w.)) (Guangce et al., 2002). The primary commercial application of these compounds is natural dyes, used in products, such as chewing gum, soft drinks, dairy products, as well as cosmetic products, such as lipstick and eyeliner (Spolaore et al., 2006). PBP also have health beneficial bioactivities, being indicated for nutraceutical products. Previous research has demonstrated their anti-oxidative, anti-inflammatory, antiviral, antitumor, neuroprotective, and hepatoprotective activities (Sekar and Chandramohan, 2008). PBP in red algae represent a big nitrogen reserve, and several studies have shown that their concentrations are correlated positively with nitrogen availability in Gracilariales (Lapointe, 1981; Vergara et al., 1995).

Lectin is a glycoprotein with more commercial interest, being the most extracted biomolecule from *Eucheuma serra* (Rhodophyta), which has been exploited for several industrial applications (Weis and Drickamer, 1996; Kawakubo et al., 1997; Hayes, 2013; Bleakley and Hayes, 2017). The bioactivities of this bio-compound comprise mitogenic, cytotoxic, antibacterial, antinociceptive, anti-inflammatory, antiviral (HIV-1), platelet aggregation inhibition, and anti-adhesion (Harnedy and Fitzgerald, 2011).

Proteins and peptides of seaweeds exhibit antioxidant and anticancer bioactivity, among others (Lakmal et al., 2014). However, the protein content can vary by season, temperature, and location (Galland-Irmouli et al., 1999; Joubert and Fleurence, 2008; Zhou et al., 2014). For example, *Gracilaria cervicornis* (Rhodophyta) and *Sargassum vulgare* (Phaeophyceae) protein content differs seasonally and it is negatively correlated with temperature and salinity (Marinho-Soriano et al., 2006). They can be used as a functional ingredient, and they have been successfully incorporated into several foods at laboratory scale. One example of it is bread including renin-inhibitory peptides obtained from *P. palmata* hydrolysates (Fitzgerald et al., 2014).

The high protein content of macroalgae can also be beneficial for use as animal feed, including aquaculture, cattle, and pets (Bleakley and Hayes, 2017). Several studies have been carried out including seaweeds, like *Laminaria* sp. (Phaeophyceae) and *Ulva* sp. (Chlorophyta), as protein sources into the diets of poultry, pigs, cattle, sheep, and rabbits, improving the health of the animals and livestock productivity (He et al., 2002; El-Deek and Brikaa, 2009). These investigations showed amazing benefits on the animals, such as decreased serum lipids, cholesterol, and uric acid concentrations (Abudabos et al., 2013; Lozano et al., 2016). The incorporation of seaweed also acts as a prebiotic to improve chicken gut health, productivity, and egg quality (Kulshreshtha et al., 2017).

11.2.3 PIGMENTS

Other compounds with potential for applications in nutraceutical and pharmaceutical products are pigments, which are different and divide the macroalgae into three taxonomic groups: Phaeophyceae (brown algae), Chlorophyta (green algae) and Rhodophyta (red algae) (Kadam et al., 2013).

Chlorophyta algae are green, since no other pigments mask the chlorophyll (Kumar et al., 2008). These seaweeds have chlorophylls (a and b) and carotenoids (β-carotene and xanthophylls) that are important in the protection against radiation and its harmful effects (Kadam et al., 2013). Carotenoids are isoprenoid compounds that are classified as carotenes (pure hydrocarbons) or xanthophylls (oxygenated carotenes) (Viera et al., 2018). They have an antioxidant role protecting chlorophyll from degradation (Franklin and Forster, 1997).

Decreased tolerance to the environmental or culture conditions would show up on macroalgae by lower pigment concentrations, reduction in photosynthetic performance and consequently growth rate (Beauchamp, 2012).

The brown color of Phaeophyceae is due to the dominance of the xanthophyll pigment, fucoxanthin, chlorophyll-a, chlorophyll-c, β-carotene and other xanthophylls (Kumar et al., 2008). In brown seaweeds, along with the chlorophylls, the major pigment is fucoxanthin, which is one of the most abundant carotenoids in nature (10% estimated total production of carotenoids) (Dembitsky and Maoka, 2007; Hosokawa et al., 2009; Pangestuti and Kim, 2011; Kadam et al., 2013b; Viera et al., 2018; Knowler et al., 2020). Different brown seaweed strains produce different compositions and profile of fucoxanthin (Pangestuti and Kim, 2011).

Studies reported the antitumoral, antioxidant, and antiobesity properties of fucoxanthin (Hosokawa et al., 1999; Yan et al., 1999; Pigmen et al., 2014) and showed that it promotes the radical scavenging activity (Sachindra et al., 2010). Protective effects in different human tissues have recently been reported for this carotenoid (Peng et al., 2011), including liver, skin, cerebrovascular, bone, and ocular well-being, in addition to the demonstrated antiangiogenic, anti-inflammatory, and anticancer activities (Martínez-Andrade et al., 2018). Some of the potential properties are the implication in the antiobesity metabolism and the antidiabetic activity (Maeda et al., 2007; Miyashita and Hosokawa, 2017).

Moreover, astaxanthin is considered as a 'super-antioxidant' with the strongest antioxidant activity compared with other carotenoids (Pashkow et al., 2008). Astaxanthin can decrease the risk of developing adverse diseases including cancer (Palozza et al., 2009), gastric ulcers caused by the bacteria *Helicobacter pylori* (Liu and Lee, 2003) and other gastrointestinal disorders and gastrointestinal health (Andersen et al., 2007; Kupcinskas et al., 2008), therapeutic agent against cardiovascular disease (Kupcinskas et al., 2008; Fassett and Coombes, 2011; Chew et al., 2013). There are evidence obtained on the inhibition of LDL oxidation by astaxanthin with the prevention of atherosclerosis (Iwamoto et al., 2000). Also reduced triacylglycerides and cholesterol in very-low-density lipoprotein (VLDL) (Karppi et al., 2007) and increased the basal artery blood flow (Miyawaki et al., 2008). In addition, astaxanthin protects against ethanol and drugs toxicity (Kim et al., 2005), enhances condition of ocular tissues (Iwasaki and Tawara, 2006; Nagaki et al., 2006) and skin (Tominaga et al., 2012; Rao et al., 2013) and improves fertility (Yuan et al., 2011; Andrisani et al., 2015). Other potential application of astaxanthin is related with the development of skincare and cosmetic products to prevent the harmful effects of exposure to UVB and UVC (Yoshihisa et al., 2014) and the treatment of pigmentation patches and marks (Arakane, 2002). Rhodophyta seaweeds present phycocyanin (PC) and phycoerythrin, in addition to chlorophyll (Kumar et al., 2008; Kadam et al., 2013b).

Pyropia orbicularis derived pigments extract is interesting due to strong anti-inflammatory, hepatoprotective, and antioxidant activities exerted by light-harvesting pigments bound to proteins that can capture solar radiation to form phycobiliproteins (PBPs) (Pina et al., 2014). Due to this, PBPs are employed as colorants in food and cosmetics (Tello-Ireland et al., 2011), with R-phycoerythrin (R-PE) and phycocyanin being the most relevant pigments in red seaweeds (Cian et al., 2014).

11.2.4 PHENOLIC COMPOUNDS (PCs)

Phenolic compounds (PCs) are also secondary metabolites of algae and they possess antioxidant, antimicrobial, and antiviral activities that are important for the protection of algal cells against biotic and abiotic stress conditions (Jiménez-Escrig et al., 2001; Karawita et al., 2005; Athukorala et al., 2006; Onofrejová et al., 2010; Francavilla et al., 2013).

Once again, the differences in content and composition of these compounds among seaweed have taxonomic, ecological, and environmental aspects (Jiménez-Escrig et al., 2012; Freile-Pelegrín and Robledo, 2013). Concentrations of macroalgae polyphenols vary according to ambient light, UV irradiation, temperature, salinity, and nutrient availability, which directly influence the chemical composition of algae through their impact on predominantly light- and temperature-controlled photosynthetic and enzymatic processes (Connan et al., 2006; Fairhead et al., 2006; Stengel et al., 2011; Pliego-Cortés et al., 2019). In fact, most polyphenols changes are linked or attributed to both irradiance and nutrient levels. Irradiance and nutrient availabilities are positively correlated to the content of polyphenolic compounds, while phenolic is negatively correlated to salinity (Cronin and

Hay, 1996; Gómez and Huovinen, 2010; Connan and Stengel, 2011; Freile-Pelegrín and Robledo, 2013).

In addition, the distribution of phenolics varies with the species. In general, green and red algae have low concentrations of phenols compared to brown algae species (Mabeau and Fleurence, 1993; Jiménez-Escrig et al., 2001; Zubia et al., 2009; Quirós et al., 2010; Jiménez-Escrig et al., 2012).

For instance, the excretion of macroalgal PCs (e.g., phlorotannins in brown algae) into the surrounding seawater is also thought to aid photoprotection (Swanson and Druehl, 2002; Pérez-Rodríguez et al., 2003).

Phenolic terpenoids are characterized in brown and red macroalgae. In addition, brown algae contain meroditerpenoids (plastoquinones, chromanols, chromenes), which are mainly limited to the Sargassaceae (brown algae) (Reddy and Urban, 2009). Rhodomelaceae (red algae) (*Laurencia* spp.) from red algae contain principally diterpenes and sesquiterpenes (Masuda et al., 2002). *Callophycus serratus* (bromophycolides) has reported to contain diterpene–benzoate macrolides (Lane et al., 2009).

Polyphenols (fucol, fucophlorethol, fucodiphloroethol G and ergosterol) and the phenolic compound phlorotannin are abundant in brown macroalgae, and have strong antioxidant effects (Holdt and Kraan, 2011). In red seaweeds, the PCs (flavonoids and phlorotannins) are potent and stable antioxidants (Namvar et al., 2012) because of their capacity to act as chelating agents with reactive oxygen species (ROS), thereby preventing oxidative stress and cell damage (Shannon and Abu-Ghannam, 2019). Polyphenols have antioxidative activity and other properties, such as radiation protective, antibiotic, and antidiabetic that make them valuable compound for cosmeceutical and pharmacological industries (Holdt and Kraan, 2011). A relation between a high intake of dietary polyphenols and lowered incidence of various chronic diseases was also reported by some epidemiological studies (Arts et al., 2001). *In vitro* and *in vivo* studies report that different classes of phenols present multifunctional antioxidant activities, antibacterial, and antifungal properties, as well as other bioactivities (Dutot et al., 2012; Hierholtzer et al., 2012; Jiménez-Escrig et al., 2012; Balboa et al., 2013).

Phlorotannins and polyphenolics are responsible for the mechanisms of antidiabetic action in seaweed (Kellogg et al., 2014; Murray et al., 2018), which inhibit hepatic gluconeogenesis, and decrease the activity of digestive enzymes, such as α-amylase, α-glucosidase, lipase, and aldose reductase (Sharifuddin et al., 2015). The collaboration of PCs to the prevention of cardiovascular diseases (CVDs) and cancers, suggesting their role in the prevention of neurodegenerative diseases and diabetes mellitus (Urquiaga and Leighton, 2000). According to Li et al. (2011) and Thomas and Kim

(2011) reviews, the marine brown algae *Ecklonia cava* subsp. *stolonifera* (formerly *Ecklonia stolonifera*), *Ecklonia cava* subsp. *kurome* (formerly *Ecklonia kurome*), *Eisenia bicyclis*, *Ishige okamurae*, *Sargassum thunbergii*, *Sargassum fusiforme* (as *Hizikia fusiformis*), *Undaria pinnatifida*, and *Saccharina japonica* (as *Laminaria japonica*) showed to have anti-HIV, anticancer, bactericidal, radio-protective, anti-allergic, and other health-beneficial biological activities related with their phlorotannin content.

Phlorotannins from Japanese Laminariaceae play a key role as bioactive ingredients in human health and nutrition (Shibata et al., 2007). Before, the same authors reported the inhibitory effect of phlorotannins on hyaluronidase (Shibata et al., 2002) and lipoxygenase (Shibata et al., 2003). In the last years, some companies have begun to produce standardized phenolic (phlorotannin) extracts from brown algae. The main goal of phenolic-containing algae supplements is in cardiovascular disease (CVD) management, due to these companies claiming arteriosclerosis prevention and an increase in protective high-density lipoprotein cholesterol (HDL-C) (Ledesma-Hernandez et al., 2013). Other purified PCs with antioxidant capacity have been reported, including bromophenols employed as "marine flavor" agents in farmed fish and prawns (Whitfield et al., 1999) and natural UV screen (Helioguard® 365, produced by Mibelle Biochemistry in Switzerland) from the red macroalga *Porphyra umbilicalis* (Cardozo et al., 2007). Bromophenols are PCs common to all major algal groups with antidiabetic, antioxidant, and anticancer effects (Liu et al., 2011). Bromophenols are mainly found in red algae, particularly in Ceramiales, Gelidiales, and Corallinales (Whitfield et al., 1999).

11.2.5 POLYSACCHARIDES

Total polysaccharide or sugar content of seaweed varies from 4 to 76% (dry mass) (Paniagua-Michel et al., 2014). Cellulose ranges from 2 to 10% of total polysaccharides in seaweed, being a structural, non-nutritive, and hard to digest component (Shannon and Abu-Ghannam, 2019). Digestible polysaccharides change between phyla. Alginate, fucoidan, and laminarin are the principal polysaccharides in brown algae (Rodrigues et al., 2015); carrageenan and agar are present in red algae; and ulvan can be found in green algae (Jiao et al., 2011b). The main commercialized polysaccharides from seaweeds are agar, carrageenan, and alginate, due to their large use in nutritional and cosmetic industries (Cardozo et al., 2007), despite polysaccharides having been tested with antioxidant, antiviral, antitumoral, and anti-coagulant properties (Mayer and Hamann, 2004; Smit, 2004). Due to their gelling, viscosifying, and emulsifying characteristics, these compounds are employed extensively, generating an increase in commercial and scientific interest (Cardozo et al., 2007).

The majority of algal polysaccharides are non-starchy fibers and this gives them functional and bioactive characteristics that are not found in terrestrial plant sources (Jiménez-Escrig and Goñi, 1999).

11.2.5.1 AGAR

Agar comes from worldly distributed red seaweed (Rhodophyceae), is the main matrix polysaccharide and is secreted by the Golgi apparatus in the cells (Imeson, 2009). Agar is present mainly in the cell wall of the order Gelidiales (*Gelidium* and *Pterocladiella*) and Gracilariales (*Gracilaria* and *Hydropuntia*). *Gracilaria tenuistipitata* is an economically important raw source for agar production because of its large amount and easier exploitation (Yarnpakdee et al., 2015).

Sulfated agarans (the principal part of the dietary fiber of *Gracilaria* species) and some minerals (e.g.: selenium) are both reported to have significant antioxidant activity (Torres et al., 2019). It was tested sulfated agarans from *Gracilaria corticata* that demonstrated antioxidant activity similar to the well-known antioxidants (ascorbic acid and butylated hydroxyanisole, BHA) (Seedevi et al., 2017).

The most important characteristic of agar is the ability to form reversible gels by cooling hot aqueous solutions. Agar is used in many practical applications as a food additive or in microbiology, biochemistry, or molecular biology besides in other industrial applications (Imeson, 2009).

Besides, agar oligosaccharides show biological activities. Antioxidant (Chen et al., 2005; Souza et al., 2012; Jin et al., 2017), antiviral (Mazumder et al., 2002), prebiotic (Bhattarai and Kashyap, 2016), antitumoral, immunomodulatory, anti-inflammatory (Liu et al., 2008; Jang et al., 2009; Hehemann et al., 2012; Higashimura et al., 2013; Coura et al., 2015; Kazłowski et al., 2015; Jin et al., 2017), glucosidase inhibitory (Chen et al., 2005), anticariogenic (Yun et al., 2017), hepatoprotective (Jin et al., 2017), and other properties of interest for skincare (Chen et al., 2005; Jang et al., 2009; Hehemann et al., 2012) are some of those activities.

Gracilaria spp. are widely employed for the production of agar and for gastrointestinal disorders' treatment (Armisen, 1995). Agar is demanded as gelling and stabilizing agent as well as a cryoprotectant for the pharmaceutical, cosmetics, and food industries (Freile-Pelegrín and Robledo, 1997;

Freile-Pelegrín et al., 2005; Marinho-Soriano and Bourret, 2005; Sousa et al., 2010; Villanueva et al., 2010). Besides commercial exploitation of agar in the hydrocolloid industry, it is used in medicinal and pharmaceutical areas, such as in a therapy against cancer cells due to its capacity to induce the apoptosis of these cells *in vitro* (Chen et al., 2004). Agar is a complex mixture of agarans, and species of *Gracilaria* genus (Gracilariales, Rhodophyta) are one of the main sources of agar, being responsible for 80% of the global production of this phycocolloids due to their abundance in the cell wall of *Gracilaria* species (FAO, 2018; Ferdouse et al., 2018;). *Gracilaria* is one of the main resources for agar production due to its fast growth and large agar content (Arbit et al., 2019).

Agar is a generally recognized as safe (GRAS) food additive in the United States and a food additive approved in Europe (E406) (Hong et al., 2017). The low-quality agar is used in food products (frozen foods, bakery icings, dessert gels or fruit juices, among others) and in industrial applications (paper sizing/coating, textile printing/dyeing, impressions, among others) (Cardozo et al., 2007). The medium quality agar is employed as the gel substrate in biological culture media and in the medical/pharmaceutical field as bulking agents, laxatives, suppositories, capsules, tablets, and anticoagulants. The highest purified and upper market types (agarose) are used for separation in molecular biology (electrophoresis, immunodiffusion, and gel chromatography) (Cardozo et al., 2007). The additive code of Agar from EU employed in nutritional industry as a natural additive is E405 (Meireles, 2013).

Over 90% of the world production of agar is used in nutritional applications (Imeson, 2009). Agar is employed as a component in tablets, capsules, and emulsions, whereas initially was used as a laxative agent in the preparation of medicines in western countries (Armisen, 1995). As alginate, the principal function of agar in the pharmaceutical industry is as an excipient (Leandro et al., 2020).

11.2.5.2 CARRAGEENAN

Carrageenan is extracted from red seaweeds harvested along the coasts of the North Atlantic, South America and the Far East. Most families belonging to the Gigartinales show superior concentrations of one particular class and thus are known as carrageenophytes (carrageenan-producers) (Pereira and Mesquita, 2003; Gurgel and Lopez-Bautista, 2007). Carrageenans are mainly obtained from the genus *Chondrus*, *Eucheuma*, *Iridae*, *Furcellaria*, *Gigartina*, and *Hypnea*. A growing demand for this compound led to the introduction of the cultivation of *Kappaphycus alvarezii* and *Eucheuma denticulatum*, with a principal content of kappa- and iota-carrageenan, respectively, available all year (Rhein-Knudsen et al., 2015). λ-carrageenan (obtained from some species from the genera *Gigartina* and *Chondrus*) (Meireles, 2013) has high potential for pharmaceutical and cosmetic industries. In the carrageenan food industry, cold-water species are unable to compete with the sub-tropical Asian carrageenophyte species. Nevertheless, some authors suggested that in medium- to long-term Asian carrageenophyte and agarophyte industries can go down due to climate change with consequences in algal flora. In addition, combining a cold-water carrageenophyte with novel market niches, such as cold-water *Chondrus crispus* λ-carrageenan, can increase the feasibility of integrated multi-trophic aquaculture (IMTA) (Meireles, 2013).

Carrageenan is a natural component, employed for many years in food applications and usually recognized as safe (GRAS) by the Food and Drug Administration; carrageenan and semi-refined carrageenan are food additives (E-407 and E407a, respectively) regulated by the European Food Safety Authority (EFSA) (Torres et al., 2019). Carrageenans are used in the pharmaceutical industry (Necas and Bartosikova, 2013; Ghanbarzadeh et al., 2018) due to their anti-inflammatory (Vanderlei et al., 2011), antioxidant (Sun et al., 2009; Gómez-Ordóñez et al., 2014; Souza et al., 2018), antiviral (Cáceres et al., 2000; Talarico et al., 2004, 2005; Chattopadhyay et al., 2008; Shi et al., 2017; Inic-Kanada et al., 2018), anticoagulant (Yuan et al., 2006; Gómez-Ordóñez et al., 2014), immunomodulatory (Liu et al., 2013), anti-tumoral (Souza et al., 2018), anti-angiogenic (Poupard et al., 2017) and neuroprotective (Souza et al., 2018) properties.

Carrageenan produced by seaweed is not assimilated by the human body, being just a fiber without nutritional value, but it gives a feature that can be employed to thicken, gel, and stabilize food products and food systems (Imeson, 2009). The food industry is the principal market for the algal hydrocolloids and around 70–80% of all carrageenan products are employed in it (Ioannou and Roussis, 2009; Bixler and Porse, 2011; Jiao et al., 2011a).

Carrageenans are more employed than agar as emulsifiers/stabilizers in many foods, especially milk-based products. kappa- and iota-carrageenans are particularly important for use in milk and meat products and pet foods due to their thickening and suspension properties (Cardozo et al., 2007). In processed meats, carrageenan is used as a water binding agent for preventing loss of moisture during cooking, which increases cooked yields and prevents an undesirable dry texture or bite. Carrageenan gives body, provides thickness, and stabilize emulsions in sauces, salad dressings and dips. The use of carrageenan has been implanted also in fluid dairy and dairy dessert products as the stabilization of cocoa. Carrageenan is responsible for whipped creams and toppings maintaining their stable form (Imeson, 2009).

11.2.5.3 ALGINATES

Alginates are present in the cell walls of brown macroalgae and have different chemical structures and characteristics depending on genera. Some of the species of macroalgae that contain alginate are *Ascophyllum*, *Durvillaea*, *Ecklonia*, *Laminaria*, *Lessonia*, *Macrocystis*, and *Sargassum* (Wiltshire et al., 2015). Fucales are also employed for nutrition and alginates, including *Scytothalia dorycarpa* (family Seirococcaceae), *Cystophora subfarcinata*, and *Sargassum linearifolium* (both Sargassaceae) (White and Wilson, 2015). Even though *Sargassum* spp. has one of the least return of alginates among brown macroalgae, it is still one of the sources of alginate (Wiltshire et al., 2015).

Alginates are recognized in the food industry due to their gel-forming characteristics given by the large portion of guluronic acid (Brownlee et al., 2009). Given their capability to stimulate the curative process, alginates are used in the pharmaceutical industry (Doyle et al., 1996; Ianova et al., 2012; Catanzano et al., 2015). Alginate may reduce blood levels of cholesterol and glucose, because it helps developing intestinal viscosity, due to being a soluble dietary fiber (Sandberg et al., 1994; Wolf et al., 2002).

The use of beverages that incorporate high guluronic acid alginate with insoluble calcium salts exposed the satiety and food intake of alginate, by assembling intragastric lumps, increasing the sensation of fullness and diminish hunger (Hoad et al., 2004). In order to benefit the effects of alginates on human health, whether the taste, stability, or oxidation, it is also used in the encapsulation of functional food oils (e.g., fish oils, wheat germ oil, primrose oils) (Chan et al., 2000). A reduction of intestinal absorption rates as well as colonic luminal toxicity, the alteration of colonic microflora, and enhancements in the colonic mucosal barrier are comprised amongst the effects (Brownlee et al., 2005).

Alginates are employed in food, cosmetic, textile, construction, and pharmaceutical/biomedical sectors for being emulsifiers, thickeners, binding, and gel-forming agents due to their capability of condensing aqueous solutions and assembling gels (Wiltshire et al., 2015). When hydrated in aqueous solution, an alginate develops viscosity which can be used in the food industry to give texturizing features acting as a thickener, a stabilizer, and a gelling agent (Imeson, 2009).

In order to maintain stability and preventing the filling bake from escaping out of the pastry while baking, alginate is commonly used in fruit fillings. It can also be used to thicken the batter. The thickening of cold-prepared bakery creams is due to the gelling abilities of alginate; in cheese sauces, alginate gives viscosity, and, in gravy, the thickening abilities can act in two distinct ways: either as temporary suspending agent or as a postponed thickening agent (Imeson, 2009).

For the purpose of keeping separation, alginate is used as a stabilizer, deepening the viscosity of the aqueous phase, and charging films at the interfaces. The first use of alginate in food was in ice cream, reducing the size of the ice crystals and giving it body and a creamy texture. In low-fat spreads, alginate is used to reach the desired texture and to sustain the emulsion. In dairy products, such as thickened and canned cream, chocolate mousse, yogurt, bakery creams, milkshake, ice cream and cheese, alginate comes to action as a gelling agent, a stabilizer, and a thickener (Imeson, 2009).

To keep moisture from the filling to spread to the rest of the cake, an alginate film can be applied. Also, in cake icings, it is applied as an anti-cracking agent and to avoid attachment to the wrapping (Imeson, 2009). In the interest of encapsulation of probiotics, flavors, and functional food oils, alginates are progressively more in use (Imeson, 2009).

Bakery goods, such as bread, muffins, cupcakes, and soft cookies, have seen their shelf life extended due to the alginate property of being a hydrophilic polymer, confining water and preserving the product moist and soft for a longer time (Butt et al., 2001). In gluten-free bakery products, such as bread, crispbread, and chocolate cakes, alginate is useful by binding and contain water, extending this way the shelf life of the products (Imeson, 2009).

11.2.5.4 ULVAN

From green seaweeds group, particularly from the Ulvaceae, could be obtained the water-soluble molecules, ulvans (Lahaye and Robic, 2007). Ulvans constitute between 8% and 29% of the dry weight of *Ulva*, depending on species and environmental growth conditions (Lahaye and Robic, 2007; Robic et al., 2009). It has been reported that this anionic polysaccharide gives *Ulva* the ability to accumulate heavy metals, removing them from contaminated waters (Webster and Gadd, 1996; Bocanegra et al., 2009; Schijf and Ebling, 2010).

Ulvans have biomedical potential, such as applications in tissue engineering, biofilm prevention and drug delivery, due to their recognition as hepatocyte membrane receptors (Lahaye and Robic, 2007; Wijesekara et al., 2011; Alves et al., 2013; Venkatesan et al., 2015; Cunha and Grenha, 2016). In addition to immunostimulatory effects, ulvans also have antiviral, antioxidant, anticoagulant, antihyperlipidemic, and anticancer activities (Lahaye and Robic, 2007; Alves et al., 2013; Kraan, 2013).

Ulvans are a high-value product with unique gelling, bioactive, and functional properties (Lahaye et al., 1996; Lahaye and Robic, 2007). In addition, *Ulva* spp. are especially suitable to mitigate impacts from anthropogenic wastewaters due to their high productivity and resilience to diverse growing conditions (Castine et al., 2013; Kraan, 2013; Lawton et al., 2013; Glasson et al., 2017). Ulvan are used in agriculture because of their capacity to improve plant immune responses (Jaulneau et al., 2010; Alves et al., 2013). These compounds are very versatile and applied as pharmaceuticals, nutraceuticals, functional foods and feed, agriculture, and bioremediation processes.

11.2.5.5 PORPHYRAN

Porphyrans are a family of agaroids polysaccharides produced by red seaweeds of the genera *Pyropia* and *Bangia* (Rhodophyta) (Brasch et al., 1981). *Porphyra/Pyropia* are a relevant food source in diverse parts of the world such as in East Asia, where they are renowned as a vegetable having power to retard aging and improve the vitality (Bhatia et al., 2008).

Porphyran present important biological activities including anticancer (Kwon and Nam, 2006; Yu et al., 2015; He et al., 2019), antioxidant (Zhao et al., 2006; Zhang et al., 2009), anti-hyperlipidemic (Ren et al., 1994; Tsuge et al., 2004; Inoue et al., 2009), anti-inflammatory (Lee et al., 2015a; Yanagido et al., 2018; Wang et al., 2020) and/or immunomodulation (Bhatia et al., 2013; Liu et al., 2017) and avoidance of illness like cardiovascular (Jimenez-Escrig and Sanchez-Muniz, 2000; Tsuge et al., 2004; Inoue et al., 2009; Rajapakse and Kim, 2011; Wijesekara et al., 2011), nervous (Liu et al., 2019), bone (Ueno et al., 2018), and diabetic disorders (Cao et al., 2016; Bito et al., 2017). In addition, porphyran has the capacity of stabilizing the tear film on the eyeball surface over a prolonged time and is employed as an artificial tear liquid (Bhatia et al., 2008). Porphyran is responsible for the releasing of histamine from the mast cells due to its great inhibitory activity against hyaluronidase (Ishihara et al., 2005). Furthermore, porphyran serve as blood anticoagulants (Zhang et al., 2005).

The modified Porphyran (with modified bioactivity and physical feature) can be obtained through the conversion of a salt of a sulfate group in Porphyran into a sulfate salt of a given salt by ion exchange. The modified Porphyran can be added and used in cosmetics, food, and drinks, having inhibitory activity against hyaluronidase activity (Toshiyoshi Araki, 1996; Sugihara et al., 2004). Oligo-porphyran (acid hydrolysis product of porphyran) can prevent and treat several pathologies, such as Parkinson's disease and acute renal failure (Liu et al., 2019).

Neopyropia yezoensis (formerly known as *Porphyra yezoensis*), *Neopyropia tenera* (formerly known as *Porphyra tenera*), *Neoporphyra haitanensis* and *Phycocalidia suborbiculata* (formerly known as *Porphyra suborbiculata*) are traditionally used as food in Japan. These algae are transformed into a sheet type of dried food, "Nori," that contains main dietary fiber and constitutes closely around 40% of mass (Ishihara et al., 2005). Nori is famous in East and Southeast Asia as well as globally, especially as a wrap for sushi (Liu et al., 2019). Porphyran is one of the main constituents of Nori (which contains 17.5% of 3,6-anhydrogalactose and free and bound sulfur 2.0% and 5.6%, respectively) (Anderson and Rees, 1965; Zhang et al., 2004, 2005).

11.2.5.6 LAMINARIN

Laminaran, called laminarin or leucosin, is a nutritional reserve of brown algae (Phaeophyceae) placed in vacuoles of the cells. Laminarin is found in the fronds of *Laminaria* and *Saccharina* species (Kadam et al., 2015). Laminarin disappears during fast growth in spring, although in autumn and winter it can reach over 35% of the dried weight of the fronds (Rinaudo, 2007). The content of laminarin in brown algae can be over levels of 35% on dry weight, which varies with the species, harvesting season, habitat, and method of extraction (Kadam et al., 2015). Some of the main sources of laminarin are: *Saccharina latissima*, *Laminaria hyperborea*, *Laminaria digitata*, *Fucus vesiculosus*, *Undaria pinnatifida* and *Ascophyllum nodosum* (Haug and Jensen, 1954; Devillé et al., 2004; MacArtain et al., 2007; Rioux et al., 2007; Je et al., 2009; Holdt and Kraan, 2011). Laminarin has been exhibited to stimulate immunity and to get antitumor effects and antibacterial activity (Kim et al., 2006; Ermakova et al., 2013; Menshova et al., 2014).

Laminarin has shown bio-functional activities, such as antitumor (Miao et al., 1999; Kuda et al., 2005; Ji et al., 2012), anti-apoptotic (Kim et al., 2006), immunostimulatory activity (Neyrinck et al., 2007; Lee et al., 2012), anticoagulant (Miao et al., 1995, 1999; Shanmugam and Mody, 2000), anti-inflammatory (Neyrinck et al., 2007; Jaswir and Monsur, 2011), and antioxidant (Tsiapali et al., 2001; Choi et al., 2012; Moroney et al., 2012;

Balboa et al., 2013) activities. Laminarin has immunostimulatory, antitumor, and wound-healing activities (Lee et al., 2012) and it was confirmed to have functional dietary fiber activity (Kadam et al., 2015). Laminarin is a possible modulator of intestinal metabolism (Devillé et al., 2004, 2007).

Laminarin decreases the levels of undesirable lipids in the liver and supplies protection against severe irradiation, decreases cholesterol levels in serum and reduces systolic blood pressure (Holdt and Kraan, 2011). One of laminarin characteristics is its low viscosity and large water solubility up to a temperature from 4 to 40°C, which facilitates processing. This is useful to inject through minimally invasive procedures or use in microfabrication of hydrogels. Photo-crosslinked laminarin hydrogels can be utilized for cell encapsulation or drug encapsulation and delivery (Custódio et al., 2016). Laminarin showed to improve natural protection responses in plants and are involved in the induction of genes encoding many pathogenesis-related (PR) proteins with antimicrobial properties (Leandro et al., 2020).

Laminarin sulfate can be used in clinical applications in different situations, such as wound healing, angiogenesis, and atherosclerosis (Kadam et al., 2015) as well as dietary fiber (Devillé et al., 2004). Also, there is a commercial application of laminarin substratum for prebiotic bacteria (Devillé et al., 2004). Nowadays, preparations containing 1→3:1→6-β-D-glucans, laminarin, and fucoidan are commercialized by the health industry due to their beneficial properties on the immune system (Holdt and Kraan, 2011).

11.2.5.7 FUCOIDAN

Brown seaweeds contain fucoidan polysaccharides in their cell walls. Fucoidans are found in several species of brown algae, such as *Fucus vesiculosus*, *Sargassum aquifolium* (formerly known as *Sargassum binderi*) and *Saccharina japonica*. Still, the chemical conformation of the majority of fucoidans is complex, and the structures of fucoidans change from species to species in distinct brown algae (Li et al., 2008).

These fucose-containing sulfated polysaccharides (FCSPs) show some bioactivities, such as antitumoral and immunomodulatory (Khilchenko et al., 2011; Anastyuk et al., 2012), antiviral (Prokofjeva et al., 2013), antithrombotic, and anticoagulant (De Zoysa et al., 2008). In addition, fucoidan has been declared to exhibit large antioxidant activity (Ananthi et al., 2010; Camara et al., 2011; Costa et al., 2011; Rodriguez-Jasso et al., 2013; Zhao et al., 2018; Wang et al., 2019). Antioxidant properties of fucoidans may be linked to their chemical structure and highly connected to the molecular weight (Xue et al., 2001; Hou et al., 2012), location, and degree of sulfation of the sulfated groups (Marudhupandi et al., 2014; Wang et al., 2019), and contents of glucuronic acid and fucose (Zhao et al., 2018). The crude fucoidan has many impurities like polyuronic acids, proteins, and PCs (Díaz-Rubio et al., 2009; Ananthi et al., 2010).

Fucoidan can be used as an anticoagulant (Chandía and Matsuhiro, 2008), antibacterial (Shannon and Abu-Ghannam, 2019) and antiviral agent (Fan et al., 1995; Mandal et al., 2007) as well as exhibit antioxidant activity (Vo and Kim, 2013; Zhao et al., 2018). *Sargassum* spp. have possible uses in the medical field as fucoidans are acknowledged for their capability to prevent tumors, alongside with anti-inflammatory and antithrombotic properties (Smit, 2004).

S. japonica is rich in fucoidan, and it is used for skincare products (anticellulite formulations), due to moisturizing, anti-aging, and anticellulite properties (Chizhov et al., 1999; Wijesinghe and Jeon, 2012; Wang et al., 2013).

U. pinnatifida is rich in fucoidan and it is employed in skincare products (aromatherapy oil; body scrub), anti-aging (antiwrinkle) and moisturizing, due to its nourishing properties (Cabarry, 2018; Osea Malibu, 2018). Interesting biological activities such as antithrombotic, anticoagulant, anticancer, antiproliferative, and anti-inflammatory, explain the study of fucoidans among polysaccharides (Chizhov et al., 1999; Kim et al., 2010; Nakazato et al., 2010).

Immuno-modulatory effects have been reported from crude fucoidan extracted from the edible brown algae *S. japonica* (Hwang et al., 2019). Different kinds of cancer like lung cancer (Han et al., 2008; Hsu et al., 2017), gastric cancer (Han et al., 2008), breast cancer (Yamasaki-Miyamoto et al., 2009) and liver cancer HepG2 cell (De Yan et al., 2019) have been described as susceptible fucoidan anticancer activity.

Fucoidans are known by the competence to inhibit tumors by decreasing serum cholesterol through their capability to prevent absorption of acids and to present anti-inflammatory and antithrombotic properties (Smit, 2004). Pharmacological activities, such as antioxidant, antiviral, anticoagulant, and immunomodulating features have been exhibited in fucoidan obtained from distinct algae species (Rupérez et al., 2002; Berteau and Mulloy, 2003; Silva et al., 2005; Li et al., 2008). Fucoidans are preeminent in the components of the genus *Padina*, that demonstrated some pharmacological activities (Ale et al., 2011; Jiao et al., 2011). Studies in fucoidan as an anticancer drug have shown that fucoidan preparations collected from different sources have induced distinct different anticancer activities *in vivo*, notwithstanding the considerable potential fucoidans have exhibited on clinical trials as a supplementary therapy. Fucoidan different structural properties seem to be linked to the distinct anticancer activities. In order to verify fucoidans' aptitude as a therapeutic agent, it is essential to establish the structural effective aspect of fucoidan for the documented *in vivo* antitumor activities (Kwak, 2014). Fucoidans from distinctive species and geographical areas present distinct structures with possible diversity in the pharmacological actions (Silva et al., 2005; Eve et al., 2009). Only recently fucoidan nutritional supplements and pharmaceuticals began to be commercialized, fucoidan studied and the compounds isolated, despite their use in traditional Chinese and Japanese folk medicines (Pádua et al., 2015). In the last years, due to the fucoidans' biological properties, they have shown a possible opportunity in cosmeceutical and nutraceutical drugs, as well as in food industries (Wijesinghe and Jeon, 2012). In some industrial sectors, such as animal feeding supplements, only lately fucoidan has been used (Fitton et al., 2015).

Figure 11.1 shows the main groups of compounds mentioned above derived from macroalgae, as well as their bioactivities with possible potential for applications in the pharmaceutical and nutraceutical industry.

11.3 AQUACULTURE: SUSTAINABLE PRODUCTION

As we demonstrated above, seaweeds are an interesting natural source of bioactive and essential compounds that can be further applied in foods (additives and nutraceutical, cosmetic, and pharmaceutical fields). Additionally, seaweeds have a structural role in the ecosystem. Being primary producers, they are the base of food webs, providing food, oxygen, and refuge to other organisms and communities (Almanza and Buschmann, 2013; Vásquez et al., 2014) and therefore they are responsible for biodiversity support (Skjermo et al., 2014). Moreover, seaweeds contribute to the coastal defense, change the environment by modifying light, sedimentation rates and reduce the hydrodynamic energy from waves and protect tidal areas from erosion (Reisewitz et al., 2006; Leclerc et al., 2013; Smale et al., 2013; Ondiviela et al., 2014; Bertocci et al., 2015).

Wild seaweed harvesting needs to be carefully carried out from wild stocks (Figure 11.2(a) and (b)), and avoid the overexploitation and environmental damaging. Thus, it is crucial that seaweeds need to be sustainably used and managed, due to the given pressure of the increasing harvesting for several applications.

Thus, to mitigate environmental pressure caused by the growing demand on seaweeds, there is a need to develop efficient aquaculture systems

(Jung et al., 2013), which could be developed offshore, onshore, and be applied in an aquaculture integrated system. The cultivation methodology should be chosen according to the seaweed species, place of the farm and cultivation facilities.

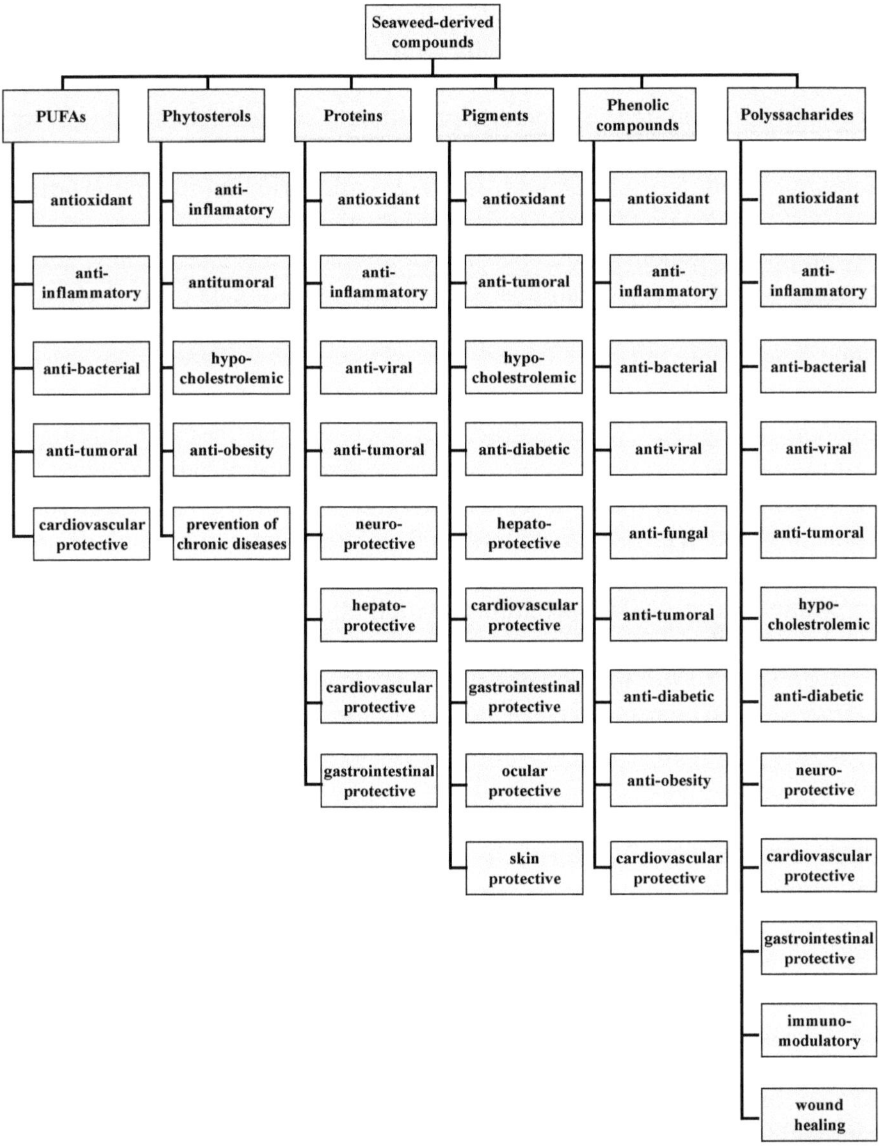

FIGURE 11.1 Scheme of the seaweed-derived compounds and their bioactive proprieties with potential for application in pharmaceutical and nutraceutical products.

Global seaweed aquaculture production is over 30 million tons and, in 2016, had an annual value of $11.7 billion. China, Indonesia, Philippines, Korea, and Japan play a major part in seaweed cultivation, accounting for 97% of global seaweed production (Ferdouse et al., 2018). Seaweed aquaculture is an old research area, in which there have been multiple new developments in the recent years, due to a continuing necessity for adaptations and the use of various technologies to transform the seaweed production into a stable and with low risk of failure activity. Furthermore, seaweed aquaculture is extremely important to suppress the need to feed the growing population, especially because it is a resource that does not compete for arable land, fertilizers, and freshwater with agricultural crops (Hafting et al., 2012; Charrier et al., 2017; Fernand et al., 2017).

Moreover, genomics area has been explored and is gaining new attention to improve and develop strains with higher growth capability, better taste, darker color, and higher resistance to diseases (Patwary and van der Meer, 1992; Chen et al., 2015).

There are different ways of seaweed farming. It could be performed offshore, using implemented structures, such as long-lines and floating rafts, in the open sea. Usually, these culture systems are placed in coastal sites which have strong water movement and are abundant in inorganic nutrient concentrations (Harrison and Hurd, 2001). Therefore, it is suitable for robust species, especially for kelps, and it has become an innovative research field to promote large scale aquaculture (Troell et al., 2009; Jansen et al., 2016; Buck and Langan, 2017).

Opposing to the offshore, seaweed could also be produced onshore (or land-based), which means that they are cultivated in closed or semi-open systems (e.g., tanks, raceways, ponds, or lagoons) (Figure 11.2(c–e)). Here, seawater is pumped on shore to the cultivation systems and maintained under agitation to keep seaweeds suspended and exposed to light (Hafting et al., 2012; Currie, 2018; Reid et al., 2020). Seaweed growth is influenced by environmental conditions, such as temperature, solar radiation, salinity, pH, and nutrient availability. So, the land-based cultivation systems have the advantage to have a better monitoring of the cultivation conditions and real-time adjustments (Hafting et al., 2012).

Within the land-based, new methodologies have been developed, taking advantage of infrastructures and natural systems for the cultivation of macroalgae species (Figure 11.2(f) and (g)). One of these most recent innovations is the use of saltpans and saline unused sites for the production of algae. In land saline aquaculture uses the ephemeral or permanent saline lakes and saline groundwater extracted from aquifers, allowing farmers to take advantage of these natural and local resources, with little expense for them (Allan et al., 2009; Bui, 2018). Seaweed production could also be conjugated in IMTA system which is breeding species from different trophic levels in proximity (Silva et al., 2015; Granada et al., 2018). In these cultures, organic and inorganic wastes of a species (for instance fish) are recycled to serve as nutritional input for others (e.g., seaweeds). Seaweeds are a solution for environmental issues of animal aquaculture, solving the eutrophication of the water because of feed supplementation and excretion (Knowler et al., 2020).

Research on aquaculture is increasing and leading to the application of the industry 4.0. Engineering and computer science allied with multisensory schemes that are being incorporated in aquaculture systems associated to online servers with logarithmic and artificial intelligence software to manage and control the system in every aspect and change of different factors in the aquaculture system, providing a better aquaculture productivity and efficiency, reducing the costs (Martin, 2019).

FIGURE 11.2 *Ulva* sp. early cultivation method from the seaweed aquaculture prototype of Lusalgae start-up in Figueira da Foz (Portugal), associated to a scientific research project to the development of environmental and efficient seaweed cultivation systems for food purposes; (a) and (b) collection of wild *Ulva* sp. in Praia da Tamargueira, Buarcos, Figueira da Foz, Portugal; (c) initial stage of *Ulva* sp. cultivation, at the laboratory; (d) and (e) second and last stage of cultivation in the acclimatization room at the laboratory level; (f) *Ulva* sp. biomass transfer for the outdoor cultivation tanks; (g) final biomass collection from the outdoor tanks for physicochemical analysis.

11.4 CONCLUSION

This chapter highlights the wide range of molecules produced by seaweeds and their explored bioactivities with beneficial effects for human health. The roadmap of extraction and characterization of seaweed-derived compounds is increasing our understanding of the potential applications of these molecules, from the selection of the best seaweed's species to the compound's isolation. Indeed, seaweeds are a potential natural raw ingredient source to be applied in many sectors, as the nutraceutical and pharmaceutical industries. Further studies and clinical trials for each compound or extract are needed to a product to be approved and safely characterized and then to be commercially available.

The interest on these resources is rising. However, to ensure the sustainability of seaweed usage and to prevent the wild beds to collapse, there is a need to resort to aquaculture production of these organisms. It is also necessary to fully understand the variable conditions that affect seaweed growth and optimize the producing methodologies. For this topic, the industry 4.0 can be an advanced key to develop the current aquaculture and techniques to a new level that can be the strategy to new findings in the pharmaceutical area.

ACKNOWLEDGMENTS

This work is financed by national funds through FCT – Foundation for Science and Technology, IP, within the scope of the projects UIDB/04292/2020 – MARE – Marine and Environmental Sciences Center and UIDP/50017/2020+UIDB/50017/2020 (by FCT/MTCES) granted to CESAM – Center for Environmental and Marine Studies. Sara García-Poza thanks to the project MENU – Marine Macroalgae: Alternative recipes for a daily nutritional diet (FA_05_2017_011) which co-financed this research, funded by the Blue Fund under Public Notice No. 5 – Blue Biotechnology. Adriana Leandro thanks FCT for the financial support provided through the doctoral grant SFRH/BD/143649/2019 funded by National Funds and Community Funds through FSE. João Cotas thanks MARE, UC, and Universidade Nova de Lisboa, through its Chemistry Department of the Faculty of Sciences and Technology within the scope of the research project "Algae in Gastronomy – Development of innovative techniques for conservation and use – Alga4Food" (MAR-01.03.01-FEAMP-0016). Diana Pacheco thanks to PTDC/BIA-CBI/31144/2017 – POCI-01 project – 0145-FEDER-031144 – MARINE INVADERS, co-financed by the ERDF through POCI

(Operational Program Competitiveness and Internationalization) and by the Foundation for Science and Technology (FCT, IP). Ana M. M. Gonçalves acknowledges the University of Coimbra for the contract IT057-18-7253.

KEYWORDS

- **anti-inflammatory**
- **antiobesity**
- **antioxidant**
- **antitumoral**
- **bioactivity**
- **fatty acids**
- **harvesting**
- **macroalgae**
- **nutraceutical**
- **pharmaceutical**
- **seaweed**
- **sustainability**

REFERENCES

Abudabos, A. M., Okab, A. B., Aljumaah, S. R., Samara, E. M., Abdoun, K. A., & Al-Haidary, A. A., (2013). Nutritional value of green seaweed (*Ulva lactuca*) for broiler chickens. *Italian Journal of Animal Science*, *12*(2), 177–181.

Ale, M. T., Mikkelsen, J. D., & Meyer, A. S., (2011). Important determinants for fucoidan bioactivity: A critical review of structure-function relations and extraction methods for fucose-containing sulfated polysaccharides from brown seaweeds. *Marine Drugs*, *9*(10), 2106–2130.

Allan, G. L., Fielder, D. S., Fitzsimmons, K. M., Applebaum, S. L., & Raizada, S., (2009). Inland saline aquaculture. In: Burnell, G., &. Allan, G., (eds.), *New Technologies in Aquaculture: Improving Production Efficiency, Quality and Environmental Management* (p. 29). Sawston: Woodhead Publishing.

Almanza, V., & Buschmann, A. H., (2013). The ecological importance of *Macrocystis pyrifera* (Phaeophyta) forests towards a sustainable management and exploitation of Chilean coastal benthic co-management areas. *International Journal of Environment and Sustainable Development*, *12*(4), 341–360.

Alves, A., Sousa, R. A., & Reis, R. L., (2013). A practical perspective on ulvan extracted from green algae. *Journal of Applied Phycology*, *25*(2), 407–424.

Anadón, A., Martínez-Larrañaga, M. R., Ares, I., & Martínez, M. A., (2018). Regulatory aspects for the drugs and chemicals used in food-producing animals in the European Union. In: *Veterinary Toxicology: Basic and Clinical Principles* (3rd edn., pp. 103–131). Cambridge: Elsevier.

Ananthi, S., Raghavendran, H. R. B., Sunil, A. G., Gayathri, V., Ramakrishnan, G., & Vasanthi, H. R., (2010). *In vitro* antioxidant and *in vivo* anti-inflammatory potential of crude polysaccharide from *Turbinaria ornata* (Marine Brown Alga). *Food and Chemical Toxicology*, *48*(1), 187–192.

Anastyuk, S. D., Shevchenko, N. M., Ermakova, S. P., et al., (2012). Anticancer activity *in vitro* of a fucoidan from the brown alga *Fucus evanescens* and its low-molecular fragments, structurally characterized by tandem mass-spectrometry. *Carbohydrate Polymers*, *87*(1), 186–194.

Andersen, L. P., Holck, S., Limas, K. L., et al., (2007). Gastric inflammatory markers and interleukins in patients with functional dyspepsia treated with astaxanthin. *FEMS Immunology and Medical Microbiology*, *50*(2), 244–248.

Anderson, N. S., & Rees, A., (1965). Porphyran: A polysaccharide with a masked repeating structure. *Journal of Chemical Society*, *10*, 5880–5887.

Andrisani, A., Donà, G., Tibaldi, E., et al., (2015). Astaxanthin improves human sperm capacitation by inducing Lyn displacement and activation. *Marine Drugs*, *13*(9), 5533–5551.

Arakane, K., (2002). Superior skin protection via astaxanthin. *Carotenoid Science*, *5*, 21–24.

Arbit, N. I. S., Omar, S. B. A., Soekendarsi, E., et al., (2019). Morphological and genetic analysis of *Gracilaria* sp. cultured in ponds and coastal waters. *IOP Conference Series: Earth and Environmental Science*, *370*, 012018.

Armisen, R., (1995). World-wide use and importance of *Gracilaria*. *Journal of Applied Phycology*, *7*(3), 231–243.

Arts, I. C. W., Hollman, P. C. H., Bueno De, M. H. B., Feskens, E. J. M., & Kromhout, D., (2001). Dietary catechins and epithelial cancer incidence: The zutphen elderly study. *The International Journal of Cancer*, *92*(2), 298–302.

Ashkenazi, D. Y., Israel, A., & Abelson, A., (2019). A novel two-stage seaweed integrated multi-trophic aquaculture. *Reviews in Aquaculture*, *11*(1), 246–262.

Athukorala, Y., Kim, K. N., & Jeon, Y. J., (2006). Antiproliferative and antioxidant properties of an enzymatic hydrolysate from brown alga, *Ecklonia cava*. *Food and Chemical Toxicology*, *44*(7), 1065–1074.

Babb, A. C., (2015). Industrial development within the blue economy: 'What These Ithacas Mean'. *Development*, *58*(4), 587–593.

Balboa, E. M., Conde, E., Moure, A., Falqué, E., & Domínguez, H., (2013). *In vitro* antioxidant properties of crude extracts and compounds from brown algae. *Food Chemistry*, *138*(2, 3), 1764–1785.

Barry, A. N., Starkenburg, S. R., & Sayre, R. T., (2015). Strategies for optimizing algal biology for enhanced biomass production. *Frontiers in Energy Research*, *3*, 1–5.

Beauchamp, E., (2012). Effects of UV radiation and salinity on the intertidal macroalgae *Palmaria palmata* and *Ulva lactuca*: Effects on photosynthetic performance, growth and pigments. *The Plymouth Student Scientist*, *5*(1), 3–22.

Berteau, O., & Mulloy, B., (2003). Sulfated fucans, fresh perspectives: Structures, functions, and biological properties of sulfated fucans and an overview of enzymes active toward this class of polysaccharide. *Glycobiology*, *13*(6), 29–40.
Bertocci, I., Araújo, R., Oliveira, P., & Sousa-Pinto, I., (2015). Potential effects of kelp species on local fisheries. *Journal of Applied Ecology*, *52*(5), 1216–1226.
Bhatia, S., Rathee, P., Sharma, K., Chaugule, B. B., Kar, N., & Bera, T., (2013). Immunomodulation effect of sulphated polysaccharide (porphyran) from *Porphyra vietnamensis*. *International Journal of Biological Macromolecules*, *57*, 50–56.
Bhatia, S., Sharma, A., Sharma, K., et al., (2008). Novel algal polysaccharides from marine source: Porphyran. *Pharmacognosy Reviews*, *2*(4), 271–276.
Bhattarai, Y., & Kashyap, P. C., (2016). Agaro-oligosaccharides: A new frontier in the fight against colon cancer?. *American Journal of Physiology-Gastrointestinal and Liver Physiology*, *310*(6), 335–336.
Bito, T., Teng, F., & Watanabe, F., (2017). Bioactive compounds of edible purple laver *Porphyra* sp. (Nori). *Journal of Agricultural and Food Chemistry*, *65*(49), 10685–10692.
Bixler, H. J., & Porse, H., (2011). A decade of change in the seaweed hydrocolloids industry. *Journal of Applied Phycology*, *23*(3), 321–335.
Bleakley, S., & Hayes, M., (2017). Algal proteins: Extraction, application, and challenges concerning production. *Foods*, *6*(5), 33.
Bocanegra, A., Bastida, S., Benedí, J., Ródenas, S., & Sánchez-Muniz, F. J., (2009). Characteristics and nutritional and cardiovascular-health properties of seaweeds. *Journal of Medicinal Food*, *12*(2), 236–258.
Brasch, D. J., Chang, H. M., Chuah, C. T., & Melton, L. D., (1981). The galactan sulfate from the edible, red alga *Porphyra columbina*. *Carbohydrate Research*, *97*(1), 113–125.
Brown, E. M., Allsopp, P. J., Magee, P. J., et al., (2014). Seaweed and human health. *Nutrition Reviews*, *72*(3), 205–216.
Brownlee, I. A., Allen, A., Pearson, J. P., et al., (2005). Alginate as a source of dietary fiber. *Critical Reviews in Food Science and Nutrition*, *45*(6), 497–510.
Brownlee, I. A., Seal, C. J., Wilcox, M., Dettmar, P. W., & Pearson, J. P., (2009). Application of alginates in food. In: Bernd, H. A. R., (ed.), *Alginates: Biology and Applications* (Vol. 13, pp. 221–228). Berlin, Heidelberg: Springer.
Buck, B. H., & Langan, R., (2017). *Aquaculture Perspective of Multi-Use Sites in the Open Ocean: The Untapped Potential for Marine Resources in the Anthropocene* (p. 404). Switzerland: Springer International Publishing.
Bui, H. T., (2018). *Technical Feasibility of Cultivating Local Seaweed Species in Inland Saline Water of Western Australia*. PhD Thesis, Curtin University, Bentley, Australia.
Butt, M. S., Anjum, F. M., Samad, A., Kausar, T., & Mukhtar, M. T., (2001). Effect of different gums on the quality and shelf life of bread. *International Journal of Agriculture and Biology*, *3*(4), 482, 483.
Cabarry, C., (2018). *SpecialChem – Connect, Innovate, Accelerate.* The Universal Selection Source: Cosmetics Ingredients. https://cosmetics.specialchem.com/ (accessed on 7 April 2022).
Cáceres, P. J., Carlucci, M. J., Damonte, E. B., Matsuhiro, B., Elisa, A., & Zúñiga, E. A., (2000). Carrageenans from Chilean samples of *Stenogramme interrupta* (Phyllophoraceae): Structural analysis and biological activity. *Phytochemistry*, *53*(1), 81–86.
Calder, P. C., (2010). Omega-3 fatty acids and inflammatory processes. *Nutrients*, *2*(3), 355–374.

Camara, R. B. G., Costa, L. S., Fidelis, G. P., et al., (2011). Heterofucans from the brown seaweed *Canistrocarpus cervicornis* with anticoagulant and antioxidant activities. *Marine Drugs*, *9*(1), 124–138.

Cao, J., Wang, J., Wang, S., & Xu, X., (2016). *Porphyra* species: A mini-review of its pharmacological and nutritional properties. *Journal of Medicinal Food*, *19*(2), 111–119.

Cardozo, K. H. M., Guaratini, T., Barros, M. P., et al., (2007). Metabolites from algae with economical impact. *Comparative Biochemistry and Physiology Part C: Toxicology and Pharmacology*, *146*(1, 2), 60–78.

Castine, S. A., McKinnon, A. D., Paul, N. A., Trott, L. A., & De Nys, R., (2013). Wastewater treatment for land-based aquaculture: Improvements and value-adding alternatives in model systems from Australia. *Aquaculture Environment Interactions*, *4*(3), 285–300.

Catanzano, O., D'Esposito, V., Acierno, S., et al., (2015). Alginate-hyaluronan composite hydrogels accelerate wound healing process. *Carbohydrate Polymers*, *131*, 407–414.

Chan, L. W., Lim, L. T., & Heng, P. W. S., (2000). Microencapsulation of oils using sodium alginate. *Journal of Microencapsulation*, *17*(6), 757–766.

Chandía, N. P., & Matsuhiro, B., (2008). Characterization of a fucoidan from *Lessonia vadosa* (Phaeophyta) and its anticoagulant and elicitor properties. *International Journal of Biological Macromolecules*, *42*(3), 235–240.

Charrier, B., Abreu, M. H., Araujo, R., et al., (2017). Furthering knowledge of seaweed growth and development to facilitate sustainable aquaculture. *New Phytologist*, *216*(4), 967–975.

Chattopadhyay, K., Ghosh, T., Carlos, A., Pujol, C. A., Carlucci, M. J., Damonte, E. B., & Ray, B., (2008). Polysaccharides from *Gracilaria corticata*: Sulfation, chemical characterization and anti-HSV activities. *International Journal of Biological Macromolecules*, *43*(4), 346–351.

Chen, H. M., Zheng, L., & Yan, X. J., (2005). The preparation and bioactivity research of agaro-oligosaccharides. *Food Technology and Biotechnology*, *43*(1), 29–36.

Chen, T. T., Lin, C. M., Mark, M. J., et al., (2015). Transgenic technology in marine organisms. In: *Springer Handbook of Marine Biotechnology* (pp. 387–412). Berlin: Springer Berlin Heidelberg.

Chen, Y. H., Tu, C. J., & Wu, H. T., (2004). Growth-inhibitory effects of the red alga *Gelidium amansii* on cultured cells. *Biological and Pharmaceutical Bulletin*, *27*(2), 180–184.

Chew, W., Mathison, B., Kimble, L., Mixter, P., & Chew, B., (2013). Astaxanthin decreases inflammatory biomarkers associated with cardiovascular disease in human umbilical vein endothelial cells. *American Journal of Advanced Food Science and Technology*, *1*, 1–17.

Chizhov, A. O., Dell, A., Morris, H. R., et al., (1999). A study of fucoidan from the brown seaweed *Chorda filum*. *Carbohydrate Research*, *320*(1, 2), 108–119.

Choi, J. I., Kim, H. J., Kim, J. H., & Lee, J. W., (2012). Enhanced biological activities of laminarin degraded by gamma-ray irradiation. *Journal of Food Biochemistry*, *36*(4), 465–469.

Choudhary, M., & Grover, K., (2012). Development of functional food products in relation to obesity. *Functional Foods in Health and Disease*, *2*(6), 188–197.

Cian, R. E., Caballero, M. S., Sabbag, N., González, R. J., & Drago, S. R., (2014). Bio-accessibility of bioactive compounds (ACE inhibitors and antioxidants) from extruded maize products added with a red seaweed *Porphyra columbina*. *LWT-Food Science and Technology*, *55*(1), 51–58.

Connan, S., & Stengel, D. B., (2011). Impacts of ambient salinity and copper on brown algae: 1. Interactive effects on photosynthesis, growth, and copper accumulation. *Aquatic Toxicology*, *104*(1, 2), 94–107.

Connan, S., Delisle, F., Deslandes, E., & Ar Gall, E., (2006). Intra-thallus phlorotannin content and antioxidant activity in Phaeophyceae of temperate waters. *Botanica Marina*, *49*(1), 39–46.

Costa, L. S., Telles, C. B. S., Oliveira, R. M., et al., (2011). Heterofucan from *Sargassum filipendula* induces apoptosis in HeLa cells. *Marine Drugs*, *9*(4), 603–614.

Cotas, J., Leandro, A., Pacheco, D., Gonçalves, A. M. M., & Pereira, L., (2020a). A comprehensive review of the nutraceutical and therapeutic applications of red seaweeds (Rhodophyta). *Life*, *10*(3), 19.

Cotas, J., Marques, V., Afonso, M. B., Rodrigues, C. M. P., & Pereira, L., (2020b). Antitumor potential of *Gigartina pistillata* carrageenans against colorectal cancer stem cell-enriched tumor spheres. *Marine Drugs*, *18*(1), 50.

Council of the European Union, (2020). *EU Action to Restrict Plastic Pollution: Council Agrees Its Position-Consilium*. https://www.consilium.europa.eu/en/press/press-releases/2018/10/31/eu-acts-to-restrict-plastic-pollution-council-agrees-its-stance/ (accessed on 7 April 2022).

Coura, C. O., Souza, R. B., Rodrigues, J. A. G., et al., (2015). Mechanisms involved in the anti-inflammatory action of a polysulfated fraction from *Gracilaria cornea* in rats. *PLoS One*, *10*(3), e0119319.

Cronin, G., & Mark, E. H., (1996). Effects of light and nutrient availability on the growth, secondary chemistry, and resistance to herbivory of two brown seaweeds. *Oikos*, *77*(1), 93–106.

Cunha, L., & Grenha, A., (2016). Sulfated seaweed polysaccharides as multifunctional materials in drug delivery applications. *Marine Drugs*, *14*(3), 42.

Currie, M. E., (2018). The growing sustainable seaweed industry: A comparison of Australian state governance directing current and future seaweed cultivation. *Independent Study Project (ISP) Collection*. https://digitalcollections.sit.edu/isp_collection (accessed on 7 April 2022).

Custódio, C. A., Reis, R. L., & Mano, J. F., (2016). Photo crosslinked laminarin based hydrogels for biomedical applications. *Biomacromolecules*, *17*(5), 1602–1609.

Darcy-Vrillon, B., (1993). Nutritional aspects of the developing use of marine macroalgae for the human food industry. *International Journal of Food Sciences and Nutrition*, *44*, 23–35.

Das, M., Zuniga, E., & Ojima, I., (2009). Novel taxoid-based tumor-targeting drug conjugates. *Chimica Oggi*, *27*(6), 54–56.

Dawczynski, C., Schubert, R., & Jahreis, G., (2007). Amino acids, fatty acids, and dietary fiber in edible seaweed products. *Food Chemistry*, *103*(3), 891–899.

De Zoysa, M., Nikapitiya, C., Jeon, Y. J., Jee, Y., & Lee, J., (2008). Anticoagulant activity of sulfated polysaccharide isolated from fermented brown seaweed *Sargassum fulvellum*. *Journal of Applied Phycology*, *20*(1), 67–74.

Dembitsky, V. M., & Maoka, T., (2007). Allenic and cumulenic lipids. *Progress in Lipid Research*, *46*(6), 328–375.

Devillé, C., Damas, J., Forget, P., Dandrifosse, G., & Peulen, O., (2004). Laminarin in the dietary fiber concept. *Journal of the Science of Food and Agriculture*, *84*(9), 1030–1038.

Devillé, C., Gharbi, M., Dandrifosse, G., & Peulen, O., (2007). Study on the effects of laminarin, a polysaccharide from seaweed, on gut characteristics. *Journal of the Science of Food and Agriculture*, *87*(9), 1717–1725.

Diamahesa, W. A., Masumoto, T., Jusadi, D., & Setiawati, M., (2017). Growth and protein content of *Ulva prolifera* maintained at different flow rates in integrated aquaculture system. *Jurnal Ilmu Dan Teknologi Kelautan Tropis*, *9*(2), 429–441.

Díaz-Rubio, M. E., Pérez-Jiménez, J., & Saura-Calixto, F., (2009). Dietary fiber and antioxidant capacity in *Fucus vesiculosus* products. *International Journal of Food Sciences and Nutrition*, *60*(2), 23–34.

Ding, Y., Wang, L., Im, S., et al., (2019). Anti-obesity effect of diphlorethohydroxycarmalol isolated from brown alga *Ishige okamurae* in high-fat diet-induced obese mice. *Marine Drugs*, *17*(11), 637.

Domínguez, H., (2013). Algae as a source of biologically active ingredients for the formulation of functional foods and nutraceuticals. In: Domínguez, H., (ed.), *Functional Ingredients from Algae for Foods and Nutraceuticals* (pp. 1–19). Sawston: Woodhead Publishing.

Doyle, J. W., Roth, T. P., Smith, R. M., Li, Y. Q., & Dunn, R. M., (1996). Effect of calcium alginate on cellular wound healing processes modeled *in vitro*. *Journal of Biomedical Materials Research*, *32*(4), 561–568.

Du, J., (2015). Research on functionality sports nutrition and health food security issues based on circulation. *The Open Cybernetics and Systemics Journal*, *9*(1), 1945–1949.

Dutot, M., Fagon, R., Hemon, M., & Rat, P., (2012). Antioxidant, anti-inflammatory, and anti-senescence activities of a phlorotannin-rich natural extract from brown seaweed *Ascophyllum nodosum*. *Applied Biochemistry and Biotechnology*, *167*(8), 2234–2240.

El-Deek, A. A., & Brikaa, M. A., (2009). Nutritional and biological evaluation of marine seaweed as a feedstuff and as a pellet binder in poultry diet. *International Journal of Poultry Science*, *8*(9), 875–881.

Ermakova, S., Men shova, R., Vishchuk, O., et al., (2013). Water-soluble polysaccharides from the brown alga *Eisenia bicyclis*: Structural characteristics and antitumor activity. *Algal Research*, *2*(1), 51–58.

European Chemicals Agency. (2020). *Position Paper on EU Restriction of Microplastic – EEB: The European Environmental Bureau*. https://eeb.org/library/ngo-position-on-eu-restriction-of-microplastic/ (accessed on 7 April 2022).

European Food Safety Authority, (2008). Consumption of food and beverages with added plant sterols. *EFSA Journal*, *6*(3), 133r.

Fairhead, V. A., Amsler, C. D., McClintock, J. B., & Baker, B. J., (2006). Lack of defense or phlorotannin induction by UV radiation or mesograzers in *Desmarestia anceps* and *D. menziesii* (Phaeophyceae). *Journal of Phycology*, *42*(6), 1174–1183.

Fan, L., Tongchun, T., Yanchun, S., Yang, L., & Shaolun, Z., (1995). Study on anti-virus effect of fucoidan *in vitro*. *Baiqiuen Yike Daxue Xuebao*, *21*(3), 255–257.

FAO, (2018). *The State of the World Fisheries and Aquaculture – Meeting the Sustainable Development Goals* (Vol. 3). Rome: FAO.

FAO, (2020). The state of world fisheries and aquaculture. *Sustainability in Action*. Rome: FAO.

Fassett, R. G., & Coombes, J. S., (2011). Astaxanthin: A potential therapeutic agent in cardiovascular disease. *Marine Drugs*, *9*(3), 447–465.

Fattorusso, E., Magno, S., & Mayol, L., (1980). Sterols of Mediterranean *Chlorophyceae*. *Experientia*, *36*(10), 1137, 1138.

Ferdouse, F., Holdt, S. L., Smith, R., Murúa, P., & Yang, Z., (2018). The global status of seaweed production, trade and utilization. *FAO Globefish Research Program*, *124*, 120. Rome: FAO.

Fernand, F., Israel, A., Skjermo, J., Wichard, T., Timmermans, K. R., & Golberg, A., (2017). Offshore macroalgae biomass for bioenergy production: Environmental aspects, technological achievements and challenges. *Renewable and Sustainable Energy Reviews*, *75*, 35–45.

Field, C. J., & Schley, P. D., (2004). Evidence for potential mechanisms for the effect of conjugated linoleic acid on tumor metabolism and immune function: Lessons from n-3 fatty acids. *The American Journal of Clinical Nutrition*, *79*(6), 1190–1198.

Fitton, J., Stringer, D., & Karpiniec, S., (2015). Therapies from fucoidan: An update. *Marine Drugs*, *13*(9), 5920–5946.

Fitzgerald, C., Gallagher, E., Doran, L., Auty, M., Prieto, J., & Hayes, M., (2014). Increasing the health benefits of bread: Assessment of the physical and sensory qualities of bread formulated using a renin inhibitory *Palmaria palmata* protein hydrolysate. *LWT-Food Science and Technology*, *56*(2), 398–405.

Fleurence, J., (1999). Seaweed proteins: Biochemical, nutritional aspects and potential uses. *Trends in Food Science and Technology*, *10*(1), 25–28.

Fleury, B. G., Figueiredo, L., Marconi, M. I., Teixeira, V. L., Ferreira, A. B. B., & Pinto, A. C., (2011). Fatty acids as chemotaxonomic markers of marine macrophytes from Rio de Janeiro State, Brazil. *Natural Product Communications*, *6*(5), 667–672.

Francavilla, M., Franchi, M., Monteleone, M., & Caroppo, C., (2013). The red seaweed *Gracilaria gracilis* as a multi products source. *Marine Drugs*, *11*(10), 3754–3776.

Franklin, L. A., & Forster, R. M., (1997). The changing irradiance environment: Consequences for marine macrophyte physiology, productivity and ecology. *European Journal of Phycology*, *32*(3), 207–232.

Freile-Pelegrín, Y., & Murano, E., (2005). Agars from three species of *Gracilaria* (Rhodophyta) from Yucatán Peninsula. *Bioresource Technology*, *96*(3), 295–302.

Freile-Pelegrín, Y., & Robledo, D., (1997). Influence of alkali treatment on agar from *Gracilaria cornea* from Yucatan, Mexico. *Journal of Applied Phycology*, *9*(6), 533–539.

Freile-Pelegrín, Y., & Robledo, D., (2013). Bioactive phenolic compounds from algae. In: Hérnandez-Ledesma, B., & Herrero, M., (eds.), *Bioactive Compounds from Marine Foods: Plant and Animal Sources* (pp. 113–29). Hoboken: Wiley.

FSAI. (2015). *Guidance on Food Additives*. Food Safety Authority of Ireland. https://www.fsai.ie/guidanceonfoodadditivesrevision12011.html (accessed on 7 April 2022).

Funk, C. D., (2001). Prostaglandins and leukotrienes: Advances in eicosanoid biology. *Science*, *294*(5548), 1871–1875.

Galland-Irmouli, A. V., Fleurence, J., Lamghari, R., et al., (1999). Nutritional value of proteins from edible seaweed *Palmaria palmata* (Dulse). *Journal of Nutritional Biochemistry*, *10*(6), 353–359.

Galloway, A. W. E., Britton-Simmons, K. H., Duggins, D. O., Gabrielson, P. W., & Brett, M. T., (2012). Fatty acid signatures differentiate marine macrophytes at ordinal and family ranks. *Journal of Phycology*, *48*(4), 956–965.

Ghanbarzadeh, M., Golmoradizadeh, A., & Homaei, A., (2018). Carrageenans and carrageenases: Versatile polysaccharides and promising marine enzymes. *Phytochemistry Reviews*, *17*(3), 535–571.

Glasson, C. R. K., Sims, I. M., Carnachan, S. M., De Nys, R., & Magnusson, M., (2017). A cascading biorefinery process targeting sulfated polysaccharides (ulvan) from *Ulva ohnoi*. *Algal Research*, *27*, 383–391.

Godfray, H. C. J., Beddington, J. R., Crute, I. R., et al., (2010). Food security: The challenge of feeding 9 billion people. *Science*, *327*(5967), 812–818.

Gómez, I., & Huovinen, P., (2010). Induction of phlorotannins during UV exposure mitigates inhibition of photosynthesis and DNA damage in the kelp *Lessonia nigrescens*. *Photochemistry and Photobiology*, *86*(5), 1056–1063.

Gómez-Ordóñez, E., Jiménez-Escrig, A., & Rupérez, P., (2014). Bioactivity of sulfated polysaccharides from the edible red seaweed *Mastocarpus stellatus*. *Bioactive Carbohydrates and Dietary Fiber*, *3*(1), 29–40.

Granada, L., Lopes, S., Novais, S. C., & Lemos, M. F. L., (2018). Modelling integrated multi-trophic aquaculture: Optimizing a three trophic level system. *Aquaculture*, *495*, 90–97.

Gressler, V., Sumie, N., Toyota, M., et al., (2010). Lipid, fatty acid, protein, amino acid and ash contents in four Brazilian red algae species. *Food Chemistry, 120*(2), 585–590.

Guangce, W., Haibao, S., Xiao, F., & Chengkui, T., (2002). Large-scale isolation and purification of R-phycoerythrin from red alga *Palmaria palmata* using the expanded bed adsorption method. *Journal of Integrative Plant Biology, 44*(5), 541–546.

Guiry, M. D., & Dawes, C. J., (1992). Daylength, temperature and nutrient control of tetrasporogenesis in *Asparagopsis armata* (Rhodophyta). *Journal of Experimental Marine Biology and Ecology, 158*(2), 197–217.

Gurgel, C. F. D., & Lopez-Bautista, J., (2007). Red algae. In: *Encyclopedia of Life Sciences* (pp. 1–5). Chichester, UK: John Wiley and Sons, Ltd.

Hafting, J. T., Critchley, A. T., Cornish, M. L., Hubley, S. A., & Archibald, A. F., (2012). On-land cultivation of functional seaweed products for human usage. *Journal of Applied Phycology, 24*(3), 385–392.

Han, J. G., Syed, A. Q., Kwon, M., Ha, J. H., & Lee, H. Y., (2008). Antioxidant, immunomodulatory and anticancer activity of fucoidan isolated from *Fucus vesiculosus*. *Journal of Biotechnology, 136*, S571.

Harnedy, P. A., & Fitzgerald, R. J., (2011). Bioactive proteins, peptides, and amino acids from macroalgae. *Journal of Phycology, 47*(2), 218–232.

Harrison, P. J., & Hurd, C. L., (2001). Nutrient physiology of seaweeds: Application of concepts to aquaculture. *Cahiers de Biologie Marine, 42*(1, 2), 71–82.

Haug, A., & Jensen, A., (1954). Seasonal variation in the chemical composition of *Alaria esculenta, Laminaria saccharina, Laminaria hyperborea* and *Laminaria digitata* from Northern Norway. Oslo: Norwegian Institute of Seaweed Research.

Hayes, M., (2013). Biological activities of proteins and marine-derived peptides from byproducts and seaweeds. In: Se-Kwon, K., (ed.), *Marine Proteins and Peptides: Biological Activities and Applications* (pp. 139–165). Hoboken: John Wiley and Sons, Ltd.

He, D., Wu, S., Yan, L., et al., (2019). Antitumor bioactivity of porphyran extracted from *Pyropia yezoensis* Chonsoo2 on human cancer cell lines. *Journal of the Science of Food and Agriculture, 99*(15), 6722–6730.

He, M. L., Hollwich, W., & Rambeck, W. A., (2002). Supplementation of algae to the diet of pigs: A new possibility to improve the iodine content in the meat. *Journal of Animal Physiology and Animal Nutrition, 86*(3, 4), 97–104.

Hehemann, J. H., Correc, G., Thomas, F., et al., (2012). Biochemical and structural characterization of the complex agarolytic enzyme system from the marine bacterium *Zobellia galactanivorans*. *Journal of Biological Chemistry, 287*(36), 30571–30584.

Hierholtzer, A., Chatellard, L., Kierans, M., Akunna, J. C., & Collier, P. J., (2012). The impact and mode of action of phenolic compounds extracted from brown seaweed on mixed anaerobic microbial cultures. *Journal of Applied Microbiology, 114*(4), 964–973.

Higashimura, Y., Naito, Y., Takagi, T., et al., (2013). Oligosaccharides from agar inhibit murine intestinal inflammation through the induction of heme oxygenase-1 expression. *Journal of Gastroenterology, 48*(8), 897–909.

Hoad, C. L., Rayment, P., Spiller, R. C., et al., (2004). *In vivo* imaging of intragastric gelation and its effect on satiety in humans. *The Journal of Nutrition, 134*(9), 2293–2300.

Holdt, S. L., & Kraan, S., (2011). Bioactive compounds in seaweed: Functional food applications and legislation. *Journal of Applied Phycology, 23*(3), 543–597.

Hong, S. J., Lee, J. H., Kim, E. J., Yang, H. J., Park, J. S., & Hong, S. K., (2017). Toxicological evaluation of neoagarooligosaccharides prepared by enzymatic hydrolysis of agar. *Regulatory Toxicology and Pharmacology, 90*, 9–21.

Hosokawa, M., Okada, T., Mikami, N., Konishi, I., & Miyashita, K., (2009). Biofunctions of marine carotenoids. *Food Science and Biotechnology*, *18*(1), 1–11.

Hosokawa, M., Wanezaki, S., Miyauchi, K., et al., (1999). Apoptosis-inducing effect of fucoxanthin on human leukemia cell line HL-60. *Food Science and Technology Research*, *5*(3), 243–246.

Hou, Y., Wang, J., Jin, W., Zhang, H., & Zhang, Q., (2012). Degradation of *Laminaria japonica* fucoidan by hydrogen peroxide and antioxidant activities of the degradation products of different molecular weights. *Carbohydrate Polymers*, *87*(1), 153–159.

Hsu, H. Y., Lin, T. Y., Lu, M. K., Leng, P. J., Tsao, S. M., & Wu, Y. C., (2017). Fucoidan induces toll-like receptor 4-regulated reactive oxygen species and promotes endoplasmic reticulum stress-mediated apoptosis in lung cancer. *Scientific Reports*, *7*(1), 1–13.

Huang, C. B., & Ebersole, J. L., (2010). A novel bioactivity of omega-3 polyunsaturated fatty acids and their ester derivatives. *Molecular Oral Microbiology*, *25*(1), 75–80.

Hwang, P. A., Lin, H. T. V., Lin, H. Y., & Lo, S. K., (2019). Dietary supplementation with low-molecular-weight fucoidan enhances innate and adaptive immune responses and protects against *Mycoplasma pneumoniae* antigen stimulation. *Marine Drugs*, *17*(3), 175.

Ianova, O. B., Kim, V. A., & Treiman, E. V., (2012). Alginates in the treatment of post-gastroresection reflux esophagitis. *Terapevticheskii Arkhivc*, *84*(10), 48–50.

Im, J. H., Choi, C. H., Mun, F., et al., (2017). A polycaprolactone/fish collagen/alginate biocomposite supplemented with phlorotannin for hard tissue regeneration. *RSC Advances*, *7*(4), 2009–2018.

Imeson, A., (2009). In: Imeson, A., (ed.), *Food Stabilizers, Thickeners and Gelling Agents* (p. 354). Oxford, UK: Wiley-Blackwell.

Inic-Kanada, A., Stein, E., Stojanovic, M., et al., (2018). Effects of iota-carrageenan on ocular *Chlamydia trachomatis* infection *in vitro* and *in vivo*. *Journal of Applied Phycology*, *30*(4), 2601–2610.

Inoue, N., Yamano, N., Sakata, K., Nagao, K., Hama, Y., & Yanagita, T., (2009). The sulfated polysaccharide porphyran reduces apolipoprotein B100 secretion and lipid synthesis in HepG2 cells. *Bioscience, Biotechnology and Biochemistry*, *73*(2), 447–449.

Ioannou, E., & Roussis, V., (2009). Natural products from seaweeds. In: Osbourn, A. E., & Lanzotti, V., (eds.), *Plant-Derived Natural Products: Synthesis, Function, and Application*. New York: Springer-Verlag.

Ishihara, K., Oyamada, C., Matsushima, R., Murata, M., & Muraoka, T., (2005). Inhibitory effect of porphyran, prepared from dried 'Nori', on contact hypersensitivity in mice. *Bioscience, Biotechnology and Biochemistry*, *69*(10), 1824–1830.

Iwamoto, T., Hosoda, K., Hirano, R., et al., (2000). Inhibition of low-density lipoprotein oxidation by astaxanthin. *Journal of Atherosclerosis and Thrombosis*, *7*(4), 216–222.

Iwasaki, T., & Tawara, A., (2006). Effects of astaxanthin on eyestrain induced by accommodative dysfunction. *Journal of the Eye*, *23*(6), 829–834.

Jang, M. K., Lee, D. G., Kim, N. Y., et al., (2009). Purification and characterization of neoagarotetraose from hydrolyzed agar. *Journal of Microbiology and Biotechnology*, *19*(10), 1197–1200.

Jansen, H. M., Van, D. B. S., Bolman, B., et al., (2016). The feasibility of offshore aquaculture and its potential for multi-use in the North Sea. *Aquaculture International*, *24*(3), 735–756.

Jaswir, I., & Monsur, H. A., (2011). Anti-inflammatory compounds of macro algae origin: A review. *Journal of Medicinal Plant Research*, *5*(33), 7146–7154.

Jaulneau, V., Lafitte, C., Jacquet, C., et al., (2010). Ulvan, a sulfated polysaccharide from green algae, activates plant immunity through the jasmonic acid signaling pathway. *Journal of Biomedicine and Biotechnology, 2010*, 1–11.

Je, J. Y., Park, P. J., Kim, E. K., et al., (2009). Antioxidant activity of enzymatic extracts from the brown seaweed *Undaria pinnatifida* by electron spin resonance spectroscopy. *LWT-Food Science and Technology, 42*(4), 874–878.

Ji, Y. B., Ji, C. F., & Zhang, H., (2012). Laminarin induces apoptosis of human colon cancer LOVO cells through a mitochondrial pathway. *Molecules, 17*(8), 9947–9960.

Jiao, G., Yu, G., Zhang, J., & Ewart, E., (2011). Chemical structures and bioactivities of sulfated polysaccharides from marine algae. *Marine Drugs, 9*(2), 196–223.

Jiménez-Escrig, A., & Goñi, I., (1999). Evaluación nutricional y efectos fisiológicos de macroalgas marinas comestibles. *Archivos Latinoamericanos de Nutrición, 49*(2), 114–120.

Jimenez-Escrig, A., & Sanchez-Muniz, F. J., (2000). Dietary fiber from edible seaweeds: Chemical structure, physicochemical properties and effects on cholesterol metabolism. *Nutrition Research, 20*(4), 585–598.

Jiménez-Escrig, A., Gómez-Ordóñez, E., & Rupérez, P., (2012). Brown and red seaweeds as potential sources of antioxidant nutraceuticals. *Journal of Applied Phycology, 24*(5), 1123–1132.

Jiménez-Escrig, A., Jiménez-Jiménez, I., Pulido, R., & Saura-Calixto, F., (2001). Antioxidant activity of fresh and processed edible seaweeds. *Journal of the Science of Food and Agriculture, 81*(5), 530–534.

Jin, M., Liu, H., Hou, Y., et al., (2017). Preparation, characterization and alcoholic liver injury protective effects of algal oligosaccharides from *Gracilaria lemaneiformis*. *Food Research International, 100*, 186–195.

Johns, R. B., Nichols, P. D., & Perry, G. J., (1979). Fatty acid composition of ten marine algae from Australian waters. *Phytochemistry, 18*(5), 799–802.

Joubert, Y., & Fleurence, J., (2008). Simultaneous extraction of proteins and DNA by an enzymatic treatment of the cell wall of *Palmaria palmata* (Rhodophyta). *Journal of Applied Phycology, 20*(1), 55–61.

Jung, K. A., Lim, S., Kim, Y., & Park, J. M., (2013). Potentials of macroalgae as feedstocks for biorefinery. *Bioresource Technology, 135*, 182–190.

Kadam, S. U., Álvarez, C., Tiwari, B. K., & O'Donnell, C. P., (2017). Extraction and characterization of protein from Irish brown seaweed *Ascophyllum nodosum*. *Food Research International, 99*, 1021–1027.

Kadam, S. U., Tiwari, B. K., & O'Donnell, C. P., (2013). Application of novel extraction technologies for bioactives from marine algae. *Journal of Agricultural and Food Chemistry, 61*(20), 4667–4675.

Kadam, S. U., Tiwari, B. K., & O'Donnell, C. P., (2015). Extraction, structure and biofunctional activities of laminarin from brown algae. *International Journal of Food Science and Technology, 50*(1), 24–31.

Kakinuma, M., Park, C. S., & Amano, H., (2001). Distribution of free L-cysteine and glutathione in seaweeds. *Fisheries Science, 67*(1), 194–196.

Kamenarska, Z. G., Dimitrova-Konaklieva, S. D., Stefanov, K. L., & Popov, S. S., (2003). A comparative study on the sterol composition of some brown algae from the Black Sea. *Journal of the Serbian Chemical Society, 68*(4, 5), 269–275.

Kapetanović, R., Sladić, D., Popov, S., Zlatović, M., Kljajić, Z., & Gašić, M. J., (2005). Sterol composition of the Adriatic Sea algae *Ulva lactuca*, *Codium dichotomum*, *Cystoseira adriatica* and *Fucus virsoides*. *Journal of the Serbian Chemical Society, 70*(12), 1395–1400.

Karawita, R., Siriwardhana, N., Lee, K. W., et al., (2005). Reactive oxygen species scavenging, metal chelation, reducing power and lipid peroxidation inhibition properties of different solvent fractions from *Hizikia fusiformis*. *European Food Research and Technology, 220*, 363–371.

Karppi, J., Rissanen, T. H., Nyyssönen, K., et al., (2007). Effects of astaxanthin supplementation on lipid peroxidation. *International Journal for Vitamin and Nutrition Research, 77*(1), 3–11.

Kawakubo, A., Makino, H., Ohnishi, J., Hirohara, H., & Hori, K., (1997). The marine red alga *Eucheuma serra* J. Agardh, a high yielding source of two isolectins. *Journal of Applied Phycology, 9*(4), 331–338.

Kayama, M., Iijima, N., Kuwahara, M., Sado, T., Araki, S., & Sakurai, T., (1985). Effect of water temperature on the fatty acid composition of *Porphyra*. *Bulletin of the Japanese Society of Scientific Fisheries, 51*(4), 687–691.

Kazłowski, B., Pan, C. L., & Ko, Y. T., (2015). Monitoring and preparation of neoagaro- and agaro-oligosaccharide products by high performance anion exchange chromatography systems. *Carbohydrate Polymers, 122*, 351–358.

Kellogg, J., Grace, M. H., & Lila, M. A., (2014). Phlorotannins from Alaskan seaweed inhibit carbolytic enzyme activity. *Marine Drugs, 12*(10), 5277–5294.

Khalil, H. P. S. A., Saurabh, C. K., Tye, Y. Y., et al., (2017). Seaweed based sustainable films and composites for food and pharmaceutical applications: A review. *Renewable and Sustainable Energy Reviews, 77*(1), 353–362.

Khilchenko, S. R., Zaporozhets, T. S., Shevchenko, N. M., et al., (2011). Immunostimulatory activity of fucoidan from the brown alga *Fucus evanescens*: Role of sulfates and acetates. *Journal of Carbohydrate Chemistry, 30*(4–6), 291–305.

Khotimchenko, S. V., (1991). Fatty acid composition of seven *Sargassum* apecies. *Phytochemistry, 30*(8), 2639–2641.

Khotimchenko, S. V., (1993). Fatty acids of green macrophytic algae from the Sea of Japan. *Phytochemistry, 32*(5), 1203–1207.

Khotimchenko, S. V., (1998). Fatty acids of brown algae from the Russian far east. *Phytochemistry, 49*(8), 2363–2369.

Khotimchenko, S. V., Vaskovsky, V. E., & Przhemenetskaya, V. F., (1991). Distribution of eicosapentaenoic and arachidonic acids in different species of *Gracilaria*. *Phytochemistry, 30*(1), 207–209.

Khotimchenko, S. V., Vaskovsky, V. E., & Titlyanova, T. V., (2002). Fatty acids of marine algae from the Pacific coast of North California. *Botanica Marina, 45*(1), 17–22.

Kim, E. J., Park, S. Y., Lee, J. Y., & Jung, P. J. H. Y., (2010). Fucoidan present in brown algae induces apoptosis of human colon cancer cells. *BMC Gastroenterology, 10*, 1–11.

Kim, J. H., Choi, S. K., Choi, S. Y., Kim, H. K., & Chang, H. I., (2005). Suppressive effect of astaxanthin isolated from the *Xanthophyllomyces dendrorhous* mutant on ethanol-induced gastric mucosal injury in rats. *Bioscience, Biotechnology and Biochemistry, 69*(7), 1300–1305.

Kim, K. H., Kim, Y. W., Kim, H. B., Lee, B. J., & Lee, D. S., (2006). Anti-apoptotic activity of laminarin polysaccharides and their enzymatically hydrolyzed oligosaccharides from *Laminaria japonica*. *Biotechnology Letters, 28*(6), 439–446.

Kim, S. K., (2011). In: Se-Kwon, K., (ed.), *Handbook of Marine Macroalgae: Biotechnology and Applied Phycology* (p. 567). Chichester, UK: John Wiley and Sons, Ltd.

Knowler, D., Chopin, T., Martínez-Espiñeira, R., et al., (2020). The economics of integrated multi-trophic aquaculture: Where are we now and where do we need to go?. *Reviews in Aquaculture, 12*(3), 1579–1594.

Kolb, N., Vallorani, L., & Stocchi, V., (1999). Chemical composition and evaluation of protein quality by amino acid score method of edible brown marine algae Arame (*Eisenia bicyclis*) and Hijiki (*Hijikia fusiforme*). *Acta Alimentaria*, *28*(3), 213–222.

Kraan, S., (2013). Mass-cultivation of carbohydrate rich macroalgae, a possible solution for sustainable biofuel production. *Mitigation and Adaptation Strategies for Global Change*, *18*(1), 27–46.

Kuda, T., Yano, T., Matsuda, N., & Nishizawa, M., (2005). Inhibitory effects of laminaran and low molecular alginate against the putrefactive compounds produced by intestinal microflora *in vitro* and in rats. *Food Chemistry*, *91*(4), 745–749.

Kulshreshtha, G., Rathgeber, B., MacIsaac, J., et al., (2017). Feed supplementation with red seaweeds, *Chondrus crispus* and *Sarcodiotheca gaudichaudii*, reduce *Salmonella enteritidis* in laying hens. *Frontiers in Microbiology*, *8*, 1–12.

Kumar, C. S., Gasesan, P., Suresh, P. V., & Bhaskar, N., (2008). Seaweeds as a source of nutritionally beneficial compounds – A review. *Journal of Food Science and Technology*, *45*(1), 1–13.

Kumari, P., Kumar, M., Gupta, V., Reddy, C. R. K., & Jha, B., (2010). Tropical marine macroalgae as potential sources of nutritionally important PUFAs. *Food Chemistry*, *120*(3), 749–757.

Kupcinskas, L., Lafolie, P., Lignell, Å., et al., (2008). Efficacy of the natural antioxidant astaxanthin in the treatment of functional dyspepsia in patients with or without *Helicobacter pylori* infection: A prospective, randomized, double blind, and placebo-controlled study. *Phytomedicine*, *15*(6, 7), 391–399.

Kwak, J. Y., (2014). Fucoidan as a marine anticancer agent in preclinical development. *Marine Drugs*, *12*(2), 851–870.

Kwon, M. J., & Nam, T. J., (2006). Porphyran induces apoptosis related signal pathway in AGS gastric cancer cell lines. *Life Sciences*, *79*(20), 1956–1962.

Lahaye, M., & Robic, A., (2007). Structure and function properties of ulvan, a polysaccharide from green seaweeds. *Biomacromolecules*, *8*, 1765–1774.

Lahaye, M., Ray, B., Baumberger, S., Quemener, B., & Axelos, M. A. V., (1996). Chemical characterization and gelling properties of cell wall polysaccharides from species of *Ulva* (*Ulvales, Chlorophyta*). *Hydrobiologia*, 326, 327, 473–480.

Lakmal, H. C., Samarakoon, K. W., Lee, W., et al., (2014). Anticancer and antioxidant effects of selected Sri Lankan marine algae. *Journal of the National Science Foundation of Sri Lanka*, *42*(4), 315–323.

Lane, A. L., Stout, E. P., Lin, A. S., et al., (2009). Antimalarial bromophycolides J-Q from the Fijian red alga *Callophycus serratus*. *The Journal of Organic Chemistry*, *74*(7), 2736–2742.

Lapointe, B. E., (1981). The effects of light and nitrogen on growth, pigment content, and biochemical composition of *Gracilaria foliifera* V. *angustissima* (*Gigartinales, Rhodophyta*). *Journal of Phycology*, *17*(1), 90–95.

Lawton, R. J., Mata, L., De Nys, R., & Paul, N. A., (2013). Algal bioremediation of waste waters from land-based aquaculture using *Ulva*: Selecting target species and strains. *PLoS One*, *8*(10), e77344.

Leandro, A., Pereira, L., & Gonçalves, A. M. M., (2020). Diverse applications of marine macroalgae. *Marine Drugs*, *18*(1), 1–17.

Leclerc, J. C., Riera, P., Leroux, C., Lévêque, L., & Davoult, D., (2013). Temporal variation in organic matter supply in kelp forests: Linking structure to trophic functioning. *Marine Ecology Progress Series*, *494*, 87–105.

Ledesma-Hernandez, B., & Herrero, M., (2013). In: Hernández-Ledesma, B., &. Herrero, M., (eds.), *Bioactive Compounds from Marine Foods* (p. 464). Chichester, UK: John Wiley & Sons Ltd.

Lee, H. A., Kim, I. H., & Nam, T. J., (2015). Bioactive peptide from *Pyropia yezoensis* and its anti-inflammatory activities. *International Journal of Molecular Medicine, 36*(6), 1701–1706.

Lee, H. S., Park, H. J., & Kim, M. K., (2008). Effect of *Chlorella vulgaris* on lipid metabolism in Wistar rats fed high fat diet. *Nutrition Research and Practice, 2*(4), 204–210.

Lee, J. Y., Kim, Y. J., Kim, H. J., Kim, Y. S., & Park, W., (2012). Immunostimulatory effect of laminarin on RAW 264.7 mouse macrophages. *Molecules, 17*(5), 5404–5411.

Li, B., Lu, F., Wei, X., & Zhao, R., (2008). Fucoidan: Structure and bioactivity. *Molecules, 13*(8), 1671–1395.

Li, X., Fan, X., Han, L., & Lou, Q., (2002). Fatty acids of some algae from the Bohai Sea. *Phytochemistry, 59*(2), 157–161.

Li, Y. X., Wijesekara, I., Li, Y., & Kim, S. K., (2011). Phlorotannins as bioactive agents from brown algae. *Process Biochemistry, 46*(12), 2219–2224.

Liu, B. H., & Lee, Y. K., (2003). Effect of total secondary carotenoids extracts from *Chlorococcum* sp. on *Helicobacter pylori*-infected BALB/c mice. *International Immunopharmacology, 3*(7), 979–986.

Liu, J., Hafting, J., Critchley, A. T., Banskota, A. H., & Prithiviraj, B., (2013). Components of the cultivated red seaweed *Chondrus crispus* enhance the immune response of *Caenorhabditis elegans* to *Pseudomonas aeruginosa* through the pmk-1, daf-2/daf-16, and skn-1 pathways. *Applied and Environmental Microbiology, 79*(23), 7343–7350.

Liu, M. Y., Mei, J. F., Yi, Y., Chen, J. S., & Ying, G. Q., (2008). Advances in study on biological activities of agaro-oligosaccharides. *Pharmaceutical Biotechnology, 15*, 493–496.

Liu, M., Hansen, P. E., & Lin, X., (2011). Bromophenols in marine algae and their bioactivities. *Marine Drugs, 9*(79), 1273–1292.

Liu, Q. M., Xu, S. S., Li, L., et al., (2017). *In vitro* and *in vivo* immunomodulatory activity of sulfated polysaccharide from *Porphyra haitanensis*. *Carbohydrate Polymers., 165*, 189–196.

Liu, Y. J., Deng, Z., Geng, L., Wang, J., & Zhang, Q., (2019). *In vitro* evaluation of the neuroprotective effect of oligo-porphyran from *Porphyra yezoensis* in PC12 cells. *Journal of Applied Phycology, 31*(4), 2559–1571.

Liu, Z., Gao, T., Ying, Y. Y., et al., (2019). Anti-cancer activity of porphyran and carrageenan from red seaweeds. *Molecules, 24*(23), 1–14.

Lopes, G., (2014). *Seaweeds from the Portuguese Coast: Chemistry, Antimicrobial and Anti-Inflammatory Capacity.* PhD Thesis, Universidade do Porto, Porto, Portugal.

Lopes, G., Sousa, C., Bernardo, J., et al., (2011). Sterol profiles in 18 macroalgae of the Portuguese coast. *Journal of Phycology, 47*(5), 1210–1218.

Lopes, G., Sousa, C., Valentão, P., & Andrade, P. B., (2013). Sterols in algae and health. In: Hernández-Ledesma, B., & Herrero, M., (eds.), *Bioactive Compounds from Marine Foods: Plant and Animal Sources* (pp. 173–191). Chichester, UK: John Wiley & Sons Ltd.

Lordan, S., Ross, R. P., & Stanton, C., (2011). Marine bioactives as functional food ingredients: Potential to reduce the incidence of chronic diseases. *Marine Drugs, 9*(6), 1056–1100.

Lozano, I., Wacyk, J. M., Carrasco, J., & Cortez-San, M. M. A., (2016). Red macroalgae *Pyropia columbina* and *Gracilaria chilensis*: Sustainable feed additive in the *Salmo salar* diet and the evaluation of potential antiviral activity against infectious salmon anemia virus. *Journal of Applied Phycology, 28*(2), 1343–1351.

Mabeau, S., & Fleurence, J., (1993). Seaweed in food products: Biochemical and nutritional aspects. *Trends in Food Science and Technology*, *4*(4), 103–107.

MacArtain, P., Gill, C. I. R., Brooks, M., Campbell, R., & Rowland, I. R., (2007). Nutritional value of edible seaweeds. *Nutrition Reviews*, *65*(12), 535–543.

Maeda, H., Hosokawa, M., Sashima, T., Funayama, K., & Miyashita, K., (2007). Effect of medium-chain triacylglycerols on anti-obesity effect of fucoxanthin. *Journal of Oleo Science*, *56*(12), 615–621.

Magalhaes, K. D., Costa, L. S., Fidelis, G. P., et al., (2011). Anticoagulant, antioxidant and antitumor activities of heterofucans from the seaweed *Dictyopteris delicatula*. *International Journal of Molecular Sciences*, *12*(5), 3352–3365.

Makkar, H. P. S., Tran, G., Heuzé, V., et al., (2016). Seaweeds for livestock diets: A review. *Animal Feed Science and Technology*, *212*, 1–17.

Mandal, P., Mateu, C. G., Chattopadhyay, K., Pujol, C. A., Damonte, E. B., & Ray, B., (2007). Structural features and antiviral activity of sulphated fucans from the brown seaweed *Cystoseira indica*. *Antiviral Chemistry and Chemotherapy*, *18*(3), 153–162.

Marinho-Soriano, E., & Bourret, E., (2005). Polysaccharides from the red seaweed *Gracilaria dura* (*Gracilariales, Rhodophyta*). *Bioresource Technology*, *96*(3), 379–382.

Marinho-Soriano, E., Fonseca, P. C., Carneiro, M. A. A., & Moreira, W. S. C., (2006). Seasonal variation in the chemical composition of two tropical seaweeds. *Bioresource Technology*, *97*(18), 2402–2406.

Martin, H., (2019). *Aquaculture 4.0: Applying Industry Strategy to Fisheries Management*. https://www.innovationnewsnetwork.com/aquaculture-4-0/596/ (accessed on 26 July 2022).

Martínez, B., & Rico, J. M., (2008). Changes in nutrient content of *Palmaria palmata* in response to variable light and upwelling in Northern Spain. *Journal of Phycology*, *44*(1), 50–59.

Martínez-Andrade, K. A. M., Lauritano, C., Romano, G., & Ianora, A., (2018). Marine microalgae with anti-cancer properties. *Marine Drugs*, *16*(5), 165.

Marudhupandi, T., Ajith, K. T. T., Senthil, S. L., & Devi, K. N., (2014). *In vitro* antioxidant properties of fucoidan fractions from *Sargassum tenerrimum*. *Pakistan Journal of Biological Sciences*, *17*(3), 402–407.

Masuda, M., Kawaguchi, S., Abe, T., Kawamoto, T., & Suzuki, M., (2002). Additional analysis of chemical diversity of the red algal genus *Laurencia* (*Rhodomelaceae*) from Japan. *Phycological Research*, *50*(2), 135–144.

Mayer, A. M. S., & Hamann, M. T., (2004). Marine pharmacology in 2000: Marine compounds with antibacterial, anticoagulant, antifungal, anti-inflammatory, antimalarial, antiplatelet, antituberculosis, and antiviral activities; affecting the cardiovascular, immune, and nervous systems and other miscellaneous mechanisms of action. *Marine Biotechnology*, *6*, 37–52.

Mazumder, S., Ghosal, P. K., Pujol, C. A., Carlucci, M. J., Damonte, E. B., & Ray, B., (2002). Isolation, chemical investigation and antiviral activity of polysaccharides from *Gracilaria corticata* (*Gracilariaceae, Rhodophyta*). *International Journal of Biological Macromolecules*, *31*(1–3), 87–95.

Meireles, F., (2013). Carrageenan yield and quality of *Chondrus crispus* Stackhouse (Rhodophyta) cultivated in an integrated multi-trophic aquaculture (IMTA) system. MSc Thesis, University of Coimbra, Coimbra, Portugal.

Mekinić, I. G., Skroza, D., Šimat, V., Hamed, I., Čagalj, M., & Perković, Z. P., (2019). Phenolic content of brown algae (*Pheophyceae*) species: Extraction, identification, and quantification. *Biomolecules*, *9*(6), 244.

Menshova, R. V., Ermakova, S. P., Anastyuk, S. D., et al., (2014). Structure, enzymatic transformation and anticancer activity of branched high molecular weight laminaran from brown alga *Eisenia bicyclis*. *Carbohydrate Polymers*, *99*, 101–109.

Miao, H. Q., Elkin, M., Aingorn, E., Ishai-Michaeli, R., Stein, C. A., & Vlodavsky, I., (1999). Inhibition of heparanase activity and tumor metastasis by laminarin sulfate and synthetic phosphorothioate oligodeoxynucleotides. *International Journal of Cancer*, *83*(3), 424–431.

Miao, H. Q., Ishai-Michaeli, R., Peretz, T., & Vlodavsky, I., (1995). Laminarin sulfate mimics the effects of heparin on smooth muscle cell proliferation and basic fibroblast growth factor-receptor binding and mitogenic activity. *Journal of Cellular Physiology*, *164*(3), 482–490.

Michalak, I., & Chojnacka, K., (2014). Algal extracts: Technology and advances. *Engineering in Life Sciences*, *14*(6), 581–591.

Michalak, I., & Chojnacka, K., (2015). Algae as production systems of bioactive compounds. *Engineering in Life Sciences*, *15*(2), 160–176.

Milovanović, M., Banjac, N., & Vucelić-Radović, B., (2009). Functional food: Rare herbs, seeds and vegetable oils as sources of flavors and phytosterols. *Journal of Agricultural Sciences*, *54*(1), 81–94.

Mišurcová, L., Buňka, F., Ambrožová, J. V., Machů, L., Samek, D., & Kráčmar, S., (2014). Amino acid composition of algal products and its contribution to RDI. *Food Chemistry*, *151*, 120–125.

Miyashita, K., & Hosokawa, M., (2017). Fucoxanthin in the management of obesity and its related disorders. *Journal of Functional Foods*, *36*, 195–202.

Miyawaki, H., Takahashi, J., Tsukahara, H., & Takehara, I., (2008). Effects of astaxanthin on human blood rheology. *Journal of Clinical Biochemistry and Nutrition*, *43*(2), 69–74.

Moreau, R. A., Whitaker, B. D., & Hicks, K. B., (2002). Phytosterols, phytostanols, and their conjugates in foods: Structural diversity, quantitative analysis, and health-promoting uses. *Progress in Lipid Research*, *41*(6), 457–500.

Moroney, N. C., O'Grady, M. N., O'Doherty, J. V., & Kerry, J. P., (2012). Addition of seaweed (*Laminaria digitata*) extracts containing laminarin and fucoidan to porcine diets: Influence on the quality and shelf-life of fresh pork. *Meat Science*, *92*(4), 423–429.

Mozaffarian, D., & Wu, J. H., (2011). Omega-3 fatty acids and cardiovascular disease: Effects on risk factors, molecular pathways, and clinical events. *Journal of the American College of Cardiology*, *58*(20), 2047–2067.

Murray, M., Dordevic, A. L., Ryan, L., & Bonham, M. P., (2018). The impact of a single dose of a polyphenol-rich seaweed extract on postprandial glycaemic control in healthy adults: A randomized cross-over trial. *Nutrients*, *10*(3), 270.

Nagaki, Y., Mihara, M., Tsukuhara, H., & Ono, S., (2006). The supplementation effect of astaxanthin on accommodation and asthenopia. *Journal Clinic Medicine and Therapeutics*, *22*(1), 41–54.

Nakazato, K., Takada, H., Iha, M., & Nagamine, T., (2010). Attenuation of N-nitrosodiethyl-amine-induced liver fibrosis by high-molecular-weight fucoidan derived from *Cladosiphon okamuranus*. *Journal of Gastroenterology and Hepatology*, *25*(10), 1692–1701.

Namvar, F., Mohamed, S., Fard, S. G., et al., (2012). Polyphenol-rich seaweed (*Eucheuma cottonii*) extract suppresses breast tumor via hormone modulation and apoptosis induction. *Food Chemistry*, *130*(2), 376–382.

Narayan, B., Miyashita, K., & Hosokawa, M., (2004). Comparative evaluation of fatty acid composition of different *Sargassum* (*Fucales*, *Phaeophyta*) species harvested from temperate and tropical waters. *Journal of Aquatic Food Product Technology*, *13*(4), 53–70.

Necas, J., & Bartosikova, L., (2013). Carrageenan: A review. *Veterinarni Medicina, 58*(4), 187–205.
Neyrinck, A. M., Mouson, A., & Delzenne, N. M., (2007). Dietary supplementation with laminarin, a fermentable marine β (1-3) glucan, protects against hepatotoxicity induced by LPS in rat by modulating immune response in the hepatic tissue. *International Immunopharmacology, 7*(12), 1497–1506.
Nwosu, F., Morris, J., Lund, V. A., Stewart, D., Ross, H. A., & McDougall, G. J., (2011). Anti-proliferative and potential anti-diabetic effects of phenolic-rich extracts from edible marine algae. *Food Chemistry, 126*(3), 1006–1012.
O'Sullivan, L., Murphy, B., McLoughlin, P., et al., (2010). Prebiotics from marine macroalgae for human and animal health applications. *Marine Drugs, 8*(7), 2038–2064.
Ondiviela, B., Losada, I. J., Lara, J. L., et al., (2014). The role of seagrasses in coastal protection in a changing climate. *Coastal Engineering, 87*, 158–168.
Onofrejová, L., Vašíčková, J., Klejdus, B., et al., (2010). Bioactive phenols in algae: The application of pressurized-liquid and solid-phase extraction techniques. *Journal of Pharmaceutical and Biomedical Analysis, 51*(2), 464–470.
Osea Malibu, (2018). *Non-Toxic Seaweed Skin Care*. https://oseamalibu.com/ (accessed on 7 April 2022).
Pádua, D., Rocha, E., Gargiulo, D., & Ramos, A. A., (2015). Bioactive compounds from brown seaweeds: Phloroglucinol, fucoxanthin and fucoidan as promising therapeutic agents against breast Cancer. *Phytochemistry Letters, 14*, 91–98.
Palozza, P., Torelli, C., Boninsegna, A., et al., (2009). Growth-inhibitory effects of the astaxanthin-rich alga *Haematococcus pluvialis* in human colon cancer cells. *Cancer Letters, 283*(1), 108–117.
Pangestuti, R., & Kim, S. K., (2011). Biological activities and health benefit effects of natural pigments derived from marine algae. *Journal of Functional Foods, 3*(4), 255–266.
Paniagua-Michel, J. J., Olmos-Soto, J. J., & Morales-Guerrero, E. R., (2014). Algal and microbial exopolysaccharides: New insights as biosurfactants and bioemulsifiers. In: Kim, S. K., (ed.), *Advances in Food and Nutrition Research* (pp. 221–257). Amsterdam: Elsevier.
Pashkow, F. J., Watumull, D. G., & Campbell, C. L., (2008). Astaxanthin: A novel potential treatment for oxidative stress and inflammation in cardiovascular disease. *American Journal of Cardiology, 101*(10), 58–68.
Patarra, R. F., Paiva, L., Neto, A. I., Lima, E., & Baptista, J., (2011). Nutritional value of selected macroalgae. *Journal of Applied Phycology, 23*(2), 205–208.
Patch, C. S., Tapsell, L. C., Williams, P. G., & Gordon, M., (2006). Plant sterols as dietary adjuvants in the reduction of cardiovascular risk: Theory and evidence. *Vascular Health and Risk Management, 2*(2), 157–162.
Patterson, G., (1971). The distribution of sterols in algae. *Lipids, 6*(2), 120–127.
Patwary, M. U., & Van, D. M. J. P., (1992). Genetics and breeding of cultivated seaweeds. *The Korean Journal of Phycology, 7*(2), 281–318.
Peinado, I., Girón, J., Koutsidis, G., & Ames, J. M., (2014). Chemical composition, antioxidant activity and sensory evaluation of five different species of brown edible seaweeds. *Food Research International, 66*, 36–44.
Peng, J., Yuan, J. P., Wu, C. F., & Wang, J. H., (2011). Fucoxanthin, a marine carotenoid present in brown seaweeds and diatoms: Metabolism and bioactivities relevant to human health. *Marine Drugs, 9*(10), 1806–1828.

Pereira, H., Barreira, L., Figueiredo, F., et al., (2012). Polyunsaturated fatty acids of marine macroalgae: Potential for nutritional and pharmaceutical applications. *Marine Drugs, 10*(9), 1920–1935.
Pereira, L., & Mesquita, J. F., (2003). Carrageenophytes of occidental Portuguese coast, 1-spectroscopic analysis in eight carrageenophytes from Buarcos Bay. *Biomolecular Engineering, 20*(4–6), 217–222.
Pereira, L., (2016). *Edible Seaweeds of the World*, 463. Boca Raton: CRC Press.
Pérez-Rodríguez, E., Aguilera, J., & Figueroa, F. L., (2003). Tissular localization of coumarins in the green alga *Dasycladus vermicularis* (Scopoli) krasser: A photoprotective role? *Journal of Experimental Botany, 54*(384), 1093–1100.
Pigmen, K., Yip, W. H., Lim, S. J., et al., (2014). Characterization and stability of pigments extracted from *Sargassum binderi* obtained from Semporna, Sabah. *Sains Malaysiana, 43*(9), 345–1354.
Piironen, V., Lindsay, D. G., Miettinen, T. A., Toivo, J., & Lampi, A. M., (2000). Plant sterols: Biosynthesis, biological function and their importance to human nutrition. *Journal of the Science of Food and Agriculture, 80*(7), 939–966.
Pina, A. L., Costa, A. R., Lage-Yusty, M. A., & López-Hernández, J., (2014). An evaluation of edible red seaweed (*Chondrus crispus*) components and their modification during the cooking process. *LWT-Food Science and Technology, 56*(1), 175–180.
Platt, D., Pelled, D., & Shulman, A., (2004). *Oils Enriched with Diacylglycerols and Phytosterol Esters for use in the Reduction of Cholesterol and Triglycerides*. WO/2004/069150.
Pliego-Cortés, H., Bedoux, G., Boulho, R., eta al., (2019). Stress tolerance and photoadaptation to solar radiation in *Rhodymenia pseudopalmata* (Rhodophyta) through mycosporine-like amino acids, phenolic compounds, and pigments in an integrated multi-trophic aquaculture system. *Algal Research, 41*, 101542.
Poudyal, H., Panchal, S. K., Ward, L. C., & Brown, L., (2013). Effects of ALA, EPA and DHA in high-carbohydrate, high-fat diet-induced metabolic syndrome in rats. *Journal of Nutritional Biochemistry, 24*(6), 1041–1052.
Poupard, N., Badarou, P., Fasani, F., et al., (2017). Assessment of heparanase-mediated angiogenesis using microvascular endothelial cells: Identification of λ-carrageenan derivative as a potent anti angiogenic agent. *Marine Drugs, 15*(5), 134.
Prokofjeva, M. M., Imbs, T. I., Shevchenko, N. M., et al., (2013). Fucoidans as potential inhibitors of HIV-1. *Marine Drugs, 11*(18), 3000–3014.
Radwan, S. S., (1991). Sources of C20-polyunsaturated fatty acids for biotechnological use. *Applied Microbiology and Biotechnology, 35*(4), 421–430.
Rajak, R. C., Jacob, S., & Kim, B. S., (2020). A holistic zero waste biorefinery approach for macroalgal biomass utilization: A review. *Science of the Total Environment., 716*, 137067.
Rajapakse, N., & Kim, S. K., (2011). Nutritional and digestive health benefits of seaweed. In: Kim, S. K., (ed.), *Advances in Food and Nutrition Research* (1st edn. Vol. 64). Amsterdam: Elsevier Inc.
Rao, A. R., Sindhuja, H. N., Dharmesh, S. M., Sankar, K. U., Sarada, R., & Ravishankar, G. A., (2013). Effective inhibition of skin cancer, tyrosinase, and antioxidative properties by astaxanthin and astaxanthin esters from the green alga *Haematococcus pluvialis*. *Journal of Agricultural and Food Chemistry, 61*(16), 3842–3851.
Reddy, P., & Urban, S., (2009). Meroditerpenoids from the southern Australian marine brown alga *Sargassum fallax*. *Phytochemistry, 70*(2), 250–255.

Reid, G. K., Lefebvre, S., Filgueira, R., et al., (2020). Performance measures and models for open-water integrated multi-trophic aquaculture. *Reviews in Aquaculture*, *12*(1), 47–75.
Reisewitz, S. E., Estes, J. A., & Simenstad, C. A., (2006). Indirect food web interactions: Sea otters and kelp forest fishes in the Aleutian archipelago. *Oecologia*, *146*(4), 623–631.
Ren, D., Noda, H., Amano, H., Nishino, T., & Nishizawa, K., (1994). Study on antihypertensive and antihyperlipidemic effects of marine algae. *Fisheries Science*, *60*(1), 83–88.
Rhein-Knudsen, N., Ale, M. T., & Meyer, A. S., (2015). Seaweed hydrocolloid production: An update on enzyme assisted extraction and modification technologies. *Marine Drugs*, *13*(6), 3340–3359.
Rinaudo, M., (2007). Seaweed Polysaccharides. In: Kamerling, H., (ed.), *Comprehensive Glycoscience: From Chemistry to Systems Biology* (pp. 691–735). Amsterdam: Elsevier.
Rioux, L. E., Turgeon, S. L., & Beaulieu, M., (2007). Characterization of polysaccharides extracted from brown seaweeds. *Carbohydrate Polymers*, *69*(3), 530–537.
Rioux, L. E., Turgeon, S. L., & Beaulieu, M., (2009). Effect of season on the composition of bioactive polysaccharides from the brown seaweed *Saccharina longicruris*. *Phytochemistry*, *70*(8), 1069–1075.
Robic, A., Sassi, J. F., Dion, P., Lerat, Y., & Lahaye, M., (2009). Seasonal variability of physicochemical and rheological properties of ulvan in two *Ulva* species (Chlorophyta) from the Brittany coast. *Journal of Phycology*, *45*(4), 962–973.
Rodrigues, D., Freitas, A. C., Pereira, L., et al., (2015). Chemical composition of red, brown and green macroalgae from Buarcos bay in central west coast of Portugal. *Food Chemistry*, *183*, 197–207.
Rodríguez-Bernaldo, D. Q. A., Lage-Yusty, M. A., & López-Hernández, J., (2010). Determination of phenolic compounds in macroalgae for human consumption. *Food Chemistry*, *121*, 634–638.
Rodriguez-Jasso, R., Solange, R. M., Mussatto, I., Pastrana, L., Aguilar, C. N., & Teixeira, J. A., (2013). Chemical composition and antioxidant activity of sulphated polysaccharides extracted from *Fucus vesiculosus* using different hydrothermal processes. *Chemical Papers*, *68*(2), 203–209.
Rupérez, P., Ahrazem, O., & Leal, A. J., (2002). Potential antioxidant capacity of sulfated polysaccharides from the edible marine brown seaweed *Fucus vesiculosus*. *Journal of Agricultural and Food Chemistry*, *50*(4), 840–845.
Sachindra, N. M., Airanthi, M. K. W. A., Hosokawa, M., & Miyashita, K., (2010). Radical scavenging and singlet oxygen quenching activity of extracts from Indian seaweeds. *Journal of Food Science and Technology*, *47*(1), 94–99.
Sánchez-Machado, D. I., López-Cervantes, J., López-Hernández, J., & Paseiro-Losada, P., (2004). Fatty acids, total lipid, protein and ash contents of processed edible seaweeds. *Food Chemistry*, *85*(3), 439–444.
Sandberg, A. S., Anderson, H., Bosaeus, I., Carlsson, N. G., Hasselblad, K., & Härröd, M., (1994). Alginate, small bowel sterol excretion, and absorption of nutrients in ileostomy subjects. *The American Journal of Clinical Nutrition*, *60*(5), 751–756.
Sato, S., (1975). Effects of environmental factors on the lipid components of *Porphyra* species. *Bulletin of the Japanese Society for the Science of Fish*, *41*, 326–339.
Schijf, J., & Ebling, A. M., (2010). Investigation of the ionic strength dependence of *Ulva lactuca* acid functional group pK_as by manual alkalimetric titrations. *Environmental Science and Technology*, *44*(5), 1644–1649.
Schmitz, G., & Ecker, J., (2008). The opposing effects of n-3 and n-6 fatty acids. *Progress in Lipid Research*, *47*(2), 147–155.

Schoschina, E. V., (1996). Seasonal and age dynamics of growth and reproduction of *Phycodrys rubens* (*Rhodophyta*) in the Barents and white seas. *Aquatic Botany*, *55*(1), 13–30.
Schwager, J., Mohajeri, M. H., Fowler, A., & Weber, P., (2008). Challenges in discovering bioactives for the food industry. *Current Opinion in Biotechnology*, *19*(2), 66–72.
Seedevi, P., Moovendhan, M., Viramani, S., & Shanmugam, A., (2017). Bioactive potential and structural characterization of sulfated polysaccharide from seaweed (*Gracilaria corticata*). *Carbohydrate Polymers*, *155*, 516–524.
Sekar, S., & Chandramohan, M., (2008). Phycobiliproteins as a commodity: Trends in applied research, patents and commercialization. *Journal of Applied Phycology*, *20*(2), 113–136.
Shanmugam, M., & Mody, K. H., (2000). Heparinoid-active sulphated polysaccharides from marine algae as potential blood anticoagulant agents. *Current Science*, *79*(12), 1672–1683.
Shannon, E., & Abu-Ghannam, N., (2019). Seaweeds as nutraceuticals for health and nutrition. *Phycologia*, *58*(5), 563–577.
Sharifuddin, Y., Chin, Y. X., Lim, P. E., & Phang, S. M., (2015). Potential bioactives from seaweed for diabetes management. *Marine Drugs*, *13*, 5447–5491.
Shi, Q., Wang, A., Lu, Z., Qin, C., Hu, J., & Yin, J., (2017). Overview on the antiviral activities and mechanisms of marine polysaccharides from seaweeds. *Carbohydrate Research*, *453*, *454*, 1–9.
Shibata, T., Ishimaru, K., Kawaguchi, S., Yoshikawa, H., & Hama, Y., (2007). Antioxidant activities of phlorotannins isolated from Japanese Laminariaceae. *Journal of Applied Phycology*, *20*(5), 255–261.
Shibata, T., Nagayama, K., Tanaka, R., & Yamaguchi, K., (2003). Inhibitory effects of brown algal phlorotannins on secretory phospholipase A_2s, lipoxygenases and cyclooxygenases. *Journal of Applied Phycology*, *15*(1), 61–66.
Shibata, T., Yamaguchi, K., Nagayama, K., Kawaguchi, S., & Nakamura, T., (2002). Inhibitory activity of brown algal phlorotannins against glycosidases from the viscera of the turban shell *Turbo cornutus*. *European Journal of Phycology*, *37*(4), 493–500.
Silva, D. M., Valente, L. M. P., Sousa-Pinto, I., Pereira, R., Pires, M. A., Seixas, F., & Rema, P., (2015). Evaluation of IMTA-produced seaweeds (*Gracilaria*, *Porphyra*, and *Ulva*) as dietary ingredients in *Nile tilapia*, *Oreochromis niloticus* L., juveniles. Effects on growth performance and gut histology. *Journal of Applied Phycology*, *27*(4), 1671–1680.
Silva, T. M. A., Alves, L. G., De Queiroz, K. C. S., et al., (2005). Partial characterization and anticoagulant activity of a heterofucan from the brown seaweed *Padina gymnospora*. *Brazilian Journal of Medical and Biological Research*, *38*, 523–533.
Singh, I. P., & Sidana, J., (2013). Phlorotannins. In: *Functional Ingredients from Algae for Foods and Nutraceuticals* (pp. 181–204). Amsterdam: Elsevier Ltd.
Skjermo, J., Aasen, I. M., Arff, J., et al., (2014). Report A new Norwegian bioeconomy based on cultivation and processing of seaweeds: Opportunities and R&D needs. *SINTEF Fisheries and Aquaculture*, 46.
Smale, D. A., Burrows, M. T., Pippa, M. P., O'Connor, N., & Hawkins, S. J., (2013). Threats and knowledge gaps for ecosystem services provided by kelp forests: A northeast Atlantic perspective. *Ecology and Evolution*, *3*(11), 4016–4038.
Smit, A. J., (2004). Medicinal and pharmaceutical uses of seaweed natural products: A review. *Journal of Applied Phycology*, *16*, 245–262.
Sousa, A. M. M., Alves, V. D., Morais, S., Delerue-Matos, C., & Gonçalves, M. P., (2010). Agar extraction from integrated multitrophic aquacultured *Gracilaria vermiculophylla*:

Evaluation of a microwave-assisted process using response surface methodology. *Bioresource Technology*, *101*(9), 3258–3267.
Sousa, O. V. E., Araújo, I. W. F., Quinderé, A. L. G., et al., (2011). The involvement of the HO-1 pathway in the anti-inflammatory action of a sulfated polysaccharide isolated from the red seaweed *Gracilaria birdiae*. *Inflammation Research*, *60*(12), 1121–1130.
Souza, B. W. S., Cerqueira, M. A., Bourbon, A. I., et al., (2012). Chemical characterization and antioxidant activity of sulfated polysaccharide from the red seaweed *Gracilaria birdiae*. *Food Hydrocolloids*, *27*(2), 287–292.
Souza, M. P., Vaz, A. F. M., Costa, T. B., et al., (2018). Construction of a biocompatible and antioxidant multilayer coating by layer-by-layer assembly of κ-carrageenan and quercetin nanoparticles. *Food and Bioprocess Technology*, *11*(5), 1050–1060.
Spolaore, P., Joannis-Cassan, C., Duran, E., & Isambert, A., (2006). Commercial applications of microalgae. *Journal of Bioscience and Bioengineering*, *101*(2), 87–96.
Stengel, D. B., Connan, S., & Popper, Z. A., (2011). Algal chemodiversity and bioactivity: Sources of natural variability and implications for commercial application. *Biotechnology Advances*, *29*, 483–501.
Sugihara, Y., Taniguchi, A., Masatsugu, K., Kazuhiro, N., Miho, T., & Takashi, A., (2004). *Polysaccharide Containing Composition and Tear Film Stabilizing Ophthalmic Solution.* 20070077278.
Sun, L., Wang, S., Gong, X., Zhao, M., Fu, X., & Wang, L., (2009). Isolation, purification and characteristics of R-phycoerythrin from a marine macroalga *Heterosiphonia japonica*. *Protein Expression and Purification*, *64*(2), 146–154.
Suzuki, K., Shimizu, T., & Nakata, T., (1998). The cholesterol metabolite cholest-4-en-3-one and its 3-oxo derivatives suppress body weight gain, body fat accumulation and serum lipid concentration in mice. *Bioorganic and Medicinal Chemistry Letters*, *8*(16), 2133–2138.
Swanson, A. K., & Druehl, L. D., (2002). Induction, exudation and the UV protective role of kelp phlorotannins. *Aquatic Botany*, *73*, 241–253.
Talarico, L. B., Pujol, C. A., Zibetti, R. G. M., Faría, P. C. S., Noseda, M. D., Duarte, M. E. R., & Damonte, E. B., (2005). The antiviral activity of sulfated polysaccharides against dengue virus is dependent on virus serotype and host cell. *Antiviral Research*, *66*(2, 3), 103–110.
Talarico, L. B., Zibetti, R. G. M., Faria, P. C. S., et al., (2004). Anti-herpes simplex virus activity of sulfated galactans from the red seaweeds *Gymnogongrus griffithsiae* and *Cryptonemia crenulata*. *International Journal of Biological Macromolecules*, *34*(1, 2), 63–71.
Tandeau De, M. N., (2003). Phycobiliproteins and phycobilisomes: The early observations. *Photosynthesis Research*, *76*(1–3), 197–203.
Tello-Ireland, C., Lemus-Mondaca, R., Vega-Gálvez, A., López, J., & Di Scala, K., (2011). Influence of hot-air temperature on drying kinetics, functional properties, color, phycobiliproteins, antioxidant capacity, texture and agar yield of alga *Gracilaria chilensis*. *LWT-Food Science and Technology*, *44*, 2112–2118.
Thomas, N. V., & Kim, S. K., (2011). Potential pharmacological applications of polyphenolic derivatives from marine brown algae. *Environmental Toxicology and Pharmacology*, *32*(3), 325–335.
Thompson, G. A., (1996). Lipids and membrane function in green algae. *Biochimica et Biophysica Acta*, *1302*(1), 17–45.
Tominaga, K., Hongo, N., Karato, M., & Yamashita, E., (2012). Cosmetic benefits of astaxanthin on humans subjects. *Acta Biochimica Polonica*, *59*(1), 43–47.

Torres, M. D., Flórez-Fernández, N., & Domínguez, H., (2019). Integral utilization of red seaweed for bioactive production. *Marine Drugs*, *17*(6), 314.

Torres, P., Nagai, A., Teixeira, D. I. A., Marinho-Soriano, E., Chow, F., & Dos, S. D. Y. A. C., (2019). Brazilian native species of *Gracilaria* (Gracilariales, Rhodophyta) as a source of valuable compounds and as nutritional supplements. *Journal of Applied Phycology*, *31*(5), 3163–3173.

Toshiyoshi, A., (1996). *Novel Beta-Agarase and Microorganism Generating the Same*. EP0743363A3.

Troell, M., Joyce, A., Chopin, T., Neori, A., Buschmann, A. H., & Fang, J. G., (2009). Ecological engineering in aquaculture-potential for integrated multi-trophic aquaculture (IMTA) in marine offshore systems. *Aquaculture*, *297*(1–4), 1–9.

Tsiapali, E., Whaley, S., Kalbfleisch, J., Ensley, H. E., William, B. I., & Williams, D. L., (2001). Glucans exhibit weak antioxidant activity, but stimulate macrophage free radical activity. *Free Radical Biology and Medicine*, *30*(4), 393–402.

Tsuge, K., Okabe, M., Yoshimura, T., Sumi, T., Tachibana, H., & Yamada, K., (2004). Dietary effects of porphyran from *Porphyra yezoensis* on growth and lipid metabolism of Sprague-Dawley rats. *Food Science and Technology Research*, *10*(2), 147–151.

Turna, I. I., Ertan, Ö. O., Cormaci, M., & Furnari, G., (2002). Seasonal variations in the biomass of macro-algal communities from the gulf of Antalya (North-Eastern Mediterranean). *Turkish Journal of Botany*, *26* (1), 19–29.

Ueno, M., Cho, K., Isaka, S., Nishiguchi, T., Yamaguchi, K., Kim, D., & Oda, T., (2018). Inhibitory effect of sulphated polysaccharide porphyran (isolated from *Porphyra yezoensis*) on RANKL-induced differentiation of RAW264.7 cells into osteoclasts. *Phytotherapy Research*, *32*(3), 452–458.

Urquiaga, I., & Leighton, F., (2000). Plant polyphenol antioxidants and oxidative stress. *Biological Research*, *33*(2), 55–64.

Valado, A., Pereira, M., Caseiro, A., et al., (2020). Effect of carrageenans on vegetable jelly in humans with hypercholesterolemia. *Marine Drugs*, *18*(1), 19.

Van, G. V. J. T., Helsper, J. P. F., De Visser, W., Van, K. H., & Brandenburg, W., (2011). Polyunsaturated fatty acids in various macroalgal species from north Atlantic and tropical seas. *Lipids in Health and Disease*, *10*(1), 1–8.

Van, K. M., Bikker, P., Van, D. M. I. M., Van Der Peet-Schwering, C. M. C., & Vereijken, J. M., (2013). *Cultivation, Processing and Nutritional Aspects for Pigs and Poultry of European Protein Sources as Alternatives for Imported Soybean Products*. Report/Wageningen UR Livestock Research 662.

Vásquez, J. A., Zuñiga, S., Tala, F., Piaget, N., Rodríguez, D. C., & Alonso, V. J. M., (2014). Economic valuation of kelp forests in Northern Chile: Values of goods and services of the ecosystem. *Journal of Applied Phycology*, *26*, 1081–1088.

Venkatesan, J., Lowe, B., Sukumar, A. S., Manivasagan, P., Al Kheraif, A. A., Kang, K. H., & Kim, S. K., (2015). Seaweed polysaccharides and their potential biomedical applications. *Starch/Staerke*, *67*(5, 6), 381–390.

Vergara, J. J., Bird, K. T., & Niell, F. X., (1995). Nitrogen assimilation following NH_4^+ pulses in the red alga *Gracilariopsis lemaneiformis*: Effect on C metabolism. *Marine Ecology Progress Series*, *122*(1–3), 253–263.

Vieira, E. F., Soares, C., Susana, M. S., et al., (2018). Seaweeds from the Portuguese coast as a source of proteinaceous material: Total and free amino acid composition profile. *Food Chemistry*, *269*, 264–275.

Viera, I., Pérez-Gálvez, A., & Roca, M., (2018). Bioaccessibility of marine carotenoids. *Marine Drugs, 16*(10), 397.

Villanueva, M. J., Morcillo, M., Tenorio, M. D., Mateos-Aparicio, I., Andrés, V., & Redondo-Cuenca, A., (2014). Health-promoting effects in the gut and influence on lipid metabolism of *Himanthalia elongata* and *Gigartina pistillata* in hypercholesterolaemic Wistar rats. *European Food Research and Technology, 238*(3), 409–416.

Villanueva, R. D., Sousa, A. M. M., Gonçalves, M. P., Nilsson, M., & Hilliou, L., (2010). Production and properties of agar from the invasive marine alga, *Gracilaria vermiculophylla* (Gracilariales, Rhodophyta). *Journal of Applied Phycology, 22*(2), 211–220.

Vo, T. S., & Kim, S. K., (2013). Fucoidans as a natural bioactive ingredient for functional foods. *Journal of Functional Foods, 5*, 16–27.

Volkmann, H., Imianovsky, U., Oliveira, J. L. B., & Sant'Anna, E. S., (2008). Cultivation of *Arthrospira* (*Spirulina*) *platensis* in desalinator wastewater and salinated synthetic medium: Protein content and amino-acid profile. *Brazilian Journal of Microbiology, 39*(1), 98–101.

Wang, J., Jin, W., Hou, Y., Niu, X., Zhang, H., & Zhang, Q., (2013). Chemical composition and moisture-absorption/retention ability of polysaccharides extracted from five algae. *International Journal of Biological Macromolecules, 57*, 26–29.

Wang, Y., Hwang, J. U., Park, H. B., Yadav, D., Oda, T., & Jin, J. O., (2020). Porphyran isolated from *Pyropia yezoensis* inhibits lipopolysaccharide-induced activation of dendritic cells in mice. *Carbohydrate Polymers, 229*, 115457.

Wang, Y., Xing, M., Cao, Q., Ji, A., Liang, H., & Song, S., (2019). Biological activities of fucoidan and the factors mediating its therapeutic effects: A review of recent studies. *Marine Drugs, 17*(3), 183.

Ward, O. P., & Singh, A., (2005). Omega-3/6 fatty acids: Alternative sources of production. *Process Biochemistry, 40*(12), 3627–3652.

Webster, E. A., & Gadd, G. M., (1996). Cadmium replaces calcium in the cell wall of *Ulva lactuca*. *BioMetals, 9*(3), 241–244.

Weis, W. I., & Drickamer, K., (1996). Structural basis of lectin-carbohydrate recognition. *Annual Review of Biochemistry, 65*, 441–473.

Wenhai, L., Cusack, C., Baker, M., et al., (2019). Successful blue economy examples with an emphasis on international perspectives. *Frontiers in Marine Science, 6*, 261.

White, W. L., & Wilson, P., (2015). *World Seaweed Utilization. Seaweed Sustainability*. Amsterdam: Elsevier Inc.

Whitfield, F. B., Helidoniotis, F., Shaw, K. J., & Svoronos, D., (1999). Distribution of bromophenols in species of marine algae from eastern Australia. *Journal of Agricultural and Food Chemistry, 47*, 2367–2373.

Wijesekara, I., Pangestuti, R., & Kim, S. K., (2011). Biological activities and potential health benefits of sulfated polysaccharides derived from marine algae. *Carbohydrate Polymers, 84*(1), 14–21.

Wijesinghe, W. A. J. P., & Jeon, Y., (2012). Biological activities and potential industrial applications of fucose rich sulfated polysaccharides and fucoidans isolated from brown seaweeds: A review. *Carbohydrate Polymers, 88*(1), 13–20.

Wiltshire, K. H., Tanner, J. E., Gurgel, C. F. D., & Deveney, M. R., (2015). *Feasibility Study for Integrated Multitrophic Aquaculture in Southern Australia*. Report to the Fisheries Research and Development Corporation. SARDI Publication No. F2015/000786-1. SARDI Research Report Series No. 883.

Wolf, B. W., Lai, C. S., Mark, S., Kipnes, M. S., et al., (2002). Glycemic and insulinemic responses of nondiabetic healthy adult subjects to an experimental acid-induced viscosity complex incorporated into a glucose beverage. *Nutrition, 18*(7, 8), 621–626.

Xue, C. H., Fang, Y., Lin, H., LChen, L., Li, Z., Deng, D., & Lu, C. X., (2001). Chemical characters and antioxidative properties of sulfated polysaccharides from *Laminaria japonica*. *Journal of Applied Phycology, 13*, 67–70.

Yamasaki-Miyamoto, Y., Yamasaki, M., Tachibana, H., & Yamada, K., (2009). Fucoidan induces apoptosis through activation of caspase-8 on human breast cancer MCF-7 cells. *Journal of Agricultural and Food Chemistry, 57*, 8677–8682.

Yan, M. D., Lin, H. Y., & Hwang, P. A., (2019). The anti-tumor activity of brown seaweed oligo-fucoidan via LncRNA expression modulation in HepG2 cells. *Cytotechnology, 71*(1), 363–374.

Yan, X., Yoshihiro, C. Y., Suzuki, M., & Nagata, T., (1999). Fucoxanthin as the major antioxidant in *Hijikia fusiformis*, a common edible seaweed. *Bioscience, Biotechnology and Biochemistry, 63*(3), 605–607.

Yanagido, A., Ueno, M., Jiang, Z., Cho, K., Yamaguchi, K., Kim, D., & Oda, T., (2018). Increase in anti-inflammatory activities of radical-degraded porphyrans isolated from discolored Nori (*Pyropia yezoensis*). *International Journal of Biological Macromolecules., 117*, 78–86.

Yarnpakdee, S., Benjakul, S., & Kingwascharapong, P., (2015). Physico-chemical and gel properties of agar from *Gracilaria tenuistipitata* from the lake of Songkhla, Thailand. *Food Hydrocolloids, 51*, 217–226.

Yoshihisa, Y., Rehman, M., & Shimizu, T., (2014). Astaxanthin, a xanthophyll carotenoid, inhibits ultraviolet-induced apoptosis in keratinocytes. *Experimental Dermatology, 23*, 178–183.

Yu, X., Zhou, C., Yang, H., Huang, X., Ma, H., Qin, X., & Hu, J., (2015). Effect of ultrasonic treatment on the degradation and inhibition cancer cell lines of polysaccharides from *Porphyra yezoensis*. *Carbohydrate Polymers, 117*, 650–656.

Yuan, H., Song, J., Li, X., Li, N., & Dai, J., (2006). Immunomodulation and antitumor activity of κ-carrageenan oligosaccharides. *Cancer Letters, 243*(2), 228–234.

Yuan, J. P., Peng, J., Yin, K., & Wang, J. H., (2011). Potential health-promoting effects of astaxanthin: A high-value carotenoid mostly from microalgae. *Molecular Nutrition and Food Research, 55*(1), 150–165.

Yun, E. J., Yu, S., & Kim, K. H., (2017). Current knowledge on agarolytic enzymes and the industrial potential of agar-derived sugars. *Applied Microbiology and Biotechnology, 101*(14), 5581–5589.

Zhang, Q., Li, N., Liu, X., Zhao, Z., Li, Z., & Xu, Z., (2004). The structure of a sulfated galactan from *Porphyra haitanensis* and its *in vivo* antioxidant activity. *Carbohydrate Research, 339*(1), 105–111.

Zhang, Q., Qi, H., Zhao, T., Deslandes, E., Ismaeli, N. M., Molloy, F., & Critchley, A. T., (2005). Chemical characteristics of a polysaccharide from *Porphyra capensis* (*Rhodophyta*). *Carbohydrate Research, 340*(15), 2447–2450.

Zhang, Z., Zhang, Q., Wang, J., Zhang, H., Niu, X., & Li, P., (2009). Preparation of the different derivatives of the low-molecular-weight porphyran from *Porphyra haitanensis* and their antioxidant activities *in vitro*. *International Journal of Biological Macromolecules, 45*(1), 22–26.

Zhao, D., Xu, J., & Xu, X., (2018). Bioactivity of fucoidan extracted from *Laminaria japonica* using a novel procedure with high yield. *Food Chemistry*, *245*(59), 911–918.

Zhao, T., Zhang, Q., Qi, H., Zhang, H., Niu, X., Xu, Z., & Li, Z., (2006). Degradation of porphyran from *Porphyra haitanensis* and the antioxidant activities of the degraded porphyrans with different molecular weight. *International Journal of Biological Macromolecules*, *38*(1), 45–50.

Zhou, A. Y., Robertson, J., Hamid, N., Ma, Q., & Lu, J., (2014). Changes in total nitrogen and amino acid composition of New Zealand *Undaria pinnatifida* with growth, location and plant parts. *Food Chemistry*, *186*, 319–325.

Zubia, M., Fabre, M. S., Kerjean, V., et al., (2009). Antioxidant and antitumoral activities of some Phaeophyta from Brittany coasts. *Food Chemistry*, *116*(3), 693–701.

CHAPTER 12

BIOACTIVE SECONDARY METABOLITES FROM MARINE MACROALGAE

SABEENA FARVIN KODUVAYUR HABEEBULLAH and SURENDRARAJ ALAGARSAMY

Ecosystem-Based Management of Marine Resources Program, Environment and Life Sciences Research Center, Kuwait Institute for Scientific Research, P.O. Box–1638, Salmiya–22017, Kuwait

ABSTRACT

Seaweeds or macroalgae are autotrophic macroscopic marine algae which come under three major groups, viz. Chlorophyta, Rhodophyta, and Phaeophyta. As the macroalgae are sessile and live in a very dynamic marine environment where they are exposed to variation in salinity, temperature, gracing pressure, and desiccation, they use chemical defense mechanisms for their survival. They produce a variety of secondary metabolites in order to save themselves from the adverse environmental conditions, interspecies competition, fouling, pathogens, and grazing. The plethora of compounds produced by the macroalgae include phenolics, alkaloids, terpenoids, steroids, polysaccharides, fatty acids, amino acids, and peptides. Many of these secondary metabolites are reported to possess biological activity of medicinal value and find application in food, pharma, and cosmetics industries. This chapter deals with the important secondary metabolites that can be isolated from marine macroalgae and their importance.

Algal Metabolites: Biotechnological, Commercial, and Industrial Applications. Jeyabalan Sangeetha, Devarajan Thangadurai, Saher Islam, & Ravichandra Hospet (Eds.)

12.1 INTRODUCTION

Marine macroalgae (seaweed) represent a group of macroscopic autotrophic plant-like organisms, which play vital ecological roles in marine communities as they provide home and food for many different sea animals. Unlike plants, they don't have root systems and they generally live attached to rock or hard substratum by holdfast. They can absorb nutrients into the tissue directly from surrounding water and can photosynthesis in all their tissue whereas, most of the plant's photosynthesis mainly in leaves (Van den Hoek et al., 1995). Based on the pigment, storage substances and composition of the cell walls, seaweeds are grouped into three categories, viz. green seaweeds (Phylum: Chlorophyta), brown seaweeds (Phylum: Phaeophyta), and red seaweeds (Phylum: Rhodophyta) (Van den Hoek et al., 1995). Seaweeds are traditionally exploited throughout the world for industrially important high volume low value polysaccharide compounds, such as agar, alginic acid and carrageenan (Ferdouse et al., 2018) and as food for coastal communities.

As the macroalgae are sessile and live in a very dynamic marine environment, they are exposed to variation in salinity, temperature, gracing pressure, fouling, and desiccation. They use chemical defense mechanisms for their survival and produce a number of chemical compounds called secondary metabolites (Paul et al., 2011). Secondary metabolites are so termed as because they do not take part directly in any primary process involving growth and development but rather have a role in interacting with abiotic and biotic factors (Dias et al., 2012). These chemicals produced by seaweeds play an important role in their survival. For example, some of these chemical compounds produced by seaweeds have been shown to inhibit the growth of other species of seaweeds (Kim et al., 2004; Kumler, 2017), seagrasses (Raniello et al., 2007), bacteria, diatoms, and other foulers (Hellio et al., 2001; Lam and Harder, 2007; Lu et al., 2011). Some seaweed compounds have UV absorbing properties which protect them from serious damage caused by sunlight (Karsten and West, 2000; Hoyer et al., 2001). In addition to their role in survival, secondary metabolites from seaweeds also have some ecological effects (Sudatti et al., 2020). For example, some of the compounds produced by algae have been shown to cause bleaching in corals (Rasher et al., 2011) and to affect the survival and settlement of coral larvae (Fong et al., 2019). As these metabolite exhibits different biological properties, such as antimicrobial, antioxidant, anti-inflammatory, anticancer, and immune-modulating properties, researchers started exploiting these natural resources for high value food, pharma, and cosmetic applications. The present chapter will give an insight into the various secondary metabolites that can be obtained from macroalgae and their bioactive properties.

12.2 THE BIOACTIVE SECONDARY METABOLITES OF IMPORTANCE FROM MACROALGAE

The different classes of secondary metabolites isolated from macroalgae and their bioactivities are described in subsections.

12.2.1 PHENOLIC COMPOUNDS (PCs)

Phenolic compounds (PCs) are a class of chemical compounds consisting of a hydroxyl group (–OH) bonded directly to an aromatic hydrocarbon group (Balasundram et al., 2006). Marine macroalgae contains a range of PCs from simple phenolics to more complex polyphenols. In general, the phenolic content of macroalgae varied from less than 1% to 14% on a dry weight basis (Holdt and Kraan, 2011). Among the three phyla, brown algae have been reported to contain a high concentration of PCs including complex phenolics, such as phlorotannins (Li et al., 2011). Environmental factors (light, depth, salinity, and nutrients), herbivory, seasonality, and intrinsic factors (age, length, and type of tissues) have a significant influence in the phenolic contents of macroalgae (Ilvessalo and Tumi, 1989; Connan et al., 2004, 2006). These biotic and abiotic factors can regulate the expression of PCs, making a marked quantitative and qualitative variation among individuals together with intra-individual variations (Connan et al., 2004; Amsler and Fairhead, 2005). Simple phenolic acids reported from various macroalgae include gallic, protocatechuic, gentisic, hydroxybenzoic, chlorogenic, vanillic, syringic, caffeic, salicylic, coumaric, and ferulic acids (Farvin and Jacobsen, 2013). Flavonoids, such as morin, hesperidin, catechol, catechin, and epigallocatechin were also reported from green, red, and brown seaweeds (Yoshie-Stark et al., 2003; Kazłowska et al., 2010).

Among the different classes of algal phenolics, phlorotannins have gained the attention of researchers because of their pharmacological properties. Phlorotannins, polymers of phloroglucinol (1,3,5-trihydroxybenzene) were mainly isolated from brown algae (Singh and Bharate, 2006). They were produced via acetate-malonate or polyketide pathway in brown algae as a defense to cellular damage (Arnold and Targett, 2002). Based on the structural linkages between phloroglucinol and number of hydroxyl groups, phlorotannins were classified as fuhalols, phlorethols (with ether linkages), fucols (with phenyl linkages), carmalols, fucophlorethols (with both ether and phenyl linkages), and eckols (with a dibenzodioxin linkage) (Figure 12.1).

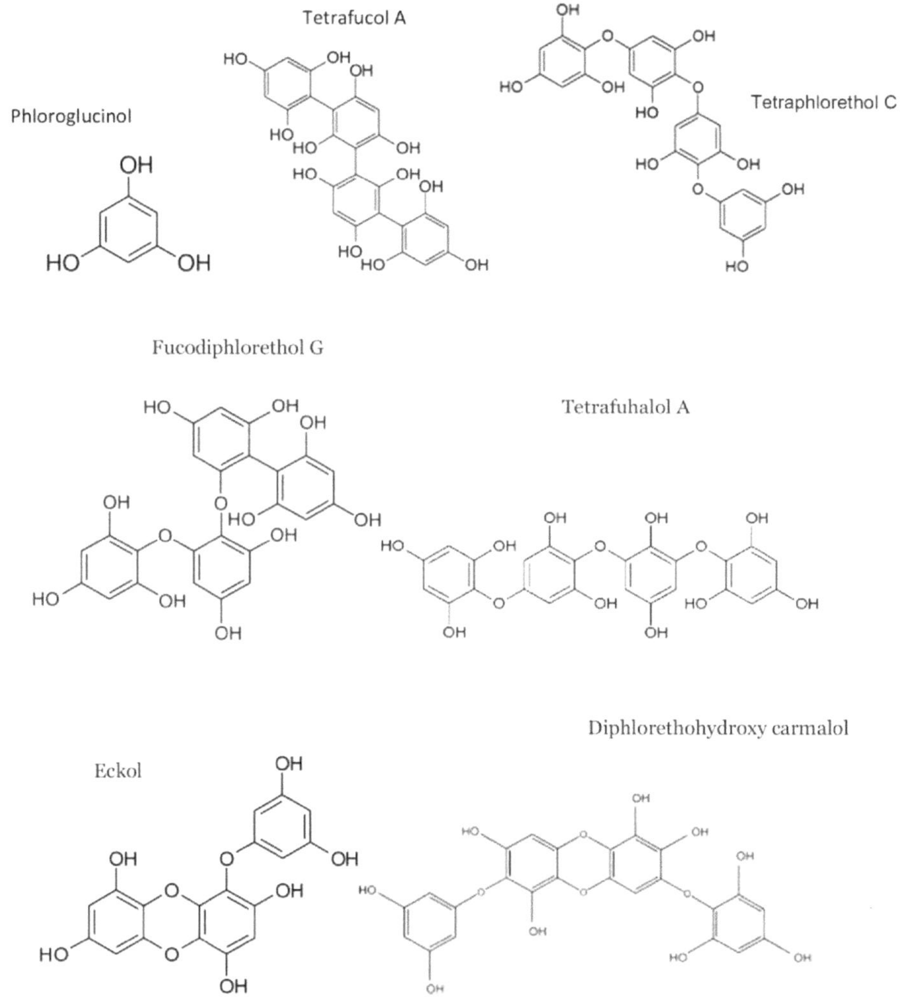

FIGURE 12.1 Phloroglucinol and its derivatives isolated from macroalgae.

The molecular weights of the most commonly found phlorotannins were in the range of 10 to 100 kDa though several studies reported a range between 126 Da and 650 kDa (Boettcher and Targett, 1993; McClintock and

Baker, 2001). Depending upon the structural heterogeneity and polymerization, phlorotannins exhibited a broad range of potential bioactivity, such as antioxidant, anti-inflammatory, antiallergic, antimicrobial, anticancer, neuroprotective, antihypertensive, and antidiabetic properties (Table 12.1).

Another interesting group of PCs present in marine macroalgae is bromophenols. Bromophenols consist of one or more benzene rings with varying degrees of bromine and hydroxyl substituents (Flodin and Whitfield, 1999). They were synthesized from phenols by an enzyme bromoperoxidase in the presence of hydrogen peroxide and bromide (Collén et al., 1994; Flodin and Whitfield, 1999). The bromophenol content and bromoperoxidase activity were reported to be high in summer for protecting from strong sun light (Pedersen et al., 1996; Flodin and Whitfield, 1999; Flodin et al., 1999). The first marine bromophenol was isolated from the red algae, *Rhodomela larix* (Katsui et al., 1967). Since then, a number of novel bromophenols were isolated from different genera belonging to all the three phyla of marine macroalgae (Liu et al., 2011). Several kinds of bromophenols were reported from the red algal family Rhodomelaceae, especially from the genera *Rhodomela*, *Odonthalia*, and *Polysiphonia* (Fujimoto et al., 1985). Bromophenols acts as a chemical defense against their predators (Kicklighter et al., 2004). Marine algal bromophenols were reported to have several beneficial biological activities, such as antioxidant, antiviral, antibacterial, antifungal, antidiabetic, anti-inflammatory, antithrombotic, and antitumor properties (Table 12.1).

12.2.2 ALKALOIDS

Alkaloids are nitrogen containing heterocyclic compounds, which include several biological amines and halogenated cyclic nitrogen containing compounds. Halogenated cyclic alkaloids were reported explicitly for marine organisms and were absent in terrestrial counterparts (Güven et al., 2013; Pérez et al., 2016). Some of the alkaloids in plants were oxidation products of amino acids, whereas the origin of halogenated alkaloids from marine algae remains unknown. As of now, about 44 alkaloids were reported from marine macroalgae. They were dominantly reported from red algae than green algae and rarely from brown algae (Güven et al., 2013). Alkaloids in marine macroalgae were classified as phenylethylamine (PEA) derivatives, indole derivatives, halogenated indole derivatives, and 2,7-Naphthyridine derivatives. 2-phenylethylamine and indole derivatives were the most widely isolated groups of alkaloids from marine algae (Güven et al., 2010; Barbosa et al., 2014).

TABLE 12.1 The Bioactive Phenolic Compounds (Simple Phenols, Phlorotannins, and Bromophenols) Isolated from Different Marine Macroalgae

Algal Species	Phenolic Compounds	Bioactivity	References
Green Algae			
Avrainvillea rawsoni	Rawsonol	Hypocholesterolemic	Carte et al. (1989)
Cymopolia barbata	Cymobarbatol and 4-isocymbarbatol	Antimutagenic agent	Wall et al. (1989)
Red Algae			
Vertebrata lanosa	2,3-dibromo-4,5-dihydroxybenzylaldehyde; 2,2′,3-tribromo-3′,4,4′,5-tetrahydroxy-6′-hydroxymethyldiphenylmethane; bis(2,3-dibromo-4,5-dihydroxylbenzyl) ether; and 5,5″-oxybis(methylene)bis (3-bromo-4-(2′,3′-dibromo-4′,5′-dihydroxylbenzyl)benzene-1,2-diol)	Antioxidant activity	Olsen et al. (2013)
Symphyocladia latiuscula	Brominated Polyphenol: methyl 4-{(2,5-dibromo-3,4-dihydroxybenzyl) [(2,3,6-tribromo-4,5-dihydroxybenzyl) carbamoyl] amino}butanoate	Antifungal	Xu et al. (2014)
	2,5-dibromo-3,4-dihydroxy-6-(2,3,6-tribromo-4,5-dihydroxybenzyl) benzyl methyl ether	Anti-diabetic	Kurihara et al. (1999a)
	2,3,6-tribromo-4,5-dihydroxybenzyl methyl ether	Antiviral activity	Park et al. (2005)
	Bromophenols	Antioxidant activity	Duan et al. (2007)
Porphyra dentata	Catechol and rutin	Anti-inflammatory	Kazlowska et al. (2010)
Grateloupia elliptica	2,4,6-tribromophenol and 2,4-dibromophenol	Antidiabetic	Kim et al. (2008a)
Odonthalia corymbifera	Bromophenols	Antidiabetic	Kurihara et al. (1999b)
		Antifungal	Lee et al. (2007a)
Polysiphonia morrowii	Bromophenols	Antidiabetic	Kurihara et al. (1999a)
	3-bromo-4,5-dihydroxybenzyl methyl ether and 3-bromo-4,5-dihydroxybenzaldehyde	Antiviral	Kim et al. (2011)
Polyopes lancifolia	Bis (2,3-dibromo-4,5-dihydroxybenzyl) ether	Antidiabetic	Kim et al. (2010a)
Polysiphonia urceolata	3-(3-bromo-4,5-dihydroxyphenyl)-2-(3,5-dibromo-4-hydroxy-phenyl)propionic acid, (E)-4-(3-bromo-4,5-dihydroxyphenyl)-but-3-en-2-one, and (3,5-dibromo-4-hydroxyphenyl) acetic acid butyl ester	Antioxidant activity	Li et al. (2008a)

TABLE 12.1 *(Continued)*

Algal Species	Phenolic Compounds	Bioactivity	References
Vidalia obtusaloba	Bromophenols Vidalols A and B	Anti-inflammatory	Wiemer et al. (1991)
Rhodomela confervoides	3-bromo-4,5-dihydroxy benzoic acid methyl ester; bis (2,3-dibromo-4,5-dihydroxybenzyl) ether; 3,4-dibromo-5-(methoxymethyl)-1,2-benzenediol and 3-bromo-4,5-dihydroxy-benzaldehyde	Antibacterial and cytotoxic	Xu et al. (2003); Lijun et al. (2005)
		Antidiabetic	Shi et al. (2008)
Tichocarpus crinitus	Bromophenol	Antioxidant activity	Lee et al. (2007b)
Brown Algae			
Leathesia nana	Bromophenol derivatives	Antitumor effect	Shi et al. (2009)
	(+)-3-(2,3-dibromo-4,5-dihydroxy-phenyl)-4-bromo-5,6-dihydroxy-1,3-dihydroiso-benzofuran	Antithrombotic effects	Shi et al. (2010)
Ecklonia cava	Phloroglucinol, Dieckol, Phlorofucofuroeckol, Eckol, Eckstolonol, Triphlorethol A, phlorofucofuroeckol A	Neuroprotective	Myung et al. (2005); Jung et al. (2009); Kang et al. (2012)
	Phloroglucinol, 7-phloro eckol, 6,6'-bieckol, eckol, fucodiphlo-roethol, phlorofucofuroeckol A, dieckol, triphlorethol-A, eckstolonol	Antioxidant activity	Kang et al. (2005a, b, 2006, 2012); Shibata et al. (2008); Li et al. (2009)
	Fucodiphloroethol G, dieckol, 6,6′-bieckol, 7-phloroeckol and phlorofucofuroeckol A	Matrix metal-loproteinase inhibitor, peroxinitrite scavenging activity, tyroninase inhibitory effect, UV protection	Jung et al. (2006); Heo et al. (2009a) Lee et al. (2009a); Yoon et al. (2009a)
	Phrolotannins	Matrix metallopro-teinase inhibitor	Kim et al. (2006)

TABLE 12.1 *(Continued)*

Algal Species	Phenolic Compounds	Bioactivity	References
	Fucodiphloroethol G, dieckol, 6,6′-bieckol, 7-phloroeckol and phlorofucofuroeckol A	Antidiabetic activity	Lee et al. (2009b, 2010)
	8,8′-bieckol, 8,4‴-dieckol, 6,6′-bieckol	HIV-I reverse transcriptase inhibitory activity	Ahn et al. (2004); Artan et al. (2008)
		Anticancer	Kong et al. (2009)
	Dioxinodihydroeckol	Cytotoxicity	Li et al. (2011)
	Phlorethol and fucophlorethal derivatives	Antiallergic	Li et al. (2008b); Le et al. (2009)
	Fucodiphloroethol-G, phlorofucofuroeckol, dieckol, 6,6′-bieckol, phloroglucinol	Antiarthritic	Ryu et al. (2009)
	Dieckol, and phlorotannin derivatives	Antidiabetic	Kang et al. (2013)
Ecklonia maxima	Dieckol, Eckol, Dibenzo [1,4] dioxine-2,4,7,9-tetraol	Neuroprotective	Kang et al. (2012); Kannan et al. (2013)
Ecklonia radiata	Phlorotannins	Antifouling	Jennings and Steinberg (1997)
Ecklonia stolonifera	Phlorofucofuroeckol A, Eckol, 2-Phloroeckol, 7-Phloroeckol, Eckstolonol	Neuroprotective	Yoon et al. (2008)
	Eckol, Phlorofuroeckol, Dieckol, Eckstolonol	Antioxidant activity	Kang et al. (2003)
		Peroxinitrite scavenging activity	Kang et al. (2004a)
		Ach enzyme inhibitory activity	Jung et al. (2006)
		Tyrosinase inhibitory activity	Kang et al. (2004b)
		α-glucosidase inhibitory activity	Moon et al. (2011)

TABLE 12.1 *(Continued)*

Algal Species	Phenolic Compounds	Bioactivity	References
Ecklonia kurome	Phloroglucinol, Eckol, Phlorofucofuroeckol A, Dieckol	Antioxidant activity	Shibata et al. (2008)
	Crude phlorotannins	Antibacterial, hyaluronidase inhibitor	Nagayama et al. (2002); Shibata et al. (2002)
Eisenia bicyclis	Phlorofucofuroeckol A, 7-Phloroeckol, Diekol, Phloroglucinol	Neuroprotective, antioxidant activity	Shibata et al. (2008); Ahn et al. (2012)
	Phlorotannin rich extract	Antibacterial activity	Eom et al. (2012)
	Phlorofucofuroeckol-A	Anticancer	Lee et al. (2012)
	Crude phlorotannins	Hyaluronidase inhibitor	Shibata et al. (2002)
Ishige okamurae	6,6′-Bieckol; Diphlorethohydroxy camalol (DPHC)	Neuroprotective	Yoon et al. (2009b); Heo et al. (2012)
		Antidiabetic	Heo et al. (2009b)
Ishige foliacea	Octaphlorethol A	Antidiabetic	Lee et al. (2012); Kang et al. (2014)
		Antihypertensive	Ko et al. (2015)
Fucus vesiculosus	Phloroglucinol	Antithrombotic and profibrinolytic activities	Bae (2011)
	Pholorotannins	Inhibits advanced glycation end products	Liu and Gu (2012)
Hizikia fusiformis and *Cystoseira trinodis*	Phlorotannin extracts	Antioxidant activity	Siriwardhana et al. (2005); Sathya et al. (2013)
Pelvetia siliquosa and *Undaria pinnatifida*	Phloroglucinol, Eckstolonol, Eckol, Phlorofucofuroeckol A, Dieckol, Triphlorethol-A	Peroxinitrite scavenging activity and enzyme inhibitory activity	Jung et al. (2006)

TABLE 12.1 *(Continued)*

Algal Species	**Phenolic Compounds**	**Bioactivity**	**References**
Eisenia arborea	Eckol, 8,8′-bieckol, Phlorofucofuroeckol (PFF)-A and PFF-B	Anti-inflammatory	Sugiura et al. (2013)
		Antiallergenic effect	Sugiura et al. (2006, 2009)
Sargassum natans and *Ascophyllum nodosum*	Phlorotannins	Antifouling	Sieburth and Conover (1965); Hellio et al. (2001)
Fucus spiralis	Fucol and fucophlorethol	Antioxidant activity	Cérantola et al. (2006)

PEA is a decarboxylation product of phenylalanine and was found initially in terrestrial plants and later in marine algae. This group includes simple phenylamine (tyramine, hordenine) and catecholamines (dopamine) (Figure 12.2) (Güven et al., 2013; Pérez et al., 2016). These compounds are reported mainly from brown and red algae, and only a few reports are available from green algae (Percot et al., 2009a, b). Recently, 3,4-dihydroxyphenethylamine (DOPA) was found in green algae *Ulvaria obscura*, an important member of green tide blooms (Alystyne et al., 2011). PEA has been reported to exert a lot of pharmacological properties. It acted as a neuromodulator and neurotransmitter in the human brain and was shown to relieve 60% of depression in depressed patients (Barroso and Rodriguez, 1996; Güven et al., 2010; Kaleagasioglu et al., 2013).

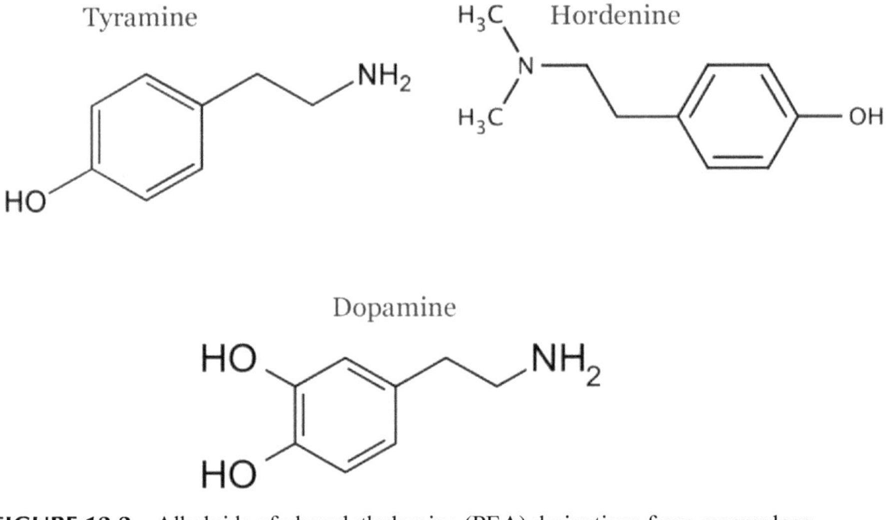

FIGURE 12.2 Alkaloids of phenylethylamine (PEA) derivatives from macroalgae.

Indole derivatives contain a benzylpyrrole, which are derived from tryptophan. They include caulerpin, caulersin, fragilamide, martensine, martefragine, denticine, almazolone, and rasmosines (Kirkup and Moore, 1983; Blunt et al., 2006; Güven et al., 2013; Kaleagasıoglu et al., 2013; Pérez et al., 2016) (Figure 12.3). Rhodophyta and Chlorophyta are the main source of indole group alkaloids with a wide spectrum of pharmacological properties, viz. antioxidant, antibacterial, antifungal, antiviral, antispasmodic, antitumor, anti-inflammatory, and neuroprotective properties (Table 12.2).

Caulerpin

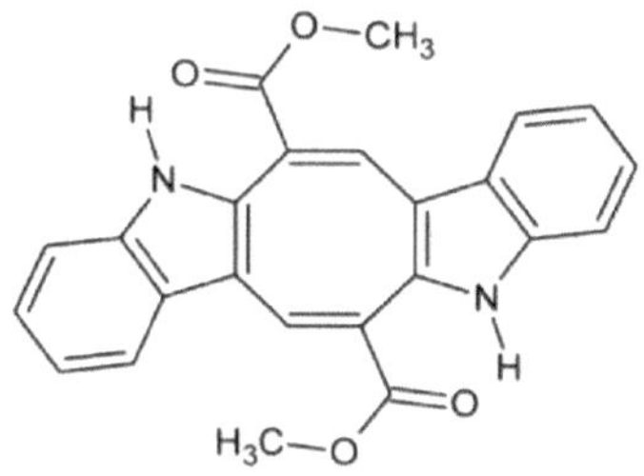

Caulersin

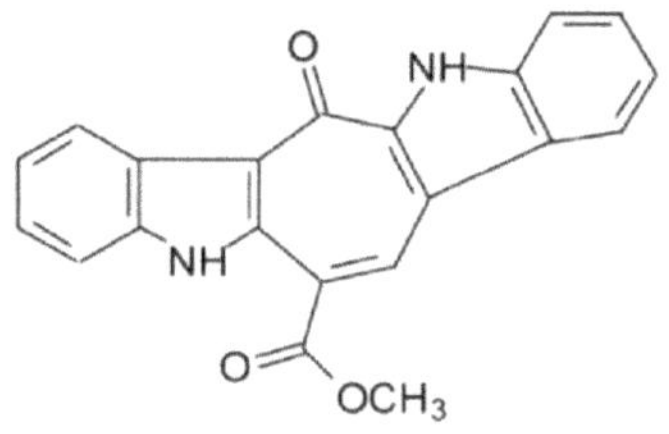

Fragilamide

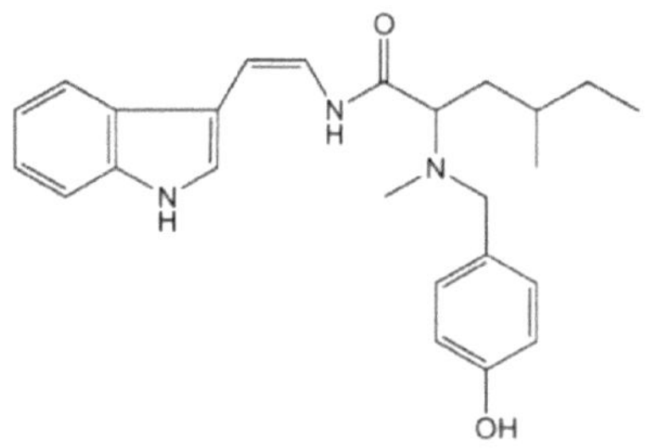

Martefragine

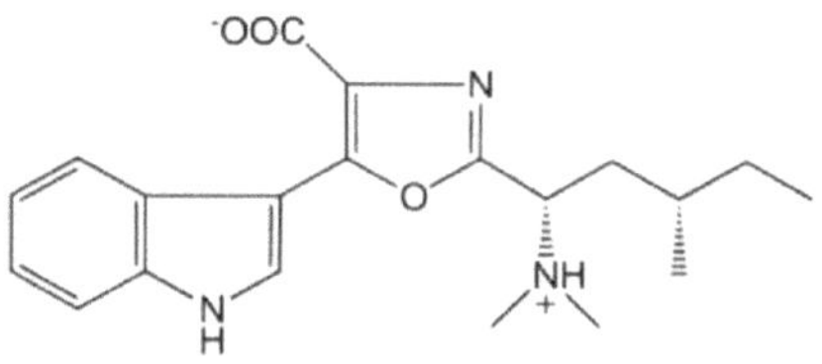

Martensin A

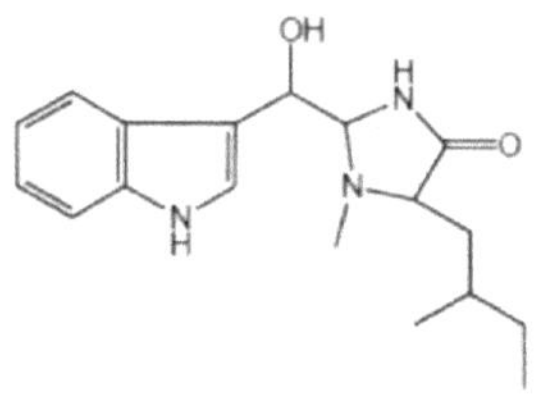

Racemosine B

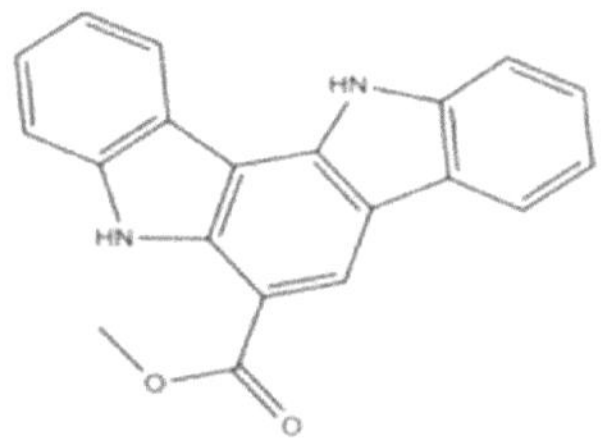

E-almazolone

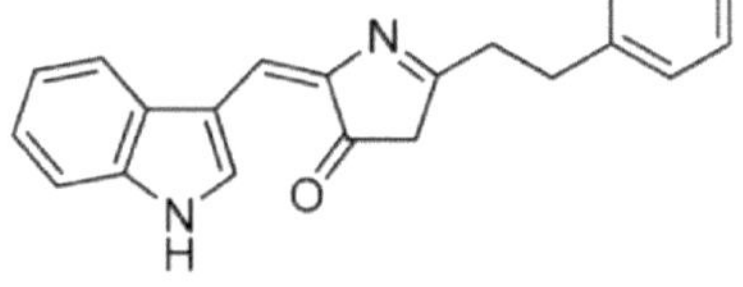

FIGURE 12.3 Some of the indol alkaloids isolated from macroalgae.

TABLE 12.2 Bioactive Alkaloids Isolated from Different Seaweeds

Algal Species	Alkaloids	Bioactivity	References
		Green Algae	
Caulerpa racemosa	Caulerpine	Antitumor activity	Ayyad and Badria (1994)
		Antioxidant activity	Takamatsu et al. (2003)
		Cytotoxic	Rocha et al. (2007); Liu et al. (2009a)
		Antinociceptive and anti-inflammatory	de Souza et al. (2009); Éverton et al. (2009); Jaswir and Monsur (2011)
		Anticorrosive activity	Kamal and Sethuraman (2012)
		Antiviral	Macedo et al. (2012)
		Antituberculosis activity	Chay et al. (2014)
		Insecticidal activity	Alarif et al. (2010)
	Caulerpinic acid	Antifungal	Liu et al. (2012)
	Rasmosin A and Rasmosin B	Neuroprotective	Liu et al. (2013)
	Caulerchlorin, Caulerpine	Spasmolytic	Cavalcante-Silva et al. (2013)
Caulerpa lentilifera, *Caulerpa microphysa*, *Caulerpa sertularoides*	Caulerpine, phytol	Antibacterial	Vairappan (2004)
		Red Algae	
Martensia fragilis	Martefragin A	Antioxidant activity	Takahashi et al. (1997); Nishida et al. (1998)
	Fragilamide	Antioxidant activity	Takahashi et al. (1998)
	Martensine A	Antibacterial activity	Kirkup and Moore (1983); Guven et al. (2010)
Martensia denticulata	Denticins (DTC)	Antiphotooxidative activity	Murakami et al. (1994)

TABLE 12.2 *(Continued)*

Algal Species	Alkaloids	Bioactivity	References
Laurencia brongniartii	Tetrabromoindole, brominated indoles	Antibacterial activity	Dembitsky and Tolstikov (2003); Carter et al. (1978)
Rhodophyllis membranacea	Tetrahalogenated indoles, bromochloroiodoindoles, Polyhalogenated indoles	Cytotoxic and antifungal activities	Brennan and Erickson (1978); Guven et al. (2010); Woolner et al. (2016)
Lophocladia sp.	Lophocladine A	δ-opioid receptor antagonist	Gross et al. (2006)
	Lophocladine B	Cytotoxic	–
Laurencia majuscula	Caulerpin I	Antifungal	Xu et al. (2006)

Marine organisms are the only reported source of halogenated indole alkaloids. Red algae are the primary source of these alkaloids wherein an indole group substituted by bromine and chlorine atoms (Güven et al., 2010). Also, there were reports of isolation of sulfur containing bromoalkaloids from red algae (Tanaka et al., 1989). Halogenated indole alkaloids were known for its antibacterial and antifungal activity (Dembitsky and Tolstikov, 2003; Woolner et al., 2016). 2,7-Naphthyridine derivatives have two members, viz., Lophocladines A and B which were isolated from red algae *Lophocladia* sp. (Figure 12.4) (Gross et al., 2006) and were reported to have cytotoxicity against cancer cells (Gross et al., 2006; Güven et al., 2010) (Table 12.2).

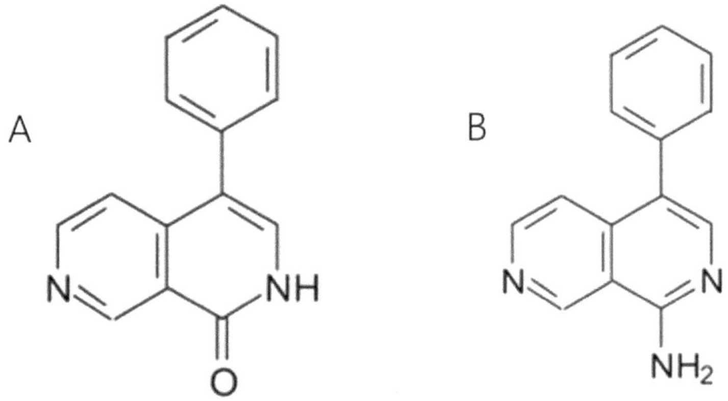

FIGURE 12.4 Lophocladines A and B isolated from red algae *Lophocladia* sp.

12.2.3 TERPENES AND TERPENOIDS

Terpenes are other important classes of secondary metabolites produced by marine macroalgae (Grosso et al., 2011). Terpenes existed in diverse chemical forms either as linear functionalized hydrocarbons or as chiral, carbocyclic skeletons with various chemical modifications, such as hydroxyl, carbonyl, ketone, aldehyde, and peroxide groups (Ajikumar et al., 2008). These functionally modified terpenes were called terpenoids or isoprenoids. These were derived from the five-carbon precursor isopentenyl pyrophosphate (IPP) via two major pathways, namely, mevalonate (MVA) and 1-deoxyxylulose 5-phosphate/2-C-methylerithrytol 4-phosphate (DOXP/MEP) (Brahmkshatriya and Brahmkshatriya, 2013; Barbosa et al., 2014). Terpenes could be categorized into hemiterpenes (C5), monoterpenes (C10), sesquiterpenes

(C15), diterpenes (C20), sesterterpenes (C25), triterpenes (C30) and polyterpenes (>C30) (Gross and König, 2006; de las Heras and Hortelano, 2009). Several terpenes and terpenoids isolated from macroalgae were shown to have anticancer, anti-inflammatory, antimicrobial, anthelmintic, neuroprotective, and gastroprotective properties (Table 12.3).

12.2.4 PIGMENTS

The light-harvesting process of all macroalgae involves chlorophylls (all groups possess chlorophyll α) and various groups of accessory pigments, which are specific to particular algal groups (Table 12.4) (Falkowski and Raven, 2007). Of the different pigment classes, chlorophylls, carotenoids, and phycobiliproteins (PBP) are the important ones.

Chlorophylls have a porphyrin tetrapyrrole esterified to a diterpene alcohol, phytol (Figure 12.5) (Ferruzzi and Blakeslee, 2007). Marine algae are found to contain four kinds of chlorophylls. The most important one is chlorophyll α, which is an essential molecule for photosynthesis. The other chlorophylls are chlorophylls b and c and the fourth type of chlorophyll discovered is chlorophyll d, which is found in red algae (Larkum and Kühl, 2005). When exposed to extreme temperature and pH conditions chlorophylls form several derivatives, such as pheophytins, pyropheophytins, and pheophorbides (Pangestuti and Kim, 2011). Chlorophyll derivatives have more potent bioactivity than chlorophyll itself (Cahyana et al., 1993). Chlorophyll α and its derivatives like pheophytins (Figure 12.5) have a number of biological properties, such as antioxidant, anti-inflammatory, antimutagenic, and neuroprotective properties (Table 12.5).

Carotenoids, the most widespread pigments in nature, are accessory pigments as they augment the light-harvesting properties of algae by passing on light excitation to chlorophyll (Larkum and Kühl, 2005). It can be classified into carotenes and xanthophylls, depending upon their polarity. The major carotenoids that occur in seaweeds include β-carotene, lutein, violaxanthin, zeaxanthin, and fucoxanthin (Table 12.5) (Takaichi, 2011). Among these, fucoxanthin (Figure 12.6) has attracted the interest of researchers due to its potential pharmacological properties (Zhang et al., 2015). Fucoxanthin, a non-provitamin A carotenoid, accounts for >10% of the total biogenic carotenoids in brown seaweeds (Moghadamtousi et al., 2014). Fucoxanthin has been shown to have a number of biological properties, such as antioxidant, antimalarial, anticancer, antiobesity, anti-inflammatory, antiangiogenic, and antidiabetic properties (Table 12.5).

TABLE 12.3 Bioactive Terpenes and Terpenoids Isolated from Different Marine Macroalgae

Algal Species	Triterpenes	Bioactivity	References
		Red Algae	
Laurencia snackeyi	Aplysistatin, a brominated sesquiterpene	Anti-inflammatory	Vairappan et al. (2013)
Portieria hornemannii	Halomin	Cytotoxic against renal, brain, colon, and lung cancer cell lines	Fuller et al. (1994)
Sphaerococcus coronopifolius	Bromoditerpene, sphaerococcenol A	Antimalarial activity	Etahiri et al. (2001)
	Bromosphaerone and 12S-hydroxybromosphaerodiol	Antibacterial	Cafieri et al. (1987); Etahiri et al. (2001)
Laurencia claviformis	Pacifenol	Antimicrobial	Dembitsky et al. (2004)
Laurencia dendroidea	Debromoelatol, Obtusane	Anti-inflammatory	Machado et al. (2014)
	Dendroidiol, Debromoelatol, Obtusane	Antimycobacterial	Machado et al. (2014)
	Elatol	Antileishmanial	Santos et al. (2010)
	Elatol	Antiparasitic-antitrypanosomal	Veiga-Santos et al. (2010)
Laurencia okamurai	Laurane-, cyclolaurane-, and cuparane-type sesquiterpenes	Antibacterial activity/Brine shrimp lethality	Liang et al. (2014)
	Laurinterol	Anticarcinoma	Kim et al. (2008b)
	Seco-laurane sesquiterpene and related laurane derivatives	Antifungal and cytotoxic	Yu et al. (2014)
Laurencia obtusa	Elatol-halogenated sesquiterpene	Antiherbivory and antifouling property	Pereira et al. (2003)
	Brominated sesquiterpenes	Antimalarial	Topcu et al. (2003)
	Kahukuen-10-ol, Laur-2-ene-3,12-diol and cuparene-3,12-diol	Anticancer property/cytotoxic activity	Angawi et al. (2014)
Laurencia venusta	Triterpene tetracyclic ether	Antiviral	Sakemi et al. (1986)

TABLE 12.3 *(Continued)*

Algal Species	Triterpenes	Bioactivity	References
Laurencia sp.	Elatol and iso-obtusol	Antibacterial activity	Vairappan et al. (2001)
	Phenolic sesquiterpenes and a polybrominated indole, laurinterol, and allolaurinterol	Antibacterial activity	Vairappan et al. (2004)
Laurencia glandulifera	Neorogioltriol-diterpene	Anti-inflammatory, analgesic properties	Chatter et al. (2011)
Laurencia obtusa and *Laurencia microcladia*	Sesquiterpenes	Anticarcinoma	Kladi et al. (2006)
Laurencia scorparia	Sesquiterpenes	Anthelmintic activity	Davyt et al. (2001)
Plocamium telfairiae	Aplysiaterpenoid A and telfairine	Insecticidal activity	Watanabe et al. (1989)
Laurencia catarinensis	Halogenated metabolites	Anticancer/cytotoxic	Lhullier et al. (2010)
Brown Algae			
Sargassum macrocarpum	Sargachromenol-plastoquinone	Neuroprotective effect/antioxidant	Tsang and Kamei (2004); Tsang et al. (2005)
	Sargaquinoic acid-Quinonic compound	Neuroprotective effect	Kamei and Tsang (2003)
Sargassum sagamianum	Sargaquinoic acid and Sargachomenol	Anticholinesterase activity/Alzheimer's disease therapy	Choi et al. (2007)
	Farnesylacetone derivatives	Cholinesterase inhibitory activity	Ryu et al. (2003
Stypopodium flabelliforme	Epitaondiol	Antiinflammatory/cardioprotective effect	Gil et al. (1995); Martinez et al. (1997)
		Gastroprotective effect	Areche et al. (2011)
		Antimicrobial activity	Rovirosa et al. (1997)
		Antiproliferative activity	Pereira et al. (2011)
		Antiviral activity	Soares et al. (2007)
	2-beta-3-alpha-epitaondiol	Cytotoxic activity	Al Sabi et al. (2006)

TABLE 12.3 *(Continued)*

Algal Species	Triterpenes	Bioactivity	References
Stypopodium zonale	Metabolites of atomaric acid	Cytotoxic activity against human lung and colon carcinoma	Dorta et al. (2002)
	Meroditerpenoids	Anti-HMPV activity	Mendes et al. (2011)
Dictyota dichotoma	Crenulacetal C – a diterpene	Inhibits Annalids – *Polydora websterii*, a harmful lugworm damaging pearl cultivation	Takikawa et al. (1998)
Cystoseira brachycarpa var. *balearica*	Biterpenes eleganolone and elegandiol	Enzyme inhibitors – acetylcholine and histamine	Pieta et al. (1993, 1995)
Green Algae			
Caulerpa taxifolia	Caulerpenyne	Toxic to Sea urchin eggs	Pesando et al. (1996)
Caulerpa sp.	Caulerpenyne – a sesquiterpene	Neuroprotective/Alzheimer's disease therapy	Cengiz et al. (2011)
Ulva fasciata	Guaiane sesquiterpenes, labdane diterpenoids	Antibacterial	Chakraborty et al. (2010a, b)

TABLE 12.4 Distribution of Main Photosynthetic Pigments, Auxiliary Pigments and Minor Pigments Associated with Three Major Classes of Macroalgae

Algal Group	**Main Pigment**	**Main Accessory Pigments**	**Minor Pigments/Pigments with Limited Distribution Within the Algal Groups**
Chlorophyceae	Chlorophyll α and b	α-, β- and γ-carotenes, lutein, siphonaxanthin, siphonein	Antheraxanthin, astaxanthin, canthaxanthin, echinone, neoxanthin, violaxanthin, zeaxanthin (Stengel et al., 2011)
Phaeophyceae	Chlorophyll α, c_1, c_2	β-carotene, fucoxanthin, zeaxanthin, violaxanthin	Antheraxanthin-like, cryptoxanthin-like, cryptoxanthin-5,6-epoxide-like, latoxanthin like, and mactraxanthin-like carotenoids, neoxanthin (Haugan and Liaaen-Jensen, 1994; Stengel et al., 2011)
Rhodophyceae	Chlorophyll α, (d)	r-Phycocyanin, b-, c-, and r-phycoerythrin, allophycocyanin, α and β-carotene	Antheraxanthin, cryptoxanthin, lutein, violaxanthin, zeaxanthin (Schubert et al., 2006)

Source: Haugan and Liaaen-Jensen (1994); Schubert et al. (2006); South and Whittick (1996); Falkowski and Raven (2007); Stengel et al. (2011).

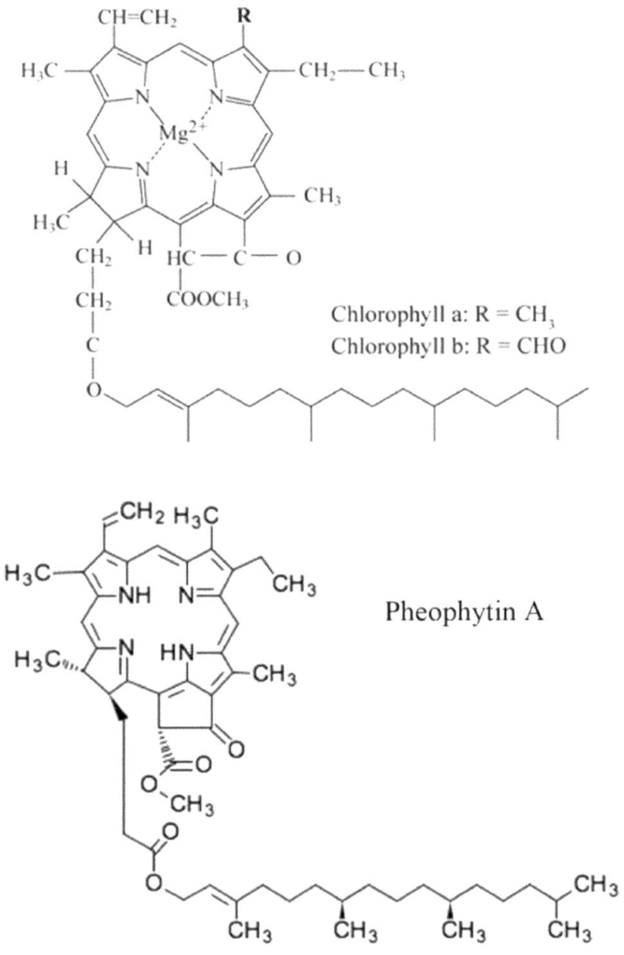

FIGURE 12.5 Chemical structure of chlorophyll α, chlorophyll β, and pheophytin A (chlorophyll α without Mg atom).

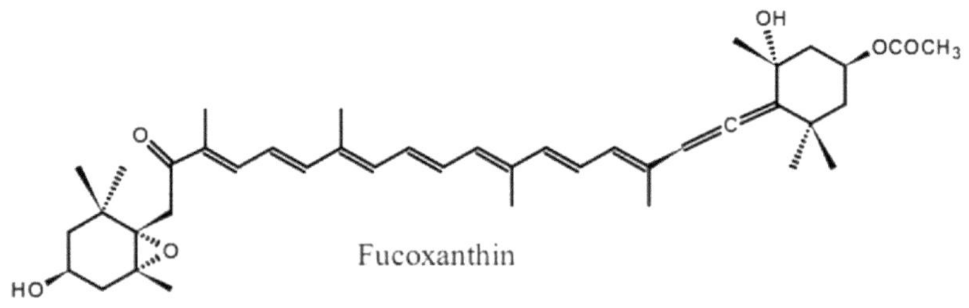

FIGURE 12.6 Structure of fucoxanthin isolated from brown seaweeds.

TABLE 12.5 The Bioactive Pigments Isolated from Different Marine Macroalgae

Algal Species	Pigment	Bioactivity	References
		Green Algae	
Enteromorpha prolifera	Pheophytin A	Anti-inflammatory	Okai and Hiqashi-Okai (1997)
		Antimutagenic	Okai and Hiqashi Okai (1997); Hiqashi-Okai et al. (1999)
		Antioxidant activity	Cho et al. (2011)
Codium fragile	Siphonaxanthin and siphonein	Anticancer	Ganesan et al. (2011)
	Siphonaxanthin	Antiangiogenic	Ganesan et al. (2010)
		Red Algae	
Odonthalia corymbifera	Pheophytin A	Potent lipoxigenase (LOX) inhibitor	Kurihara et al. (2014)
Porphyra tenera	β-carotene, chlorophyll α and lutein	Antimutagenic activity	Okai et al. (1996)
		Brown Algae	
Sargassum fulvellum	Pheophytin A	Neuroprotective	Ina and Kamei (2006); Ina et al. (2007)
Sargassum siliquastrum	Fucoxanthin	UV protection	Heo and Jeon (2009)
		Cytoprotective effect	Heo et al. (2008)
Sargassum heterophyllum	Fucoxanthin	Antimalarial	Afolayan et al. (2008)
Eisenia bicyclis	Pyropheophytin A	Antioxidative effect	Cahyana et al. (1992)
Ishige okamurae	Fucoxanthin	Induce apoptosis in human leukemia HL-60 cells	Kim et al. (2010b)
		Anti-inflammatory	Kim et al. (2010c)
Myagropsis myagroides	Fucoxanthin	Anti-inflammatory	Heo et al. (2010)

TABLE 12.5 *(Continued)*

Algal Species	Pigment	Bioactivity	References
Undaria pinnatifida	Fucoxanthin	Antiobesity effect	Maeda et al. (2005)
		Anticancer effect	Hosokawa et al. (2004); Kotake-Nara et al. (2005); Liu et al. (2009b); Nakazawa et al. (2009); Ganesan et al. (2011)
		Antiobesity	Woo et al. (2009)
		Antiangiogenic	Sugawara et al. (2006)
		Antiobesity and antidiabetic	Maeda et al. (2009)
		Neuroprotective	Ikeda et al. (2003)
Hijikia fusiformis	Fucoxanthin	Antioxidant, neuroprotective	Nomura et al. (1997); Yan et al. (1999)
Laminaria japonica	Fucoxanthin	Anticancer	Das et al. (2008); Zhang et al. (2008)
		Prevent osteoporosis	Das et al. (2010)
		Photoprotective	Shimoda et al. (2010)
Halocynthia roretzi	Halocynthiaxanthin and fucoxanthinol	Anticancer	Konishi et al. (2006)
Fucus serratus	Fucoxanthin	Antioxidant	Sasaki et al. (2008)

12.2.5 PROTEINS, NONRIBOSOMAL PEPTIDES, AND OTHER AMINO ACIDS

12.2.5.1 PROTEINS

In macroalgae, PBP and lectins are the two major categories of macroproteins with immense bioactivity. PBP are water-soluble, colored, and highly fluorescent compounds consisting of a protein backbone to which prosthetic billin chromophores (tetrapyrrolic open chain) are covalently bound (Harnedy and FitzGerald, 2011; Ibañez et al., 2012). These include phycocyanins, allophycocyanins (APCs), and phycoerythrins. Phycoerythrins were most abundant PBP in Rhodophyta (Pangestuti and Kim, 2011; Barbosa et al., 2014). PBPs made up to 1–10% of the dry weight of algal biomass (Lordan et al., 2011; Barbosa et al., 2014). Variations in nitrogen availability, light irradiation, and time of sampling have been shown to affect the ratio of each PBP produced (Martínez and Rico, 2008; Harnedy and FitzGerald, 2011). Species variation of PBPs were also reported, for example, *P. palmata* was reported to contain 12% phycoerythrin (w/w) per dry weight, while *Gracilaria tikuahiae* contained only 0.5% w/w per dry weight (Chronakis, 2000; Fleurence, 2004; Harnedy and FitzGerald, 2011, 2013). PBPs are commonly used in fluorescence-based biotechnological assays as these pigments can fluoresce *in vitro* and *in vivo* (Aneiros and Garateix, 2004; Fleurence, 2004; Sekar and Chandramohan, 2008; Harnedy and Fitzgerald, 2013). PBPs are also widely used as natural colorants in food (i.e., chewing gum and dairy products) and cosmetic products (i.e., lipsticks and eyeliners) (Sekar and Chandramohan, 2008). Pharmacological properties, such as antioxidant, anti-inflammatory, antidiabetic, neuroprotective, hypocholesterolemic, hepatoprotective, antiviral, antitumor, liver-protecting, atherosclerosis treatment, serum-lipid-reducing, and lipase-inhibition activities have been reported for PBPs (Sekar and Chandramohan, 2008; Soni et al., 2009; Yabuta et al., 2010; Harnedy and FitzGerald, 2011).

12.2.5.2 LECTINS

Lectins are nonimmune carbohydrate-binding proteins, found in both terrestrial and marine organisms (Hamid et al., 2013; Barbosa et al., 2014). Lectins interact with specific glycan structures in cell membranes and were involved in several biological processes such as host-pathogen interactions, cell-cell communication, cancer metastasis, and induction of apoptosis (Conde et al., 2013; De Schutter and Van Damme, 2015). Study on biochemical and structural properties of macroalgal lectins are limited, and they were sequentially different from plant lectins (Calvete et al., 2000; Hori et al., 2000; Cardozo et al., 2007; Harnedy and FitzGerald, 2011). Macroalgal lectins are low-molecular-weight monomeric thermostable proteins with affinity for oligosaccharides or glycoproteins (Barbosa et al., 2014; Praseptiangga, 2016). They did not require divalent cations for structural integrity or biological activity (Praseptiangga, 2016). Hemagglutinating activity against erythrocytes was the main biological activity associated with lectins, and hence they were often referred to as "hemagglutinins" or "agglutinins" (Sharon and Lis, 2004; Harnedy and FitzGerald, 2011). In addition to this, macroalgal lectins exhibited a number of biological properties, such as mitogenic, antimicrobial, anti-HIV, anti-inflammatory, and antinociceptive properties (Table 12.6). Griffithsin, a lectin isolated from *Griffithsia* sp. is a broad-spectrum antiviral agent which is in clinical trials against HIV virus (Lee, 2019). It is of the most potent antiviral discovered to date with broad spectrum of antiviral activity. It was reported to have antiviral activity against a number of enveloped viruses, such as SARS-CoV (O'Keefe et al., 2010), MERS-CoV (Millet et al., 2016), HCV (Takebe et al., 2013), HSV (Nixon et al., 2013), JEV (Ishag et al., 2016), and PEDV (Li et al., 2019). Activity stemmed from its ability to bind terminal mannoses present in high-mannose oligosaccharides and crosslink these glycans on the surface of the viral envelope glycoproteins. Even human papillomavirus (HPV), a non-enveloped virus, was inactivated by GRFT via a glycosylation-independent mechanism (Lee, 2019).

12.2.5.3 PEPTIDES

Major classes of bioactive peptides reported from macroalgae are linear peptides, cyclic peptides, and depsipeptides (Table 12.6). Dipeptides, such as carnosine and glutathione (GSH), with antioxidant property, have been isolated from red seaweed *Ancanthophora delilei* and *Ulva faciata* (Fleurence, 2004; Shiu and Lee, 2005). A tridecapeptide, IRLIIVLMPILMA isolated from the *Palmaria palmata* and dipeptide Phe-Tyr isolated from *Undaria pinnatifida* are the most promising antihypertensive compounds identified from seaweed so far (Rosa et al., 2020). N'Diaye et al. (1996) reported the isolation of a novel antibacterial peptide, Almazole D, from red algae of the Senegal coast. The other peptides reported having antiproliferative activity against cancer cells

TABLE 12.6 Bioactive Proteins/Peptides/Amino Acids Isolated from Marine Macroalgae

Algal Species	Protein/Peptides/Amino Acids	Bioactivity	References
Green Algae			
Codium fragile	Lectins-carbohydrate binding glycoproteins	Agglutinins	Smit (2004)
Bryopsis	Cyclic depsipeptides Kahalalide A and F	Antibacterial, Anti TB, Anti-HIV	El Sayed et al. (2000)
	Kahalalide F	Anticancer property	Faircloth et al. (2000); Sparidans et al. (2001); Suarez et al. (2003)
Ulva spp.	Hexapeptide, SECMA 1, with the sequence EDRLKP	Mitogenic activity	Ennamany et al. (1998)
Red Algae			
Palmaria palmata	Peptide – IRLIIVLMPILMA	Antihypertensive-renin inhibition	Fitzgerald et al., 2012
	MAAs – Usujirene, Palythine, and asterine-330	Antiproliferative and anti-inflammatory	Athukorale et al. (2016)
	MAAs – Usujirene	UV protection	Sekikawa et al. (1986)
Porphyra rosengurttii	Mycosporine-like amino acids	Antioxidant activity	De la Coba et al. (2009)
Porphyra yezoensis	Peptide AKYSU	ACE inhibitory and Antihypertensive	Suetsuna (1998); Saito and Hagino (2005)
Gracilaria cornea	MAAs-porphyra-334 and shinorine	UV protection	Sinha et al. (2000)
Porphyra columbina, Grateloupia doryphora, Mazzaella laminarioides	MAAs-shinorine	UV protection	Huovinen et al. (2004)
Gelidium corneum, Ahnfeltiopsis devoniensis	MAAs	Antioxidant activity	De la Coba et al. (2009)
Griffithsia sp.	Lectins-Griffithsin	Anti-HIV activity	Mori et al. (2005); Ziółkowska and Wlodawer (2006)
Hypnea japonica	Lectins-Hypnins A-D	Agglutinins	Hori et al. (1986)

TABLE 12.6 *(Continued)*

Algal Species	Protein/Peptides/Amino Acids	Bioactivity	References
Soleria robusta	Isoagglutinin – a glycoproteins	Mitogenic and cytotoxic	Hori et al. (1988)
Eucheuma serra, Eucheuma amakusaensis, and *Eucheuma cottonii*	Lectins/agglutinins	Mitogenic activity	Kawakubo et al. (1997, 1999); Sugahara et al. (2001)
Eucheuma serra	ESA-2-agglutinin	Anticancer activity	Hori et al. (2007)
Galaxaura marginata	Lectins	Antibacterial	Liao et al. (2003)
Galaxaura filamentous	Cyclic pentapeptide, Galaxamide (1)	Antiproliferative activity	Xu et al. (2008)
Amansia multifida	Lectin	Antinociceptive properties	Neves et al. (2007)
Hypnea cervicornis	Mucin-binding agglutinin	Antinociceptive and anti-inflammatory activity	Bitencourt et al. (2008)
Bryothamnion triquetrum and *Bryothamnion seaforthii*	Lectins	Prevention of dental plaque formation	Teixeira et al. (2007)
Porphyra umbilicalis	MAAs-Porphyra335 and shinorine	UV protection-sun screen lotion	Schmid et al. (2006); Cardozo et al. (2007)
Ancanthophora delilei	Carnosine (β-alanyl-L-histidine), a histidyl dipeptide	Antioxidant activity	Fleurence (2004)
Chondrus crispes, Mastocarpus stellatus	MAAs-Usujirene, Palythine, and asterine-330	Antiproliferative and anti-inflammatory	Athukorale et al. (2016)
Delesseriacean seaweed	Almazole D, a novel dipeptide derivative	Antibacterial activity	N'Diaye et al. (1996)
Brown Algae			
U. pinnatifida	Dipeptides and tetra peptides	Antihypertensive	Suetsuna and Nakano (2000); Sato et al. (2002); Suetsuna et al. (2004)

isolated from macroalgae include pentapeptide, galaxamide (1) from *Galaxaura filamentosa* (Xu et al., 2008), and hexapeptide SECMA1 from *Ulva* sp. (Ennamany et al., 1998). Cyclic depsipeptides, Kahalalide A and F, isolated from *Bryopsis* sp. (Hamann and Scheuer, 1993; Hamann et al., 1996; Suárez et al., 2003) were reported to have antimicrobial and anticancer properties (Sparidans et al., 2001; Suárez et al., 2003). Morse (1991) isolated different oligopeptides, which acted like analogs of the neurotransmitter γ-aminobutyric acid (GABA), from red algae and were useful in the diagnosis and therapy of brain and central nervous system disorders (Harnedy and FitzGerald, 2011; Lazcano-Pérez et al., 2012). Some macroalgal derived peptides are sold in the Japanese market with antihypertensive claims, for example, Nori peptide AKYSY is sold under the trade name Peptide Nori S (Harnedy and Fitzgerald, 2013; Holdt and Kraan, 2011).

12.2.5.4 FREE AMINO ACIDS

The bioactive free amino acids found in macroalgae include taurine, laminine, kainoids, and mycosporine-like amino acids (Harnedy and Fitzgerald, 2013).

12.2.5.4.1 Taurine

Taurine is a naturally occurring sulfonic β-amino acid derived from cysteine and methionine (Lourenço and Camilo, 2002). As it was not utilized in protein synthesis, taurine was found free or in simple peptides (Houston, 2005). Macroalgal species were reported to contain low concentrations of taurine (Dawczynski et al., 2007; Harnedy and Fitzgerald, 2013). Among the three algal phyla, red algae contained higher levels of taurine (Kataoka and Ohnishi, 1986; Dawczynski et al., 2007; Harnedy and Fitzgerald, 2013). Though, taurine was not utilized in protein synthesis, they were regarded as a conditionally essential amino acid (EAA) (Lourenço and Camilo, 2002; Harnedy and Fitzgerald, 2013). Taurine is essential for vital functions, such as bile-acid conjugation, retinal, and neurological development, osmoregulation, modulation of cellular calcium level, and immune function (Lourenço and Camilo, 2002; Harnedy and Fitzgerald, 2013). Several authors have reported that taurine has health-promoting properties such as antihypertensive, hypocholesterolemic, and antioxidant activities (Lourenço and Camilo, 2002; Militante and Lombardin, 2002; Zhang et al., 2004; Houston, 2005).

12.2.5.4.2 Laminine

Laminine (trimethyl (5-amino-5-carboxy pentyl) ammonium hydroxide) is a choline-like basic amino acid widely distributed in the brown algae of the Laminariaceae family (Takemoto et al., 1964; Girard et al., 1988; Holdt and Kraan, 2011). Laminine isolated from *Laminaria augustata* was shown to have hypotensive action at high doses through the ganglion-blocking effect (Girard et al., 1988) (Table 12.6).

12.2.5.4.3 Kainoid Amino Acids

Kainoid amino acids are a group of unusual amino acids that are structurally and functionally related to aspartic and glutamic acids (Holdt and Kraan, 2011). Kainic and domoic acids known for insecticidal, anthelmintic, and neuroexcitatory properties (Parsons, 1996) were widely reported from different algal species (Smit, 2004; Harnedy and FitzGerald, 2011). Since long the extracts from *Digenea simplex* and *Chondria armata* containing domoic acid and kainic acid have been used in the treatment of ascariasis in Japan (Parsons, 1996; Smit, 2004; Bhakuni and Rawat, 2005). These compounds were used in the study of neurophysiological disorders, such as Parkinson's disease, epilepsy, and Alzheimer's disease (AD) (Smit, 2004). Kainoids were difficult to synthesize due to intricate molecular structure and were extracted from macroalgae as and when required (Smit, 2004; Harnedy and FitzGerald, 2011). "Digenin" and "Helminal" are some of the commercially available preparations of kainic acid sold in the market (Holdt and Kraan, 2011).

12.2.5.4.4 Mycosporine-Like Amino Acids

Mycosporine-like amino acids (MAAs) are low-molecular-weight (<400 Da) water-soluble molecules, composed of either an aminocyclohexenone or an aminocycloheximine ring conjugated with one or more amino acids residue (Harnedy and FitzGerald, 2011). These are produced in the aquatic organism to protect them against UV radiation (Oren and Gunde-Cimerman, 2007; Harnedy and FirzGerald, 2011). The primary MAA, mycosporine–glycine, synthesized by the shikimate pathway is converted to secondary MAAs through chemical and biochemical conversions (Shick and Dunlap, 2002;

Singh et al., 2008) and hence different MAAs are reported in a wide variety of marine macroalgae (Oren and Gunde-Cimerman, 2007). These include mycosporineglycine, shinorine, porphyra-334, palythine, asterina-330, palythinol, and palythene (Figure 12.7).

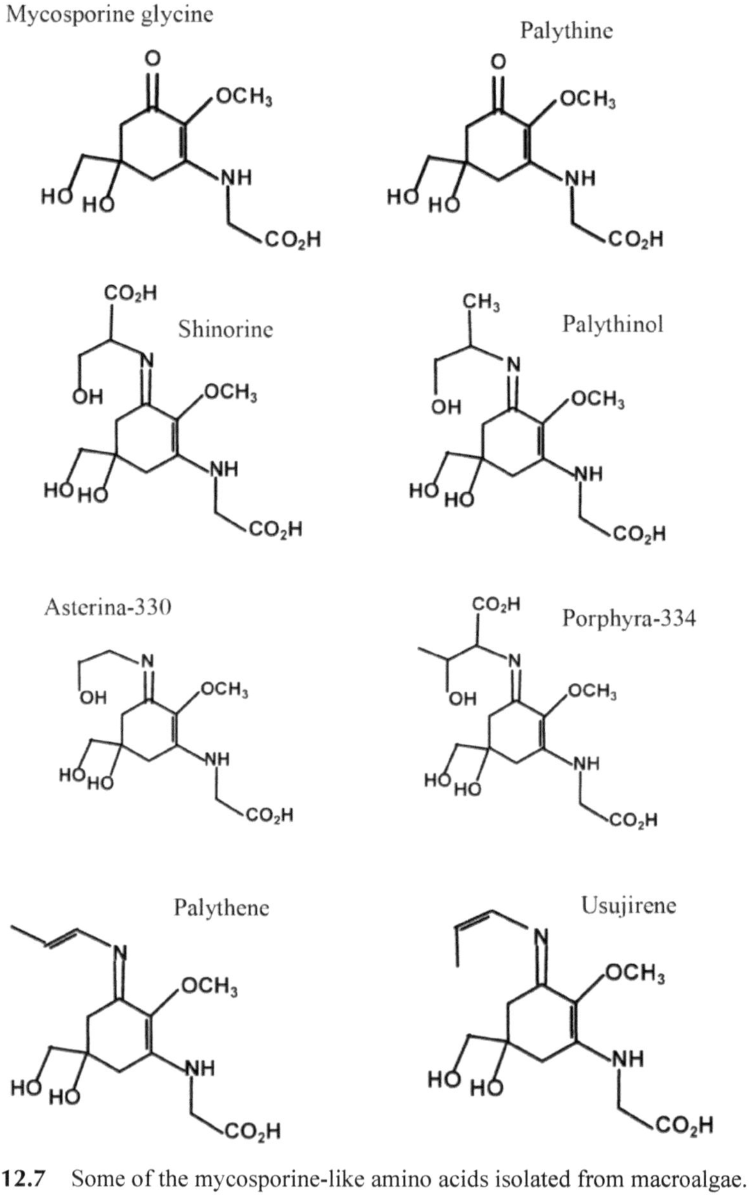

FIGURE 12.7 Some of the mycosporine-like amino acids isolated from macroalgae.

Varied levels of MAAs were reported from different classes of macroalgae (Hoyer et al., 2001; Singh et al., 2008). Rhodophyceae are the primary producers of MAAs followed by Phaeophyceae and some species of Chlorophyceae (Karsten et al., 1999, 2005; Kräbs et al., 2002; Carreto et al., 2005; Singh et al., 2008; Yuan et al., 2009; Cardozo et al., 2011). The depth and latitude of collection site also played an essential role in the levels of MAAs from macroalgae (Karsten et al., 1998; Hoyer et al., 2001; Aguilera et al., 2002; Harnedy and FirzGerald, 2011).

Red algae could be classified into three groups based on MAAs synthesis (Hoyer et al., 2001; Singh et al., 2008). The first group lives in sublittoral zones and completely lacks MAAs due to poor light penetration. In the second group, some algae living in supralittoral zones with limited exposure of UV, have varying quantities of MAAs, and in the third category, most of the algae living in supralittoral zones with higher doses of UV exposure are reported to have the highest MAAs. MAAs content in macroalgae was directly proportional to the solar radiation, and hence macroalgae with high MAAs content were distributed in lower, high solar latitudes and vice versa (Karsten and West, 2000). Shinorine and porphyra-334 are the most common MAAs reported being isolated from macroalgae of all latitudes (Banaszak and Lesser, 1995; Karsten et al., 1998; Sinha et al., 2007; Barbosa et al., 2014). However, structurally more complex MAAs were isolated from some intertidal red macroalgae of Antarctic waters which includes palythine, asterina-330, palythinol, palythene, usujirene, and the unusual M335/360 (Kräbs et al., 2002; Carreto et al., 2005). The MAAs isolated from different macroalgae were shown to have UV protection, antioxidant, antiproliferative, and anti-inflammatory activities (Table 12.6).

12.2.6 SULFATED POLYSACCHARIDES

Marine macroalgae contain a variety of polysaccharides for both structural and protective functions. Cell-wall structural polysaccharides, such as alginates (brown algae), carrageenan, and agar (red algae), are highly exploited by the hydrocolloid industry (de Jesus Raposo et al., 2015). Other minor polysaccharides of interest that could be found as part of cell walls with important functional properties are sulfated polysaccharides, such as fucoidans (brown algae), galactans (certain red and green algae), agaroids, such as porphyran (red algae) and ulvans (green algae) (Wijesekara et al., 2011; de Jesus Raposo et al., 2015).

12.2.6.1 FUCOIDANS

Fucoidans are fucose-rich sulfated polysaccharides present in brown macroalgae (Figure 12.8). These polysaccharides differ in their structure, sugar backbone composition and sulfation patterns according to the species of origin and extraction techniques used (Barbosa et al., 2014; Fitton et al., 2015). Fucoidans are made of sulfated α-L-fucose residues, but may also contain mannose, xylose, galactose, acetyl groups and uronic acids (Jhamandas et al., 2005; Li et al., 2008c). Fucoidans were known to have a number of biological properties, such as antioxidant (Wang et al., 2015), hypocholesterolemic (Li et al., 1999; Thomesa et al., 2010), anticancer (Aisa et al., 2005; Alekseyenko et al., 2007), anticoagulant, antiangiogenic (Cumashi et al., 2007), antiviral (Prokofjeva et al., 2013) and anti-inflammatory activities (Park et al., 2011). The biological effects of fucoidans were reported to be species-dependent as the structure and composition of the fucoidan differs with the species from which it was isolated (Cumashi et al., 2007; Barbosa et al., 2014).

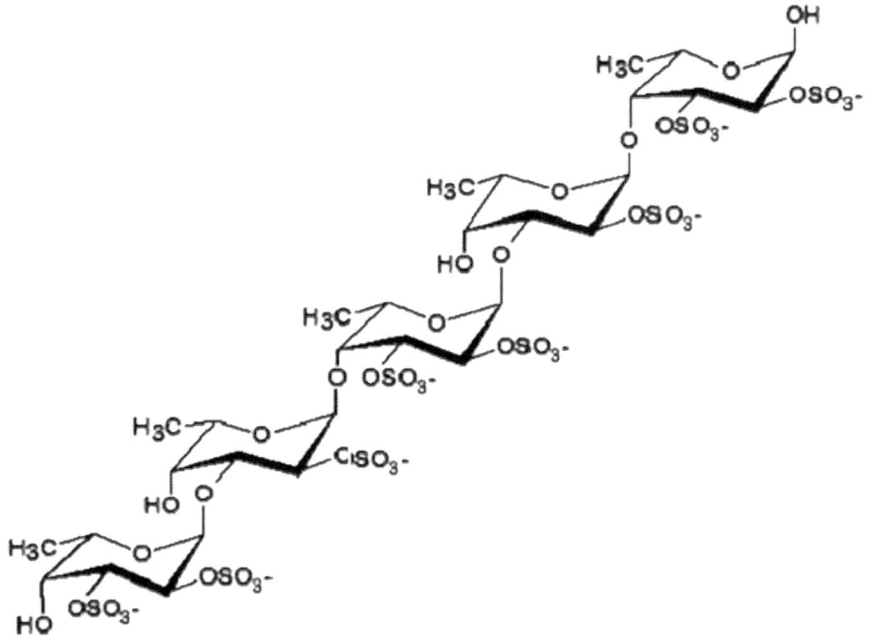

FIGURE 12.8 Structure of fucoidan reported from *Ascophyllum nodosum* and *Fucus vesiculosus*.

12.2.6.2 GALACTANS

Marine sulfated galactans are abundant in red algae (Pomin and Mourão, 2008). They are made up of linear chains of β-D-galactopyranose residues, linked through positions 1 and 3 (A units) and α-D/L-galactopyranose residues, linked through positions 1 and 4 (B units), arranged in an alternated sequence (AB)n (Figure 12.9) (Pomin and Mourão, 2008; Delattre et al., 2011). However, there were considerable structural variations reported in the sulfated galactans obtained from the red algae depending on the species, location, and season (Pereira et al., 2005; Pomin and Mourão, 2008).

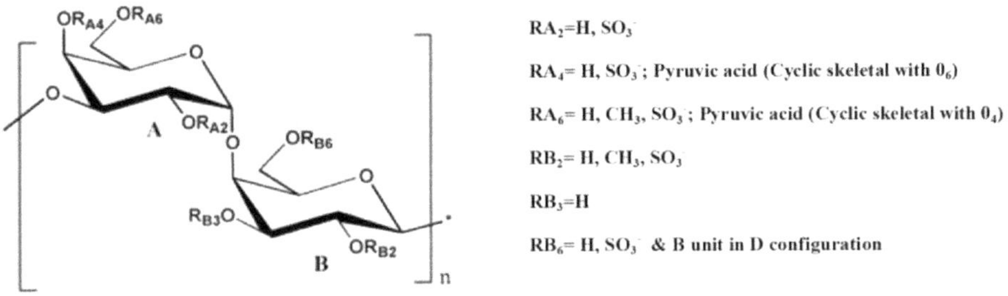

FIGURE 12.9 General structure of sulfated galactans from red seaweeds.

The most common sulfated galactans reported from macroalgae are carrageenans and agarans which are homogeneous galactans. Some complex heterogeneous galactans have also been isolated from red algae of the order Halymeniales, especially from the genus *Grateloupia* and *Pachymenia* (Hirase et al., 1967; Miller et al., 1995). Apart from Halymeniales, the other orders that have been reported to contain heterogenous galactans are Gigartinales (Genus: *Cryptonemia*, *Gymnogongrus*, *Gloiopeltis*, and *Kapaphycus*), Bangiales (Genus: *Porphyra*), and Bonnemaisoniales (Genus: *Asparagopsis*) (Delattre et al., 2011). The primary structural variations in these polysaccharides depend on the sulfation pattern (Pomin and Mourão, 2008). The complex nature of these polysaccharides was attributed to the substitution of hydroxyl groups at various positions of Unit A and B by a sulfate ester, a methyl group, or pyruvic acid (Usov, 1998; Pomin and Mourão, 2008).

There were reports of sulfated galactans from some green algae, mainly from *Codium* species (Matsubara et al., 2001; Farias et al., 2008). Sulfated arabinogalactan and sulfated glucogalactan are the complex and heterogeneous galactans isolated from *C. fragile* and *C. cylindricum* respectively

(Love and Percival, 1964; Matsubara et al., 2001). The sulfated galactans from green algae exhibited a backbone composed mainly of 4-sulfated, 3-linked β-D-galactopyranose units compared to the repetitive disaccharides found in the sulfated galactans from red algae (Bilan et al., 2007; Farias et al., 2008; Pomin and Mourão, 2008). Polysaccharides containing sulfated galactans were isolated as minor components from other green seaweeds including *Caulerpa* and *Ulva* (Love and Percival, 1964; Mao et al., 2006; Bilan et al., 2007; Shevchenko et al., 2009). The marine sulfated galactans were known for their antiviral (Duarte et al., 2001; Duarte et al., 2004; Talarico et al., 2004), antitumor (Zhou et al., 2004), immune modulation (Zhou et al., 2004; Wongprasert et al., 2014), antiangiogenic, anti-inflammatory (Bürgermeister et al., 2002), anticoagulant (Farias et al., 2000; Pereira et al., 2005), and antithrombotic properties (Fonseca et al., 2008).

12.2.6.3 AGAROIDS

Agaroids are sulfated polysaccharides and have a structure similar to agarobiose (Figure 12.10). In Agaroids some of the 3,6-anhydrogalactose are replaced by 4-α-L-galactopyranose and C6 of α-L-galactopyranose carries sulfate groups up to 20% (Pomin and Mourão, 2008; Delattre et al., 2011).

Funorans and porphyrans are examples of sulfated polysaccharides isolated from red algae (*Gloiopeltis* sp. and *Porphyra umbilicalis*), whereas the ulvan is from green algae (*Ulva* and *Enteromorpha* sp.) (Delattre et al., 2011; Tuvikene et al., 2015). In funorans, the C6 and C2 hydroxyl of 3-D-galactopyranose and 4-α-L-galactopyranose are substituted by sulfates or methyl groups (Figure 12.10) (Delattre et al., 2011). Porphyrans, highly substituted agarose-type, constitute of 4-linked 6-O-sulfo-α-L-galactopyranose (the precursor of the 3,6-anhydro unit) and 3-linked 6-O-methyl-β-D-galactopyranose (Figure 12.10) (Morrice et al., 1984; Delattre et al., 2011). Porphyrans have been reported to have anticancer (Kwon and Nam, 2006), antihyperlipidemic (Tsuge et al., 2004; Inoue et al., 2009), antioxidant, and anticoagulant properties (Zhang et al., 2010).

The main constituents of ulvan are sulfate, rhamnose, xylose, iduronic, and glucuronic acids (Figure 12.10) (Percival and McDowell, 1967; Lahaye and Ray, 1996). Kaeffer et al. (1999) reported the antitumor and immune modulation activities of ulvan in normal or tumoral colonic epithelial cells. Ulvan exhibited strain-specific anti-influenza activities (Ivanova et al., 1994) and anticoagulant activities (Mao et al., 2006). Ulvan has also been reported to have antioxidant activities (Qi et al., 2005, 2006) and reduced the hepatotoxicity against acetaminophen-induced hepatotoxicity in experimental albino rats (Rao et al., 2004).

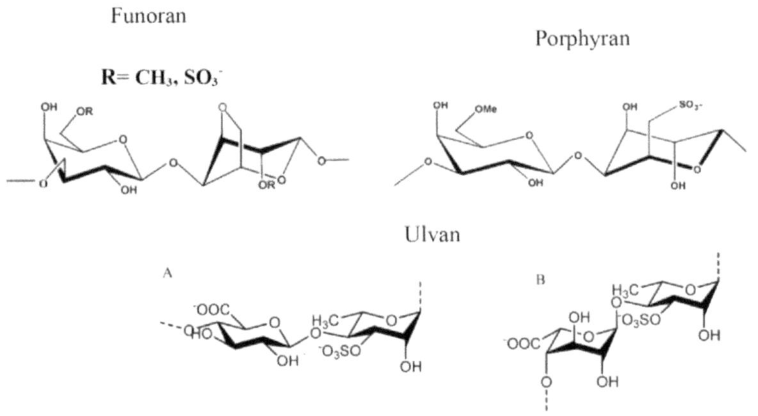

The main repeating disaccharide units of ulvan. A. [→4)-β-D-Glc*p*-(1→4)-α-L-Rha*p* 3S-(1→]$_n$; B. [→4)-α-L-Ido*p*-A (1→4)-α-L-Rha*p* 3S-(1→]$_n$

FIGURE 12.10 Structure of funoran, porphyran, and ulvan isolated from marine macroalgae.

12.3 LIPID COMPONENTS

In general, Macroalgae contain up to 1–5% of lipids on dry weight basis (Lordan et al., 2011) and glycolipids (GL) are the major lipid class in all algae, followed by neutral lipids and phospholipids (PLs) with some exceptions (Narayan et al., 2005; Holdt and Kraan, 2011).

12.3.1 FATTY ACIDS

The polar lipids like glyco- or PLs form the main building blocks of lipid bilayers of cellular membranes and the fatty acid composition of the bilayer strongly influence the membrane fluidity (Allakhverdiev et al., 2009). Among the different fatty acids, polyunsaturated fatty acids (PUFAs) mainly eicosapentaenoic acid (EPA; 20:5n-3) and docosahexaenoic acid (DHA; 22:6n-3) are an important category of fatty acids with a wide range of biological activity. The PUFA content in marine seaweeds varies with species, area of collection, and level of exposure to other environmental factors (Kumari et al., 2010; Pereira et al., 2012; Schmid et al., 2014). For example, the low temperature increases the proportion of polyunsaturated fatty acids (PUFA) to maintain the membrane fluidity and to facilitate the electron transport between the photosystems in the thylakoid membranes (Allakhverdiev et al., 2009). Macroalgae, especially Phaeophyta and Rhodophyta species, are valuable sources of PUFAs (Colombo et al., 2006; Kumari et al., 2010; Pereira et al., 2012; Barbosa et al., 2014). EPA and DHA are proven to provide cardiovascular protection and have anti-inflammatory and anticancer properties (Swanson et al., 2012).

12.3.2 GLYCOLIPIDS (GL)

Glycolipids (GL) are an important category of seaweed lipids, which attracted the attention of researchers. The isolation and structural characterization of macroalgal GL were reviewed by Plouguerné et al. (2014). There are three major types of GL present in seaweeds, viz., monogalactosyldiacylglycerides (MGDGs), digalactosyldiacylglycerides (DGDGs), and sulfoquinovosyldiacylglycerides (SQDGs) (Figure 12.11) (Plouguerné et al., 2014). They play an important role in the photosynthesis of eukaryotic algae (Hölzl and Dörmann, 2007). Several authors reported the antibacterial (Amiguet et al., 2011), antitumor, and antiviral activities (Mattos et al., 2011; de Souza et al., 2012) of the seaweed GL establishing their pharmacological potential. GL from macroalgae were also reported to have antifouling and anti-herbivory properties (Deal et al., 2003; Plouguerné et al., 2010).

12.3.3 STEROLS

Sterols are amphipathic compounds that are biosynthesized by MVA or DOXP/MEP (Lopes et al., 2011; Barbosa et al., 2014). Sterols structures are similar to cholesterol except for an alkyl substitution at C-24 (Lopes et al., 2013). The predominant sterols in Phaeophyta are Fucosterol, whereas in Rhodophyta, it is cholesterol (Figure 12.12).

Sterols varied greatly among Chlorophyta (Lopes et al., 2011; Mendis and Kim, 2011). Isofucocholesterol, cholesterol, 24-methylene cholesterol, and β-sitosterol were the frequently reported sterols from Chlorophyta (Mendis and Kim, 2011). Prolonged consumption of sterols from marine algae was reported to reduce fatty liver and excess fat deposition in the human heart by inhibiting cholesterol absorption (Ikeda et al., 1988). Specifically, fucosterol from macroalgae was reported to have several beneficial properties, such as anti-inflammatory, UV protective (Hwang et al., 2014), anti-osteoporotic

(Bang et al., 2011), anticancerous (Huh et al., 2012; Khanavi et al., 2012), antioxidative (Lee et al., 2003) and antidiabetic effects (Lee et al., 2004).

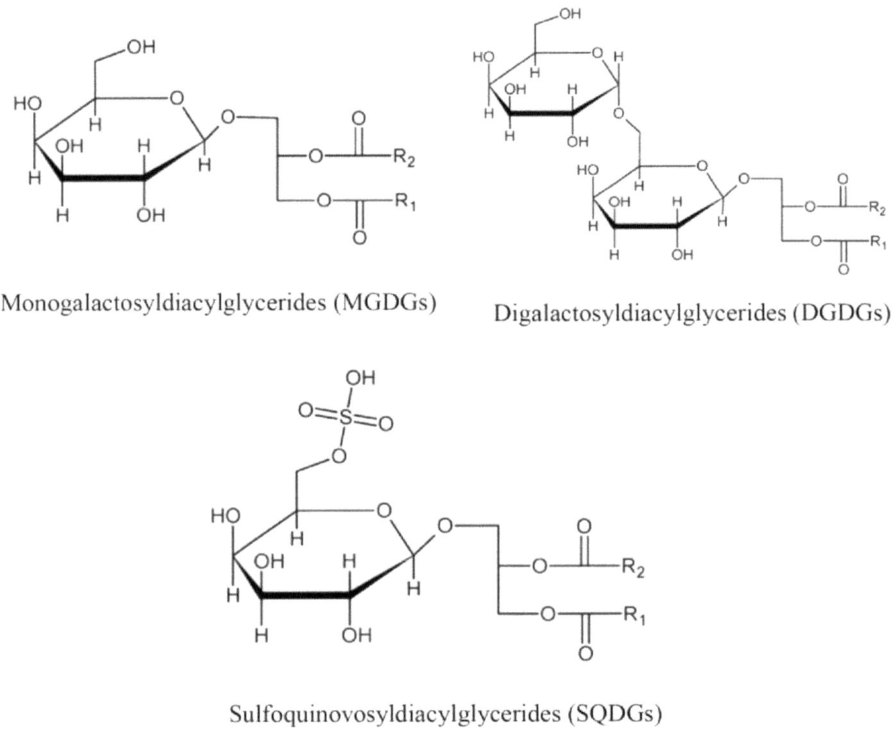

FIGURE 12.11 Structure of the three main glycoglycerolipids from seaweeds.

12.4 CONCLUSION

Marine macroalgae are potential sources of chemically diverse secondary metabolites, varying from simple linear peptides and fatty acids to complex alkaloids, terpenes, and polyketides. These biologically active secondary metabolites serve as an invaluable resource in the ongoing search for natural products. Recent advances in extraction, isolation, and characterization techniques, together with advancements in organic synthesis and chemical structure elucidation, accelerated the sustainable exploitation of these unique resources for obtaining commercially useful molecules. The immense chemical diversity and unique metabolic pathways of macroalgae due to adaptation to broad environmental conditions provide prospects for the potential application of "omic" techniques to identify species with the highest potential for newer molecules of interest for biotechnological and biomedical applications.

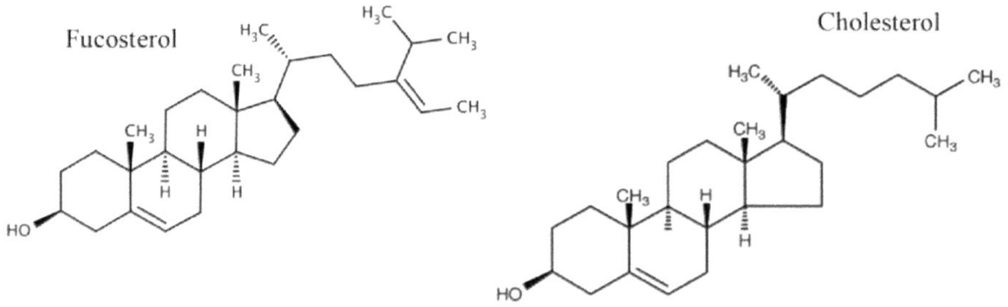

FIGURE 12.12 Structure of fucosterol and cholesterol.

KEYWORDS

- **alkaloids**
- **lipids**
- **macroalgae**
- **natural products**
- **phenolics**
- **polysaccharides**
- **proteins**
- **secondary metabolites**
- **terpenoids**

REFERENCES

Afolayan, A. F., Bolton, J. J., Lategan, C. A., Smith, P. J., & Beukes, D. R., (2008). Fucoxanthin, tetraprenylated toluquinone and toluhydroquinone metabolites from *Sargassum heterophyllum* inhibit the *in vitro* growth of the malaria parasite *Plasmodium falciparum. Zeitschrift Für Naturforschung. C. Journal of Bioscience*, *63*(11, 12), 848–852.

Aguilera, J., Bischof, K., Karsten, U., & Hanelt, D., (2002). Seasonal variation in ecophysiological patterns in macroalgae from an Arctic fjord. II. Pigment accumulation and biochemical defense systems against high light stress, *Marine Biology*, *140*, 1087–1095.

Ahn, B. R., Moon, H. E., Kim, H. R., Jung, H. A., & Choi, J. S., (2012). Neuroprotective effect of edible brown alga *Eisenia bicyclis* on amyloid beta peptide-induced toxicity in PC12 cells. *Archives of Pharmacological Research, 35*(11), 1989–1998.

Ahn, M. J., Yoon, K. D., Min, S. Y., et al., (2004). Inhibition of HIV-1 reverse transcriptase and protease by phlorotannins from the brown alga *Ecklonia cava. Biological and Pharmaceutical Bulletin, 27*(4), 544–547.

Aisa, Y., Miyakawa, Y., Nakazato, T., et al., (2005). Fucoidan induces apoptosis of human HS-Sultan cells accompanied by activation of caspase-3 and down-regulation of ERK pathways. *American Journal of Hematology, 78*(1), 7–14.

Ajikumar, P. K., Tyo, K., Carlsen, S., Mucha, O., Phon, T. H., & Stephanopoulos, G., (2008). Terpenoids: Opportunities for biosynthesis of natural product drugs using engineered microorganisms. *Molecular Pharmacology, 5*(2), 167–190.

Al Sabi, A., McArthur, J., Ostroumov, V., & French, R. J., (2006). Marine toxins that target voltage-gated sodium channels. *Marine Drugs, 4*(3), 157–192.

Alarif, W. M., Abou-Elnaga, Z. S., Ayyad, S. E. N., & Al-Lihaibi, S. S., (2010). Insecticidal metabolites from the green alga *Caulerpa racemosa. Clean - Soil, Air, Water, 38*(5, 6), 548–557.

Alekseyenko, T. V., Zhanayeva, S. Y., Venediktova, A. A., et al., (2007). Antitumor and antimetastatic activity of fucoidan, a sulfated polysaccharide isolated from the Okhotsk Sea *Fucus evanescens* brown alga. *Bulletin of Experimental Biology and Medicine, 143*(6), 730–732.

Allakhverdiev, S. I., Los, D. A., & Murata, N., (2009). Regulatory roles in photosynthesis of unsaturated fatty acids in membrane lipids. In. Wada, H., & Murata, N., (eds.), *Lipids in Photosynthesis: Essential and Regulatory Functions* (pp. 373–388). Springer, Dordrecht.

Alystyne, K. L., Anderson, K. J., Winans, A. K., & Gifford, S., (2011). Dopamine release by the green alga *Ulvaria obscura* after simulated immersion by incoming tides. *Marine Biology, 158*(9), 2087–2094.

Amiguet, V. T., Jewell, L. E., Mao, H., et al., (2011). Antibacterial properties of a glycolipid-rich extract and active principle from Nunavik collections of the macroalgae *Fucus evanescens* C. Agardh (*Fucaceae*). *Canadian Journal of Microbiology, 57*(9), 745–749.

Amsler, C. D., & Fairhead, V. A., (2005). Defensive and sensory chemical ecology of brown algae. *Advances in Botanical Research, 43*(2005), 1–91.

Aneiros, A., & Garateix, A., (2004). Bioactive peptides from marine sources: Pharmacological properties and isolation procedures. *Journal of Chromatography B. Analytical Technologies in the Biomedical and Life Sciences, 803*(1), 41–53.

Angawi, R. F., Alarif, W. M., Hamza, R. I., Badria, F. A., & Ayyad, S.N. (2014). Few cytotoxic laurene-, cuparene-, and laurokamurene-type sesquiterpenes from the red alga *Laurencia obtusa. Helvetica Chimica Acta, 97*(10), 1388–1395.

Areche, C., San-Martín, A., Rovirosa, J., & Sepúlveda, B., (2011). Gastroprotective activity of epitaondiol and sargaol. *Natural Product Communication, 6*(8), 1073, 1074.

Arnold, T. M., & Targett, N. M., (2002). Marine tannins: The importance of a mechanistic framework for predicting ecological roles. *Journal of Chemical Ecology, 28*(10), 1919–1934.

Artan, M., Li, Y., Karadeniz, F., et al., (2008). Anti-HIV-1 activity of phloroglucinol derivative, 6,6'-bieckol, from *Ecklonia cava. Bioorganic and Medicinal Chemistry, 16*(17), 7921–7926.

Athukorala, Y., Trang, S., Kwok, C., & Yuan, Y., (2016). Antiproliferative and antioxidant activities and mycosporine-like amino acid profiles of wild-harvested and cultivated edible Canadian marine red macroalgae. *Molecules*, *21*(1), 119.

Ayyad, S. E. N., & Badria, F. A., (1994). Caulerpine: An antitumor indole alkaloid from *Caulerpa racemosa. Alexandria Journal of Pharmacological Sciences*, *8*(3), 217–219.

Bae, J. S., (2011). Antithrombotic and profibrinolytic activities of phloroglucinol. *Food and Chemical Toxicology*, *49*(7), 1572–1577.

Balasundram, B., Sundram, K., & Samman, S., (2006). Phenolic compounds in plants and agri-industrial by-products: Antioxidant activity, occurrence, and potential uses. *Food Chemistry*, *99*(1), 191–203.

Banaszak, A. T., & Lesser, M. P., (1995). Survey of mycosporine-like amino acids in macrophytes of Kaneohe Bay. In: Gulko, D., & Jokiel, P. L., (eds.), *Ultraviolet Radiation and Coral Reefs* (pp. 171–179). Hawaii Institute of Marine Biology Technical Report 41, University of Hawaii, Honolulu, HI, USA.

Bang, M. H., Kim, H. H., Lee, D. Y., Han, M. W., Baek, Y. S., & Chung, D. K., (2011). Anti-osteoporotic activities of fucosterol from sea mustard (*Undaria pinnatifida*). *Food Science and Biotechnology*, *20*(2), 343–347.

Barbosa, M., Valentão, P., & Andrade, P., (2014). Bioactive compounds from macroalgae in the new millennium: Implications for neurodegenerative diseases. *Marine Drugs*, *12*(9), 4934–4972.

Barroso, N., & Rodriguez, M., (1996). Action of beta-phenylethylamine and related amines on nigrostriatal dopamine neurotransmission. *European Journal of Pharmacology*, *297*(3), 195–203.

Bhakuni, D. S., & Rawat, D. S., (2005). Bioactive metabolites of marine algae, fungi and bacteria. In: Bhakuni, D. S., & Rawat, D. S., (eds.), *Bioactive Marine Natural Products* (pp. 1–19). Anamaya Publishers, New Delhi, India.

Bilan, M. I., Vinogradova, E. V., Shashkov, A. S., & Usov, A. I., (2007). Structure of a highly pyruvylated galactan sulfate from the Pacific green alga *Codium yezoense* (Bryopsidales, Chlorophyta). *Carbohydrate Research*, *342*(3, 4), 586–596.

Bitencourt, F. D. S., Figueiredo, J. G., Mota, M. R. L., et al., (2008). Antinociceptive and anti-inflammatory effects of a mucin-binding agglutinin isolated from the red marine alga *Hypnea cervicornis. Naunyn-Schmiedeberg's Archives of Pharmacology*, *377*(2), 139–148.

Blunt, J. W., Copp, B. R., Munro, M. H. G., Northcote, P. T., & Prinsep, M. R., (2006). Marine natural products. *Natural Product Reports*, *23*, 26–78.

Boettcher, A. A., & Targett, N. M., (1993). Role of polyphenolic molecular size in reduction of assimilation efficiency in *Xiphister mucosus. Ecology*, *74*(3), 891–903.

Brahmkshatriya, P. P., & Brahmkshatriya, P. S., (2013). Terpenes: Chemistry, biological role, and therapeutic applications. In: Ramawat, K. G., & Mérillon, J. M., (eds.), *Natural Products: Phytochemistry, Botany and Metabolism of Alkaloids, Phenolics and Terpenes* (pp. 2665–2691). Springer, Berlin, Germany.

Brennan, M. R., & Ericson, K. L., (1978). Polyhalogenated indoles from the marine alga *Rhodophyllis membranacea* Harvey. *Tetrahedron Letters*, *19*(19), 1637–1640.

Bürgermeister, J., Paper, D. H., Vogl, H., Linhardt, R. J., & Franz, G., (2002). LaPSvS1, a (1→3)-β-galactan sulfate and its effect on angiogenesis *in vivo* and *in vitro*. *Carbohydrate Research*, *337* (16), 1459–1466.

Cafieri, F., De Napoli, L., Fattorusso, E., & Santacroce, C., (1987). Diterpenes from the red alga *Sphaerococcus coronopifolius*. *Phytochemistry*, *22*(8), 471–473.

Cahyana, A. H., Shuto, Y., & Kinoshita, Y., (1992). Pyropheophytin A as an antioxidative substance from the marine alga, Arame (*Eisenia bicyclis*). *Bioscience, Biotechnology and Biochemistry*, *56*(10), 1533–1535.

Cahyana, H., Shuto, Y., & Kinoshita, Y., (1993). Antioxidative activity of porphyrin derivatives. *Bioscience, Biotechnology and Biochemistry*, *57*(4), 680, 681.

Calvete, J. J., Costa, F. H. F., Saker-Sampaio, S., et al., (2000). The amino acid sequence of the agglutinin isolated from the red marine alga *Bryothamnion triquetrum* defines a novel lectin structure. *Cellular and Molecular Life Sciences*, *57*(2), 343–350.

Cardozo, K. H. M., Marques, L. G., Carvalho, V. M., et al., (2011). Analyses of photoprotective compounds in red algae from the Brazilian coast. *Brazilian Journal of Pharmacognosy*, *21*(2), 202–208.

Cardozo, K. H., Guaratini, T., Barros, M. P., et al., (2007). Metabolites from algae with economical impact. *Comparative Biochemistry and Physiology C Toxicology and Pharmacology*, *146*(1, 2), 60–78.

Carreto, J. I., Carignan, M. O., & Montoya, N. G., (2005). A high-resolution reverse-phase liquid chromatography method for the analysis of mycosporine-like amino acids (MAAs) in marine organisms. *Marine Biology*, *146*, 237–252.

Carte, B. K., Troupe, N., Chan, J. A., Westley, J. W., & Faulkner, D. J., (1989). Rawsonol, an inhibitor of HMG-CoA reductase from the tropical green alga *Avrainvillea rawsoni*. *Phytochemistry*, *28*(11), 2917–2919.

Carter, G. T., Rinehart, K. L., Li, L. H., Kuentzel, S. L., & Connor, J. L., (1978). Brominated indoles from *Laurencia brongniartii*. *Tetrahedron Letters*, *46*, 4479–4482.

Cavalcante-Silva, L., De Carvalho, C. A., Barbosa-Filho, J., et al., (2013). Spasmolytic effect of caulerpine involves blockade of Ca^{2+} influx on guinea pig ileum. *Marine Drugs*, *11*(5), 1553–1564.

Cengiz, S., Cavas, L., Yurdakoc, K., & Pohnert, G., (2011). The sesquiterpene caulerpenyne from *Caulerpa* spp. is a lipoxygenase inhibitor. *Marine Biotechnology*, *13*(2), 321–326.

Cérantola, S., Breton, F., Gall, E. A., & Deslandes, E., (2006). Co-occurrence and antioxidant activities of fucol and fucophlorethol classes of polymeric phenols in *Fucus spiralis*. *Botanica Marina*, *49*(4), 347–351.

Chakraborty, K., Lipton, A. P., Paul, R. R., & Vijayan, K. K., (2010a). Antibacterial labdane diterpenoids of *Ulva fasciata* Delile from southwestern coast of the Indian Peninsula. *Food Chemistry*, *119*(4), 1399–1408.

Chakraborty, K., Lipton, A. P., Paulraj, R., & Chakraborty, R. D., (2010b). Guaianesesquiterpenes from seaweed *Ulva fasciata* Delile and their antibacterial properties. *European Journal of Medicinal Chemistry*, *45*(6), 2237–2244.

Chatter, R., Ben, O. R., Rabhi, S., et al., (2011). *In vivo* and *in vitro* anti-inflammatory activity of neorogioltriol, a new diterpene extracted from the red algae *Laurencia glandulifera*. *Marine Drugs*, *9*(7), 1293–1306.

Chay, C. I. C., Gómez, C. R., Espitia, P. C. I., Torres-Ochoa, R. O., & Martínez, R., (2014). Synthesis and anti-tuberculosis activity of the marine natural product caulerpin and its analogue. *Marine Drugs*, *12*(4), 1757–1772.

Cho, M., Lee, H. S., Kang, I. J., Won, M. H., & You, S., (2011). Antioxidant properties of extract and fractions from *Enteromorpha prolifera*, a type of green seaweed. *Food Chemistry*, *127*(3), 999–1006.

Choi, B. W., Ryu, G., Park, S. H., et al., (2007). Anticholinesterase activity of plastoquinones from *Sargassum sagamianum*: Lead compounds for Alzheimer's disease therapy. *Phytotherapy Research*, *21*(5), 423–426.

Chronakis, I. S., (2000). Biosolar proteins from aquatic algae. In: Doxastakis, G., & Kiosseoglou, V., (eds.), *Novel Macromolecules in Food Systems* (pp. 39–72). Elsevier Applied Science, Amsterdam.

Collén, J., Ekdahl, A., Abrahamsson, K., & Pedersén, M., (1994). The involvement of hydrogen peroxide in the production of volatile halogenated compounds by *Meristiella gelidium*. *Phytochemistry*, *36*(5), 1197–1202.

Colombo, M. L., Risè, P., Giavarini, F., De Angelis, L., Galli, C., & Bolis, C. L., (2006). Marine macroalgae as sources of polyunsaturated fatty acids. *Plant Foods for Human Nutrition*, *61*(2), 64–69.

Conde, E., Balboa, E. M., Parada, M., & Falqué, E., (2013). Algal proteins, peptides and amino acids. In: Dominguez, H., (ed.), *Functional Ingredients from Algae for Foods and Nutraceuticals* (pp. 135–180). Woodhead Publishing, Oxford, UK.

Connan, S., Delisle, F., Deslandes, E., & Ar Gall, E., (2006). Intra-thallus phlorotannin content and antioxidant activity in phaeophyceae of temperate waters. *Botanica Marina*, *49*, 39–46.

Connan, S., Goulard, F., Stiger, V., Deslandes, E., & Ar Gall, E., (2004). Interspecific and temporal variation in phlorotannin levels in an assemblage of brown algae. *Botanica Marina*, *47*(4), 410–416.

Cumashi, A., Ushakova, N. A., Preobrazhenskaya, M. E., et al., (2007). A comparative study of the anti-inflammatory, anticoagulant, antiangiogenic, and antiadhesive activities of nine different fucoidans from brown seaweeds. *Glycobiology*, *17*(5), 541–552.

Das, S. K., Hashimoto, T., & Kanazawa, K., (2008). Growth inhibition of human hepatic carcinoma HepG2 cells by fucoxanthin is associated with down-regulation of cyclin D. *Biochimica et Biophysica Acta – General Subjects*, *1780*(4), 743–749.

Das, S. K., Ren, R., Hashimoto, T., & Kanazawa, K., (2010). Fucoxanthin induces apoptosis in osteoclast-like cells differentiated from RAW264.7 cells. *Journal of Agricultural and Food Chemistry*, *58*(10), 6090–6095.

Davyt, D., Fernandez, R., Suescun, L., et al., (2001). New sesquiterpene derivatives from the red alga *Laurencia scoparia*. Isolation, structure determination, and anthelmintic activity. *Journal of Natural Products*, *64*(12), 1552–1555.

Dawczynski, C., Schubert, R., & Jahreis, G., (2007). Amino acids, fatty acids, and dietary fiber in edible seaweed products. *Food Chemistry*, *103*(3), 891–899.

De Jesus, R. M. F., De Morais, A. M. B., & De Morais, R. M. S. C., (2015). Marine polysaccharides from algae with potential biomedical applications. *Marine Drugs*, *13*(5), 2967–3028.

De La Coba, F., Aguilera, J., Figueroa, F. L., De Gálvez, M. V., & Herrera, E., (2009). Antioxidant activity of mycosporine-like amino acids isolated from three red macroalgae and one marine lichen. *Journal Applied Phycology*, *21*(2), 161–169.

De Las Heras, B., & Hortelano, S., (2009). Molecular basis of the anti-inflammatory effects of terpenoids. *Inflammation and Allergy Drug Targets*, *8*(1), 28–39.

De Schutter, K., & Van, D. E. J. M., (2015). Protein-carbohydrate interactions as part of plant defense and animal immunity. *Molecules*, *20*(5), 9029–9053.

De Souza, E. T., De Lira, D. P., De Queiroz, A. C., et al., (2009). The antinociceptive and anti-inflammatory activities of caulerpin, a bisindole alkaloid isolated from seaweeds of the genus *Caulerpa*. *Marine Drugs*, *7*(4), 689–704.

De Souza, L. M., Sassaki, G. L., Romanos, M. T. V., & Barreto-Bergter, E., (2012). Structural characterization and anti-HSV-1 and HSV-2 activity of glycolipids from the marine algae *Osmundaria obtusiloba* isolated from south-eastern Brazilian coast. *Marine Drugs*, *10*(4), 918–931.

Deal, M. S., Hay, M. E., Wilson, D., & Fenical, W., (2003). Galactolipids rather than phloro tannins as herbivore deterrents in the brown seaweed *Fucus vesiculosus*. *Oecologia, 136*(1), 107–114.

Delattre, C., Fenoradosoa, T. A., & Michaud, P., (2011). Galactans: An overview of their most important sourcing and applications as natural polysaccharides. *Brazilian Archives of Biology and Technology, 54*(6), 1075–1092.

Dembitsky, V. M., & Tolstikov, G. A., (2003). Natural halogenated alkaloids. *Chemistry for Sustainable Development, 11*, 451–466.

Dembitsky, V. M., & Tolstkov, G. A., (2004). Natural halogenated sesquiterpenes from marine organisms. *Chemistry for Sustainable Development, 12*, 1–12.

Dias, D. A., Urban, S., & Roessner, U., (2012). A historical overview of natural products in drug discovery. *Metabolites, 2*(2), 303–336.

Dorta, E., Cueto, M., Brito, I., & Darias, J., (2002). New terpenoids from the brown alga *Stypopodium zonale*. *Journal Natural Products, 65*(11), 1727–1730.

Duan, X. J., Li, X. M., & Wang, B. G., (2007). Highly brominated mono- and bis-phenols from the marine red alga *Symphyocladia latiuscula* with radical-scavenging activity. *Journal of Natural Products, 70*(7), 1210–1213.

Duarte, M. E. R., Cauduro, J. P., Noseda, D. G., et al., (2004). The structure of the agaran sulfate from *Acanthophora spicifera* (Rhodomelaceae, Ceramiales) and its antiviral activity. Relation between structure and antiviral activity in agarans. *Carbohydrate Research, 339*(2), 335–347.

Duarte, M. E., Noseda, D. G., Noseda, M. D., Tulio, S., Pujol, C. A., & Damonte, E. B., (2001). Inhibitory effect of sulfated galactans from the marine alga *Bostrychia montagnei* on herpes simplex virus replication *in vitro*. *Phytomedicine, 8*(1), 53–58.

El Sayed, K. A., Bartyzel, P., Shen, X., Perry, T. L., Zjawiony, J. K., & Hamann, M. T., (2000). Marine natural products as antituberculosis agents. *Tetrahedron, 56*(7), 949–953.

Ennamany, R., Saboureau, D., Mekideche, N., & Creppy, E. E., (1998). SECMA 1, a mitogenic hexapeptide from *Ulva algeae* modulates the production of proteoglycans and glycosaminoglycans in human foreskin fibroblast. *Human and Experimental Toxicology, 17*(1), 18–22.

Eom, S. H., Kim, Y. M., & Kim, S. K., (2012). Antimicrobial effect of phlorotannins from marine brown algae. *Food and Chemical Toxicology, 50*(9), 3251–3255.

Etahiri, S., Bultel-Poncé, V., Caux, C., & Guyot, M., (2001). New bromoditerpenes from the red alga *Sphaerococcus coronopifolius*. *Journal of Natural Products, 64*(8), 1024–1027.

Éverton, T. S., Daysianne, P. L., Aline, C. Q., et al., (2009). The antinociceptive and anti-inflammatory activities of caulerpin, a bisindole alkaloid isolated from seaweeds of the genus *Caulerpa*. *Marine Drugs, 7*(4), 689–704.

Faircloth, G. T., Grant, W., Smith, B., et al., (2000). Preclinical development of Kahalalide F: A new marine compound selected for clinical studies. *Proceeding of American Association of Cancer Research, 41*, 600, 601.

Falkowski, G., & Raven, J. A., (2007). *Aquatic Photosynthesis* (pp. 1–42). Princeton and Oxford: Princeton University Press, Oxford.

Farias, E. H. C., Pomin, V. H., Valente, A. P., Nader, H. B., Rocha, H. A. O., & Mourão, P. A. S., (2008). A preponderantly 4-sulfated, 3-linked galactan from the green alga *Codium isthmocladum*. *Glycobiology, 18*(3), 250–259.

Farias, W. R., Valente, A. P., Pereira, M. S., & Mourão, P. A., (2000). Structure and anticoagulant activity of sulfated galactans; Isolation of a unique sulfated galactan from the red algae *Botryocladia occidentalis* and comparison of its anticoagulant action with that of sulfated galactans from invertebrates. *Journal of Biological Chemistry*, *275*(38), 29299–29307.

Farvin, K. H. S., & Jacobsen, C., (2013). Phenolic compounds and antioxidant activities of selected species of seaweeds from Danish coast. *Food Chemistry*, *138*(2, 3), 1670–1681.

Ferdouse, F., Holdt, S. L., Smith, R., Murúa, P., & Yang, Z., (2018). *The Global Status of Seaweed Production, Trade and Utilization* (Vol. 124). FAO Globefish Research Program. FAO, Rome.

Ferruzzi, M. G., & Blakeslee, J., (2007). Digestion, absorption, and cancer preventative activity of dietary chlorophyll derivatives. *Nutrition Research*, *27*(1), 1–12.

Fitton, J., Dell'Acqua, G., Gardiner, V. A., Karpiniec, S., Stringer, D., & Davis, E., (2015). Topical benefits of two fucoidan-rich extracts from marine macroalgae. *Cosmetics*, *2*(2), 66–81.

Fitzgerald, C., Mora-Soler, L., Gallagher, E., et al., (2012). Isolation and characterization of bioactive pro-peptides with *in vitro* renin inhibitory activities from the macroalga *Palmaria palmata*. *Journal Agricultural and Food Chemistry*, *60*(30), 7421–7427.

Fleurence, J., (2004). Seaweed proteins. In: Yada, R. Y., (ed.), *Proteins in Food Processing* (pp. 197–213). Woodhead Publishing Limited, Cambridge, UK.

Flodin, C., & Whitfield, F., (1999). Biosynthesis of bromophenols in marine algae. *Water Science and Technology*, *40*(6), 53–58.

Flodin, C., Helidoniotis, F., & Whitfield, F. B., (1999). Seasonal variation in bromophenol content and bromoperoxidase activity in *Ulva lactuca*. *Phytochemistry*, *51*(1), 135–138.

Fong, J., Lim, Z. W., Bauman, A. G., et al., (2019). Allelopathic effects of macroalgae on *Pocillopora acuta* coral larvae. *Marine Environmental Research*, *151*, 104745.

Fonseca, R. J., Oliveira, S. N. M. C. G., Melo, F. R., Pereira, M. G., Benevides, N. M., & Mourão, P. A., (2008). Slight differences in sulfation of algal galactans account for differences in their anticoagulant and venous antithrombotic activities. *Thrombosis and Haemostasis*, *99*(3), 539–545.

Fujimoto, K., Ohmura, H., & Kaneda, T., (1985). Screening for antioxygenic compounds in marine algae and bromophenols as effective principles in a red alga *Polysiphonia urceolata*. *Nippon Suisan Gakk*, *51*(7), 1139–1143.

Fuller, R. W., Cardellina, J. H., Jurek, J., et al., (1994). Isolation and structure/activity features of halomon-related antitumor monoterpenes from the red alga *Portieria hornemannii*. *Journal of Medicinal Chemistry*, *37*(25), 4407–4411.

Ganesan, P., Matsubara, K., Ohkubo, T., et al., (2010). Anti-angiogenic effect of siphonaxanthin from green alga, *Codium fragile*. *Phytomedicine*, *17*(14), 1140–1144.

Ganesan, P., Noda, K., Manabe, Y., et al., (2011). Siphonaxanthin, a marine carotenoid from green algae, effectively induces apoptosis in human leukemia (HL-60) cells. *Biochimca et Biophysica Acta (BBA)-General Subjects*, *1810*(5), 497–503.

Gil, B., Ferrándiz, M. L., Sanz, M. J., et al., (1995). Inhibition of inflammatory responses by epitaondiol and other marine natural products. *Life Science*, *57*(2), PL25–30.

Girard, J. P., Marion, C., Liutkus, M., et al., (1988). Hypotensive constituents of marine algae; 1. Pharmacological studies of laminine. *Planta Medica*, *54*(3), 193–196.

Gross, H., & König, G. M., (2006). Terpenoids from marine organisms: Unique structures and their pharmacological potential. *Phytochemical Reviews*, *5*(1), 115–141.

Gross, H., Goeger, D. E., Hills, P., et al., (2006). Lophocladines, bioactive alkaloids from the red alga *Lophocladia* sp. *Journal of Natural Products*, *69*(4), 640–644.

Grosso, C., Vinholes, J., Valentão, P., & Andrade, P. B., (2011). Halogenated compounds from seaweed, a biological overview. In: Pomin, V. H., (ed.), *Seaweed: Ecology, Nutrient Composition and Medicinal Uses* (pp. 163–184). Nova Science Publishers, New York, USA.

Güven, K. C., Coban, B., Sezik, E., Erdugan, H., & Kaleağasioğlu, F., (2103). Alkaloids of marine macro algae. In: Ramawat, K. G., & Merillon, J. M., (eds.), *Natural Products* (pp. 27–37). Springer-Verlag, Berlin, Germany.

Güven, K. C., Percot, A., & Sezik, E., (2010). Alkaloids in marine algae. *Marine Drug, 8*(2), 269–284.

Hamann, M. T., & Scheuer, P. J., (1993). Kahalalide F: A bioactive depsipeptide from the sacoglossan mollusk *Elysia rufescens* and the green alga *Bryopsis* sp. *Journal of American Chemical Society, 115*(13), 5825–5826.

Hamann, M. T., Otto, C. S., Scheuer, P. J., & Dunbar, D. C., (1996). Kahalalides: Bioactive peptides from a marine mollusk *Elysia rufescens* and its algal diet *Bryopsis* sp. *Journal of Organic Chemistry, 61*(19), 6594–6600.

Hamid, R., Masood, A., Wani, I. H., & Rafiq, S., (2013). Lectins: Proteins with diverse applications. *Journal of Applied Pharmaceutical Sciences, 3*(4), 93–103.

Harnedy, P. A., & FitzGerald, R. J., (2011). Bioactive proteins, peptides, and amino acids from macroalgae. *Journal of Phycology, 47*(2), 218–232.

Harnedy, P. A., & FitzGerald, R. J., (2013). *In vitro* assessment of the cardioprotective, anti-diabetic and antioxidant potential of *Palmaria palmata* protein hydrolysates. *Journal of Applied Phycology, 25*(6), 1793–1803.

Haugan, J. A., & Liaaen-Jensen, S., (1994). Algal carotenoids 54. Carotenoids of brown algae (Phaeophyceae). *Biochemical Systematics and Ecology, 22*(1), 31–41.

Hellio, C., De La Broise, D., Dufossé, L., Le Gal, Y., & Bourgougnon, N., (2001). Inhibition of marine bacteria by extracts of macroalgae: Potential use for environmentally friendly antifouling paints. *Marine Environmental Research, 52*(3), 231–247.

Heo, S. J., & Jeon, Y. J., (2009). Protective effect of fucoxanthin isolated from *Sargassum siliquastrum* on UV-B induced cell damage. *Journal of Photochemistry and Photobiology B., 95*(2), 101–107.

Heo, S. J., Cha, S. H., Kim, K. N., et al., (2012). Neuroprotective effect of phlorotannin isolated from *Ishige okamurae* against H_2O_2-induced oxidative stress in murine hippocampal neuronal cells, HT22. *Applied Biochemistry and Biotechnology, 166*(6), 1520–1532.

Heo, S. J., Hwang, J. Y., Choi, J. I., Han, J. S., Kim, H. J., & Jeon, Y. J., (2009b). Diphlorethohydroxycarmalol isolated from *Ishige okamurae*, a brown algae, a potent alpha-glucosidase and α-amylase inhibitor, alleviates postprandial hyperglycemia in diabetic mice. *European Journal of Pharmacology, 615*(1–3), 252–256.

Heo, S. J., Ko, S. C., Cha, S. H., et al., (2009a). Effect of phlorotannins isolated from *Ecklonia cava* on melanogenesis and their protective effect against photo-oxidative stress induced by UV-B radiation. *Toxicology In Vitro, 23*(6), 1123–1130.

Heo, S. J., Ko, S. C., Kang, S. M., et al., (2008). Cytoprotective effect of fucoxanthin isolated from brown algae *Sargassum siliquastrum* against H_2O_2-induced cell damage. *European Food Research and Technology, 228*(1), 145–151.

Heo, S. J., Yoon, W. J., Kim, K. N., et al., (2010). Evaluation of anti-inflammatory effect of fucoxanthin isolated from brown algae in lipopolysaccharide-stimulated RAW 264.7 macrophages. *Food and Chemical Toxicology, 48*(8, 9), 2045–2051.

Hiqashi-Okai, K., Otani, S., & Okai, Y., (1999). Potent suppressive effect of a Japanese edible seaweed, *Enteromorpha prolifera* (Sujiao-nori) on initiation and promotion phases of chemically induced mouse skin tumorigenesis. *Cancer Letters, 140*(1, 2), 21–25.

Hirase, S., Araki, C., & Watanabe, K., (1967). Component sugars of the polysaccharide of the red seaweed *Grateloupia elliptica*. *Bulletin of Chemical Society of Japan*, *40*(6), 1445–1448.

Holdt, S. L., & Kraan, S., (2011). Bioactive compounds in seaweed: Functional food applications and legislation. *Journal of Applied Phycology*, *23*(3), 543–597.

Hölzl, G., & Dörmann, P., (2007). Structure and function of glycoglycerolipids in plants and bacteria. *Progress in Lipid Research*, *46*(5), 225–243.

Hori, K., Ikegami, S., Miyazawa, K., & Ito, K., (1988). Mitogenic and antineoplastic isoagglutinins from the red alga *Solieria robusta*. *Phytochemistry*, *27*(7), 2063–2067.

Hori, K., Matsubara, K., & Miyazawa, K., (2000). Primary structures of two hemagglutinins from the marine red alga, *Hypnea japonica*. *Biochimica et Biophysica Acta*, *1474*(2), 226–236.

Hori, K., Miyazawa, K., & Ito, K., (1986). Isolation and characterization of glycoconjugate-specific isoagglutinins from a marine green alga *Boodlea coacta* (Dickie) Murray et De Toni. *Botanica Marina*, *29*(4), 323–328.

Hori, K., Sato, Y., Ito, K., et al., (2007). Strict specificity for high-mannose type N-glycans and primary structure of a red alga *Eucheuma serra* lectin. *Glycobiology*, *17*(5), 479–491.

Hosokawa, M., Kudo, M., Maeda, H., Kohno, H., Tanaka, T., & Miyashita, K., (2004). Fucoxanthin induces apoptosis and enhances the antiproliferative effect of the PPARgamma ligand, troglitazone, on colon cancer cells. *Biochemical and Biophysical Acta*, *1675*(1–3), 113–119.

Houston, M. C., (2005). Nutraceuticals, vitamins, antioxidants, and minerals in the prevention and treatment of hypertension. *Progress in Cardiovascular Diseases*, *47*(6), 396–449.

Hoyer, K., Karsten, U., & Wiencke, C., (2001). Photoprotective substances in Antarctic macroalgae and their variations with respect to depth distributions, different tissues and developmental stages. *Marine Ecology Progress Series*, *211*, 117–129.

Huh, G. W., Lee, D. Y., In, S. J., et al., (2012). Fucosterols from *Hizikia fusiformis* and their proliferation activities on osteosarcoma-derived cell MG63. *Journal of Korean Society of Applied Biological Chemistry*, *55*(4), 551–555.

Huovinen, P., Gómez, I., Figueroa, F. L., Ulloa, N., Morales, V., & Lovengreen, C., (2004). Ultraviolet-absorbing mycosporine-like amino acids in red macroalgae from Chile. *Botanica Marina*, *47*(1), 21–29.

Hwang, E., Park, S. Y., Shin, H. S., Lee, D. G., & Yi, T. H., (2014). Effect of oral administration of fucosterol from *Hizikia fusiformis* on DNCB-induced atopic dermatitis in NC/Nga mice. *Food Science and Biotechnology*, *23*(2), 593–599.

Ibañez, E., Herrero, M., Mendiola, J., & Castro-Puyana, M., (2012). Extraction and characterization of bioactive compounds with health benefits from marine resources: Macro and micro algae, cyanobacteria, and invertebrates. In: Hayes, M., (ed.), *Marine Bioactive Compounds: Sources, Characterization and Applications* (pp. 55–98). Springer, New York, USA.

Ikeda, I., Tanaka, K., Sugano, M., Vahouny, G. V., & Gallo, L. L., (1988). Inhibition of cholesterol absorption in rats by plant sterols. *Journal of Lipid Research*, *29*(12), 1573–1582.

Ikeda, K., Kitamura, A., Machida, H., et al., (2003). Effect of *Undaria pinnatifida* (Wakame) on the development of cerebrovascular diseases in stroke prone spontaneously hypertensive rats. *Clinical and Experimental Pharmacology and Physiology*, *30*(1–2), 44–48.

Ilvessalo, H., & Tuomi, J., (1989). Nutrient availability and accumulation of phenolic compounds in the brown alga *Fucus vesiculosus*. *Marine Biology*, *101*(1), 115–119.

Ina, A., & Kamei, Y., (2006). Vitamin B(12), a chlorophyll-related analog to pheophytin a from marine brown algae, promotes neurite outgrowth and stimulates differentiation in PC12 cells. *Cytotechnology, 52*(3), 181–187.

Ina, A., Hayashi, K. I., Nozaki, H., & Kamei, Y., (2007). Pheophytin a, a low molecular weight compound found in the marine brown alga *Sargassum fulvellum*, promotes the differentiation of PC12 cells. *International Journal of Developmental Neuroscience, 25*(1), 63–68.

Inoue, N., Yamano, N., Sakata, K., Nagao, K., Hama, Y., & Yanagita, T., (2009). The sulfated polysaccharide porphyran reduces apolipoprotein B100 secretion and lipid synthesis in HepG2 cells. *Bioscience, Biotechnology, and Biochemistry, 73*(2), 447–449.

Ishag, H. Z., Li, C., Wang, F., & Mao, X., (2016). Griffithsin binds to the glycosylated proteins (E and prM) of Japanese encephalitis virus and inhibit its infection. *Virus Research, 215*, 50–54.

Ivanova, V., Rouseva, R., Kolarova, M., Serkedjieva, J., Rachev, R., & Manolova, N., (1994). Isolation of a polysaccharide with antiviral effect from *Ulva lactuca. Preparative Biochemistry, 24*(2), 83–97.

Jaswir, I., & Monsur, H., (2011). Anti-inflammatory compounds of macro algae. *Journal of Medicinal Plant Research, 5*(33), 7146–7154.

Jennings, J. G., & Steinberg, P. D., (1997). Phlorotannins versus other factors affecting epiphyte abundance on the kelp *Ecklonia radiata. Oecologia, 109*(3), 461–473.

Jhamandas, J. H., Wie, M. B., Harris, K., MacTavish, D., & Kar, S., (2005). Fucoidan inhibits cellular and neurotoxic effects of beta-amyloid (A beta) in rat cholinergic basal forebrain neurons. *European Journal of Neuroscience, 21*(10), 2649–2659.

Jung, H. A., Hyun, S. K., Kim, H. R., & Choi, J. S., (2006). Angiotensin-converting enzyme I inhibitory activity of pholorotannins from *Ecklonia stolonifera. Fisheries Science, 72*(6), 1292–1299.

Jung, W. K., Heo, S. J., Jeon, Y. J., et al., (2009). Inhibitory effects and molecular mechanism of dieckol isolated from marine brown alga on COX-2 and iNOS in microglial cells. *Journal of Agricultural and Food Chemistry, 57*(10), 4439–4446.

Kaeffer, B., Bénard, C., Lahaye, M., Blottière, H. M., & Cherbut, C., (1999). Biological properties of ulvan, a new source of green seaweed sulfated polysaccharides, on cultured normal and cancerous colonic epithelial cells. *Planta Medica, 65*(6), 527–531.

Kaleagasioglu, F., Guven, K. C., Sezik, Erdugan, H., & Coban, B., (2013). Pharmacology of macroalgae alkaloids. In: Ramawat, K. G., & Merillon, J. M., (eds.), *Natural Products – Phytochemistry, Botany and Metabolism of Alkaloids, Phenolics and Terpenes, Part 2* (pp. 1203–1216). Springer Verlag, Berlin, Heidelberg.

Kamal, C., & Sethuraman, M. G., (2012). Caulerpin-A bis-indole alkaloid as a green inhibitor for the corrosion of mild steel in 1 M HCl solution from the marine alga *Caulerpa racemosa. Industrial and Engineering Chemistry Research, 51*, 10399–10407.

Kamei, Y., & Tsang, C. K., (2003). Sargaquinoic acid promotes neurite outgrowth via protein kinase A and MAP kinases-mediated signaling pathways in PC12D cells. *International Journal of Developmental Neuroscience, 21*(5), 255–262.

Kang, H. S., Chung, H. Y., Jung, J. H., Son, B. W., & Choi, J. S., (2003). A new phlorotannin from the brown alga *Ecklonia stolonifera. Chemical and Pharmacological Bulletin, 51*(8), 1012–1014.

Kang, H. S., Chung, H. Y., Kim, J. Y., Son, B. W., Jung, H. A., & Choi, J. S., (2004a). Inhibitory phlorotannins from the edible brown alga *Ecklonia stolonifera* on total reactive oxygen species (ROS) generation. *Archives of Pharmocological Research, 27*(2), 194–198.

Kang, H. S., Kim, H. R., Byun, D. S., Son, B. W., Nam, T. J., & Choi, J. S., (2004b). Tyrosinase inhibitors isolated from the edible brown alga *Ecklonia stolonifera. Archives in Pharmacological Research, 27*(12), 1226–1232.

Kang, K. A., Lee, K. H., Chae, S., et al., (2005a). Triphlorethol-A from *Ecklonia cava* protects V79-4 lung fibroblast against hydrogen peroxide induced cell damage. *Free Radical Research, 39*(8), 883–892.

Kang, K. A., Lee, K. H., Chae, S., et al., (2005b). Eckol isolated from *Ecklonia cava* attenuates oxidative stress induced cell damage in lung fibroblast cells. *FEBS Letters, 579*, 6295–6304.

Kang, K. A., Lee, K. H., Chae, S., et al., (2006). Cytoprotective effect of phloroglucinol on oxidative stress induced cell damage via catalase activation. *Journal of Cellular Biochemistry, 97*(3), 609–620.

Kang, M. C., Kim, K. N., Lakmal, H. H. C., et al., (2014). Octaphlorethol A isolated from *Ishige foliacea* prevents and protects against high glucose-induced oxidative damage *in vitro* and *in vivo*. *Environmental Toxicology and Pharmacology, 38*(2), 607–615.

Kang, M. C., Wijesinghe, W. A. J. P., Lee, S. H., et al., (2013). Dieckol isolated from brown seaweed *Ecklonia cava* attenuates type II diabetes in db/db mouse model. *Food and Chemical Toxicology, 53*, 294–298.

Kang, S. M., Cha, S. H., Ko, J. Y., et al., (2012). Neuroprotective effects of phlorotannins isolated from a brown alga, *Ecklonia cava*, against H_2O_2-induced oxidative stress in murine hippocampal HT22 cells. *Environmental Toxicology and Pharmacology, 34*(1), 96–105.

Kannan, R. R. R., Aderogba, M. A., Ndhlala, A. R., Stirk, W. A., & Van, S. J., (2013). Acetylcholinesterase inhibitory activity of phlorotannins isolated from the brown alga, *Ecklonia maxima* (Osbeck) Papenfuss. *Food Research International, 54*(1), 1250–1254.

Karsten, U., & West, J. A., (2000). Living in the intertidal zone-seasonal effects on heterosides and sun-screen compounds in the red alga *Bangia atropurpurea* (Bangiales). *Journal of Experimental Marine Biology and Ecology, 254*(2), 221–234.

Karsten, U., Bischof, K., Hanelt, D., Tug, H., & Wiencke, C., (1999). The effect of ultraviolet radiation on photosynthesis and ultraviolet-absorbing substances in the endemic Arctic macroalga *Devaleraea ramentacea* (*Rhodophyta*). *Physiologia Plantarum, 105*(1), 58–66.

Karsten, U., Friedl, T., Schumann, R., Hoyer, K., & Lembcke, S., (2005). Mycosporine-like amino acids and phylogenies in green algae: Prasiola and its relatives from the *Trebouxiophyceae* (*Chlorophyta*). *Journal of Phycology, 41*(3), 557–566.

Karsten, U., Sawall, T., Hanelt, D., et al., (1998). An inventory of UV-absorbing mycosporine-like amino acids in macroalgae from polar to warm-temperate regions. *Botanica Marina, 41*(1–6), 443–453.

Kataoka, H., & Ohnishi, N., (1986). Occurrence of taurine in plants. *Agricultural and Biological Chemistry, 50*(7), 1887–1888.

Katsui, N., Suzuki, Y., Kitamura, S., & Irie, T., (1967). 5,6-dibromoprotocatechualdehyde and 2,3-dibromo-4,5-dihydroxybenzyl methyl ether: New dibromophenols from *Rhodomela larix. Tetrahedron, 23*(3), 1185–1188.

Kawakubo, A., Makino, H., Ohnishi, J., Hirohara, H., & Hori, K., (1999). Occurrence of highly yielded lectins homologous within the genus *Eucheuma. Journal of Applied Phycology, 11*(2), 149–156.

Kawakubo, A., Makino, H., Ohnishi, J., Hirohara, H., & Hori, K., (1997). The marine red alga *Eucheuma serra* J. Agardh, a high yielding source of two isolectins. *Journal of Applied Phycology, 9*(4), 331–338.

Kazłowska, K., Hsu, T., Hou, C. C., Yang, W. C., & Tsai, G. J., (2010). Anti-inflammatory properties of phenolic compounds and crude extract from *Porphyra dentata*. *Journal of Ethnopharmacology*, *128*(1), 123–130.

Khanavi, M., Gheidarloo, R., Sadati, N., et al., (2012). Cytotoxicity of fucosterol containing fraction of marine algae against breast and colon carcinoma cell line. *Pharmacognosy Magazine*, *8*(29), 60–64.

Kicklighter, C. E., Kubanek, J., & Hay, M. E., (2004). Do brominated natural products defend marine worms from consumers? Some do, most don't. *Limnology and Oceanography*, *49*(2), 430–441.

Kim, K. N., Heo, S. J., Kang, S. M., Ahn, G., & Jeon, Y. J., (2010c). Fucoxanthin induces apoptosis in human leukemia HL-60 cells through a ROS-mediated Bcl-xL pathway. *Toxicology In Vitro*, *24*(6), 1648–1654.

Kim, K. N., Heo, S. J., Yoon, W. J., et al., (2010b). Fucoxanthin inhibits the inflammatory response by suppressing the activation of NF-κB and MAPKs in lipopolysaccharide-induced RAW 264.7 macrophages. *European Journal of Pharmacology*, *649*(16), 369–375.

Kim, K. Y., Nam, K. A., Kurihara, H., & Kim, S. M., (2008a). Potent alpha-glucosidase inhibitors purified from the red alga *Grateloupia elliptica*. *Phytochemistry*, *69*(16), 2820–2825.

Kim, K. Y., Nguyen, T. H., Kurihara, H., & Kim, S. M., (2010a). Alpha-glucosidase inhibitory activity of bromophenol purified from the red alga *Polyopes lancifolia*. *Journal of Food Science*, *75*(5), H145–150.

Kim, M. J., Choil, J. S., Kang, S. E., et al., (2004). Multiple allelopathic activity of the crustose coralline alga *Lithophyllum yessoense* against settlement and germination of seaweed spores. *Journal of Applied Phycology*, *16*(3), 175–179.

Kim, M. M., Mendis, E., & Kim, S. K., (2008b). *Laurencia okamurai* extract containing laurinterol induces apoptosis in melanoma cells. *Journal of Medicinal Food*, *11*(2), 260–266.

Kim, M. M., Van, T. Q., Mendis, E., et al., (2006). Phlorotannins in *Ecklonia cava* extract inhibit matrix metalloproteinase activity. *Life Sciences*, *79*(15), 1436–1443.

Kim, S. Y., Kim, S. R., Oh, M. J., Jung, S. J., & Kang, S. Y., (2011). *In vitro* antiviral activity of red alga, *Polysiphonia morrowii* extract and its bromophenols against fish pathogenic infectious hematopoietic necrosis virus and infectious pancreatic necrosis virus. *Journal of Microbiology*, *49*(1), 102–106.

Kirkup, M. P., & Moore, R. E., (1983). Indole alkaloids from the marine red alga *Martensia fragilis*. *Tetrahedron Letters*, *24*(20), 2087–2090.

Kladi, M., Xenaki, H., Vagias, C., Papazafiri, P., & Roussis, V., (2006). New cytotoxic sesquiterpenes from the red algae *Laurencia obtusa* and *Laurencia microcladia*. *Tetrahedron*, *62*(1), 182–189.

Ko, S. C., Jung, W. K., Kang, S. M., et al., (2015). Angiotensin I-converting enzyme (ACE) inhibition and nitric oxide (NO)-mediated antihypertensive effect of octaphlorethol A isolated from *Ishige sinicola*: *In vitro* molecular mechanism and *in vivo* SHR model. *Journal of Functional Foods*, *8*(Part A), 289–299.

Kong, C. S., Kim, J. A., Yoon, N. Y., & Kim, S. K., (2009). Induction of apoptosis by phloroglucinol derivative from *Ecklonia cava* in MCF-7 human breast cancer cells. *Food and Chemical and Toxicology*, *47*(7), 1653–1658.

Konishi, I., Hosokawa, M., Sashima, T., Kobayashi, H., & Miyashita, K., (2006). Halocynthiaxanthin and fucoxanthinol isolated from *Halocynthia roretzi* induce apoptosis in human leukemia, breast and colon cancer cells. *Comparative Biochemistry and Physiology C. Toxicology and Pharmacology*, *142*(1, 2), 53–59.

Kotake-Nara, E., Sugawara, T., & Nagao, A., (2005). Antiproliferative effect of neoxanthin and fucoxanthin on cultured cells. *Fishery Science*, *71*(2), 459–461.

Kräbs, G., Bischof, K., Hanelt, D., Karsten, U., & Wiencke, C., (2002). Wavelength-dependent induction of UV-absorbing mycosporine-like amino acids in the red alga *Chondrus crispus* under natural solar radiation. *Journal of Experimental Marine Biology and Ecology*, *268*(1), 69–82.

Kumari, P., Kumar, M., Gupta, V., Reddy, C. R. K., & Jha, B., (2010). Tropical marine macroalgae as potential sources of nutritionally important PUFAs. *Food Chemistry*, *120*(3), 749–757.

Kumler, W. E., (2017). Evidence against a mechanism of allelopathy in the green alga *Chlorodesmis fastigiata*. *Peer J. Preprints*, *5*, e2700v1.

Kurihara, H., Kagawa, Y., Konno, R., Kim, S. M., & Takahashi, K., (2014). Lipoxygenase inhibitors derived from marine macroalgae. *Bioorganic and Medicinal Chemistry Letters*, *24*(5), 1383–1385.

Kurihara, H., Mitani, T., Kawabata, J., & Takahashi, K., (1999a). Inhibitory potencies of bromophenols from rhodomelaceae algae against α-glucosidase activity. *Fisheries Science*, *65*(2), 300–303.

Kurihara, H., Mitani, T., Kawabata, J., & Takahashi, K., (1999b). Two new bromophenols from the red alga *Odonthalia corymbifera*. *Journal of Natural Products*, *62*(6), 882–884.

Kwon, M. J., & Nam, T. J., (2006). Porphyran induces apoptosis related signal pathway in AGS gastric cancer cell lines. *Life Sciences*, *79*(20), 1956–1962.

Lahaye, M., & Ray, B., (1996). Cell-wall polysaccharides from the marine green alga *Ulva "rigida"* (*Ulvales, Chlorophyta*) – NMR analysis of ulvan oligosaccharides. *Carbohydrate Research*, *283*, 161–173.

Lam, L., & Harder, T., (2007). Marine macroalgae affect abundance and community richness of bacterioplankton in close proximity. *Journal of Phycology*, *43*(5), 874–881.

Larkum, A. W. D., & Kühl, M., (2005). Chlorophyll D: The puzzle resolved. *Trends in Plant Science*, *10*(8), 355–357.

Lazcano-Pérez, F., Román-González, S. A., Sánchez-Puig, N., & Arreguin-Espinosa, R., (2012). Bioactive peptides from marine organisms: A short overview. *Protein Peptide Letters*, *19*(7), 700–707.

Le, Q. T., Li, Y., Qian, Z. J., Kim, M. M., & Kim, S. K., (2009). Inhibitory effects of polyphenols isolated from marine alga *Ecklonia cava* on histamine release. *Process Biochemistry*, *44*(2), 168–176.

Lee, C., (2019). Griffithsin, a highly potent broad-spectrum antiviral lectin from red algae: From discovery to clinical application. *Marine Drugs*, *17*(10), 567.

Lee, H. S., Lee, T. H., Lee, J. H., et al., (2007a). Inhibition of the pathogenicity of *Magnaporthe grisea* by bromophenols, isocitrate lyase inhibitors, from the red alga *Odonthalia corymbifera*. *Journal of Agricultural and Food Chemistry*, *55*(17), 6923–6928.

Lee, J. H., Lee, T. K., Kang, R. S., Shin, H. J., & Lee, H. S., (2007b). The *in vitro* antioxidant activities of the bromophenols from the red alga *Tichocarpus crinitus* and phenolic derivatives. *Journal of Korean Magnetic Resonance Society*, *11*(1), 56–63.

Lee, J. Y., Kim, S. M., Jung, W. S., et al., (2012). Phlorofucofuroeckol-A, a potent inhibitor of aldo-keto reductase family 1 member B10, from the edible brown alga *Eisenia bicyclis*. *Journal of Korean Society of Applied Biological Chemistry*, *55*(6), 721–727.

Lee, S. H., Han, J. S., Heo, S. J., Hwang, J. Y., & Jeon, Y. J., (2010). Protective effects of dieckol isolated from *Ecklonia cava* against high glucose-induced oxidative stress in human umbilical vein endothelial cells. *Toxicology In Vitro, 24*(2), 375–381.

Lee, S. H., Kang, S. M., Ko, S. C., Lee, D. H., & Jeon, Y. J., (2012). Octaphlorethol A, a novel phenolic compound isolated from a brown alga, *Ishige foliacea*, increases glucose transporter 4-mediated glucose uptake in skeletal muscle cells. *Biochemistry Biophysics Research Communication, 420*(3), 576–581.

Lee, S. H., Yong-Li, F., Karadeniz, F., Kim, M. M., & Kim, S. K., (2009b). α-Glucosidase and α-amylase inhibitory activities of phloroglucinal derivatives from edible marine brown alga, *Ecklonia cava. Journal of Science of Food and Agriculture, 89*(9), 1552–1558.

Lee, S., Lee, Y. S., Jung, S. H., Kang, S. S., & Shin, K. H., (2003). Anti-oxidant activities of fucosterol from the marine algae *Pelvetia siliquosa. Archives of Pharmacal Research, 26*(9), 719–722.

Lee, Y. R., Noh, E. M., Jeong, E. Y., et al., (2009a). Cordycepin inhibits UVB-induced matrix metalloproteinase expression by suppressing the NF-κB pathway in human dermal fibroblasts. *Experimental and Molecular Medicine, 41*(8), 548–554.

Lee, Y. S., Shin, K. H., Kim, B. K., & Lee, S., (2004). Anti-diabetic activities of fucosterol from *Pelvetia siliquosa. Archives of Pharmacal Research, 27*(11), 1120–1122.

Lhullier, C., Falkenberg, M., Ioannou, E., et al., (2010). Cytotoxic halogenated metabolites from the Brazilian red alga *Laurencia catarinensis. Journal of Natural Products, 73*(1), 27–32.

Li, B., Lu, F., Wei, X., & Zhao, R., (2008c). Fucoidan: Structure and bioactivity. *Molecules, 13*(8), 1671–1695.

Li, D. Y., Xu, Z., & Zhang, S. H., (1999). Prevention and cure of fucoidan of *L. japonica* on mice with hypercholesterolemia. *Food Science, 20*, 45–46.

Li, K., Li, X. M., Ji, N. Y., & Wang, B. G., (2008a). Bromophenols from the marine red alga *Polysiphonia urceolata* with DPPH radical scavenging activity. *Journal of Natural Products, 71*(1), 28–30.

Li, L., Yu, X., Zhang, H., et al., (2019). *In vitro* antiviral activity of griffithsin against porcine epidemic diarrhea virus. *Virus Genes, 55*(2), 174–181.

Li, Y. X., Wijesekara, I., Li, Y., & Kim, S. K., (2011). Phlorotannins as bioactive agents from brown algae. *Process Biochemistry, 46*, 2219–2224.

Li, Y., Lee, S. H., Le, Q. T., Kim, M. M., & Kim, S. K., (2008b). Anti-allergic effects of phlorotannins on histamine release via binding inhibition between IgE and FcεRI. *Journal of Agricultural and Food Chemistry, 56*(24), 12073–12080.

Li, Y., Qian, Z. J., Ryu, B., Lee, S. H., Kim, M. M., & Kim, S. K., (2009). Chemical components and its antioxidant properties *in vitro*: An edible marine brown alga, *Ecklonia cava. Bioorganic and Medicinal Chemistry, 17*(5), 1963–1973.

Liang, Y., Li, X. M., Li, C. S., Sun, H., & Wang, B. G., (2014). Laurane-, cyclolaurane-, and cuparane-type sesquiterpenes from the marine red alga *Laurencia okamurai. Natural Product Communications, 9*(3), 323–324.

Liao, W. R., Lin, J. Y., Shieh, W. Y., Jeng, W. L., & Huang, R., (2003). Antibiotic activity of lectins from marine algae against marine vibrios. *Journal of Industrial Microbiology and Biotechnology, 30*(7), 433–439.

Lijun, H., Nianjun, X., Jiangong, S., Xiaojun, Y., & Chengkui, Z., (2005). Isolation and pharmacological activities of bromophenols from *Rhodomela confervoides. Chinese Journal of Oceanology and Limnology, 23*(2), 226–229.

Liu, C. L., Huang, Y. S., Hosokawa, M., Miyashita, K., & Hu, M. L., (2009b). Inhibition of proliferation of a hepatoma cell line by fucoxanthin in relation to cell cycle arrest and enhanced gap junctional intercellular communication. *Chemical and Biological Interactions*, *182*(2, 3), 165–172.

Liu, D. Q., Mao, S. C., Yu, X. Q., Feng, L. H., & Lai, X. P., (2012). Caulerchlorin, a novel chlorinated bisindole alkaloid with antifungal activity from the Chinese green alga *Caulerpa racemosa*. *Heterocycles*, *85*(3), 661–666.

Liu, D. Q., Mao, S. C., Zhang, H. Y., et al., (2013). Racemosins A and B, two novel bisindole alkaloids from the green alga *Caulerpa racemosa*. *Fitoterapia*, *91*, 15–20.

Liu, H., & Gu, L., (2012). Phlorotannins from brown algae (*Fucus vesiculosus*) inhibited the formation of advanced glycation end products by scavenging reactive carbonyls. *Journal of Agricultural and Food Chemistry*, *60*(5), 1326–1334.

Liu, M., Hansen, P. E., & Lin, X., (2011). Bromophenols in marine algae and their bioactivities. *Marine Drugs*, *9*(7), 1273–1292.

Liu, Y., Morgan, J. B., Coothankandaswamy, V., et al., (2009a). The *Caulerpa* pigment caulerpin inhibits HIF-1 activation and mitochondrial respiration. *Journal of Natural Products*, *72*(12), 2104–2109.

Lopes, G., Sousa, C., Bernardo, J., et al., (2011). Sterol profiles in 18 macroalgae of the Portuguese Coast. *Journal of Phycology*, *47*(5), 1210–1218.

Lopes, G., Sousa, C., Valentão, P., & Andrade, P. B., (2013). Sterols in algae and health. In: Hernández-Ledesma, B., & Herrero, M., (eds.), *Bioactive Compounds from Marine Foods* (pp. 173–191). John Wiley and Sons Ltd, Oxford, UK.

Lordan, S., Ross, R. P., & Stanton, C., (2011). Marine bioactives as functional food ingredients: Potential to reduce the incidence of chronic diseases. *Marine Drug*, *9*(6), 1056–1100.

Lourenço, R., & Camilo, M. E., (2002). Taurine: A conditionally essential amino acid in humans? An overview in health and disease. *Nutrition Hospitalaria*, *17*(6), 262–270.

Love, J., & Percival, E., (1964). The polysaccharides of the green seaweed codium fragile. Part II. The water-soluble sulphated polysaccharides. *Journal of Chemical Society*, *632*, 3338.

Lu, H., Xie, H., Gong, Y., Wang, Q., & Yang, Y., (2011). Secondary metabolites from the seaweed *Gracilaria lemaneiformis* and their allelopathic effects on *Skeletonema costatum*. *Biochemical Systematics and Ecology*, *39*(4–6), 397–400.

Macedo, N. R. P. V., Ribeiro, M. S., Villaca, R. C., et al., (2012). Caulerpin as a potential antiviral drug against herpes simplex virus type 1. *Brazilian Journal of Pharmacognosy*, *22*(4), 861–867.

Machado, F. L. S., Ventura, T. L. B., De Gestinari, L. M. S., et al., (2014). Sesquiterpenes from the Brazilian red alga *Laurencia dendroidea* J. Agardh. *Molecules*, *19*(3), 3181–3192.

Maeda, H., Hosokawa, M., Sashima, T., Funayama, K., & Miyashita, K., (2005). Fucoxanthin from edible seaweed, *Undaria pinnatifida*, shows antiobesity effect through UCP1 expression in white adipose tissues. *Biochemical and Biophysical Research Communications*, *332*(2), 392–397.

Maeda, H., Hosokawa, M., Sashima, T., Murakami-Funayama, K., & Miyashita, K., (2009). Anti-obesity and anti-diabetic effects of fucoxanthin on diet-induced obesity conditions in a murine model. *Molecular Medical Reports*, *2*(6), 897–902.

Mao, W., Zang, X., Li, Y., & Zhang, H., (2006). Sulfated polysaccharides from marine green algae *Ulva conglobata* and their anticoagulant activity. *Journal of Applied Phycology*, *18*(1), 9–14.

Martínez, B., & Rico, J. M., (2008). Changes in nutrient content of *Palmaria palmata* in response to variable light and upwelling in Northern Spain. *Journal of Phycology*, *44*(1), 50–59.

Martinez, J. L., Sepulveda, S. P., Rovirosa, J., & San, M. A., (1997). Efectos en aorta y auricula aisladas de rata de diacetil epitaondiol, diterpenoide del alga *Stypopodium flabelliforme*. *The Journal of Argentine Chemical Society* (*Anales de la Asociación Química Argentina*) 85, 69–75.

Matsubara, K., Matsuura, Y., Bacic, A., Liao, M., Hori, K., & Miyazawa, K., (2001). Anticoagulant properties of a sulfated galactan preparation from a marine green alga, *Codium cylindricum*. *International Journal of Biological Macromolecules*, *28*(5), 395–399.

Mattos, B. B., Romanos, M. T. V., De Souza, L. M., Sassaki, G., & Barreto-Bergter, E., (2011). Glycolipids from macroalgae: Potential biomolecules for marine biotechnology. *Revista Brasileira de Farmacognosia*, *21*, 244–247.

McClintock, J. B., & Baker, B. J., (2001). *Marine Chemical Ecology*. CRC Press, Florida.

Mendes, G., Soares, A. R., Sigiliano, L., et al., (2011). *In vitro* anti-HMPV activity of meroditerpenoids from marine alga *Stypopodium zonale* (Dictyotales). *Molecules*, *16*(10), 8437–8450.

Mendis, E., & Kim, S. K., (2011). Present and future prospect of seaweeds in developing functional foods. In: Kim, S. K., (ed.). *Advances in Food and Nutrition Research – Marine Functional Foods: Implication and Applications, Macro and Micro Algae* (Vol. 64, p. 11). Academic press, USA.

Militante, J. D., & Lombardini, J. B., (2002). Taurine: Evidence of physiological function in the retina. *Nutrional Neuroscience*, *5*(2), 75–90.

Miller, I. J., Falshaw, R., & Furneaux, R. H., (1995). Structural analysis of the polysaccharide from *Pachymenia lusoria* (Cryptonemiaceae, Rhodophyta). *Carbohydrate Research*, *268*(2), 219–232.

Millet, J. K., Seron, K., Labitt, R. N., et al., (2016). Middle east respiratory syndrome coronavirus infection is inhibited by griffithsin. *Antiviral Research*, *133*, 1–8.

Moghadamtousi, S. Z., Karimian, H., Khanabdali, R., et al., (2014). Anticancer and antitumor potential of fucoidan and fucoxanthin, two main metabolites isolated from brown algae. *Science World Journal*, *2014*, 1–10.

Moon, H. E., Islam, N., Ahn, B. R., et al., (2011). Protein tyrosine phosphatase 1B and α-glucosidase inhibitory phlorotannins from edible brown algae, *Ecklonia stolonifera* and *Eisenia bicyclis*. *Bioscience, Biotechnology and Biochemistry*, *75*(8), 1472–1480.

Mori, T., O'Keefe, B. R., Sowder, R. C., et al., (2005). Isolation and characterization of griffithsin, a novel HIV-inactivating protein, from the red alga *Griffithsia* sp. *Journal of Biological Chemistry*, *280*(10), 9345–9353.

Morrice, L. M., McLean, M. W., Long, W. F., & Williamson, F. B., (1984). Porphyran primary structure. In: *Eleventh International Seaweed Symposium* (pp. 572–575). Springer Netherlands, Dordrecht.

Morse, A. N. C., (1991). GABA-mimetic peptides from marine algae and cyanobacteria as potential diagnostic and therapeutic agents. In: Thompson, M. F., Sarojini, R., & Nagabhushanam, R., (eds.), *Bioactive compounds from marine organisms-with emphasis on the Indian Ocean* (pp. 167–172). Oxford & IBH Publishing Co., New Delhi.

Murakami, H., Kato, T., Mimura, A., & Takahara, Y., (1994). New indole derivatives from *Martensia denticulata* seaweed. *Bioscience, Biotechnology and Biochemistry*, *58*(3), 535–538.

Myung, C. S., Shin, H. C., Bao, H. Y., Yeo, S. J., Lee, B. H., & Kang, J. S., (2005). Improvement of memory by dieckol and phlorofucofuroeckol in ethanol-treated mice:

Possible involvement of the inhibition of acetylcholinesterase. *Archieves of Pharmacal Research, 28*(6), 691–698.

N'Diaye, I., Guella, G., Mancini, I., & Pietra, F., (1996). Almazole D, a new type of antibacterial 2,5-disubstituted oxazolic dipeptide from a red alga of the coast of Senegal. *Tetrahedron Letters, 37*(17), 3049–3050.

Nagayama, K., Iwamura, Y., Shibata, T., Hirayama, I., & Nakamura, T., (2002). Bactericidal activity of phlorotannins from the brown alga *Ecklonia kurome*. *Journal of Antimicrobial Chemotheraphy, 50*(6), 889–893.

Nakazawa, Y., Sashima, T., Hosokawa, M., & Miyashita, K., (2009). Comparative evaluation of growth inhibitory effect of stereoisomers of fucoxanthin in human cancer cell lines. *Journal of Functional Foods, 1*(1), 88–97.

Narayan, B., Miyashita, K., & Hosakawa, M., (2005). Comparative evaluation of fatty acid composition of different *Sargassum* (*Fucales*, *Phaeophyta*) species harvested from temperate and tropical waters. *Journal of Aquatic Food Product Technology, 13*(4), 53–70.

Neves, S. A., Freitas, A. L. P., Sousa, B. W., et al., (2007). Antinociceptive properties in mice of a lectin isolated from the marine alga *Amansia multifida* Lamouroux. Brazilian Journal of Medical and Biological Research, *40*(1), 127–134.

Nishida, A., Fuwa, M., Fujikawa, Y., Nakahata, E., Furuno, A., & Nakagawa, M., (1998). First total synthesis of martefragin A, a potent inhibitor of lipid peroxidation isolated from sea alga. *Tetrahedron Letters, 39*(33), 5983–5986.

Nixon, B., Stefanidou, M., Mesquita, P. M., et al., (2013). Griffithsin protects mice from genital herpes by preventing cell-to-cell spread. *Journal of Virology, 87*(11), 6257–6269.

Nomura, T., Kikuchi, M., Kubodera, A., & Kawakami, Y., (1997). Proton-donative antioxidant activity of fucoxanthin with 1,1-diphenyl-2-picrylhydrazyl (DPPH). *IUBMB Life, 42*(2), 361–370.

O'Keefe, B. R., Giomarelli, B., Barnard, D. L., et al., (2010). Broad-spectrum *in vitro* activity and *in vivo* efficacy of the antiviral protein griffithsin against emerging viruses of the family coronaviridae. *Journal of Virology, 84*(5), 2511–2521.

Okai, Y., & Higashi-Okai, K., (1997). Potent anti-inflammatory activity of pheophytin a derived from edible green alga, *Enteromorpha prolifera* (Sujiao-nori). *International Journal of Immunopharmacology, 19*(6), 355–358.

Okai, Y., Higashi-Okai, K., Yano, Y., & Otani, S., (1996). Identification of antimutagenic substances in an extract of edible red alga, *Porphyra tenera* (Asakusa-nori). *Cancer Letters, 100*(1–2), 235–240.

Olsen, E. K., Hansen, E., Isaksson, J., & Andersen, J. H., (2013). Cellular antioxidant effect of four bromophenols from the red algae, *Vertebrata lanosa*. *Marine Drugs, 11*(8), 2769–2784.

Oren, A., & Gunde-Cimerman, N., (2007). Mycosporines and mycosporine-like amino acids: UV protectants or multipurpose secondary metabolites. *FEMS Microbiological Letters, 269*(1), 1–10.

Pangestuti, R., & Kim, S. K., (2011). Biological activities and health benefit effects of natural pigments derived from marine algae. *Journal of Functional Foods, 3*(4), 255–266.

Park, H. J., Kurokawa, M., Shiraki, K., Nakamura, N., Choi, J. S., & Hattori, M., (2005). Antiviral activity of the marine alga *Symphyocladia latiuscula* against herpes simplex virus (HSV-1) *in vitro* and its therapeutic efficacy against HSV-1 infection in mice. *Biological and Pharmaceutical Bulletin, 28*(12), 2258–2262.

Park, H. Y., Han, M. H., Park, C., et al., (2011). Anti-inflammatory effects of fucoidan through inhibition of NF-κB, MAPK and Akt activation in lipopolysaccharide-induced BV2 microglia cells. *Food and Chemical Toxicology*, *49*(8), 1745–1752.

Parsons, A. F., (1996). Recent developments in kainoid amino acid chemistry. *Tetrahedron*, *52*(12), 4149–4174.

Paul, V. J., Ritson-Williams, R., & Sharp, K., (2011). Marine chemical ecology in benthic environments. *Natural Product Reports*, *28*(2), 345–387.

Pedersen, M., Collen, J., Abrahamsson, K., & Ekdahl, A., (1996). Production of halocarbons from seaweeds: An oxidative stress reaction. *Marine Science*, *60*, 257–263.

Percival, E., & McDowell, R. H., (1967). *Chemistry and Enzymology of Marine Algal Polysaccharides* (p. 219). Academic Press, New York, USA.

Percot, A., Guven, K. C., Aysel, V., Erdugan, H., & Gezgin, T., (2009b). *N*-acetyltyramine from *Phyllophora crispa* (Hudson) P.S. Dixon and N-acetylphenylethylamine from *Gelidium crinale* (Hare ex Turner) graillon. Acta Pharmaceutica Sciencia, *51*(1), 9–14.

Percot, A., Yalçın, A., Aysel, V., Erdugan, H., Dural, B., & Güven, K. C., (2009a). β-Phenylethylamine content in marine algae around Turkish coasts. *Botanica Marina*, *52*(1), 87–90.

Pereira, D. M., Cheel, J., Areche, C., et al., (2011). Anti-proliferative activity of meroditerpenoids isolated from the brown alga *Stypopodium flabelliforme* against several cancer cell lines. *Marine Drugs*, *9*(5), 852–862.

Pereira, H., Barreira, L., Figueiredo, F., et al., (2012). Polyunsaturated fatty acids of marine macroalgae: Potential for nutritional and pharmaceutical applications. *Marine Drugs*, *10*(9), 1920–1935.

Pereira, M. G., Benevides, N. M. B., Melo, M. R. S., Valente, A. P., Melo, F. R., & Mourão, P. A. S., (2005). Structure and anticoagulant activity of a sulfated galactan from the red alga, *Gelidium crinale*. Is there a specific structural requirement for the anticoagulant action. *Carbohydrate Research*, *340*(12), 2015–2023.

Pereira, R. C., Da Gama, B. A. P., Teixeira, V. L., & Yoneshigue-Valentin, Y., (2003). Ecological roles of natural products of the Brazilian red seaweed *Laurencia obtusa. Brazilian Journal of Biology*, *63*(4), 665–672.

Pérez, M. J., Falqué, E., & Domínguez, H., (2016). Antimicrobial action of compounds from marine seaweed. *Marine Drugs*, *14*(3), 52.

Pesando, D., Lemée, R., Ferrua, C., Amade, P., & Girard, J. P., (1996). Effects of caulerpenyne, the major toxin from *Caulerpa taxifolia* on mechanisms related to sea urchin egg cleavage. *Aquatic Toxicology*, *35*(3–4), 139–155.

Pietá, F. D., Bilia, A. R., Breschi, M. C., Cinelli, F., Morelli, I., & Scatizzi, R., (1993). Crude extracts and two linear diterpenes from *Cystoseira balearica* and their activity. *Planta Medica*, *59*(2), 135–138.

Pietà, F. D., Breschi, M. C., Scatizzi, R., & Cinelli, F., (1995). Relaxing activity of two linear diterpenes from *Cystoseira brachycarpa* var. *balearica* on the contractions of intestinal preparations. *Planta Medica*, *61*(6), 493–496.

Plouguerne, E., Da Gama, B. A. P., Pereira, R. C., & Barreto-Bergter, E., (2014). Glycolipids from seaweeds and their potential biotechnological applications. *Frontiers in Cellular and Infection Microbiology*, *4*, 1–5.

Plouguerné, E., Ioannou, E., Georgantea, P., et al., (2010). Anti-microfouling activity of lipidic metabolites from the invasive brown alga *Sargassum muticum* (Yendo) Fensholt. *Marine Biotechnology*, *12*(1), 52–61.

Pomin, V. H., & Mourão, P. A., (2008). Structure, biology, evolution, and medical importance of sulfated fucans and galactans. *Glycobiology, 18*(12), 1016–1027.
Praseptiangga, D., (2016). Algal lectins and their potential uses. *Squalen Bulletin of Marine Fisheries Postharvest Biotechnology, 10*(2), 89–98.
Prokofjeva, M. M., Imbs, T. I., Shevchenko, N. M., et al., (2013). Fucoidans as potential inhibitors of HIV-1. *Marine Drugs, 11*(8), 3000–3014.
Qi, H., Zhang, Q., Zhao, T., et al., (2005). Antioxidant activity of different sulfate content derivatives of polysaccharide extracted from *Ulva pertusa* (*Chlorophyta*) *in vitro*. *International Journal of Biological Macromolecules, 37*(4), 195–199.
Qi, H., Zhang, Q., Zhao, T., Hu, R., Zhang, K., & Li, Z., (2006). *In vitro* antioxidant activity of acetylated and benzoylated derivatives of polysaccharide extracted from *Ulva pertusa* (*Chlorophyta*). *Bioorganic and Medicinal Chemistry Letters, 16*(9), 2441–2445.
Raniello, R., Mollo, E., Lorenti, M., Gavagnin, M., & Buia, M. C., (2007). Phytotoxic activity of caulerpenyne from the Mediterranean invasive variety of *Caulerpa racemosa*: A potential allelochemical. *Biological Invasions, 9*(4), 361–368.
Rao, H. B. R., Sathivel, A., & Devaki, T., (2004). Antihepatotoxic nature of *Ulva reticulata* (*Chlorophyceae*) on acetaminophen-induced hepatoxicity in experimental rats. *Journal of Medicinal Food, 7*(4), 495–497.
Rasher, D. B., Stout, E. P., Engel, S., Kubanek, J., & Hay, M. E., (2011). Macroalgal terpenes function as allelopathic agents against reef corals. *Proceedings of the National Academy of Sciences USA, 108*(43), 17726–17731.
Rocha, F. D., Soares, A. R., Houghton, P. J., Pereira, R. C., Kaplan, M. A. C., & Teixeira, V. L., (2007). Potential cytotoxic activity of some Brazilian seaweeds on human melanoma cells. *Phytotheraphy Research, 21*(2), 170–175.
Rosa, G. P., Tavares, W. R., Sousa, P. M. C., Pages, A. K., Seca, A. M. L., & Pinto, D. C. G. A., (2020). Seaweed secondary metabolites with beneficial health effects: An overview of successes in *in vivo* studies and clinical trials. *Marine Drugs, 18*(1), 8.
Rovirosa, J., & Martin, A. S., (1997). Antimicrobial activity of the brown alga *Stypopodium flabelliforme* constituents. *Fitoterapia, 68*(5), 473–475.
Ryu, B., Li, Y., Qian, Z. J., Kim, M. M., & Kim, S. K., (2009). Differentiation of human osteosarcoma cells by isolated phlorotannins is subtly linked to COX-2, iNOS, MMPs, and MAPK signaling: Implication for chronic articular disease. *Chemico Biological Interactions, 179*(2–3), 192–201.
Ryu, G., Park, S. H., Kim, E. S., Choi, B. W., Ryu, S. Y., & Lee, B. H., (2003). Cholinesterase inhibitory activity of two farnesylacetone derivatives from the brown alga *Sargassum sagamianum*. *Archives in Pharmaceutical Research, 26*(10), 796–799.
Saito, M., & Hagino, H., (2005). Antihypertensive effect of oligopeptides derived from nori (*Porphyra yezoensis*) and Ala-Lys-Tyr-Ser-Tyr in rats. *Journal of Japanese Society for Nutrition and Food Science, 5*, 177–184.
Sakemi, S., Higa, T., Jefford, C. W., & Bernardinelli, G., (1986). Venustatriol. A new, anti-viral, triterpene tetracyclic ether from *Laurencia venusta*. *Tetrahedron Letters, 27*, 4287–4290.
Santos, A. O., Veiga-Santos, P., Ueda-Nakamura, T., et al., (2010). Effect of elatol, isolated from red seaweed *Laurencia dendroidea*, on *Leishmania amazonensis*. *Marine Drugs, 8*(11), 2733–2743.
Sasaki, K., Ishihara, K., Oyamada, C., et al., (2008). Effects of fucoxanthin addition to ground chicken breast meat on lipid and color stability during chilled storage, before and after cooking. *Asian-Australasian Journal of Animal Sciences, 21*(7), 1067–1072.

Sathya, R., Kanaga, N., Sankar, P., & Jeeva, S., (2013). Antioxidant properties of phlorotannins from brown seaweed *Cystoseira trinodis* (Forsskål) C. Agardh. *Arabian Journal of Chemistry, 10*, S2608–S2614.

Sato, M., Hosokawa, T., Yamaguchi, T., et al., (2002). Angiotensin I-converting enzyme inhibitory peptides derived from wakame (*Undaria pinnatifida*) and their antihypertensive effect in spontaneously hypertensive rats. *Journal of Agricultural and Food Chemistry, 50*(21), 6245–6452.

Schmid, D., Schürch, C., & Zülli, F., (2006). Mycosporine-like amino acids from red algae protect against premature skin-aging. *European Cosmetics, 9*, 1–4.

Schmid, M., Guihéneuf, F., & Stengel, D. B., (2014). Fatty acid contents and profiles of 16 macroalgae collected from the Irish Coast at two seasons. *Journal of Applied Phycology, 26*(1), 451–463.

Schubert, N., García-Mendoza, E., & Pacheco-Ruiz, I., (2006). Carotenoid composition of marine red algae. *Journal of Phycology, 42*(6), 1208–1216.

Sekar, S., & Chandramohan, M., (2008). Phycobiliproteins as a commodity: Trends in applied research, patents and commercialization. *Journal of Applied Phycology, 20*(2), 113–136.

Sekikawa, I., Kubota, C., Hiraoki, T., & Tsujino, I., (1986). Isolation and structure of a 357 nm UV-absorbing substance usujirene from the red alga *Palmaria palmata* (L. O. Kuntze). *Japanese Journal of Phycology, 34*, 185–188.

Sharon, N., & Lis, H., (2004). History of lectins: From hemagglutinins to biological recognition molecules. *Glycobiology, 14*(11), 53R–62R.

Shevchenko, N. M., Burtseva, Y. V., Zvyagintseva, T. N., et al., (2009). Polysaccharides and sterols from green algae *Caulerpa lentillifera* and *C. sertularioides*. *Chemistry of Natural Compounds, 45*(1), 1–5.

Shi, D., Li, J., Guo, S., Su, H., & Fan, X., (2009). The antitumor effect of bromophenol derivatives *in vitro* and *Leathesia nana* extract *in vivo*. *Chinese Journal of Oceanology and Limnology, 27*(2), 277–282.

Shi, D., Li, X., Li, J., Guo, S., Su, H., & Fan, X., (2010). Antithrombotic effects of bromophenol, an alga-derived thrombin inhibitor. *Chinese Journal of Oceanology and Limnology, 28*(1), 96–98.

Shi, D., Xu, F., He, J., Li, J., Fan, X., & Han, L., (2008). Inhibition of bromophenols against PTP1B and anti-hyperglycemic effect of *Rhodomela confervoides* extract in diabetic rats. *Science Bulletin, 53*(16), 2476–2479.

Shibata, T., Fujimoto, K., Nagayama, K., Yamaguchi, K., & Nakamura, T., (2002). Inhibitory activity of brown algal phlorotannins against hyaluronidase. *International Journal of Food Science and Technology, 37*(6), 703–709.

Shibata, T., Ishimaru, K., Kawaguchi, S., Yoshikawa, H., & Hama, Y., (2008). Antioxidant activities of phlorotannins isolated from Japanese *Laminariaceae*. *Journal of Applied Phycology, 20*(5), 705–711.

Shick, J. M., & Dunlap, W. C., (2002). Mycosporine-like amino acids and related gadusols: Biosynthesis, acumulation, and UV-protective functions in aquatic organisms. *Annual Reviews on Physiology, 64*(1), 223–262.

Shimoda, H., Tanaka, J., Shan, S. J., & Maoka, T., (2010). Anti-pigmentary activity of fucoxanthin and its influence on skin mRNA expression of melanogenic molecules. *Journal Pharm. Pharmacol., 62*(9), 1137–1145.

Shiu, C. T., & Lee, T. M., (2005). Ultraviolet-B-induced oxidative stress and responses of the ascorbate-glutathione cycle in a marine macroalga *Ulva fasciata*. *Journal of Experimental Botany*, *56*(421), 2851–2865.

Sieburth, J. M., & Conover, J. T., (1965). Sargassum tannin, an anti-biotic which retards fouling. *Nature*, *208*(5005), 52, 53.

Singh, I. P., & Bharate, S. B., (2006). Phloroglucinol compounds of natural origin. *Natural Products Reports*, *23*(4), 558–591.

Singh, S. P., Kumari, S., Rastogi, R. P., Singh, K. L., & Sinha, R. P., (2008). Mycosporine-like amino acids (MAAs): Chemical structure, biosynthesis and significance as UV-absorbing/screening compounds. *Indian Journal of Experimental Biology*, *46*(1), 7–17.

Sinha, R. P., Klisch, M., Gröniger, A., & Häder, D. P., (2000). Mycosporine-like amino acids in the marine red alga *Gracilaria cornea*-effects of UV and heat. *Environmental and Experimental Botany*, *43*(1), 33–43.

Sinha, R. P., Singh, S. P., & Häder, D. P., (2007). Database on mycosporines and mycosporine-like amino acids (MAAs) in fungi, cyanobacteria, macroalgae, phytoplankton and animals. *Journal of Photochemistry and Photobiology B*, *89*(1), 29–35.

Siriwardhana, N., Lee, K. W., & Jeon, Y. J., (2005). Radical scavenging potential of hydrophilic phlorotannins of *Hizikia fusiformis*. *Algae*, *20*(1), 69–75.

Smit, A. J., (2004). Medicinal and pharmaceutical uses of seaweed natural products: A review. *Journal of Applied Phycology*, *16*(4), 245–262.

Soares, A. R., Abrantes, J. L., Lopes, S. T. M., et al., (2007). *In vitro* antiviral effect of meroditerpenes isolated from the Brazilian seaweed *Stypopodium zonale* (Dictyotales). *Planta Medica*, *73*(11), 1221–1224.

Soni, B., Visavadiya, N. P., & Madamwar, D., (2009). Attenuation of diabetic complications by C-phycoerythrin in rats: Antioxidant activity of C-phycoerythrin including copper-induced lipoprotein and serum oxidation. *Brazilian Journal of Nutrition*, *102*(1), 102–109.

South, G. R., & Whittick, A., (2009). *An Introduction to Phycology* (p. 354). John Wiley & Sons, Vancouver.

Sparidans, R. W., Stokvis, E., Jimeno, J. M., López-Lázaro, L., Schellens, J. H., & Beijnen, J. H., (2001). Chemical and enzymatic stability of a cyclic depsipeptide, the novel, marine-derived, anti-cancer agent kahalalide F. *Anticancer Drugs*, *12*(7), 575–582.

Stengel, D. B., Connan, S., & Popper, Z. A., (2011). Algal chemodiversity and bioactivity: Sources of natural variability and implications for commercial application. *Biotechnological Advances*, *29*(5), 483–501.

Suárez, Y., González, L., Cuadrado, A., Berciano, M., Lafarga, M., & Muñoz, A., (2003). Kahalalide F, a new marine-derived compound, induces oncosis in human prostate and breast cancer cells. *Molecular Cancer Therapy*, *2*(9), 863–872.

Sudatti, D. B., Duarte, H. M., Soares, A. R., Salgado, L. T., & Pereira, R. C., (2020). New ecological role of seaweed secondary metabolites as autotoxic and allelopathic. *Frontiers in Plant Science*, *11*, 347.

Suetsuna, K., & Nakano, T., (2000). Identification of an antihypertensive peptide from peptic digest of wakame (*Undaria pinnatifida*). *Journal of Nutritional Biochemistry*, *11*(9), 450–454.

Suetsuna, K., (1998). Purification and identification of angiotensin I-converting enzyme inhibitors from the red alga *Porphyra yezoensis*. *Journal of Marine Biotechnology*, *6*(3), 163–167.

Suetsuna, K., Maekawa, K., & Chen, J. R., (2004). Antihypertensive effects of *Undaria pinnatifida* (wakame) peptide on blood pressure in spontaneously hypertensive rats. *Journal of Nutritional Biochemistry*, *15*(5), 267–272.

Sugahara, T., Ohama, Y., Fukuda, A., Hayashi, M., Kawakubo, A., & Kato, K., (2001). The cytotoxic effect of *Eucheuma serra* agglutinin (ESA) on cancer cells and its application to molecular probe for drug delivery system using lipid vesicles. *Cytotechnology*, *36*(1), 93–99.

Sugawara, T., Matsubara, K., Akagi, R., Mori, M., & Hirata, T., (2006). Antiangiogenic activity of brown algae fucoxanthin and its deacetylated product, fucoxanthinol. *Journal of Agricultural and Food Chemistry*, *54*(26), 9805–9810.

Sugiura, Y., Matsuda, K., Okamoto, T., et al., (2009). The inhibitory effects of components from a brown alga, *Eisenia arborea*, on degranulation of mast cells and eicosanoid synthesis. *Journal of Functional Foods*, *1*(4), 387–393.

Sugiura, Y., Tanaka, R., Katsuzaki, H., Imai, K., & Matsushita, T., (2013). The anti-inflammatory effects of phlorotannins from *Eisenia arborea* on mouse ear edema by inflammatory inducers. *Journal of Functional Food*, *5*(1), 2019–2023.

Swanson, D., Block, R., & Mousa, S. A., (2012). Omega-3 fatty acids EPA and DHA: Health benefits throughout life. *Advances in Nutrition*, *3*(1), 1–7.

Takahashi, S., Matsunaga, T., Hasegawa, C., et al., (1997). Studies on bioactive substances in natural resources. VI. Isolation and structure elucidation of martefragin A. *Toyama-ken Yakuji Kenkyusho Nenpo*, *24*, 53–57.

Takahashi, S., Matsunaga, T., Hasegawa, C., et al., (1998). Martefragin A, a novel indole alkaloid isolated from red alga, inhibits lipid peroxidation. *Chemical Pharmaceutical Bulletin (Tokyo)*, *46*(10), 1527–1529.

Takaichi, S., (2011). Carotenoids in algae: Distributions, biosyntheses and functions. *Marine Drugs*, *9*(6), 1101–1118.

Takamatsu, S., Hodges, T. W., Rajbhandari, I., Gerwick, W. H., Hamann, M. T., & Nagle, D. G., (2003). Marine natural products as novel antioxidant prototypes. *Journal of Natural Products*, *66*(5), 605–608.

Takebe, Y., Saucedo, C. J., Lund, G., et al., (2013). Antiviral lectins from red and blue-green algae show potent *in vitro* and *in vivo* activity against hepatitis C virus. *PLoS One*, *8*(5), e64449.

Takemoto, T., Daigo, K., & Takagi, S., (1964). Hypotensive constituents of marine algae. I. A new basic amino acid 'laminine' and other basic constituents isolated from *Laminaria angustata*. *Yakugaku Zasshi*, *84*(3), 1176–1179.

Takikawa, M., Uno, K., Ooi, T., et al., (1998). Crenulacetal C, a marine diterpene, and its synthetic mimics inhibiting *Polydora websterii*, a harmful lugworm damaging pearl cultivation. *Chemical and Pharmcological Bulletin*, *46*(3), 462–466.

Talarico, L. B., Zibetti, R. G. M., Faria, P. C. S., et al., (2004). Anti-herpes simplex virus activity of sulfated galactans from the red seaweeds *Gymnogongrus griffithsiae* and *Cryptonemia crenulata*. *International Journal of Biological Macromolecules*, *34*(1), 63–71.

Tanaka, J., Higa, T., Bernardinelli, G., & Jefford, C. W., (1989). Sulfur containing polybromoindoles from the red alga *Laurencia brongniartii*. *Tetrahedron Letters*, *45*(23), 7301–7310.

Teixeira, E. H., Napimoga, M. H., Carneiro, V. A., et al., (2007). *In vitro* inhibition of oral *Streptococci* binding to the acquired pellicle by algal lectins. *Journal of Applied Microbiology*, *103*(4), 1001–1006.

Thomesa, P. B., Murugan, R., Balu, P., & Ramasamy, R., (2010). Cardioprotective activity of *Cladosiphon okamuranus* fucoidan against isoproterenol induced myocardial infarction in rats. *Phytomedicine*, *18*(1), 52–57.

Topcu, G., Aydogmus, Z., Imre, S., et al., (2003). Brominated sesquiterpenes from the red alga *Laurencia obtusa. Journal of Natural Products*, *66*(11), 1505–1508.

Tsang, C. K., & Kamei, Y., (2004). Sargaquinoic acid supports the survival of neuronal PC12D cells in a nerve growth factor-independent manner. *European Journal of Pharmacology*, *488*(1–3), 11–18.

Tsang, C. K., Ina, A., Goto, T., & Kamei, Y., (2005). Sargachromenol, a novel nerve growth factor-potentiating substance isolated from *Sargassum macrocarpum*, promotes neurite outgrowth and survival via distinct signaling pathways in PC12D cells. *Neuroscience*, *132*(3), 633–643.

Tsuge, K., Okabe, M., Yoshimura, T., Sumi, T., Tachibana, H., & Yamada, K., (2004). Dietary effect of porphyran from *Porphyra yezoensis* on growth and lipid metabolism of Sprague-Dawley rats. *Food Science and Technology Research*, *10*(2), 147–151.

Tuvikene, R., Robal, M., Fujita, D., et al., (2015). Funorans from *Gloiopeltis* species. Part I. Extraction and structural characteristics. *Food Hydrocolloids*, *43*(2), 481–492.

Usov, A., (1998). Structural analysis of red seaweed galactans of agar and carrageenan groups. *Food Hydrocolloids*, *12*(3), 301–308.

Vairappan, C. S., (2004). Antibacterial activity of major secondary metabolites found in four species of edible green macroalgae genus *Caulerpa*. *Asian Journal of Microbiology, Biotechnology and Environmental Sciences*, *6*(2), 197–201.

Vairappan, C. S., Daitoh, M., Suzuki, M., Abe, T., & Masuda, M., (2001). Antibacterial halogenated metabolites from the Malaysian *Laurencia* species. *Phytochemistry*, *58*(2), 291–297.

Vairappan, C. S., Kamada, T., Lee, W. W., & Jeon, Y. J., (2013). Anti-inflammatory activity of halogenated secondary metabolites of *Laurencia snackeyi* (Weber-van Bosse) Masuda in LPS-stimulated RAW 264.7 macrophages. *Journal of Applied Phycology*, *25*(6), 1805–1813.

Vairappan, C. S., Kawamoto, T., Miwa, H., & Suzuki, M., (2004). Potent antibacterial activity of halogenated compounds against antibiotic-resistant bacteria. *Planta Medica*, *70*(11), 1087–1090.

Van, D. H. C., Mann, D. G., & Jahns, H. M., (1995). *Algae: An Introduction to Phycology* (pp. 1–641). Cambridge University Press, Cambridge, UK.

Veiga-Santos, P., Pelizzaro-Rocha, K. J., Santos, A. O., et al., (2010). *In vitro* anti-trypanosomal activity of elatol isolated from red seaweed *Laurencia dendroidea*. *Parasitology*, *137*(11), 1661–1670.

Wall, M. E., Wani, M. C., Manikumar, G., et al., (1989). Plant antimutagenic agents, 7. Structure and antimutagenic properties of cymobarbatol and 4-Isocymbarbatol, new cymopols from green alga (*Cymopolia barbata*). *Journal of Natural Products*, *52*(5), 1092–1099.

Wang, C. Y., Wu, T. C., Hsieh, S. L., Tsei, Y. H., Yeh, C. W., & Huang, C. Y., (2015). Antioxidant activity and growth inhibition of human colon cancer cells by crude and purified fucoidan preparations extracted from *Sargassum cristaefolium*. *Journal of Food Drug Analysis*, *23*(4), 766–777.

Watanabe, K., Miyakado, M., Ohno, N., Okada, A., Yanagi, K., & Moriguchi, K., (1989). A polyhalogenated insecticidal monoterpene from the red alga *Plocamium telfairiae*. *Phytochemistry*, *28*(1), 77, 78.

Wiemer, D. F., Idler, D. D., & Fenical, W., (1991). Vidalols A and B, new anti-inflammatory bromophenols from the Caribbean marine red alga *Vidalia obtusaloba*. *Experientia*, *47*(1991), 851–853.

Wijesekara, I., Pangestuti, R., & Kim, S. K., (2011). Biological activities and potential health benefits of sulfated polysaccharides derived from marine algae. *Carbohydrate Polymers*, *84*(1), 14–21.

Wongprasert, K., Rudtanatip, T., & Praiboon, J., (2014). Immunostimulatory activity of sulfated galactans isolated from the red seaweed *Gracilaria fisheri* and development of resistance against white spot syndrome virus (WSSV) in shrimp. *Fish and Shellfish Immunology*, *36*(1), 52–60.

Woo, M. N., Jeon, S. M., Shin, Y. C., Lee, M. K., Kang, M. A., & Choi, M. S., (2009). Anti-obese property of fucoxanthin is partly mediated by altering lipid-regulating enzymes and uncoupling proteins of visceral adipose tissue in mice. *Molecular Nutrition and Food Research*, *53*(12), 1603–1611.

Woolner, V. H., Jones, C. M., Field, J. J., et al., (2016). Polyhalogenated indoles from the red alga *Rhodophyllis membranacea*: The first isolation of bromo-chloro-iodo secondary metabolites. *Journal of Natural Products*, *79*(3), 463–469.

Xu, N., Fan, X., Yan, X., Li, X., Niu, R., & Tseng, C. K., (2003). Antibacterial bromophenols from the marine red alga *Rhodomela confervoides*. *Phytochemistry*, *62*(8), 1221–1224.

Xu, W. J., Liao, X. J., Xu, S. H., et al., (2008). Isolation, structure determination, and synthesis of galaxamide, a rare cytotoxic cyclic pentapeptide from a marine algae *Galaxaura filamentosa*. *Organic Letters*, *10*(20), 4569–4572.

Xu, X. H., Zhu, Y. D., Gao, W., & Kong, C. H., (2006). Agricultural lead compounds from *Laurencia majuscula*. In: *Proceedings of International Forum on Green Chemical Science and Engineering and Process Systems Engineering* (Vol. 1, pp. 538–541). Tianjin.

Xu, X., Yin, L., Gao, J., Gao, L., & Song, F., (2014). Antifungal bromophenols from marine red alga *Symphyocladia latiuscula*. *Chemical Biodiversity*, *11*(5), 807–811.

Yabuta, Y., Fujimura, H., Kwak, C. S., Enomoto, T., & Watanabe, F., (2010). Antioxidant activity of the phycoerythrobilin compound formed from a dried Korean purple laver (*Porphyra* sp.) during *in vitro* digestion. *Food Science and Technology Research*, *16*(4), 347–352.

Yan, X., Chuda, Y., Suzuki, M., & Nagata, T., (1999). Fucoxanthin as the major antioxidant in *Hijikia fusiformis*, a common edible seaweed. *Bioscience, Biotechnology and Biochemistry*, *63*(3), 605–607.

Yoon, N. Y., Chung, H. Y., Kim, H. R., & Choi, J. S., (2008). Acetyl- and butyrylcholinesterase inhibitory activities of sterols and phlorotannins from *Ecklonia stolonifera*. *Fisheries Science*, *74*(200), 200–207.

Yoon, N. Y., Eom, T. K., Kim, M. M., & Kim, S. K., (2009a). Inhibitory effect of phlorotannins isolated from *Ecklonia cava* on mushroom tyrosinase activity and melanin formation in mouse B16F10 melanoma cells. *Journal of Agricultural and Food Chemistry*, *57*(10), 4124–4129.

Yoon, N. Y., Lee, S. H., Yong-Li, & Kim, S. K., (2009b). Phlorotannins from *Ishige okamurae* and their acetyl- and butyrylcholinesterase inhibitory effects. *Journal of Functional Foods*, *1*(4), 331–335.

Yoshie-Stark, Y., Hsieh, Y. P., & Suzuki, T., (2003). Distribution of flavonoids and related compounds from seaweeds in Japan. *Journal of Tokyo University of Fisheries*, *89*, 1–6.

Yu, X. Q., He, W. F., Liu, D. Q., et al., (2014). A seco-laurane sesquiterpene and related laurane derivatives from the red alga *Laurencia okamurai* Yamada. *Phytochemistry*, *103*, 162–170.

Yuan, Y. V., Westcott, N. D., Hu, C., & Kitts, D. D., (2009). Mycosporine-like amino acid composition of the edible red alga, *Palmaria palmata* (Dulse) harvested from the west and east coasts of grand Manan Island, New Brunswick. *Food Chemistry*, *112*(2), 321–328.

Zhang, H., Tang, Y., Zhang, Y., et al., (2015). Fucoxanthin: A promising medicinal and nutritional ingredient. *Evidence Based Complementary and Alternative Medicine*, *2015*, 1–10.

Zhang, M., Izumi, I., Kagamimori, S., et al., (2004). Role of taurine supplementation to prevent exercise-induced oxidative stress in healthy young men. *Amino Acids*, *26*(2), 203–207.

Zhang, Z., Zhang, P., Hamada, M., et al., (2008). Potential chemoprevention effect of dietary fucoxanthin on urinary bladder cancer EJ-1 cell line. *Oncology Reports*, *20*(5), 1099–1103.

Zhang, Z., Zhang, Q., Wang, J., Song, H., Zhang, H., & Niu, X., (2010). Regioselective syntheses of sulfated porphyrans from *Porphyra haitanensis* and their antioxidant and anticoagulant activities *in vitro*. *Carbohydrate Polymers*, *79*(4), 1124–1129.

Zhou, G., Sun, Y., Xin, H., Zhang, Y., Li, Z., & Xu, Z., (2004). *In vivo* antitumor and immunomodulation activities of different molecular weight lambda-Carrageenans from *Chondrus ocellatus*. *Pharmacological Research*, *50*(1), 47–53.

Ziółkowska, N. E., & Wlodawer, A., (2006). Structural studies of algal lectins with anti-HIV activity. *Acta Biochemica Polonica*, *53*(4), 617–626.

CHAPTER 13

MARINE MACROALGAE FOR INDUSTRIAL EXTRACTION OF VALUABLE BIOFUNCTIONAL COMPOUNDS USING BIOREFINERY

NUNES NUNO[1,2], BARRETO MARIA CARMO,[3] and PINHEIRO DE CARVALHO MIGUEL ÂNGELO ALMEIDA[1]

[1]*ISOPlexis, Center for Sustainable Agriculture and Food Technology, University of Madeira, Penteada Campus, Funchal–9050-290, Madeira, Portugal*

[2]*UBQ II, Biochemistry Unit, Lda. Rua Visconde de Anadia, Anadia Building 5° Floor CC, Funchal–9050-020, Madeira, Portugal*

[3]*cE3c–Center for Ecology, Evolution, and Environmental Changes/ Azorean Biodiversity Group and Faculty of Sciences and Technology, University of Azores, Ponta Delgada–9501-801, Portugal*

ABSTRACT

According to the latest data from Food and Agriculture Organization (FAO, 2017), 730,575 tons of macroalgae were yearly harvested worldwide. Chile alone was responsible for 36% of the worldwide macroalgae harvesting. Norway followed with 20% of the macroalgae world harvest, followed by Japan, Indonesia, Peru, and Canada. In 2015, macroalgae production reached 4,356,863.47 tons. Major worldwide production was achieved by China, producing almost 50%, mainly constituted by Wakame, *Gracilaria*, and indistinct macroalgae. Indonesia, the second world producer of aquaculture

Algal Metabolites: Biotechnological, Commercial, and Industrial Applications. Jeyabalan Sangeetha, Devarajan Thangadurai, Saher Islam, & Ravichandra Hospet (Eds.)

macroalgae reaching 19% of total, which mainly corresponds to the production of *Eucheuma* macroalgae for the extraction of carrageenan. Korea, Chile, and the Philippines are the following countries after China in macroalgae aquaculture production, respectively. The biorefinery concept is intrinsically connected with high-efficiency fractionation of biomass and the production of valuable biofunctional compounds. It represents a sustainable multi-process, transforming biomass into various marketable products and energy. Several strategies were developed for numerous industrial crops or biomass applications. Nowadays, chemical production or extraction using macroalgae as feedstock is mainly focused on single products, such as the extraction and purification of hydrocolloids, polysaccharides, pigments, proteins, and biofuels production, discarding the remaining biomass. Integrating sustainable strategies for cascade processing with efficient disintegration of biomass to obtain valuable biocompounds could be the key to a profitable industry. Several research works have been published using *Gracilaria* and *Gelidiella* genus for primary extraction of agar, bioethanol, and phycobiliproteins (PBP), with secondary extraction of fertilizers, lipids, bio-oil, biochar, and biogas.

13.1 INTRODUCTION

In earlier centuries, coastal populations harvested a wide variety of macroalgae that were used in multiple purposes such as food, animal feed, treatment of diseases (goiter, intestinal afflictions, and dilation of the cervix in difficult childbirths) and more recently to fertilize soil and produce gels (Buschmann et al., 2017). Nowadays, harvesting macroalgae is performed using manpower and/or recurring to equipment to facilitate this work.

According to the latest data available, in 2015, 730,575 tons of macroalgae were harvested worldwide, of which 52,208 tons were in fresh weight and the remaining in dry weight (FAO, 2017). From 2005 till 2015, Chile was the biggest harvester in the world, reaching a gather peak of 428,674 tons, in 2013, when the total world harvest was 858,842 tons. Chile contributes for almost 50% of macroalgae worldwide harvest in this year (Figure 13.1). In 2015, Chile harvesting capacity decreased to 261,246 tons per year of macroalgae, still representing 36% of the worldwide harvesting capacity. In the same period, Norway, the second biggest harvester of macroalgae worldwide, maintained its capacity, slightly variating and in 2015 harvested 147,391 tons of macroalgae, more than 20% of macroalgae world harvest. Japan, Indonesia, Peru, and Canada are the following countries in macroalgae harvesting, after Chile and Norway. Chile's major harvest was brown macroalgae (Chilean kelp and *Lessonia trabeculata*), followed by red macroalgae (*Gracilaria*) in the southeast Pacific Ocean (Figure 13.2). Norway also majorly harvests brown macroalgae, which includes *Laminaria hyperborea* and *Ascophyllum nodosum* from the northeast Atlantic Ocean. Estimated biomass of *L. hyperborea* situates between 50 and 100 million tons, with an average density of 10 to 15 kg per m^2. *A. nodosum* is estimated in almost 2 million tons along the Norwegian coast, with densities of 4 to 7 kg per m^2 (Meland and Rebours, 2012). According to the FAO (2017), in 2015, the European Union (EU) harvested a total of 45,394 tons, including fresh and dried macroalgae, corresponding to 6.21% of worldwide harvesting. Ireland, in 2015, was the largest harvester in the EU, maintaining its harvesting scores of 29,500 tons fresh macroalgae since 2005 (Figure 13.3), representing 65% of the total EU share. France occupies the 2nd position, with 10,331 tons of fresh macroalgae, 22.76% of total EU share, reaching its maximum yield in 2014, with 34,165 tons. Other countries harvesting macroalgae are Spain (2,376 tons – 5.23%), Portugal (1,574 tons – 3.47%), Italy (1,200 tons – 2.64%) and Estonia (413 tons – 0.91%). France presents a variable harvesting tendency, but between 2005 and 2015, the harvesting contribution of the remaining European countries were maintained.

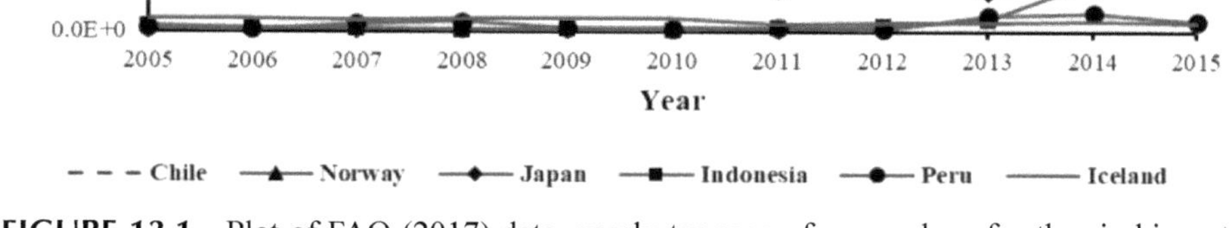

FIGURE 13.1 Plot of FAO (2017) data, yearly tonnage of macroalgae for the six biggest macroalgae harvesters worldwide, Chile, Norway, Japan, Indonesia, Peru, and Iceland, from 2005 till 2015 (harvesters still active in 2015).

In 2015, China produced almost 50% of the worldwide production of macroalgae in aquaculture, which corresponds to 2,152,239.06 tons (Figure 13.4), majorly constituted by Wakame, *Gracilaria*, and indistinct macroalgae (Figure 13.5). In the same year, Indonesia was the second world aquaculture producer, with 842,852.09 tons of macroalgae (Figure 13.4),

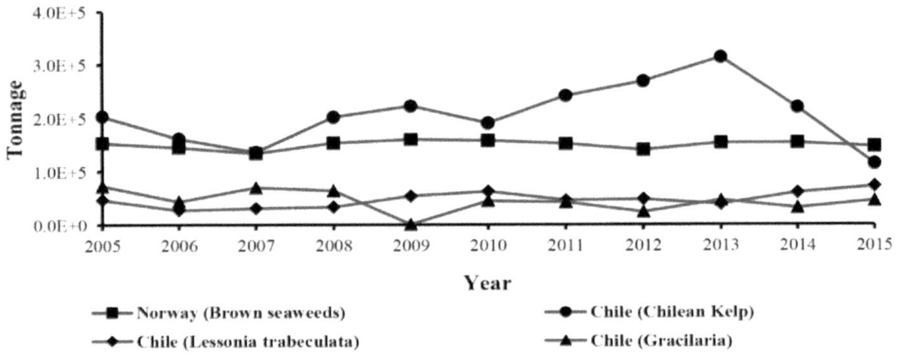

FIGURE 13.2 Plot of FAO (2017) data, tonnage per year, for the four biggest macroalgae that were harvested in Norway (brown seaweed) and Chile (Chilean kelp, *Lessonia trabeculata* and *Gracilaria*) from 2005 to 2015 (still active in 2015).

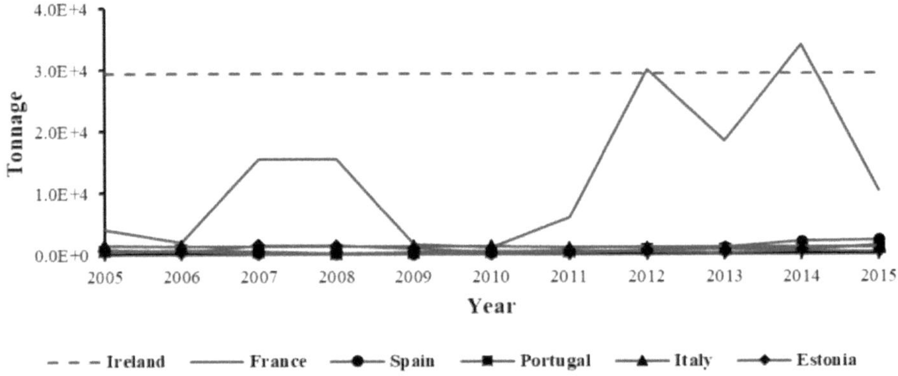

FIGURE 13.3 Plot of FAO (2017) data, yearly tonnage of macroalgae for the only six macroalgae harvesters still active in 2015 in the European Union, Ireland, France, Spain, Portugal, Italy, and Estonia, from 2005 to 2015.

corresponding to 19.35% of worldwide production, which mainly corresponds to the production of *Eucheuma* macroalgae (Figure 13.5) for the extraction of carrageenan. Since 2005, China increased macroalgae aquaculture production only by 1.12-fold, while in the same period, Indonesia increased its production by 6.61-fold, showing a distinct evolution in macroalgae aquaculture production in recent years (FAO, 2017). In 2005 Japan was the 2nd largest aquaculture producer, essentially manufacturing Laver, which are square sheet of dried macroalgae, in this case Nori, very commonly used in sushi. Over this period, we assisted a decrease in the production capacity, which has become slightly above Indonesia production in 2015. Korea, Chile, and the Philippines are the countries showing a significant macroalgae aquaculture production after China, Indonesia, and Japan (Figure 13.4) in a worldwide scenario. EU is a small player in macroalgae aquaculture production, with a total of 1,951.49 tons in 2015, representing only 0.04% of worldwide production (FAO, 2017). Nevertheless, in 2015, Spain was the largest macroalgae producer in EU, with a total share of almost 81%, representing 1,579.92 tons, followed by France that produced 149.8 tons, approximately 7.68%, Portugal 90.56 tons, 4.64%, Ireland 77.67 tons, 3.98%, and Denmark 53.54 tons, 2.74% of total EU macroalgae aquaculture production (Figure 13.6). All these countries maintain their macroalgae aquaculture productions in the northeast of the Atlantic Ocean.

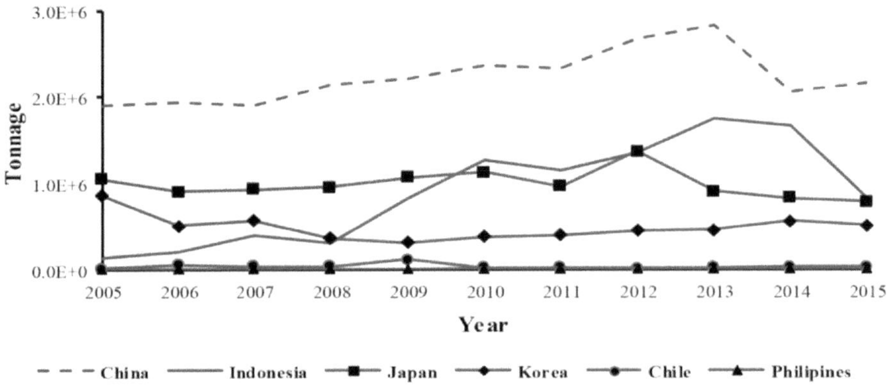

FIGURE 13.4 Plot of FAO (2017) data, yearly tonnage of macroalgae for the six largest macroalgae aquaculture producers, China, Indonesia, Japan, Korea, Chile, and the Philippines, from 2005 till 2015 (that were still active in 2015).

The aquaculture of macroalgae is often considered as an alternative to the harvest of wild macroalgae because cultivation can provide the intensification of yield and allows to control the quality of the macroalgae, crucial for some industries. At the same time, it allows enough macroalgae to be produced without inducing pressure on marine resources. Aquaculture is mainly carried out by macroalgae farmers, with the particularity of entrancing macroalgae propagules in ropes, attached to poles that are firmly secured to the shallow sea bottom. Alternatively, at major depths, reaching up to 30 to 40 meters, macroalgae propagules are attached to ropes and suspended, with the assistance of heavyweights, positioned in the sea bottom and maintained in fluctuation, together with buoys. In either case, fluctuating structures can be implemented to keep the macroalgae at surface or submerged, which can vary in depth with the high and low tide.

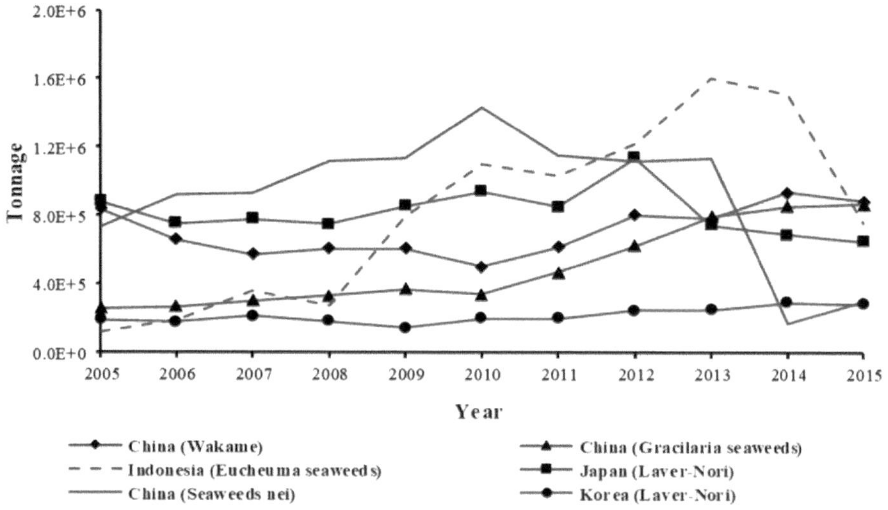

FIGURE 13.5 Plot of FAO (2017) data, tonnage per year, for the six largest aquaculture macroalgae produced, China (Wakame, Gracilaria, and indistinct macroalgae), Indonesia (Eucheuma macroalgae), Japan, and Korea (Laver-Nori), from 2005 to 2015 (that were still active in 2015).

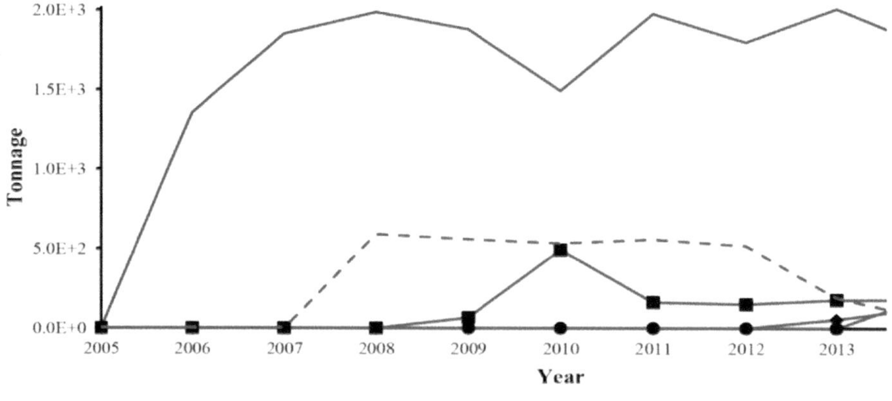

FIGURE 13.6 Plot of FAO (2017) data, tonnage per year, for the only five macroalgae aquaculture producers in the European Union, that were still active in 2015, Spain, France, Portugal, Ireland, and Denmark from 2005 till 2015.

13.2 BIOREFINERY

The global concept of biorefinery defines it as a network of facilities, equipment, and strategies, which integrates biomass for extraction and/or conversion processes to produce biofuels, energy, and commodity biochemicals (Moncada et al., 2016). The biorefinery concept is intrinsically connecting highly efficient fractionation of biomass and the production of several valuable biofunctional products. This is considered as a sustainable multi-process, able to transform biomass into various marketable products and energy (Jung et al., 2013). Several strategies were already developed for numerous industrial crops or biomass applications, aiming to achieve the circular use of biomass and waste. Chemical composition of macroalgae and their fast growth enables its use in a large set of biorefinery strategies. Nowadays, chemicals production using macroalgae as feedstock, is mainly focused on single products, such as the extraction and purification of hydrocolloids (agar, alginates or carrageenans), polysaccharides (fucoidan), pigments (fucoxanthin), proteins (phycoerythrin) and production of biofuels, discarding the remaining biomass (Van Hal et al., 2014). Macroalgae are extensively studied, envisioning the extraction or production of biofuels, which is questionable from the point of sustainability and economic feasibility (Baghel et al., 2015). Alternatively, cascade integrated strategies could efficiently disintegrate biomass allowing to obtain valuable biocompounds and possibly be the answer for a profitable industry (Gilbert-López et al., 2015). A reliable dynamic model for the biorefinery system implementation is extremely important for sustainable resource management and mitigation of climate change (Seghetta et al., 2016) promoted by traditional industry and intensive exploitation of natural resources. In summary, lignocellulosic feedstock, green, and whole crop biorefineries represent three independent systems that are in general studied and developed to determine their performance and sustainability (Gnansounou et al., 2017). Biorefinery potential is enormous, since there is a great variety of unique feedstock that can be decomposed or converted using multiple-step strategies and technologies, available nowadays. Table 13.1 resumes several biorefinery approaches to extract multiple products, using singular strategies. Furthermore, numerous methods are reported for distinct lignocellulosic feedstock. Silalertruksa et al. (2017) evaluated the potential of sugarcane bagasse in Thailand, which after primary juice extraction for sugar production, was assessed for steam and electricity, molasses to produce ethanol and vinasse to serve as fertilizer and soil conditioner. Other examples of lignocellulosic-based biorefineries are for corn stover (Wang et al., 2017) and wheat-bran (Alonso, 2017). Also, several strategies using the whole plant are already being established for other crops, such as soybean (*Glycine max*), producing refined oil, biodiesel, defatted meal, and energy (Abdulkhani et al., 2017; De Pretto et al., 2017) and more recently arugula (*Eruca sativa*), for the production of energy, biodiesel, ethanol, biomethane, and glycerol (Rahimi et al., 2017).

TABLE 13.1 Biomass Used for the Extraction of Several Commodities

Biomass	Commodities	References
Arugula seeds (*Eruca sativa*)	Methanol	Rahimi et al. (2017)
	Ethanol	
	Biodiesel	
	Biomethane	
	Glycerol	
Lignocellulose	Lignin	Özdenkçi et al. (2017)
	Syngas	
	Bio-oil	
Oleaginous yeast (*Rhodosporidium toruloides*)	Refined oil	Marques et al. (2017)
	Biogas	
Soybean	Refined oil	De Pretto et al. (2017)
	Biodiesel	
	Defatted meal	
	Glycerol	
	Biofuel dimethyl furan (DMF)	Abdulkhani et al. (2017)
	Dimethyl ether (DME)	
	Ethanol	
Sugarcane	Sugar	Mendes et al. (2017)
	Ethanol	
	Steam (electricity)	
Wheat bran	Oil	Alonso (2017)
	β-glucans	
	Dextrins	
	Glucose	
	Soluble lignin	
	Proteins	
	Xylose	
	Arabinose	
	Cello-oligosaccharides	
	Insoluble lignin	
	Alkylresorcinols	

Unicellular photosynthetic microalgae are also studied to be integrated into the biorefinery effort. These organisms have a complex cell wall, are capable of accelerating their reproduction rate, and are considered an ideal material to extract multiple products, such as proteins, lipids, carbohydrates, flavonoids, terpenoids, and alkaloids (Chua and Schenk, 2017). Table 13.2 summarizes the bibliography referring microalgae biorefinery approaches to extract various byproducts. Wiesberg et al. (2017) considered green microalgae *Chlorella pyrenoidosa* as the starting biomass to integrate in a biorefinery process (Table 13.2). The primary objective was to assess the economic feasibility of CO_2 bio-capture and storage by these microorganisms, transforming it into valuable products for mankind. Oil was the first product obtained through liquefaction processes and secondly the production of bio-syngas, applying gasification procedures. They concluded that the major expenditure in this approach, 1/5 of the total cost, was the purchase of photobioreactors. García Prieto et al. (2017) developed an integrated microalgae-based biorefinery through the formulation of a mixed-integer nonlinear programming model. They cultivated green microalgae *Haematococcus pluvialis* in open ponds and used a surfactant-chelate strategy to reduce production costs. Primarily, astaxanthin was extracted using a solid-liquid extraction with hexane. Posteriorly, the residual biomass was transformed into other valued products, such as biodiesel, through transesterification of fatty acids and production of polyhydroxybutyrate (PHB) by the fermentation with *Cupriavidus necator*. Three other valuable byproducts were obtained in the final process, glycerol, fertilizer, and energy. Additionally, they designed a detailed capital cost model for process equipments acquisition. They concluded that the extraction of astaxanthin and production of PHB from these microalgae, followed by the production of biodiesel, becomes economically feasible. Mitra et al. (2015) assessed the marine eustigmatophyte *Nannochloropsis* sp. potential to integrate in a biorefinery strategy platform in a two-stage cultivation, under photoautotrophic conditions. The cultivation of these microalgae was maintained in a photobioreactor up to the late log phase. Afterwards, they first extracted eicosapentaenoic acid (EPA), using conventional methods of solid-liquid extraction. As a second step of biorefinery, the transesterification of remaining saturated fatty acids (SFAs) and monounsaturated (MUFAs) for the production of biodiesel was performed. In two phase cultivation, the authors concluded that the temperature and light reduction in the second phase can increase EPA production by 3.4-fold, but presents the disadvantage of reducing the biomass, lipid content and productivity. Fractionation of lipids in neutral lipids, GL, and phospholipids demonstrated that the major portion of EPA is presented in

TABLE 13.2 Microalgae Used to Develop Biorefinery Strategies to Extract Several Products from This Biomass*

Biomass	Application				References
	Primary	**Methodology**	**Secondary**	**Methodology**	
Chlorella pyrenoidosa	Carbon dioxide bio-capture	–	Oil	Liquefaction	Wiesberg et al. (2017)
			Bio-syngas	Gasification	
Haematococcus pluvialis	Astaxanthin	Solid-liquid extraction	Biodiesel	Transesterification	García Prieto et al. (2017)
			Glycerol	Byproduct	
			Poly(hydroxybutyrate) (PHB)	Fermentation by *Cupriavidus necator*	
			Fertilizer	Byproduct	
			Energy	Byproduct	
Isochrysis galbana	Carotenoids	Supercritical fluid extraction	Chlorophylls	Supercritical fluid extraction	Gilbert-López et al. (2015)
			Mid and highly-polar lipids		
	Non-polar lipids		Proteins		
			Sugars		
Nannochloropsis sp.	Eicosapentaenoic acid (EPA)	Solid-liquid extraction	Biodiesel	Transesterification	Mitra et al. (2015)
Nannochloropsis sp.	Fatty acids (for biodiesel)	Supercritical fluid extraction	Carotenoids	Supercritical fluid extraction	Nobre et al. (2013)
			Chlorophylls		
			Biogas (hydrogen)	Dark fermentation with *Enterobacter aerogenes*	
Spirogyra sp.	Carotenoids	Solid-liquid extraction	Biogas (hydrogen)	Fermentation using *Clostridium butyricum*	Pacheco et al. (2015)
			Organic acids	Byproduct	
			Methanol	Chemical reaction between CO_2 and hydrogen	

*Primary and secondary applications are presented.

this last class. Nobre et al. (2013) also evaluated the potential of *Nannochloropsis* sp. as feedstock in a biorefinery strategy, primarily extracting fatty acids to produce biodiesel, secondly carotenoids and chlorophyll *a* as a high added-value product and thirdly producing biohydrogen, through a dark fermentation with *Enterobacter aerogenes*. The extraction procedures were performed using supercritical fluid extraction, variating temperature, pressure, and solvent flow rate in order to determine their effect on extraction yield. They concluded that using only CO_2, at 40°C, 300 bar with a flow-rate of 0.62 g/min, they could efficiently extract 33 g of lipids per 100 g of dry biomass. The use of 20% ethanol as co-solvent increased the lipid extraction to 45 g of lipids per 100 g of dry biomass. Gilbert-López et al. (2015) assessed the potential of *Isochrysis galbana*, also using supercritical fluid extraction within the same parameters described for *Nannochloropsis* sp. They performed multiple extracts, applying a downstream processing platform, with the assistance of various pressurized green solvents (CO_2, ethanol, and water) to achieve the isolation of bioactive compounds and compare its yield efficiency with conventional methods. Several extracts were produced, including firstly carotenoids and non-polar lipids, secondly carotenoids, chlorophylls, mid, and highly-polar lipids, and thirdly proteins and sugars. They also performed a factorial design to optimize the extraction of fucoxanthin, a primary carotenoid found in *Isochrysis galbana*, using three different percentages of ethanol, 15, 45, and 75%. Their main target was to determine the similarity between conventional and pressurized methods during the extraction of bioactive compounds, complying with the principles of green chemistry. This work presents a paradigm shift when addressing the biorefinery of microalgae, targeting only high valuable products and not bio-fuel. Furthermore, Pacheco et al. (2015) evaluated the overall energy consumption and greenhouse gas emissions of pigments extraction and biohydrogen production from green microalgae *Spyrogyra* sp.

Macroalgae biomass is a potential feedstock for biofuel production, an abundant source of renewable sugars as its growth does not require arable land, freshwater, or intense labor (Konda et al., 2015). Table 13.3 summarizes the information from the bibliography referring biorefinery methods to extract several products, using a range of strategies. Ben et al. (2016) introduced the biorefinery of green macroalgae *Chaetomorpha linum*. They investigated an eco-friendly biorefining process to co-produce bioethanol and ferment the biomass with *Saccharomyces cerevisiae*, obtaining biogas, mainly methane as a valuable byproduct. Several pre-treatments of thallus were applied to determine its disintegration potential. This included acid, neutral, and alkali treatments, with 3% NaOH giving the best results,

TABLE 13.3 Macroalgae Used to Determine Biorefinery Strategies with Primary and Secondary Applications

Biomass	Application				References
	Primary	**Methodology**	**Secondary**	**Methodology**	
Gracilariopsis longissima	Agar	Alkali treatment	Ethanol	Fermentation (*Saccharomyces cerevisiae*)	Kumar et al. (2013)
			Fertilizer	Byproduct	
Gelidiella acerosa	Bioethanol	Enzymatic hydrolysis (*Saccharomyces cerevisiae*)	Lipids	Solid-liquid extraction	Baghel et al. (2015)
			Agar	Freezing and thawing	
Gracilaria dura			R-phycoerytrin	Solid-liquid extraction	
			R-phycocyanin	Solid-liquid extraction	
Gelidium pusillum			Liquid fertilizer	Byproduct	
Gracillaria gracilis	Phycobiliproteins	Solid-liquid extraction	Bio-oil	Pyrolysis	Francavilla et al. (2014)
	–	–	Biochar	Byproduct	
Several macroalgae	–	–	Biogas (CO_2)	Byproduct	Raikova et al. (2017)
	Bio-oil	Hydrothermal liquefaction	Ashes	Byproduct	
	–	–	Aqueous phase products	Byproduct	
Chaetomorpha linum	Bioethanol	Fermentation (*Saccharomyces cerevisiae*)	Biogas (Biomethane)	Byproduct	Ben Yahmed et al. (2016)

assessed by scanning electron microscope and saccharification tests. After pre-treatment, the biomass would be recovered and subjected to hydrolytic enzymes, extracted from *Aspergillus awamori*. Bioethanol yield reached the maximum of 0.41 g per g of reducing sugars. Francavilla et al. (2014) performed a cascade approach using the red macroalga *Gracilaria gracilis*, to extract and purify phycobiliproteins (PBPs), using a solid-liquid methodology. R-phycoerytrin, allophycocyanin, and phycocyanin were the main PBP extracted, yielding 7 mg/g dry weight (dw), 3.5 mg/g dw and 2 mg/g dw, respectively. This extraction was followed by the production of bio-oil and the residual biomass submitted to a pyrolysis conversion, with optimum temperature between 400°C and 600°C, obtaining biochar as a byproduct. Bio-oil yield was high, 65 weight (wt.)% at 500°C, but unfortunately with a high content of nitrogenous compounds, preventing its use as a biofuel. Biochar yield was also high, reaching 33 wt.% at 400°C. Baghel et al. (2015) selected three macroalgae, *Gelidiella acerosa*, *Gelidium pusillum* and *Gracilaria dura* to produce bioethanol from the enzymatic hydrolysate using cellulase followed by a *Saccharomyces cerevisiae* fermentation. Lipids, R-phycoerytrin, and R-phycocyanin were extracted from residual pulp using a solid-liquid extraction. Agar was extracted following the freezing and thawing methodology and a liquid fertilizer obtained as a final byproduct. This work envisions a scalable production of valuable products with manageable integrated processes, facilitating the sequential extraction of major biocomponents of red macroalgae. This strategy shows some advantages over direct extraction, such as the improvement of the agar quality, which increased gel strength by 1.5 to 3-fold, without alkali or acid pre-treatment. The residue resulted from the sequential extractions was eliminated, and chemicals used in cellulose extraction were reduced up to 85%. Kumar et al. (2013) used a different strategy in the biorefinery of red macroalgae *Gracilariopsis longissima* (formerly *Gracilaria verrucosa*). Macroalgae was harvested at various growth times and agar was primarily extracted by an alkali treatment, yielding 27 to 33%. The leftover pulp, containing from 62 to 68% holocellulose, was subjected to an enzymatic hydrolysis, to obtain 0.87 g of sugars per g of cellulose. This hydrolysate was subjected to fermentation with *Saccharomyces cerevisiae* for the production of bio-ethanol yielding 0.43 g of ethanol per g of sugar. A single final byproduct was obtained, which can be used as a fertilizer or soil conditioner. Raikova et al. (2017) applied a hydrothermal liquefaction (HTL) process to produce a bio-crude oil, using 13 different macroalgae. This work intended to produce a cohesive assessment of the effects which the biochemical composition imprints, when converting macroalgae biomass to bio-crude oil.

Additionally, three byproducts were obtained using this methodology, biogas (mainly CO_2), aqueous phase products and ashes. *Ulva lactuca* showed the highest yield for bio-crude oil, 29.9%. Aqueous phase phosphate concentrations of 236 $mg.L^{-1}$ was obtained using *Solieria chordalis* biomass. A positive correlation between the increase of lipids in the biomass and the increase of bio-crude oil yield was obtained. Likewise, the increase of biomass nitrogen will generally contribute to bio-crude oil nitrogen content. This HTL is seen to be a promising biomass conversion method, capable of simultaneously producing fuels and aqueous fertilizers or as auxiliar in the remediation, when producing and using biomass from contaminated marine or freshwater effluents. We have successfully determined in a previous work (Nunes et al., 2018) that a nutraceutical supplement, rich in iodine, could be extracted from the red macroalga *Asparagopsis taxiformis* and using a cascade approach, also to extract lipids, carrageenan, and cellulose, respectively. These compounds were very similar to the same compounds extracted from this macroalga when performing single product extraction. This showed that multiple products could be extracted from the same biomass, maintaining its quality, reducing waste, and improving income.

13.3 BIOCOMPOUNDS

13.3.1 PHENOLIC COMPOUNDS (PCs)

These compounds are the main secondary metabolites, present in plants, lichens, and algae. Structurally, they are formed from one to multiple aromatic rings with one or more hydroxyl groups, totalizing 8,000 different known compounds (Dai and Mumper, 2010). Their main functions include the protection against epiphytes, marine grazers, pathogens, photoprotection, and in brown algae they appear cross-linked with alginates to reinforce the cell wall (Audibert et al., 2010). The composition and concentration of the phenolic compounds (PCs) tend to fluctuate according to the macroalgae location. Macroalgae exposed to higher levels of ultraviolet (UV) radiation and desiccation periods, will produce higher content of PC than submerged macroalgae (O'Sullivan et al., 2011). Flavonoids are the most consumed polyphenols within human food, containing 15 carbon atoms arranged in three rings, C6-C3-C6. They are divided into six subgroups, which include anthocyanins, flavanones, flavonols, flavones, flavanols, and isoflavones, according to the oxidation status of the central carbon ring (Dai and Mumper, 2010).

In green and red macroalgae, PC present are mostly bromophenols, but in lower quantity than PC in brown macroalgae, which are mainly phlorotannins (Connan, 2015). Phlorotannins are oligomers or polymers of monomeric units of 1,3,5-trihydroxybenzene (phloroglucinol), interconnected with various linkages, such as dibenzodioxin for eckols and carmalols, aryl-aryl bonds for fucols, ether bonds for phlorethols, hydroxyphlorethols, and fuhalols or a combination of both aryl-aryl and ether bonds for fucophlorethols (Rajauria, 2018). These compounds have a wide range of molecular sizes (0.4 to 400 kDa) and can be present in brown macroalgae at various concentrations (Balboa et al., 2013). Special vesicles, named physodes, store these compounds that can be halogenated or sulfated (Stengel and Walker, 2015) and are known to absorb in the UVB range (280–320 nm) (Yuan and Walsh, 2006). Purified phlorotannins are known to present a strong antioxidant activity, due to the presence of phenol rings and to their unique molecular structure, acting as an electron trap, scavenging peroxides, superoxide anions and hydroxyl radicals (Wang et al., 2009). Extracts of PCs properly purified, fractioned, and characterized, can represent an important source of bioactive molecules for food, cosmetics, and medicine (Audibert et al., 2010). Several epidemiologic data, as well as *in vitro* and *in vivo* studies, have reported health benefits, when consuming polyphenol rich foods, preventing cardiovascular diseases (CVDs) related to ageing, due to their antioxidant properties, antimutagenic activities and cholesterol-lowering effects (Keyrouz et al., 2011). Additionally, PCs are found to modulate the activity of a wide range of enzymes and cell receptors (Dai and Mumper, 2010). *In vitro* tests showed that polymeric phlorotannins from brown algae *Eisenia bicyclis* and *Ecklonia kurome* inhibit enzyme activity, such as hyaluronidase (Shibata et al., 2002a), phospholipase A, lipoxygenase, and cyclooxygenase-1. This last inhibitory activity was detected in phlorotannins extracted from *Eisenia bicyclis* of the Japan coast (Shibata et al., 2002, b). In the food industry, hypopigmentation of products from natural sources is progressively required. The inhibition of tyrosinase or blockade of upstream regulation of melanogenesis for active whitening, can be achieved with phlorotannins, without cytotoxic effects (Balboa et al., 2013). Additionally, these compounds have the potential to be used as natural preservatives, due to their antibacterial activity (Audibert et al., 2010). Various solvents and extraction methods are continuously developed due to the diversity of PCs and matrix features needed. The most common solvents used for PC extraction include methanol, acetone, and ethanol, in various concentrations. Several assays for PC measurement are available such as Folin-Denis, Folin-Ciocalteu, and 2,4-dimethoxybenzaldehyde assays (Connan, 2015).

Table 13.4 summarizes the total phenolic content (TPC) in macroalgae, from several worldwide places, using different standards to validate TPC content, namely phloroglucinol equivalents (PGE), gallic acid equivalents (GAE) and catechol equivalents (CE). According to Table 13.4, TPC varies from 0.1 to 24.2 g/100 g dw, using PGE as a standard. Also, the highest values for TPC were obtained when extraction from the brown macroalgae *Fucus vesiculosus* was performed with 70% acetone at room temperature during 24 h (Wang et al., 2009). TPC ranges from 20.8 g in *Grateloupia turuturu* and 92.0 g/100 g dw in *Codium tomentosum*, a green macroalgae, when using CE as a standard and extraction performed with ethyl acetate (Rodrigues et al., 2015). The TPC determined using GAE as a standard, originated values between 0.2 and 0.4 g/100 g dw, when extracted with MeOH/H_2O (60/40, v/v) at 40°C, during a period of 3 h (O'Sullivan et al., 2011). Kumar et al. (2008) determined 2.1 g/100 g of dw of TPC in *Kappaphycus alvarezii*, a red macroalgae, using $ChCl_3$/MeOH (2:1, v/v) in a Soxhlet apparatus. Although some variation can be observed in TPC, depending on the macroalgae and sampling location, the biggest source of its variation seems to be the solvent system used to obtain the primary extracts and the selected phenolic standard (PGE, GAE, or CE). The TPC in *Fucus vesiculosus* from Brittany (France) had 5.8 g PGE /100 g dw, using the MeOH/H_2O solvent system, at 40°C, during 3 h to the extraction (Connan et al., 2004). Wang et al. (2009) obtained significantly higher TPC, 24.2 g PGE/100 g dw in the same species of macroalgae collected in Hafnarfjordur (Iceland), but performing the extraction using a 70% acetone solvent system, during 24 h, at room temperature. Also, comparing the TPC in *Fucus vesiculosus*, using the same solvent system, but varying the standards, it was achieved 5.8 g PGE/100 g dw (Connan et al., 2004) or only 0.3 g CE /100 g dw (O'Sullivan et al., 2011).

13.3.2 PROTEINS

Macro and microalgae are often consumed due to their high protein content, with a similar or higher content than some plant sources (Fleurence, 1999; Phong et al., 2016). Protein quality depends mainly on its amino acid composition, content variation, proportion, availability, and digestibility. There are 20 proteinogenic amino acids and 11 of them are essential for human body functioning and should be obtained through food ingestion, due to our metabolic inability to biosynthesize. These essential amino acids (EAAs) exist in many plant sources but often in imbalanced and/or insufficient amounts, commonly absence of one or more EAA, which are histidine, isoleucine,

TABLE 13.4 Total Phenolic Content (TPC) Described for Macroalgae Collected in Several Places Worldwide*

Species	Collection Site	Extraction Methodology	Total Phenolic Content (TPC) g/100 g Equivalents (dw)	References
Ascophyllum nodosum	Portsall, Brittany (France)	$MeOH/H_2O$, 40°C, 3 h	5.8 (PGE)	Lee et al. (2016)[F]
Ascophyllum nodosum	Hvassahraun, Hafnarfjordur (Iceland)	70%, acetone, room temperature, 24 h	15.9 (PGE)	Wang et al. (2009)[L]
Ascophyllum nodosum	Galway (Ireland)	$MeOH/H_2O$ (60/40, v/v), 40°C, 3 h	0.5 (GAE)	O'Sullivan et al. (2011)[F]
Bifurcaria bifurcata	Portsall, Brittany (France)	$MeOH/H_2O$, 40°C, 3 h	3.7 (PGE)	Connan et al. (2004)[F]
Codium tomentosum	Buarcos bay, Figueira da Foz (Portugal)	Ethyl acetate	92.0 (CE)	Rodrigues et al. (2015)[AD]
Fucus serratus	Portsall, Brittany (France)	$MeOH/H_2O$, 40°C, 3 h	4.3 (PGE)	Connan et al. (2004)[F]
Fucus serratus	Hvassahraun, Hafnarfjordur (Iceland)	70%, acetone, room temperature 24 h	24.0 (PGE)	Wang et al. (2009)[L]
Fucus serratus	Galway (Ireland)	$MeOH/H_2O$ (60/40, v/v), 40°C, 3 h	0.4 (GAE)	O'Sullivan et al. (2011)[F]
Fucus spiralis	Portsall, Brittany (France)	$MeOH/H_2O$, 40°C, 3 h	3.9 (PGE)	Connan et al. (2004)[F]
Fucus vesiculosus	Portsall, Brittany (France)	$MeOH/H_2O$, 40°C, 3 h	5.8 (PGE)	Connan et al. (2004)[F]
Fucus vesiculosus	Hvassahraun, Hafnarfjordur (Iceland)	70%, acetone, room temperature, 24 h	24.2 (PGE)	Wang et al. (2009)[L]
Fucus vesiculosus	Galway (Ireland)	$MeOH/H_2O$ (60/40, v/v), 40°C, 3 h	0.3 (GAE)	O'Sullivan et al. (2011)[F]
Gracilaria gracilis	Buarcos bay, Figueira da Foz (Portugal)	Ethyl acetate	22.8 (CE)	Rodrigues et al. (2015)[AD]

TABLE 13.4 *(Continued)*

Species	Collection Site	Extraction Methodology	Total Phenolic Content (TPC) g/100 g Equivalents (dw)	References
Grateloupia turuturu	Buarcos bay, Figueira da Foz (Portugal)	Ethyl acetate	20.8 (CE)	Rodrigues et al. (2015)[AD]
Himanthalia elongata	Portsall, Brittany (France)	MeOH/H_2O, 40°C, 3 h	2.2 (PGE)	Connan et al. (2004)[F]
Kappaphycus alvarezii	Port Okha (India)	Chloroform:MeOH (2:1, v/v), Soxhlet	2.1 (GAE)	Kumar et al. (2008)[AD]
Laminaria digitata	Portsall, Brittany (France)	MeOH/H_2O 40°C 3 h	0.1 (PGE)	Connan et al. (2004)[F]
Laminaria hyperborea	Galway (Ireland)	MeOH/H_2O (60/40, v/v), 40°C, 3 h	0.2 (GAE)	O'Sullivan et al. (2011)[F]
Osmundea pinnatifida	Buarcos bay, Figueira da Foz (Portugal)	Ethyl acetate	33.7 (CE)	Rodrigues et al. (2015)[AD]
Pelvetia canaliculata	Portsall, Brittany (France)	MeOH/H_2O, 40°C, 3 h	3.4 (PGE)	Connan et al. (2004)[F]
Pelvetia canaliculata	Galway (Ireland)	MeOH/H_2O (60/40, v/v), 40°C, 3 h	0.4 (GAE)	O'Sullivan et al. (2011)[F]
Sargassum muticum	Buarcos bay, Figueira da Foz (Portugal)	Ethyl acetate	49.9 (CE)	Rodrigues et al. (2015)[AD]
Saccorhiza polyschides	Buarcos bay, Figueira da Foz (Portugal)	Ethyl acetate	22.4 (CE)	Rodrigues et al. (2015)[AD]
Ulva rigida	Sidi Mansour Sfax (Tunisia)	MeOH, 70°C, 24 h, Soxhlet	0.9 (PGE)	Trigui et al. (2013)[AD]

*TPC was extracted differently, using different solvents, temperatures, and times. Distinct standards were used to assess TPC content namely phloroglucinol equivalents (PGE), Gallic acid equivalents (GAE) and catechol equivalents (CE). After individual reference, letters in superscript indicate the state of the biomass for the extraction protocol used by different authors, Fresh (F), Lyophilized (L) or Air-Dried (AD).

lysine, leucine, phenylalanine, methionine, tryptophan, threonine, and valine (Young and Pellett, 1994). Although the EAA composition in algae fluctuates throughout the year due to seasonality (Fleurence, 1999), they are worthwhile classified as a viable and complete protein source (FAO/WHO, 1991). However, phlorotannins, and high polysaccharide content in algae can act as anti-nutrients, significantly reducing protein digestibility. Also, the application and widespread food use of algae is limited by harvesting capabilities, seasonality, geographical location, production, and methods for protein extraction and purification (Bleakley and Hayes, 2017). Cell disruption methods for protein extraction are commonly achieved through aqueous, acid, and alkaline extraction or enzymatic digestion (Galland-Irmouli et al., 1999). Enzymatic assisted protein extraction frequently implies the use of glucanases, amylases, cellulase, endoproteases or proteases. The chemical or subcritical water hydrolysis and more recently, supercritical fluid extraction, ultrahigh-pressure extraction, pressurized fluid extraction and microwave-assisted extraction were also used for macroalgae protein extraction (Kadam et al., 2017).

Table 13.5 summarizes total protein content in macroalgae collected from several worldwide places, using Kjeldahl's methodology and multiplying the percentage of nitrogen content by a factor of 6.25. Protein content ranges between 5.0 g/100 g dw with *Lobophora variegata*, a brown macroalgae from Tamil Nadu, India (Verma et al., 2017) and 24.8 g/100 g dw with *Porphyra* sp., a red macroalgae collected in Azores, Portugal (Paiva et al., 2014). The bibliographical analysis shows that green macroalgae contains between 6.9 and 18.8 g/100 g dw of protein, red macroalgae from 9.1 to 24.8 g/100 g dw and brown macroalgae from 5.0 to 16.9 g/100 g dw. Interestingly, the highest protein content was observed in red macroalgae, such as *Porphyra* sp., *Gracilaria* sp., *Grateloupia* sp. and *Osmundea* sp.

Two different kinds of bioactive algae proteins have been extracted and purified for industrial applications, i.e., PBP, and lectins, which are commonly extracted from Rhodophyta macroalgae (Bleakley and Hayes, 2017).

13.3.2.1 PHYCOBILIPROTEINS (PBP)

These water-soluble macromolecules are a light-harvesting family, existing in the photosynthetic apparatus of cyanobacteria and some groups of eukaryotic algae, such as red algae, Glaucophytes, and Cryptomonads (Apt et al., 1995). They are constituents of the phycobilisome multi-molecular complex, linked to the thylakoid membrane (Tandeau De Marsac et al., 2003). These structures

TABLE 13.5 Protein Content Described in Macroalgae Collected from Several Places Worldwide*

Species	Collection Site	Protein Content (g/100 g) dw (Conversion Factor 6.25) Kjeldahl Method	References
Caulerpa racemosa	Veraval Coast, Gujarat (India)	12.9	Kumar et al. (2011)[L]
Caulerpa scalpelliformis	Veraval Coast, Gujarat (India)	10.5	Kumar et al. (2011)[L]
Caulerpa veravalensis	Veraval Coast, Gujarat (India)	7.8	Kumar et al. (2011)[L]
Codium tomentosum	Buarcos bay, Figueira da Foz (Portugal)	18.8	Rodrigues et al. (2015)[AD]
Codium spp.	Barbate estuary, Gulf of Cádiz (Spain)	15.6	Gracía et al. (2016)[L]
Dictyota dicotoma	Port Okha, Gujarat (India)	11.9	Verma et al. (2017)[AD]
Fucus spiralis	São Miguel Island, Azores (Portugal)	9.7	Paiva et al. (2014)[AD]
Gelidiella acerosa	Port Okha, Gujarat (India)	14.6	Verma et al. (2017)[AD]
Gelidium micropterum	Mandapam, Tamil Nadu (India)	9.1	Verma et al. (2017)[AD]
Gracilaria cervicornis	Búzios beach (Brazil)	19.7	Marinho-Soriano et al. (2006)[AD]
Gracilaria changii	Santubong, Sarawak (Malaysia)	12.6	Chan and Matanjun (2017)[L]
Gracilaria fisheri	Pattani bay (Thailand)	11.6	Benjama and Masniyom (2012)[AD]
Gracilaria gracilis	Buarcos bay, Figueira da Foz (Portugal)	20.2	Rodrigues et al. (2015)[AD]
Gracilaria tenuistipitata	Pattani bay (Thailand)	21.6	Benjama and Masniyom (2012)[AD]
Grateloupia indica	Port Okha, Gujarat (India)	10.1	Verma et al. (2017)[AD]
Grateloupia turuturu	Piriac-sur-Mer (France)	22.9	Denis et al. (2010)[L]
Grateloupia turuturu	Buarcos bay, Figueira da Foz (Portugal)	22.5	Rodrigues et al. (2015)[AD]
Halymenia floresii	Barbate estuary, Gulf of Cádiz (Spain)	9.4	Gracía et al. (2016)[L]
Lobophora variegata	Mandapam, Tamil Nadu (India)	5.0	Verma et al. (2017)[AD]
Osmundea pinnatifida	São Miguel Island, Azores (Portugal)	20.8	Paiva et al. (2014)[AD]

TABLE 13.5 *(Continued)*

Species	Collection Site	Protein Content (g/100 g) dw (Conversion Factor 6.25) Kjeldahl Method	References
Osmundea pinnatifida	Buarcos bay, Figueira da Foz (Portugal)	23.8	Rodrigues et al. (2015)[AD]
Padina gymnospora	Port Okha, Gujarat (India)	11.6	Verma et al. (2017)[AD]
Padina tetrastromatica	Mandapam, Tamil Nadu (India)	12.4	Verma et al. (2017)[AD]
Porphyra sp.	São Miguel Island, Azores (Portugal)	24.8	Paiva et al. (2014)[AD]
Saccorhiza polyschides	Buarcos bay, Figueira da Foz (Portugal)	14.4	Rodrigues et al. (2015)[AD]
Saccorhiza polyschides	Barbate estuary, Gulf of Cádiz (Spain)	7.3	Gracía et al. (2016)[L]
Sargassum linearifolium	Port Okha, Gujarat (India)	8.9	Verma et al. (2017)[AD]
Sargassum muticum	Buarcos bay, Figueira da Foz (Portugal)	16.9	Rodrigues et al. (2015)[AD]
Sargassum naozhouense	Techeng Island, Guangdong province (China)	11.2	Peng et al. (2013)[AD]
Sargassum vulgare	Búzios beach (Brazil)	13.6	Marinho-Soriano et al. (2006)[AD]
Ulva lactuca	Port Okha, Gujarat (India)	12.2	Verma et al. (2017)[AD]
Ulva reticulata	Mandapam, Tamil Nadu (India)	6.9	Verma et al. (2017)[AD]
Ulva spp.	Barbate estuary, Gulf of Cádiz (Spain)	7.2	Gracía et al. (2016)[L]

*The methodology for protein content was Kjeldahl method, multiplying the nitrogen content by a factor of 6.25. After individual reference, letters in superscript indicate the state of the biomass for the extraction protocol used by different authors, Lyophilized (L) or Air-Dried (AD).

present inside the chloroplast structure, have the ability to capture the photon energy toward the photosystem (Sonani et al., 2014). PBP are divided in three classes, according to their absorption properties, comprising phycoerythrins (PEs) absorbing light at 495 nm and 540–570 nm, phycocyanins (PCs) at 610–620 nm and allophycocyanins (APCs) at 650–655 nm (Sudhakar et al., 2015). All PBPs have two subunits (α and β) while a third subunit (γ) is only found in PEs as a linker peptide. R-phycoerythrin and B-phycoerythrin (BPE), two PEs, are structurally composed and arranged in $(\alpha\beta)6\gamma$ complex. Meanwhile, PCs, and APCs have a structure that can be described as $(\alpha\beta)_3$ complex (Isailovic et al., 2004). PBPs absorption and fluorescence characteristics, in the visible spectrum, result from the presence of photosynthetic pigments named phycobilins and their interactions within polypeptide chains. Individual subunits (α, β, and γ) can possess one or more phycobilin chromophores, linked with specific cysteines, present in the polypeptide chains, with a thioester bond (Apt et al., 1995). Nowadays, there is a widespread of biotechnological applications for PBP, including several fluorescent applications such as labelling, microscopy, immunohistochemistry, and flow cytometry analysis (Aneiros and Garateix, 2004). Nonetheless, the majority of industrial applications use low purity compounds (low purity index) as they are used as natural dyes in food and cosmetics (Yamaguchi, 1997). *In vitro* and *in vivo* tests demonstrate the nutraceutical potential of these bioactive compounds, with several patents filed for anti-inflammatory, anti-tumor, anti-oxidative, anti-viral, hepatoprotective, and neuroprotective activities (Sekar and Chandramohan, 2008). Numerous extraction and purification techniques have been developed over the years, mainly using phosphate buffer, but also testing distilled water and seawater for PBP extraction, applying or not enzyme digestion, to facilitate the breakage of the cell wall (Sudhakar et al., 2015).

Table 13.6 summarizes the bibliographic and methodology information about PBP extraction, namely for PEs, PCs, and APCs, from macroalgae collected in several places of the world. Some variability in PBP, that were extracted from red macroalgae, can be observed, ranging from 81.9 to 457.5 mg of PEs /100 g dw in *Champia* spp. and *Scinaia fascicularis*, respectively. PCs were found to variate between 15.4 mg in *Grateloupia indica* and 282.60 mg/100 g dw in *Hypnea musciformis*, both collected in India. The lowest content of APCs was obtained in *Portieria hornemanni*, 15.2 mg and the highest content of 50 mg/100 g dw in *Porphyra* spp., both collected in India. These results were published by Verma et al. (2017). The extraction was performed by suspending air-dried macroalgae flour in 1 Molar (M) acetic acid-sodium acetate buffer, at pH 5.5, with 0.01% of sodium azide, during 30 minutes in dark condition. PBPs extraction could

TABLE 13.6 Phycobiliproteins Namely Phycoerythrins (PEs), Phycocyanins (PCs) and Allophycocyanins (APCs) Content, Described for Macroalgae Collected from Several Places Worldwide*

Species	Collection Site	Extraction Methodology	Phycoerythrins (PEs) mg/100g dw	Phycocyanins (PCs) mg/100g dw	Allophycocyanins (APCs) mg/100g dw	References
Botryocladia leptopoda	Port Okha, Gujarat (India)	Suspended in 1 M acetic acid–sodium acetate buffer (pH 5.5) with 0.01% of sodium azide for 30 min in the dark	351.10	117.40	28.60	Verma et al. (2017)[AD]
Champia spp.	Mandapam, Tamil Nadu (India)	Suspended in 1 M acetic acid–sodium acetate buffer (pH 5.5) with 0.01% of sodium azide for 30 min in the dark	81.90	101.90	26.40	Verma et al. (2017)[AD]
Gelidiella acerosa	Port Okha, Gujarat (India)	Suspended in 1 M acetic acid–sodium acetate buffer (pH 5.5) with 0.01% of sodium azide for 30 min in the dark	335.10	199.80	29.10	Verma et al. (2017)[AD]
Gelidium micropterum	Mandapam, Tamil Nadu (India)	Suspended in 1 M acetic acid–sodium acetate buffer (pH 5.5) with 0.01% of sodium azide for 30 min in the dark	274.80	85.70	30.10	Verma et al. (2017)[AD]
Gracilaria crassa	Rameswaram, Tamil Nadu (India)	Suspended in H_2O (pH 7.0) at room temperature	50.00 (fw)	28.00 (fw)	34.00 (fw)	Sudhakar et al. (2015)[F]
Grateloupia indica	Port Okha, Gujarat (India)	Suspended in 1 M acetic acid–sodium acetate buffer (pH 5.5) with 0.01% of sodium azide for 30 min in the dark	385.50	15.40	28.00	Verma et al. (2017)[AD]

TABLE 13.6 *(Continued)*

Species	Collection Site	Extraction Methodology	Phycoerythrins (PEs) mg/100g dw	Phycocyanins (PCs) mg/100g dw	Allophycocyanins (APCs) mg/100g dw	References
Grateloupia turuturu	Batz-sur-Mer (France)	Ultrasound-assisted enzymatic hydrolysis (UAEH) for 2 h at 22°C	360.00	–	–	Le Guillard et al. (2015)[F]
Grateloupia turuturu	Piriac-sur-Mer (France)	Suspended in phosphate buffer 20 mM (pH 7.1) at 4°C	300.00	33.00	–	Denis et al. (2010)[L]
Haloplegma duperreyi	Port Okha, Gujarat (India)	Suspended in 1 M acetic acid–sodium acetate buffer (pH 5.5) with 0.01% of sodium azide for 30 min in the dark	89.10	203.20	33.70	Verma et al. (2017)[AD]
Halymenia venusta	Port Okha, Gujarat (India)	Suspended in 1 M acetic acid–sodium acetate buffer (pH 5.5) with 0.01% of sodium azide for 30 min in the dark	377.10	104.80	31.40	Verma et al. (2017)[AD]
Hypnea musciformis	Mandapam, Tamil Nadu (India)	Suspended in 1 M acetic acid–sodium acetate buffer (pH 5.5) with 0.01% of sodium azide for 30 min in the dark	271.30	282.60	45.20	Verma et al. (2017)[AD]
Hypnea valentiae	Mandapam, Tamil Nadu (India)	Suspended in 1 M acetic acid–sodium acetate buffer (pH 5.5) with 0.01% of sodium azide for 30 min in the dark	283.20	236.80	42.70	Verma et al. (2017)[AD]
Portieria hornemannii	Mandapam, Tamil Nadu (India)	Suspended in 1 M acetic acid–sodium acetate buffer (pH 5.5) with 0.01% of sodium azide for 30 min in the dark	290.40	148.80	15.20	Verma et al. (2017)[AD]

TABLE 13.6 *(Continued)*

Species	Collection Site	Extraction Methodology	Phycoerythrins (PEs) mg/100g dw	Phycocyanins (PCs) mg/100g dw	Allophycocyanins (APCs) mg/100g dw	References
Porphyra spp.	Mandapam, Tamil Nadu (India)	Suspended in 1 M acetic acid–sodium acetate buffer (pH 5.5) with 0.01% of sodium azide for 30 min in the dark	242.30	67.20	50.00	Verma et al. (2017)[AD]
Rhodymenia dissecta	Port Okha, Gujarat (India)	Suspended in 1 M acetic acid–sodium acetate buffer (pH 5.5) with 0.01% of sodium azide for 30 min in the dark	141.70	51.10	35.60	Verma et al. (2017)[AD]
Scinaia fascicularis	Port Okha, Gujarat (India)	Suspended in 1 M acetic acid–sodium acetate buffer (pH 5.5) with 0.01% of sodium azide for 30 min in the dark	457.50	39.50	31.50	Verma et al. (2017)[AD]
Scinaia hatei	Port Okha, Gujarat (India)	Suspended in 1 M acetic acid–sodium acetate buffer (pH 5.5) with 0.01% of sodium azide for 30 min in the dark	335.50	98.30	31.70	Verma et al. (2017)[AD]

*These were assessed differently by different authors and methods and described individually in each reference. All measures are described as dry weight (dw, except when fresh weight (fw) is written after value for individual measurements. After individual reference, letters in superscript indicate the state of the biomass for the extraction protocol used by different authors, Fresh (F), Lyophilized (L) or Air-Dried (AD).

be modulated and optimized, since several researchers published different methodologies for these compounds. Sudhakar et al. (2015) extracted it from *Gracilaria crassa* collected in India, through suspension of fresh macroalgae in water (pH 7), at room temperature. Le Guillard et al. (2015) optimized an ultrasound-assisted enzymatic hydrolysis extraction, during 2 h, at 22°C, for compounds of *Grateloupia turuturu* from Batz-sur-Mer (France), yielding 360 mg of PEs /100 g dw. Denis et al. (2010) assessed these compounds in the same macroalgae from Piriac-sur-Mer, France, suspending lyophilized macroalgae flour in phosphate buffer (20 mM), pH 7.1, at 4°C, and obtained a similar PEs yield, 300 mg/100 g dw.

13.3.2.2 LECTINS

These are important bioactive glycoproteins, responsible for sugar cellular surface recognition in bacteria, plants, and animals (Weis and Drickamer, 1996). Lectins are also known as haemagglutinins or agglutinins due to their ability to agglutinate red blood cells (Smit, 2004). Lectins have a non-immune origin and reversibly bind to carbohydrates in a highly specific, non-covalent way (Goldstein and Poretz, 1986). Algae lectins are also known as phycolectins. Their physicochemical features are distinct from other plant lectins, due to its monomeric structure, with low molecular weight, showing specificity to glycoproteins more than to monosaccharides (Hori et al., 1990). Phycolectins do not require metal ions for its biological activity and possess a high content of acidic amino acids (Singh et al., 2017a). Numerous biological processes were undertaken, with lectins participation such as cell-cell communication, host-pathogen interactions, antiviral activities, cancer metastasis and apoptosis induction (Holdt and Kraan, 2011). Over the years, these glycoproteins were successfully assayed in many biotechnological applications such as cytotoxic, mitogenic, antinociceptive, antibiotic, anti-adhesion, and anti-inflammatory activities (Freitas et al., 2015). In addition, lectin-glycan interactions have been extensively studied as a new therapeutic approach for anti-HIV vaccines, assessing the adherence stage of this virus to lectins from algae sources (Huskens and Schols, 2012).

13.3.2.3 BIOACTIVE PEPTIDES

Amino acid content and physiological utility after digestion and absorption, will determine the nutritional quality of proteins. Bioactive peptides are inactive when integrated in the polypeptide chain of the protein and normally have less than 3 kDa and 2 to 30 amino acids in length. These peptides are released by proteolytic enzymes throughout gastrointestinal digestion or food processing (Vercruysse et al., 2005). The peptides with hormone-like properties exhibit a wide diversity of beneficial physiological activities, including antimicrobial, immunomodulatory, angiotensin-converting enzyme inhibition, opioid, renin inhibitory, anticoagulant, platelet-activating factor acetylhydrolase inhibitory, prolyl endopeptidase inhibitory, satiety-inducing, and α-amylase inhibitory activities (Hayes, 2013). Nowadays, milk proteins are the most common source of bioactive peptides. Nonetheless, various bioactive peptides have been identified in animal food products, such as meat, fish, eggs, blood, and plant sources including soybean, broccoli, rice, pea, garlic, and algae (Bleakley and Hayes, 2017). Algae proteins show high variability throughout the year due to season, temperature, and location, becoming a challenge the attempts to obtain bioactive peptides (Joubert and Fleurence, 2008).

13.3.3 POLYSACCHARIDES

These are polymers of monosaccharides linked by glycosidic bonds. They are the main structural compounds of macroalgae cell walls and could be involved in the recognition patterns between macroalgae and pathogens (Cardozo et al., 2007). Macroalgae represent the majority of marine organisms producing a large quantity of carbohydrates, due to their unique metabolism and physiological capability to prosper in extreme habitats conditions (Sudha et al., 2014). The term phycocolloids is used to identify the high molecular weight polysaccharides extracted from macroalgae that possess the ability to hold significant amounts of water and the capacity to form gel or viscous substances. They are the principal component of commercial macroalgae extracts, with an annual production of 100,000 tons. These extracts have been used for a variety of purposes such as food technology, pharmaceutics, cosmetics, and biotechnology (Pangestuti and Kim, 2015). The biological properties of macroalgae polysaccharides have been tested *in vitro* and *in vivo*, showing variable evidence for immunostimulatory, anticoagulant, anti-inflammatory, antiviral, antitumor, anti-apoptosis, antioxidant, and antiproliferative activities (Garcia-vaquero et al., 2017).

Table 13.7 summarizes the bibliographic information about the total carbohydrates content in macroalgae, collected in several places worldwide. The methodology used and sample condition varies among references, since

TABLE 13.7 Total Carbohydrate Content Described for Macroalgae Collected from Several Places Worldwide*

Species	Collection Site	Carbohydrate Content (g/100 g) dw	Method	References
Caulerpa racemosa	Veraval Coast, Gujarat (India)	48.95	Dubois et al. (1956)	Kumar et al. (2011)[L]
Caulerpa scalpelliformis	Veraval Coast, Gujarat (India)	38.84	Dubois et al. (1956)	Kumar et al. (2011)[L]
Caulerpa veravalensis	Veraval Coast, Gujarat (India)	37.23	Dubois et al. (1956)	Kumar et al. (2011)[L]
Codium tomentosum	Buarcos bay, Figueira da Foz (Portugal)	32.80	Difference method	Rodrigues et al. (2015)[AD]
Codium spp.	Barbate estuary, Gulf of Cádiz (Spain)	8.80	Anthrone method	Gracía et al. (2016)[L]
Dictyota dicotoma	Port Okha, Gujarat (India)	15.00	Dubois et al. (1956)	Verma et al. (2017)[AD]
Fucus spiralis	São Miguel Island, Azores (Portugal)	17.59	Dubois et al. (1956)	Paiva et al. (2014)[AD]
Gelidiella acerosa	Port Okha, Gujarat (India)	32.09	Dubois et al. (1956)	Verma et al. (2017)[AD]
Gelidium micropterum	Mandapam, Tamil Nadu (India)	34.81	Dubois et al. (1956)	Verma et al. (2017)[AD]
Gracilaria cervicornis	Búzios beach (Brazil)	63.13	Difference method	Marinho-Soriano et al. (2006)[AD]
Gracilaria changii	Santubong, Sarawak (Malaysia)	41.52	Difference method	Chan and Matanjun (2017)[L]
Gracilaria gracilis	Buarcos bay, Figueira da Foz (Portugal)	46.60	Difference method	Rodrigues et al. (2015)[AD]
Grateloupia indica	Port Okha, Gujarat (India)	22.48	Dubois et al. (1956)	Verma et al. (2017)[AD]
Grateloupia turuturu	Buarcos bay, Figueira da Foz (Portugal)	43.20	Difference method	Rodrigues et al. (2015)[AD]
Halymenia floresii	Barbate estuary, Gulf of Cádiz (Spain)	15.50	Anthrone method	Gracía et al. (2016)[L]
Lobophora variegata	Mandapam, Tamil Nadu (India)	32.31	Dubois et al. (1956)	Verma et al. (2017)[AD]
Osmundea pinnatifida	São Miguel Island, Azores (Portugal)	17.61	Dubois et al. (1956)	Paiva et al. (2014)[AD]
Osmundea pinnatifida	Buarcos bay, Figueira da Foz (Portugal)	32.40	Difference method	Rodrigues et al. (2015)[AD]
Padina gymnospora	Port Okha, Gujarat (India)	34.45	Dubois et al. (1956)	Verma et al. (2017)[AD]
Padina tetrastromatica	Mandapam, Tamil Nadu (India)	32.51	Dubois et al. (1956)	Verma et al. (2017)[AD]

TABLE 13.7 *(Continued)*

Species	**Collection Site**	**Carbohydrate Content (g/100 g) dw**	**Method**	**References**
Porphyra sp.	São Miguel Island, Azores (Portugal)	25.37	Dubois et al. (1956)	Paiva et al. (2014)[AD]
Saccorhiza polyschides	Barbate estuary, Gulf of Cádiz (Spain)	6.30	Anthrone method	Gracía et al. (2016)[L]
Saccorhiza polyschides	Buarcos bay, Figueira da Foz (Portugal)	45.60	Difference method	Rodrigues et al. (2015)[AD]
Sargassum linearifolium	Port Okha, Gujarat (India)	29.82	Dubois et al. (1956)	Verma et al. (2017)[AD]
Sargassum muticum	Buarcos bay, Figueira da Foz (Portugal)	49.30	Difference method	Rodrigues et al. (2015)[AD]
Sargassum naozhouense	Techeng Island, Guangdong province (China)	47.73	Difference method	Peng et al. (2013)[AD]
Sargassum vulgare	Búzios beach (Brazil)	61.61	Difference method	Marinho-Soriano et al. (2006)[AD]
Ulva lactuca	Port Okha, Gujarat (India)	32.61	Dubois et al. (1956)	Verma et al. (2017)[AD]
Ulva reticulata	Mandapam, Tamil Nadu (India)	33.35	Dubois et al. (1956)	Verma et al. (2017)[AD]
Ulva spp.	Barbate estuary, Gulf of Cádiz (Spain)	11.70	Anthrone method	Gracía et al. (2016)[L]

*Carbohydrates were assessed differently, using different solvents, temperatures, and time. After individual reference, letters in superscript indicate the state of the biomass for the extraction protocol used by different authors, Lyophilized (L) or Air-Dried (AD).

samples used are lyophilized or air-dried. Total carbohydrate content varies between 6.30 g/100 g dw in *Saccorhiza polyschides*, a brown macroalgae (Gracía et al., 2016) and 63.13 g/100 g dw in *Gracilaria cervicornis*, a red macroalgae (Marinho-Soriano et al., 2006). The different color in macroalgae shown specific maximum and minimum carbohydrates yields. Green macroalgae carbohydrate varies between 8.80 g/100 g of dw in *Codium* spp. and 48.95 g/100 g of dw in *Caulerpa racemosa* from Gulf of Cádiz (Spain) and Gujarat (India), respectively. Red macroalgae shows between 15.50 g/100 g dw in *Halymenia floresii* and 63.13 g/100 g dw in *Gracilaria cervicornis*. For brown macroalgae, it varies from 6.30 g/100 g dw in *Saccorhiza polyschides* and 61.61 g/100 g dw found in *Sargassum vulgare*.

13.3.3.1 ALGINATES

The function of alginate in brown macroalgae (Ochrophyta) is predominantly structural, being in the cell wall and intercellular matrix, attributing strength, and flexibility to withstand sea currents. The molecular arrangement is a linear, anionic block copolymer heteropolysaccharide, comprised of β-D-mannuronic acid (M) and α-L-glucuronic acid (G). The structure of alginates varies according to the species, season, and geographical location. It is composed by sequential distributed homogeneous M-M (M-blocks) and G-G segments (G-blocks) alternating with M-G segments (MG-blocks), which vary in relative amounts (D'Ayala et al., 2008). Furthermore, M/G ratio and block structure are crucial for the physicochemical properties of alginates since this parameter is widely used to predict the nature and behavior of gels when calcium ions are added. Moreover, two main factors influence the quality and applicability of alginate gel, since a high amount of G-blocks favorably increases gel strength and industrial application will depend on its uronic acid composition (Fertah, 2017). These gels can be synthesized in a wide range of molecular weights, ranging from 50 to 100,000 residues, to better fit different applications (D'Ayala et al., 2008). This polysaccharide is mainly applied in the food industry as a thickener, stabilizer or as an emulsifying agent (Sudha et al., 2014). Alginates extraction from brown macroalgae begins with a pre-extraction using hydrochloric acid, followed by washing, filtration, and neutralization with an alkaline product. It forms sodium alginate that is precipitated with ethanol or isopropanol, frequently being re-precipitated in the same way (Laurienzo, 2010).

Sodium alginate shows high ability to remove heavy metals, such as cadmium and copper through adsorption (Papageorgiou et al., 2008). This capability is derived from their natural condition, when they are bound with salts present in seawater, particularly Ca^{2+}, Na^{+}, Mg^{2+}, Sr^{2+}, and Ba^{2+} ions (Fertah, 2017). Pharmaceutical and medical interest for alginate applications has increased since it was demonstrated to be efficient in esophageal reflux and it produces multiquality calcium fibers for dermatology and wound healing. It is an active natural disintegrative material, a tablet binder, and an alternative for a sustained release system due to various properties, such as the formation of hydrogels under relatively mild pH and temperature. It is considered non-toxic, biodegradable, biocompatible, and abundantly available in nature (D'Ayala et al., 2008).

13.3.3.2 AGAR

This hydrocolloid is also known as agar-agar, being a collective term used to describe gelling polysaccharides. Currently, it is mostly extracted from red species of macroalgae, such as *Pterocladiella*, *Gelidium*, *Ahnfeltia*, *Gelidiella*, *Gracilariopsis*, and *Gracilaria* (Laurienzo, 2010). Galactans from red species of macroalgae were named as agaroids, agarose, agarans, and agaroptectin, depending on the conformation and side chains that different agar components can present. The term agarose is nowadays widely used due to later research, that showed a regular polymer of alternating 3-linked β-D-galactopyranose and 4-linked 3,6-anhydro-L-galactopyranose residues. The proportion of 3,6-anhydro-α-L-galactopyranose in the polysaccharides influences the strength and brittleness of the gel. It is widely used for special preparations in gel electrophoresis, gel permeation chromatography and other applications in biochemical, biomedical, and biotechnological laboratories (Knutsen et al., 1994). It is estimated that 70% of agar is comprised of agarose, which presents a linear structure having sulfate ester, methoxyl group and pyruvate ketal as side-chain substituents, which can be found in the backbone of agar, influencing its gelling properties. Purified agarose is also commercialized but considerably more expensive, evidencing a significantly lower sulfate content, better optical clarity and increased gel strength comparing with agar (Laurienzo, 2010). Agar is the most abundant cell wall polysaccharide present in red macroalgae and its biological function is to provide structural stability with high flexibility to withstand ocean waves and currents. It also provides resistance to pathogens, protect algae against extreme salinity, pH, temperatures, and desiccation (Lee et al., 2017). Agar extract is added to boiling water and jellifies at room temperature, forming thermally reversible gels without losing their mechanical and thermic properties, being suitable for multiple applications. The gel structure results from the hydrogen bonds, established in a continuous way, following a coil-double helix transition structure, interacting between themselves, and linking bundles of associated right-handed double helices (Norziah et al., 2006). The subsequent three-dimensional network is capable of restraining water molecules in its interstices (Arnott et al., 1974). Agar can be subjected to sodium hydroxide and enzymatic treatments using sulfohydrolase to improve its strength (Freile-Pelegrín and Murano, 2005; Shukla et al., 2011). Additionally, genetic manipulation in the development stages of macroalgae has been performed to minimize seasonal variations in plant growth and agar quality (Lee et al., 2016). Recently, the use of agar and agarose beads for sustained release of water-soluble drugs was investigated, consisting in dissolving the drug in a hot agar aqueous solution, with a temperature around 70°C. The beads are jellified when drops of agar with the dissolved drug are dropped in a cold bath containing a non-solvent for agar, normally acetone or ethyl acetate (Laurienzo, 2010).

13.3.3.3 CARRAGEENANS

This hydrocolloid is widely extracted from Rhodophyta species, such as *Kappaphycus*, *Betaphycus*, *Eucheuma*, *Hypnea*, *Gigartina*, and *Chondrus*. Carrageenans are mainly defined by alternating 3-linked-β-D-galactopyranose and 4-linked-α-D-galactopyranose units. Carrageenan and agar show a fundamental stereochemical difference between the diastereoisomeric polysaccharide molecules (Anderson et al., 1965). They are classified according to the degree of substitutions occurring on their free hydroxyl groups, which generally could be the addition of ester sulfate or the presence of the 3,6-anhydride on the 4-linking residues (Nanaki et al., 2010). Additionally, carrageenan can present eight different forms, Kappa (κ), Iota (τ), Lambda (λ), Beta (β), Mu (μ), Nu (ν), Xi (ξι) and Theta (θ), depending on the macroalgae source used for carrageenan extraction (Campo et al., 2009). The first three are the most commercially important, since κ produces strong and rigid gels, τ develops soft gels, and λ is used as a thickener as it does not form a gel (Sankalia et al., 2006). Carrageenan properties are primarily influenced by the number and position of ester sulfate groups as well as the content of 3,6-anhydro-galactose. The commercial carrageenan presents a molecular weight that can fluctuate between 100 and 1,000 kDa (Li et al., 2014). κ and τ gel consist of an ordered three-dimensional network of double helices cross-linked with sulfate groups, oriented externally while in λ, the sulfate groups in the 2° position are directed inward, forming the cross-link of an ordered network. Biological properties of carrageenan show potential in pharmaceutics as an anticoagulant, anticancer, antihyperlipidemic, and with immunomodulatory activity (Campo et al., 2009). Although no nutritional value could be attributed to carrageenan, it is widely used as a food additive, functioning as an emulsifier, stabilizer or thickener and has been applied to increase the quality of dairy sweets, puddings, cheese, etc. (Wijesekara et al., 2011; Zia et al., 2017). Also, carrageenan could be used in non-food products, such as firefighting foams, air freshener gels and shoe polish (Zia et al., 2017). The economic feasibility is extremely important to choose the proper methodology to extract carrageenan from macroalgae. Primarily, macroalgae is washed following a hot alkali extraction process, freeing the carrageenan from the cell. Secondly, one of three methods currently used in the industry could be used for carrageenan separation, namely the "freeze-thaw" technique, the "alcohol precipitation" method or the "KCl precipitation" process (Li et al., 2014).

13.3.3.4 CELLULOSE

This natural biopolymer is one of the most abundant polysaccharides on earth, providing physical structure and strength which can be found in cell walls of almost all plants and macroalgae (D'Ayala et al., 2008). There are four types of cellulose available, which include native cellulose (type I), native cellulose with major structural changes caused by treatment with an alkaline solution (type II) and cellulose treated with multiple reagents (type III and IV). These last cellulose derivatives can be amongst others cellulose acetate, ethylcellulose, methylcellulose, hydroxyethylcellulose, carboxymethylcellulose, and hydroxypropyl cellulose (Yang and Li, 2018). It is a versatile material, mainly used to produce paper, pharmaceutical, textile, veterinary, food, and cosmetic goods (Lakshmi et al., 2017). Its unique properties, such as high strength and modulus, high tensile strength, high surface area, lightweight, high aspect ratio, stiffness, cost-effectiveness and good biodegradability, allows several industrial applications (Singh et al., 2017b). Cellulose type I comprises the general formula $(C_6H_{10}O_5)_n$ composed by β-(1–4)-linked anhydroglucose repeating units, where n can variate from 10,000 to 15,000 depending on the source of raw material (Yang and Li, 2018). It is a homopolysaccharide whose rigid structure depends on the intermolecular and intramolecular hydrogen bonds and hydrophobic interactions (van der Waals) (Lindman et al., 2010). Cellulose type I is considered amphiphilic, because all three hydroxyl groups of the anhydroglucose units possess an equatorial alignment, making it hydrophilic, and the hydrogen atoms of C–H bonds have an axial orientation making it hydrophobic. Also, it has a highly ordered crystalline regions and disordered amorphous regions along its chain (Nagarajan et al., 2017).

13.3.3.5 FUCOIDAN

Xylofucoglycuronans, glycuronogalactofucans, and fucoidan are the main constituents of biological macromolecules classified as fucans (Phull and Kim, 2017). Fucoidan is a group of fucose-enriched sulfated polysaccharides with a high degree of complexity. These compounds originate in the fibrillary cell wall, where they act as reinforcement molecules and in the intercellular spaces of brown macroalgae, involved in the macroalgae protection against the effects of desiccation, when exposed to low tide (Senthilkumar et al., 2013; Sanjeewa et al., 2017). Its chemical structure mainly consists of L-fucose sulfate groups and in smaller extend one or more groups of D-mannose, D-xylose, arabinose, D-galactose, D-glucuronic acid, L-rhamnose, glucose, and acetyl (Wu et al., 2016). Fucoidan molecular structure will vary depending on the species, harvesting time and habitat conditions of macroalgae. Fucoidan homofucose backbone consists primarily of 2 types, type I with $\alpha(1\rightarrow3)$-L-fucopyranose residues or type II with alternating $\alpha(1\rightarrow3)$ and $\alpha(1\rightarrow4)$-linked L-fucopyranosyls (Cumashi et al., 2007; Sanjeewa et al., 2017). Initially isolated by Kylin in 1913, it was named "fucoidin" and later, according to IUPAC guidelines, the name was changed to "fucoidan." However, other names are used, such as fucan, fucosan or sulfated fucan (Senthilkumar et al., 2013). According to recent reviews by Senthilkumar et al. (2013) and Vo and Kim (2013), several macroalgae have been targeted for the extraction and isolation of fucoidan to multiple biological applications, such as *Analipus japonicus*, *Ascophyllum nodosum*, *Cladosiphon okamuranus*, *Dictyota menstrualis*, *Fucus distichus*, *Fucus evanescens*, *Fucus serratus*, *Fucus vesiculosus*, *Hizikia fusiformis*, *Kjellmaniella crassifolia*, *Pelvetia canaliculata*, *Sargassum kjellmanianum*, *Sargassum stenophyllum*, and *Sargassum thunbergii*. Fucoidan extraction and purification often begins with dehydration and milling of macroalgae to obtain a homogeneous biomass with higher surface-to-volume-ratio. A pre-treatment is normally applied for removal of lipids, terpenes, and phenols, using a ternary solvent system, composed by methanol/chloroform/water 4:2:1 (v/v), and further applying formaldehyde in an ethanolic solution or washing the macroalgae with acetone (Hahn et al., 2012). The extracted fucoidan fraction can be further cleaned with activated charcoal. Typically, the extraction continues with hot aqueous or acidic solution at temperatures ranging from 70°C to 100°C for several hours (Ale et al., 2011). Further fucoidan purification is achieved either using precipitation with ethanol or dialysis for desalting and removal of low molecular weight compounds. Recently, low pressure anionic or molecular exclusion liquid chromatography has been used to achieve high purified fucoidan fraction (Hemmingson et al., 2006; Foley et al., 2011). Fucoidan biological activity has been extensively studied; including anti-inflammatory, antioxidant, anti-virus, antithrombotic, and anticoagulant effects, as well preventing various renal, hepatic, and neuropathic disorders (Senthilkumar et al., 2013). In addition, the antitumor potential was compared with synthetic drugs, where this natural product demonstrated lower side effects (Wu et al., 2016). These results turned fucoidan into a promising substance with multiple applications in the pharmaceutical, cosmeceutical, and nutraceutical industries (Phull and Kim, 2017).

13.3.3.6 ULVANS

These are a group of major matrix soluble polysaccharides which appear equally scattered throughout the frond, predominantly within the intercellular space and in the fibrillar wall. This polysaccharide is arranged in a bead-like structure, stabilized by cell wall proteins or strong physical interactions (Alves et al., 2013). Originally, these substances were named as ulvin and ulvacin to designate different fractions of *Ulva lactuca* water-soluble sulfated polysaccharides. Later their designations were changed to ulvan (Lahaye and Robic, 2007). These anionic sulfated polysaccharides are structurally constituted by various proportions of repeating disaccharide units, which include 3-sulfated rhamnose (Rha3S), glucuronic acid (GlcA), iduronic acid (IduA) and some xylose (Xyl) units (Hehemann et al., 2014). Additionally, in their composition, two major aldobiuronic acids disaccharides, which are differentiated in type A and B are present. Type A includes ulvanobiuronic acid 3-sulfate (A3s), a 1,4-linked glucuronic acid with O-3-sulfated rhamnose, and type B, ulvanobiuronic acid 3-sulfate (B3s), a 1,4-linked iduronic acid with O-3-sulfated rhamnose (Glasson et al., 2017). Various biological activities are hypothesized for ulvan, such as osmotic functions, regulating the ionic balance and preventing tissue desiccation, due to its highly hygroscopic capability. Other possible functions are protecting macroalgae from marine bacterial attacks and mechanically regulating the spore release and adhesivity (Alves et al., 2013). Ulvan molecular structure, macromolecular characteristics and rheological properties are shown to be variable among species and seasons (Robic et al., 2009a). Classical extraction of ulvan polysaccharides are performed using sodium oxalate or ammonium oxalate aqueous solutions, with temperatures between 80°C and 90°C, to chelate the Ca^{2+} which crosslinks ulvan strands in the cell wall (Glasson et al., 2017). Although, this extraction methodology will dictate the physical and chemical properties of ulvan (Robic et al., 2009b). Several potential applications have been described for these biological active polymers, such as anti-hyperlipidemic, anti-oxidant, anti-peroxidative, and anti-viral activity (Alves et al., 2013). Additionally, it can function as a heparinoid agent, replacing heparin, treating gastric ulcer, or having anti-thrombotic activity (Lahaye and Robic, 2007).

13.3.4 LIPIDS

Macroalgae are a low-energy food, which environmental conditions influence its lipid composition and quantity, as different mechanisms are triggered to allow adaptation (Mišurcová et al., 2011). Although macroalgae have low lipid content, they are considered a potential source of functional lipids due to their large stock in coastal waters. Total lipids in macroalgae are composed of lipid classes, whose quantity varies according to species, geographical location, and environmental factors (Miyashita et al., 2013). Table 13.8 summarizes bibliographic and methodology information available about lipids, from macroalgae sampled in different places of the world. The low and high lipid values were found in different red macroalgae, 0.30 g/100 g dw in *Gracilaria changii*, Sarawak, Malaysia (Chan and Matanjun, 2017) and 12.30 g/100 g dw, found in *Halymenia floresii*, Gulf of Cádiz, Spain, reported by Gracía et al. (2016). The total lipids in green macroalgae varies between 1.45 g/100 g in dw of *Ulva lactuca*, Gujarat, India, and 7.10 g/100 g dw in *Codium* spp., Gulf of Cádiz, Spain. In the brown macroalgae, total lipids oscillate between 0.49 g/100 g dw in *Sargassum vulgare*, Búzios beach, Brazil, and 8.20 g/100 g dw in *Saccorhiza polyschides*, Gulf of Cádiz, Spain.

Total lipids can be divided into classes, which include glycolipids (GL), triacylglycerols (TAG), phospholipids (PLs), sterols, and pigments. In these classes, fatty acids comprise the largest component, which can be present in three different forms such as saturated fatty acids (SFAs), monounsaturated fatty acids (MUFAs) and polyunsaturated fatty acids (PUFAs). The PUFAs

TABLE 13.8 Total Lipid Content Described for Macroalgae Collected from Several Places Worldwide*

Species	Collection Site	Lipid Content (g/100 g) dw	Method	References
Caulerpa racemosa	Veraval Coast, Gujarat (India)	2.64	Bligh and Dyer (1959)	Kumar et al. (2011)[L]
Caulerpa scalpelliformis	Veraval Coast, Gujarat (India)	3.06	Bligh and Dyer (1959)	Kumar et al. (2011)[L]
Caulerpa veravalensis	Veraval Coast, Gujarat (India)	2.80	Bligh and Dyer (1959)	Kumar et al. (2011)[L]
Codium spp.	Barbate estuary, Gulf of Cádiz (Spain)	7.10	Soxhlet (n-hexane)	J.S. and V. (2016)[L]
Codium tomentosum	Buarcos bay, Figueira da Foz (Portugal)	3.60	Soxhlet	Rodrigues et al. (2015)[AD]
Dictyota dicotoma	Port Okha, Gujarat (India)	2.95	Bligh and Dyer (1959)	Verma et al. (2017)[AD]
Fucus spiralis	São Miguel Island, Azores (Portugal)	5.23	Folch, Lees, and Stanley (1957)	Paiva et al. (2014)[AD]
Gelidiella acerosa	Port Okha, Gujarat (India)	2.75	Bligh and Dyer (1959)	Verma et al. (2017)[AD]
Gelidium micropterum	Mandapam, Tamil Nadu (India)	2.20	Bligh and Dyer (1959)	Verma et al. (2017)[AD]
Gracilaria cervicornis	Búzios beach (Brazil)	0.43	Soxhlet (n-hexane)	Marinho-Soriano et al. (2006)[AD]
Gracilaria changii	Santubong, Sarawak (Malaysia)	0.30	Soxtec system (petroleum ether) (AOAC 991.36)	Chan and Matanjun (2017)[L]
Gracilaria fisheri	Pattani bay (Thailand)	2.20	Bligh and Dyer (1959)	Benjama and Masniyom (2012)[AD]
Gracilaria gracilis	Buarcos bay, Figueira da Foz (Portugal)	0.60	Soxhlet extraction	Rodrigues et al. (2015)[AD]
Gracilaria tenuistipitata	Pattani bay (Thailand)	2.80	Bligh and Dyer (1959)	Benjama and Masniyom (2012)[AD]
Grateloupia indica	Port Okha, Gujarat (India)	1.80	Bligh and Dyer (1959)	Verma et al. (2017)[AD]
Grateloupia turuturu	Piriac-sur-Mer (France)	2.60	Folch, Lees, and Stanley (1957)	Denis et al. (2010)[L]

TABLE 13.8 *(Continued)*

Species	Collection Site	Lipid Content (g/100 g) dw	Method	References
Grateloupia turuturu	Buarcos bay, Figueira da Foz (Portugal)	2.20	Soxhlet extraction	Rodrigues et al. (2015)[AD]
Halymenia floresii	Barbate estuary, Gulf of Cádiz (Spain)	12.30	Soxhlet (n-hexane)	Gracía et al. (2016)[L]
Lobophora variegata	Mandapam, Tamil Nadu (India)	1.25	Bligh and Dyer (1959)	Verma et al. (2017)[AD]
Osmundea pinnatifida	São Miguel Island, Azores (Portugal)	7.53	Folch, Lees, and Stanley (1957)	Paiva et al. (2014)[AD]
Osmundea pinnatifida	Buarcos bay, Figueira da Foz (Portugal)	0.90	Soxhlet extraction	Rodrigues et al. (2015)[AD]
Padina gymnospora	Port Okha, Gujarat (India)	2.28	Bligh and Dyer method (1959).	Verma et al. (2017)[AD]
Padina tetrastromatica	Mandapam, Tamil Nadu (India)	1.25	Bligh and Dyer method (1959).	Verma et al. (2017)[AD]
Porphyra sp.	São Miguel Island, Azores (Portugal)	8.88	Folch, Lees, and Stanley (1957)	Paiva et al. (2014)[AD]
Saccorhiza polyschides	Buarcos bay, Figueira da Foz (Portugal)	1.10	Soxhlet	Rodrigues et al. (2015)[AD]
Saccorhiza polyschides	Barbate estuary, Gulf of Cádiz (Spain)	8.20	Soxhlet (n-hexane)	Gracía et al. (2016)[L]
Sargassum linearifolium	Port Okha, Gujarat (India)	1.93	Bligh and Dyer method (1959).	Verma et al. (2017)[AD]
Sargassum muticum	Buarcos bay, Figueira da Foz (Portugal)	1.45	Soxhlet	Rodrigues et al. (2015)[AD]
Sargassum naozhouense	Techeng Island, Guangdong province (China)	1.06	Soxhlet	Peng et al. (2013)[AD]
Sargassum vulgare	Búzios beach (Brazil)	0.49	Soxhlet (n-hexane)	Marinho-Soriano et al. (2006)[AD]

TABLE 13.8 *(Continued)*

Species	Collection Site	Lipid Content (g/100 g) dw	Method	References
Ulva lactuca	Port Okha, Gujarat (India)	1.45	Bligh and Dyer method (1959)	Verma et al. (2017)[AD]
Ulva reticulata	Mandapam, Tamil Nadu (India)	2.50	Bligh and Dyer method (1959)	Verma et al. (2017)[AD]
Ulva spp.	Barbate estuary, Gulf of Cádiz (Spain)	3.40	Soxhlet (n-hexane)	Gracía et al. (2016)[L]

*Lipids were extracted differently, using different solvents, temperatures, and time. After individual reference, letters in superscript indicate the state of the biomass for the extraction protocol used by different authors, Lyophilized (L) or Air-Dried (AD).

are essential to human body functions and should be obtained through food intake, since they cannot be synthesized by our own metabolic processes. Nowadays, they are considered an important component of functional food and nutraceutics, promoting health benefits, since they are able to reduce the risk of CVDs, diabetes, osteoporosis, and cancer (Mišurcová et al., 2011). GL is the most common lipid class and is comprised of monogalactosyl-diacylglycerols, digalactosyl-diacylglycerol, and sulfoquinovosyldiacylglycerol (Holdt and Kraan, 2011). GL is extremely important in photosynthetic membranes of plants, algae, and bacteria. In brown macroalgae, GL is the foremost membrane lipid, with a high concentration of specific long-chain fatty acids, such as stearidonic acid (18:4n-3; SDA), EPA (20:5n-3) and arachidonic acid (AA) (20:4n-6; ARA) (Miyashita et al., 2013). SDA are produced by the desaturation of α-linolenic acid (ALA), and when consumed, it increases EPA in red blood cell membranes (Lemke et al., 2013). Epidemiological tests demonstrated that the consumption of long-chain ω3 fatty acids reduces the risk of cardiovascular disease, particularly sudden heart attack (Albert et al., 1998). The ω6 fatty acids are extremely important in biological systems. ARA participates in the immune response, prevents thrombosis, and helps brain function and together with docosahexaenoic acid (DHA), improves cognitive functions and prevents age-related brain disorders (Hoffman et al., 2009; Kiso, 2011). These unsaturated acids are the main components of cell membranes and participate in neuron structure of the central nervous system (Miyashita et al., 2013). Table 13.9 summarizes bibliographic information about saturated fatty acids (SFAs), monounsaturated fatty acids (MUFAs) and PUFAs content in total fatty acid methyl esters (FAME) of macroalgae, collected in several places of the world. In Table 13.9, it can be observed that SFAs varies between 7.53% in *Gracilaria changii* collected in Sarawak (Malaysia) and 67.4% of the total FAME, in *Grateloupia turuturu*, from Piriac-sur-Mer (France). MUFAs show a variation between 4.97% in *Caulerpa veravalensis*, Gujarat (India) and 48.7% of the total FAME in *Codium* spp., from Gulf of Cádiz (Spain). For PUFAs, these values oscillate between 19.1% of *Grateloupia turuturu* and 51.2% of the total FAME, in *Gracilaria changii*.

13.3.5 PIGMENTS

13.3.5.1 CAROTENOIDS

Among the most abundant pigments found in nature, carotenoids account for approximately 600 compounds. They are lipid soluble methylated polyene

TABLE 13.9 Total Saturated Fatty Acids (SFAs), Monounsaturated Fatty Acids (MUFAs) and Polyunsaturated Fatty Acids (PUFAs) Content in Total Fatty Acid Methyl Esters (FAME) are Described for Macroalgae Collected from Several Places Worldwide*

Species	Collection Site	SFAs (% of Total FAME)	MUFAs (% of Total FAME)	PUFAs (% of Total FAME)	References
Caulerpa racemosa	Veraval Coast, Gujarat (India)	62.15	10.68	27.17	Kumar et al. (2011)[L]
Caulerpa scalpelliformis	Veraval Coast, Gujarat (India)	50.25	7.00	42.75	Kumar et al. (2011)[L]
Caulerpa veravalensis	Veraval Coast, Gujarat (India)	55.03	4.97	40.00	Kumar et al. (2011)[L]
Codium spp.	Barbate estuary, Gulf of Cádiz (Spain)	17.0	48.7	27.2	Gracía et al. (2016)[L]
Codium tomentosum	Buarcos bay, Figueira da Foz (Portugal)	38.88	18.51	42.60	Rodrigues et al. (2015)[AD]
Fucus spiralis	São Miguel Island, Azores (Portugal)	33.59	27.10	38.97	Paiva et al. (2014)[AD]
Gracilaria changii	Santubong, Sarawak (Malaysia)	7.53	38.30	51.20	Chan and Matanjun (2017)[L]
Gracilaria gracilis	Buarcos bay, Figueira da Foz (Portugal)	63.54	15.24	21.22	Rodrigues et al. (2015)[AD]
Grateloupia turuturu	Buarcos bay, Figueira da Foz (Portugal)	42.74	11.54	45.72	Rodrigues et al. (2015)[AD]
Grateloupia turuturu	Piriac-sur-Mer (France)	67.4	11.6	19.1	Denis et al. (2010)[L]
Halymenia floresii	Barbate estuary, Gulf of Cádiz (Spain)	18.8	37.0	32.6	Gracía et al. (2016)[L]
Osmundea pinnatifida	São Miguel Island, Azores (Portugal)	56.78	20.64	22.23	Paiva et al. (2014)[AD]
Osmundea pinnatifida	Buarcos bay, Figueira da Foz (Portugal)	58.07	18.92	23.01	Rodrigues et al. (2015)[AD]
Porphyra sp.	São Miguel Island, Azores (Portugal)	54.32	17.31	24.96	Paiva et al. (2014)[AD]
Saccorhiza polyschides	Buarcos bay, Figueira da Foz (Portugal)	36.42	29.09	34.49	Rodrigues et al. (2015)[AD]
Saccorhiza polyschides	Barbate estuary, Gulf of Cádiz (Spain)	20.5	37.6	27.7	Gracía et al. (2016)[L]
Sargassum binderi	Straits of Malacca near Port Dickson (Malaysia)	31.17	13.65	36.91	Jaswir et al. (2012)[F]

TABLE 13.9 *(Continued)*

Species	Collection Site	SFAs (% of Total FAME)	MUFAs (% of Total FAME)	PUFAs (% of Total FAME)	References
Sargassum duplicatum	Straits of Malacca near Port Dickson (Malaysia)	29.49	15.39	33.10	Jaswir et al. (2012)[F]
Sargassum muticum	Buarcos bay, Figueira da Foz (Portugal)	42.17	21.13	36.70	Rodrigues et al. (2015)[AD]
Sargassum naozhouense	Techeng Island, Guangdong province (China)	33.63	10.42	18.84	Peng et al. (2013)[AD]
Ulva spp.	Barbate estuary, Gulf of Cádiz (Spain)	24.0	45.5	24.2	Gracía et al. (2016)[L]

*After individual reference, letters in superscript indicate the state of the biomass for the extraction protocol used by different authors, fresh (F), lyophilized (L), or air-dried (AD).

derivatives or unsaturated terpenoids with multiple colors, due to the delocalization of the excited state π-electron energy, stated by the variation of conjugated carbon double bonds (Ruban, 2010). The molecular structure comprises a backbone of 40 carbon polyene chain, attributing their distinctive structure, chemical properties, and light-absorbing features. Cyclic groups could exist in its extremities (one or both) including functional groups that contain oxygen (Del Campo et al., 2007). Carotenoids are essential for photosynthesis, contributing to light harvesting, quench chlorophyll triplet states, scavenge reactive oxygen species (ROS), dissipate excess energy, maintain the structure and function of photosynthetic complexes (Demmig-Adams and Adams, 2002). The contribution of carotenoids to light-harvesting is often greater in algae than in higher plants due to the smaller amount of light that reaches their photosynthetic tissue (Britton et al., 2008). The total carotenoid content (TCC) in several macroalgae is described in Table 13.10. TCC ranges between 7.42 mg, in *Gracilaria changii*, Sarawak (Malaysia) and 297.77 mg/100 g dw of *Zonaria tournefortii* from Madeira (Portugal). Green macroalgae was found to contain between 20.41 mg, in *Ulva lactuca*, Madeira (Portugal) and 141.70 mg/100 g dw in *Caulerpa veravalensis* from Tamil Nadu (India). Red macroalgae contains between 7.42 mg, in *Gracilaria changii* and 192.80 mg/100 g dw of *Botryocladia leptopoda*, from Gujarat (India). Finally, brown macroalgae shows TCC between 32.70 mg in *Stoechospermum marginatum*, Gujarat (India) and 297.77 mg/100 g dw of *Zonaria tournefortii* from Madeira (Portugal).

13.3.5.2 CHLOROPHYLLS

These are greenish pigments directly linked to photosynthesis, synthesized by phototrophic organisms in structures called chloroplasts, within flattened vesicles (thylakoids), to absorb sunlight. They are the most abundant pigments on earth and generally carry Magnesium as the central metal ion and a phytol (C_{20}) esterified with the propionic acid moiety at C_{17} (Roca et al., 2016). These pigments belong to the tetrapyrroles organic molecules group together with hems, bellins, and corrins. Tetra pyrroles are constituted by four pyrrole rings with five atoms each, one of which is nitrogen. Alone, a pyrrole molecule does not absorb light, but when four pyrroles ring merge, linked by unsaturated methane groups, it becomes a molecule rich in conjugated double bonds which can strongly absorb light. Additionally, the four nitrogen atoms are oriented towards the center of the structure and became

TABLE 13.10 Total Chlorophyll α and Carotenoid Content Described for Macroalgae Collected from Several Places Worldwide*

Species	Collection Site	Carotenoid Content (mg/100 g) dw	Chlorophyll α Content (mg/100 g) dw	Method	References
Acrosiphonia orientalis	Port Okha, Gujarat (India)	140.50	203.50	Duxbury and Yentach (1956)	Verma et al. (2017)[AD]
Asparagopsis taxiformis	Madeira (Portugal)	13.14	28.81	Wellburn (1994); Kumar et al. (2010)	Nunes et al. (2017)[L]
Boodlea composita	Port Okha, Gujarat (India)	39.80	1151.30	Duxbury and Yentach (1956)	Verma et al. (2017)[AD]
Botryocladia leptopoda	Port Okha, Gujarat (India)	192.80	374.70	Duxbury and Yentach (1956)	Verma et al. (2017)[AD]
Caulerpa vervelansis	Mandapam, Tamil Nadu (India)	141.70	2357.90	Duxbury and Yentach (1956)	Verma et al. (2017)[AD]
Champia spp.	Mandapam, Tamil Nadu (India)	73.90	177.50	Duxbury and Yentach (1956)	Verma et al. (2017)[AD]
Chondrus crispus	Madeira (Portugal)	21.25	62.51	Wellburn (1994); Kumar et al. (2010)	Nunes et al. (2017)[L]
Dictyopteris australis	Port Okha, Gujarat (India)	159.50	355.30	Duxbury and Yentach (1956)	Verma et al. (2017)[AD]
Dictyota dicotoma	Port Okha, Gujarat (India)	140.60	263.10	Duxbury and Yentach (1956)	Verma et al. (2017)[AD]
Gelidium micropterum	Mandapam, Tamil Nadu (India)	74.60	392.90	Duxbury and Yentach (1956)	Verma et al. (2017)[AD]
Gelidiella acerosa	Port Okha, Gujarat (India)	75.10	71.2	Duxbury and Yentach (1956)	Verma et al. (2017)[AD]
Gracilaria changii	Santubong, Sarawak (Malaysia)	7.42	57.79	Kumar, Ramakritinan, and Kumaraguru (2010)	Chan and Matanjun (2017)[L]
Grateloupia indica	Port Okha, Gujarat (India)	31.70	89.60	Duxbury and Yentach (1956)	Verma et al. (2017)[AD]
Grateloupia lanceola	Madeira (Portugal)	131.13	32.30	Wellburn (1994); Kumar et al. (2010)	Nunes et al. (2017)

TABLE 13.10 *(Continued)*

Species	Collection Site	Carotenoid Content (mg/100 g) dw	Chlorophyll α Content (mg/100 g) dw	Method	References
Haloplegma duperreyi	Port Okha, Gujarat (India)	178.10	344.70	Duxbury and Yentach (1956)	Verma et al. (2017)[AD]
Halymenia venusta	Port Okha, Gujarat (India)	27.10	50.40	Duxbury and Yentach (1956)	Verma et al. (2017)[AD]
Hypnea musciformis	Mandapam, Tamil Nadu (India)	123.80	268.00	Duxbury and Yentach (1956)	Verma et al. (2017)[AD]
Hypnea valentiae	Mandapam, Tamil Nadu (India)	119.20	296.40	Duxbury and Yentach (1956)	Verma et al. (2017)[AD]
Iyengaria stellata	Port Okha, Gujarat (India)	82.00	113.00	Duxbury and Yentach (1956)	Verma et al. (2017)[AD]
Lobophora variegata	Mandapam, Tamil Nadu (India)	115.00	222.50	Duxbury and Yentach (1956)	Verma et al. (2017)[AD]
Nemalion elminthoides	Madeira (Portugal)	8.99	89.09	Wellburn (1994); Kumar et al. (2010).	Nunes et al. (2017)[L]
Padina gymnospora	Port Okha, Gujarat (India)	207.10	343.90	Duxbury and Yentach (1956)	Verma et al. (2017)[AD]
Padina tetrastromatica	Mandapam, Tamil Nadu (India)	249.60	340.40	Duxbury and Yentach (1956)	Verma et al. (2017)[AD]
Porphyra spp.	Shrivardhan, Maharashtra (India)	58.00	426.20	Duxbury and Yentach (1956)	Verma et al. (2017)[AD]
Portieria hornemannii	Mandapam, Tamil Nadu (India)	41.50	89.50	Duxbury and Yentach (1956)	Verma et al. (2017)[AD]
Rhodymenia dissecta	Port Okha, Gujarat (India)	118.00	347.90	Duxbury and Yentach (1956)	Verma et al. (2017)[AD]
Sargassum linearifolium	Port Okha, Gujarat (India)	265.80	239.80	Duxbury and Yentach (1956)	Verma et al. (2017)[AD]
Scinaia fascicularis	Port Okha, Gujarat (India)	31.80	53.10	Duxbury and Yentach (1956)	Verma et al. (2017)[AD]
Scinaia hatei	Port Okha, Gujarat (India)	71.10	66.60	Duxbury and Yentach (1956)	Verma et al. (2017)[AD]

TABLE 13.10 *(Continued)*

Species	**Collection Site**	**Carotenoid Content (mg/100 g) dw**	**Chlorophyll α Content (mg/100 g) dw**	**Method**	**References**
Spatoglossum asperum	Port Okha, Gujarat (India)	37.90	409.70	Duxbury and Yentach (1956)	Verma et al. (2017)[AD]
Stoechospermum marginatum	Port Okha, Gujarat (India)	32.70	502.60	Duxbury and Yentach (1956)	Verma et al. (2017)[AD]
Turbinaria spp.	Mandapam, Tamil Nadu (India)	153.90	260.00	Duxbury and Yentach (1956)	Verma et al. (2017)[AD]
Ulva lactuca	Port Okha, Gujarat (India)	36.90	909.20	Duxbury and Yentach (1956)	Verma et al. (2017)[AD]
Ulva lactuca	Madeira (Portugal)	20.41	92.72	Wellburn (1994); Kumar et al. (2010)	Nunes et al. (2017)[L]
Ulva reticulata	Mandapam, Tamil Nadu (India)	35.70	904.50	Duxbury and Yentach (1956)	Verma et al. (2017)[AD]
Valonia utricularis	Port Okha, Gujarat (India)	51.50	687.40	Duxbury and Yentach (1956)	Verma et al. (2017)[AD]
Zonaria tournefortii	Madeira (Portugal)	297.77	244.25	Wellburn (1994); Kumar et al. (2010)	Nunes et al. (2017)[L]

*Chlorophyll α and carotenoid were assessed differently by different authors and methods are described individually in each reference. After individual reference, letters in superscript indicate the state of the biomass for the extraction protocol used by different authors, lyophilized (L) or air-dried (AD).

capable of accepting coordinating metal ions above or below the plane of the macromolecule. The stabilized oxidation states, solubility, and geometry of these metals (which can be manganese, magnesium, cobalt, iron, zinc, vanadium, and nickel), together with the ring structure of tetrapyrroles, mediates interactions with proteins in living organisms (Smith and Witty, 2002).

Three main classes of chlorophylls are nowadays known, which are distinguished by their unsaturation degree. These are phytoporphyrins, with complete unsaturated macrocycle (chlorophyll χ), bacteriochlorins with a double saturation of the C7–C8 and C17–C18 bonds (bacteriochlorophylls *a*, *b*, and *g*) and chlorins with a saturated bond between C17 and C18 (chlorophyll α, *b*, *d*, and bacteriochlorophylls *c*, *d*, *e*, *f*) (Roca et al., 2016). The biochemical pathway of chlorophyll synthesis is tightly controlled since it can become a potential cellular phototoxic. In certain circumstances, when excess of light occurs, the photosynthetic apparatus can be overexcited, resulting in additional absorbed energy transferred to oxygen, with the formation of ROS. These could also accumulate as a result of the inhibition of chlorophyll βiosynthesis or of its degradation during cellular death (Hörtensteiner and Kräutler, 2011). Additionally, it is essential that a correct stoichiometry is achieved between chlorophylls and apoproteins that naturally link with this pigment (Stenbaek and Jensen, 2010). Green macroalgae produce chlorophyll α and *b*, red macroalgae only chlorophyll α and brown macroalgae are known to produce chlorophyll α and *c* (Takaichi, 2013). Each person is estimated to consume up to 86 mg of chlorophyll per day. This consumption can provide some health benefits, such as antioxidant activity (Lanfer-Marquez et al., 2005) and good ability to chelate some chemical carcinogens or mutagens, reducing cancer risk (Chen and Roca, 2018). *Porphyra* (red macroalgae) and *Enteromorpha* (green macroalgae) species have been successfully assessed as a strong suppressor against genotoxin-induced umuC gene expression in *Salmonella typhimurium*, which is possibly correlated with carcinogenesis (Okai et al., 1996). In processing food, chlorophylls can be converted into pheophytin, pyropheophytin, and pheophorbide, known to prevent cancer (Holdt and Kraan, 2011). Pheophorbide *a*, which results from chlorophyll βreakdown, was isolated from *Grateloupia elliptica* and successfully inhibited the viability of U87MG, SK-OV-3 and HeLa cancer line cells (Cho et al., 2014). Pheophytin, a chlorophyll without the central magnesium ion (Mg^{2+}), was isolated from *Sargassum fulvellum*, a brown macroalgae that demonstrated potential to treat neurodegenerative diseases, such as Alzheimer's disease (AD), due to its capability to promote the differentiation of phaeochromocytoma (PC12) cells (Ina et al., 2007). For these reasons, chlorophylls, and its derivatives, isolated from macroalgae, have become an attractable biomass source for food, cosmetic, and pharmacology applications (Chen et al., 2017). Table 13.10 summarizes bibliographic and methodology information about total chlorophyll *a* (TChl *a*) content in macroalgae collected in several places of the world. TChl *a* varies between 28.81 mg in *Asparagopsis taxiformis*, Madeira (Portugal) and 2,357.90 mg/100 g dw in *Caulerpa veravalensis*, collected in Tamil Nadu (India) (Duxbury and Yentach, 1956). Green macroalgae varies its TChl *a* content between 92.72 mg of *Ulva lactuca* from Madeira (Portugal) and 2,357.90 mg/100 g dw in *Caulerpa veravalensis*. Red macroalgae TChl *a* content fluctuates between 28.81 mg in *Asparagopsis taxiformis* from Madeira (Portugal) and 426.20 mg/100 g dw of *Porphyra* spp., collected in Maharashtra (India). Brown macroalgae TChl *a* content varies between 113.00 mg in *Iyengaria stellata*, Gujarat (India) and 502.60 mg/100 g dw of *Stoechospermum marginatum*, collected in Gujarat (India).

13.3.5.3 FUCOXANTHIN

This is a photosynthetically active molecule that in brown algae integrates in a 'chlorophyll-carotenoid-protein complex' in the thylakoids structure (Haugan and Liaaen-Jensen, 1992). This xanthophyll is a major marine carotenoid, contributing to more than 10% of the estimated total production of carotenoids in nature (Kim et al., 2012). It contains six oxygen atoms in its structure, which are organized in epoxy, hydroxyl, carbonyl, and carboxyl moieties in its molecule, resulting in a highly polar carotenoid (Peng et al., 2011). At each end of the molecule, a cyclic polar group is located, increasing the coordination of the molecule in the membrane, and regulating the interaction patterns with protein membrane-spanning helixes. Together with violaxanthin, lutein, neoxanthin, and zeaxanthin, they are the most common xanthophylls (Ruban, 2010). It possesses an unusual highly reactive allenic bound, found rarely in carotenoids, contributing to its unique features (Peng et al., 2011). This pigment has an orange color and was first isolated from brown macroalgae *Fucus*, *Dictyota*, and *Laminaria* (Willstätter and Page, 1914). Based on its unique molecular structure, fucoxanthin is considered an important nutraceutical biocompound with various therapeutic activities (Fariman et al., 2016). Its major applications are as an antioxidant (Rajauria et al., 2016), anti-obesity (Miyashita and Hosokawa, 2017) and anti-cancer agent (Gutiérrez-Rodríguez et al., 2017; Ma et al., 2017), but costs associated with production and purification are a major drawback for its industrialization and commercialization (Guedes et al., 2011). Furthermore, we have recently published a new validated methodology, using an HPLC coupled with a PDA, to accurately quantify fucoxanthin in brown macroalgae and macroalgae beach casts, containing also brown macroalgae (Nunes et al., 2019). This study showed that brown macroalgae from Madeira archipelago retain from 1.01 mg to 85.2 mg/100 g dw of fucoxanthin, in *Halopteris scoparia* and *Zonaria tournefortii*, respectively. Macroalgae beach casts, collected in "Playa de Las Canteras" in Gran Canaria, contain from 1.32 mg to 4.94 mg/100 g dw, in June and September, respectively. Table 13.11 summarizes bibliographic and methodology information about fucoxanthin extraction and quantification, determined in brown macroalgae, collected in several places of the world. Fucoxanthin extracted and determined in fresh weight (fw), shows values ranging between 2.20 mg, in *Hizikia fusiformis*, collected in Tokushima (Japan) and 26.00 mg/100 g fw of *Eisenia bicyclis*, collected in South Korea. Dry weight (dw) fucoxanthin values varied from 17.90 mg in *Iyengaria stellata*, from Gujarat (India) and 101.00 mg/100 g dw of *Sargassum duplicatum*, from Straits of Malacca (Malaysia).

13.4 CONCLUSION

Macroalgae present itself as a new feedstock with great potential for multi-product extraction. Over the past decade, it has become more sustainable, with the development of new technologies, producing valuable marketable products and energy. The biorefinery approach increases the sustainability, efficiently extracting more compounds and reducing waste, which allows the creation of a bioeconomy, based in multi-product production from a single source. This enables an increase of the biomass conversion efficiency, reducing the amount of raw material discarded. Additionally, the implementation of a biorefinery centered in a cascade extraction, could provide higher profitability, and improve residue management of the feedstock, mitigating the environmental impacts. Several macroalgae species can be integrated in the biorefinery approach, providing enough material to a designed cascade extraction. This improves sustainability, economic growth, and market development. Numerous economic appealing products can be extracted, due to the singular properties of macroalgae. Among these products are included proteins, lipids with distinct fatty acids, sugars, and polysaccharides with bioactive properties, antioxidants, and other

TABLE 13.11 Fucoxanthin Content (mg/100 g Dry Weight (dw) or Fresh Weight (fw) When Stated After Individual Value, for Macroalgae Collected from Several Places Worldwide*

Species	Collection Site	Extraction Methodology	Fucoxanthin mg/100 g dw	References
Eisenia bicyclis	South Korea	LC-MS and HPLC (ethanol)	26.00 (fw)	Kim et al. (2012)[F]
Eisenia bicyclis	Naruto, Tokushima (Japan)	HPLC (methanol)	7.70 (fw)	Kanazawa et al. (2008)[F]
Hizikia fusiformis	Naruto, Tokushima (Japan)	HPLC (methanol)	2.20 (fw)	Kanazawa et al. (2008)[F]
Laminaria japonica	Hakodate, Hokkaido (Japan)	HPLC (methanol)	18.70 (fw)	Kanazawa et al. (2008)[F]
Sargassum binderi	Straits of Malacca near Port Dickson (Malaysia)	HPLC (methanol)	73.00	Jaswir et al. (2012)[F]
Sargassum duplicatum	Straits of Malacca near Port Dickson (Malaysia)	HPLC (methanol)	101.00	Jaswir et al. (2012)[F]
Sargassum fulvellum	Kasumi, Hyogo (Japan)	HPLC (methanol)	6.50 (fw)	Kanazawa et al. (2008)[F]
Sargassum plagiophyllum	Straits of Malacca near Port Dickson (Malaysia)	RP-HPLC (Cold acetone-methanol, 7:3 v/v)	71.00	Jaswir et al. (2013)[F]
Turbinaria turbinata	Straits of Malacca near Port Dickson (Malaysia)	RP-HPLC (Cold acetone-methanol, 7:3 v/v)	59.00	Jaswir et al. (2013)[F]
Undaria pinnatifida	Naruto, Tokushima (Japan)	HPLC (metanol)	11.10 (fw)	Kanazawa et al. (2008)[F]
Dictyopteris australis	Port Okha, Gujarat (India)	Spectrophotometer (Acetone 90%)	22.70	Verma et al. (2017)[AD]
Dictyota dicotoma	Port Okha, Gujarat (India)	Spectrophotometer (Acetone 90%)	18.40	Verma et al. (2017)[AD]
Iyengaria stellata	Port Okha, Gujarat (India)	Spectrophotometer (Acetone 90%)	17.90	Verma et al. (2017)[AD]
Lobophora variegata	Mandapam, Tamil Nadu (India)	Spectrophotometer (Acetone 90%)	22.90	Verma et al. (2017)[AD]
Padina gymnospora	Port Okha, Gujarat (India)	Spectrophotometer (Acetone 90%)	43.50	Verma et al. (2017)[AD]
Padina tetrastromatica	Mandapam, Tamil Nadu (India)	Spectrophotometer (Acetone 90%)	41.10	Verma et al. (2017)[AD]

TABLE 13.11 *(Continued)*

Species	Collection Site	Extraction Methodology	Fucoxanthin mg/100 g dw	References
Sargassum linearifolium	Port Okha, Gujarat (India)	Spectrophotometer (Acetone 90%)	37.50	Verma et al. (2017)[AD]
Spatoglossum asperum	Port Okha, Gujarat (India)	Spectrophotometer (Acetone 90%)	57.80	Verma et al. (2017)[AD]
Stoechospermum marginatum	Port Okha, Gujarat (India)	Spectrophotometer (Acetone 90%)	36.90	Verma et al. (2017)[AD]
Turbinaria spp.	Mandapam, Tamil Nadu (India)	Spectrophotometer (Acetone 90%)	43.10	Verma et al. (2017)[AD]

*Fucoxanthin content, assessed differently by different authors and methods, is described individually in each reference. After individual reference, letters in superscript indicate the state of the biomass for the extraction protocol used by different authors, Fresh (F) or Air-Dried (AD).

high-value compounds, which can be used by the biotechnology industry. The remaining residue, after biomass exhaustion, could be successfully processed to produce bioenergy, in the form of biochar and bio-oil or even integrate a bio-fertilizer for crop production. Moreover, macroalgae was not domesticated, having natural defenses against epiphytes and marine grazers, reducing the inputs during aquaculture production, and minimizing the environmental impact. Its production does not occupy land area, reducing the deforestation pressure. Furthermore, it is of extreme importance, when studying different macroalgae species to integrate in a biorefinery, that sufficient research is performed to design a specific strategy that copes with the feedstock availability and integrates technological processes and pathways available for product extraction, adapting this industry to the local conditions. Moreover, a biorefinery concept enables the interaction between different bioeconomy sectors, forming partnerships between different stakeholders, enabling higher social, environmental, and economic benefits. Still, some apprehensions exist, concerning the sustainability principle, since macroalgae availability oscillates with climate, season, local conditions, and legal requirements, posing a challenging factor. However, with the increasing development of new cost-effective bioprocessing routes and employing eco-friendly downstream processes in the biorefinery complex, could be the key factors for a worldwide successful implementation.

ACKNOWLEDGMENTS

This work was financially supported by ARDITI – Regional Agency for the Development of Research Technology and Innovation and UBQ II, Unidade de Bioquímica through the program of PhD scholarships in Business, M14-20-09-5369-FSE-000001-Doctorate. ISOPlexis, University of Madeira through the projects: CASBio "Programa Operacional Madeira 14–20," Portugal 2020 and the European Union through the European Regional Development Fund, grant number M1420-01-0145-FEDER-000011. MACBIOBLUE, MAC/1.1b/086, program INTERREG MAC 2014–2020. The authors also present their acknowledgment to DRCT (Azores Regional Government) for funding the Azorean Biodiversity Group (ABG), the FCT-Fundação para a Ciência e a Tecnologia, through Portuguese National Funds, and the FEDER through the PT2020 Partnership Agreement, UID/BIA/00329/2013, 2015–2018, UID/BIA/00329/2019 funding the cE3c center.

KEYWORDS

- **agar**
- **alginates**
- **bioactive peptides**
- **biorefinery**
- **carrageenans**
- **cellulose**
- **chlorophylls**
- **fucoidan**
- **fucoxanthin**
- **lectins**
- **lipids**
- **phenolic compounds**
- **phycobiliproteins**
- **polysaccharides**
- **ulvans**

REFERENCES

Abdulkhani, A., Alizadeh, P., Hedjazi, S., & Hamzeh, Y., (2017). Potential of soya as a raw material for a whole crop biorefinery. *Renew. Sustain. Energy Rev.*, *75*, 1269–1280. doi: https://doi.org/10.1016/j.rser.2016.10.082.

Albert, C. M., Hennekens, C. H., O'Donnell, C. J., Ajani, U. A., Carey, V. J., Willett, W. C., Ruskin, J. N., & Manson, J. A. E., (1998). Fish consumption and risk of sudden cardiac death. *J. Am. Med. Assoc.*, *279*, 23–28. doi: https://doi.org/10.1001/jama.279.1.23.

Ale, M. T., Mikkelsen, J. D., & Meyer, A. S., (2011). Important determinants for fucoidan bioactivity: A critical review of structure-function relations and extraction methods for fucose-containing sulfated polysaccharides from brown seaweeds. *Mar. Drugs*, *9*, 2106–2130. doi: https://doi.org/10.3390/md9102106.

Alonso, E., (2017). The role of supercritical fluids in the fractionation pretreatments of a wheat bran-based biorefinery. *J. Supercrit. Fluids*. doi: https://doi.org/10.1016/j.supflu.2017.09.010.

Alves, A., Sousa, R. A., & Reis, R. L., (2013). A practical perspective on ulvan extracted from green algae. *J. Appl. Phycol.*, *25*, 407–424. doi: https://doi.org/10.1007/s10811-012-9875-4.

Anderson, N. S., Dolan, T. C. S., & Rees, D. A., (1965). Evidence for a common structural pattern in the polysaccharide sulfates of the Rhodophyceae. *Nature*, *205*, 1060–1062. doi: https://doi.org/10.1038/2051060a0.

Aneiros, A., & Garateix, A., (2004). Bioactive peptides from marine sources: Pharmacological properties and isolation procedures. *J. Chromatogr. B Anal. Technol. Biomed. Life Sci.*, *803*, 41–53. doi: https://doi.org/10.1016/j.jchromb.2003.11.005.

Apt, K. E., Collier, J. L., & Grossman, A. R., (1995). Evolution of the phycobiliproteins. *J. Mol. Biol.*, *248*, 79–96. doi: https://doi.org/10.1006/jmbi.1995.0203.

Arnott, S., Fulner, A., Scott, W. E., Dea, I. C. M., Morehouse, R., & Rees, D. A., (1974). The agarose double helix and its function in agarose gel structure. *J. Mol. Biol.*, *90*, 269–284. doi: https://doi.org/10.1016/0022-2836(74)90372-6.

Audibert, L., Fauchon, M., Blanc, N., Hauchard, D., & Ar Gall, E., (2010). Phenolic compounds in the brown seaweed *Ascophyllum nodosum*: Distribution and radical-scavenging activities. *Phytochem. Anal.*, *21*, 399–405. doi: https://doi.org/10.1002/pca.1210.

Baghel, R. S., Trivedi, N., Gupta, V., Neori, A., Reddy, C. R. K., Lali, A., & Jha, B., (2015). Biorefining of marine macroalgal biomass for production of biofuel and commodity chemicals. *Green Chem.*, *17*, 2436–2443. doi: https://doi.org/10.1039/C4GC02532F.

Balboa, E. M., Conde, E., Moure, A., Falqué, E., & Domínguez, H., (2013). *In vitro* antioxidant properties of crude extracts and compounds from brown algae. *Food Chem.*, *138*, 1764–1785. doi: https://doi.org/10.1016/j.foodchem.2012.11.026.

Ben, Y. N., Jmel, M. A., Ben, A. M., Bouallagui, H., Marzouki, M. N., & Smaali, I., (2016). A biorefinery concept using the green macroalgae *Chaetomorpha linum* for the coproduction of bioethanol and biogas. *Energy Convers. Manag.*, *119*, 257–265. doi: https://doi.org/10.1016/j.enconman.2016.04.046.

Benjama, O., & Masniyom, P., (2012). Biochemical composition and physicochemical properties of two red seaweeds (*Gracilaria fisheri* and *G. tenuistipitata*) from the Pattani bay in Southern Thailand. *Songklanakarin J. Sci. Technol.*, *34*, 223–230.

Bleakley, S., & Hayes, M., (2017). Algal proteins: Extraction, application, and challenges concerning production. *Foods*, *6*, 33. doi: https://doi.org/10.3390/foods6050033.

Britton, G., (2008). Functions of intact carotenoids. In: Britton, G., Liaaen-Jensen, S., & Pfander, H., (eds.), *Carotenoids, Vol. 4: Natural Functions* (pp. 189–212). Birkhäuser Verlag, Basel.

Buschmann, A. H., Camus, C., Infante, J., Neori, A., Israel, Á., Hernández-González, M. C., Pereda, S. V., et al., (2017). Seaweed production: Overview of the global state of exploitation, farming and emerging research activity. *Eur. J. Phycol.*, *52*, 391–406. doi: https://doi.org/10.1080/09670262.2017.1365175.

Campo, V. L., Kawano, D. F., Silva, D. B., & Carvalho, I., (2009). Carrageenans: Biological properties, chemical modifications and structural analysis: A review. *Carbohydr. Polym.*, *77*, 167–180. doi: https://doi.org/10.1016/j.carbpol.2009.01.020.

Cardozo, K. H. M., Guaratini, T., Barros, M. P., Falcão, V. R., Tonon, A. P., Lopes, N. P., Campos, S., et al., (2007). Metabolites from algae with economical impact. *Comp. Biochem. Physiol.-C Toxicol. Pharmacol.*, *146*, 60–78. doi: https://doi.org/10.1016/j.cbpc.2006.05.007.

Chan, P. T., & Matanjun, P., (2017). Chemical composition and physicochemical properties of tropical red seaweed, *Gracilaria changii*. *Food Chem.*, *221*, 302–310. doi: https://doi.org/10.1016/j.foodchem.2016.10.066.

Chen, K., & Roca, M., (2018). *In vitro* bioavailability of chlorophyll pigments from edible seaweeds. *J. Funct. Foods*, *41*, 25–33. doi: https://doi.org/10.1016/j.jff.2017.12.029.

Chen, K., Ríos, J. J., Pérez-Gálvez, A., & Roca, M., (2017). Comprehensive chlorophyll χomposition in the main edible seaweeds. *Food Chem.*, *228*, 625–633. doi: https://doi.org/10.1016/j.foodchem.2017.02.036.

Cho, M. L., Park, G. M., Kim, S. N., Amna, T., Lee, S., & Shin, W. S., (2014). Glioblastoma-specific anticancer activity of pheophorbide a from the edible red seaweed *Grateloupia elliptica*. *J. Microbiol. Biotechnol.*, *24*, 346–353. doi: https://doi.org/10.4014/jmb.1308.08090.

Chua, E. T., & Schenk, P. M., (2017). A biorefinery for *Nannochloropsis*: Induction, harvesting, and extraction of EPA-rich oil and high-value protein. *Bioresour. Technol.*, *244*, 1416–1424. doi: https://doi.org/10.1016/j.biortech.2017.05.124.

Connan, S., (2015). Spectrophotometric assays of major compounds extracted from algae. In: Stengel, D. B., & Connan, S., (eds.), *Natural Products from Marine Algae* (pp. 75–101). Springer, Berlin, Heidelberg. doi: https://doi.org/0.1007/978-1-4939-2684-8.

Connan, S., Goulard, F., Stiger, V., Deslandes, E., & Gall, E. A., (2004). Interspecific and temporal variation in phlorotannin levels in an assemblage of brown algae. *Bot. Mar.*, *47*, 410–416. doi: https://doi.org/10.1515/BOT.2004.057.

Cumashi, A., Ushakova, N. A., Preobrazhenskaya, M. E., D'Incecco, A., Piccoli, A., Totani, L., Tinari, N., et al., (2007). A comparative study of the anti-inflammatory, anticoagulant, antiangiogenic, and antiadhesive activities of nine different fucoidans from brown seaweeds. *Glycobiology*, *17*, 541–552. doi: https://doi.org/10.1093/glycob/cwm014.

D'Ayala, G. G., Malinconico, M., & Laurienzo, P., (2008). Marine derived polysaccharides for biomedical applications: Chemical modification approaches. *Molecules*, *13*, 2069–2106. doi: https://doi.org/10.3390/molecules13092069.

Dai, J., & Mumper, R. J., (2010). Plant phenolics: Extraction, analysis and their antioxidant and anticancer properties. *Molecules*, *15*, 7313–7352. doi: https://doi.org/10.3390/molecules15107313.

De Pretto, C., Tardioli, P. W., & Costa, C. B. B., (2017). Assessing energetic and available fuel demands from a soybean biorefinery producing refined oil, biodiesel, defatted meal and power. *Comput. Chem. Eng.*, *104*, 259–270. doi: https://doi.org/10.1016/j.compchemeng.2017.03.025.

Del, C. J. A., García-González, M., & Guerrero, M. G., (2007). Outdoor cultivation of microalgae for carotenoid production: Current state and perspectives. *Appl. Microbiol. Biotechnol.*, *74*, 1163–1174. doi: https://doi.org/10.1007/s00253-007-0844-9.

Demmig-Adams, B., & Adams, W., (2002). Antioxidants in photosynthesis and human nutrition. *Science*, *298*, 2149–2153. doi: https://doi.org/10.1126/science.1078002.

Denis, C., Morançais, M., Li, M., Deniaud, E., Gaudin, P., Wielgosz-Collin, G., Barnathan, G., et al., (2010). Study of the chemical composition of edible red macroalgae *Grateloupia turuturu* from Brittany (France). *Food Chem.*, *119*, 913–917. doi: https://doi.org/10.1016/j.foodchem.2009.07.047.

DuBois, M., Gilles, K. A., Hamilton, J. K., Rebers, P. A., & Smith, F., (1956). Colorimetric method for determination of sugars and related substances. *Anal. Chem.*, *28*, 350–356. doi: https://doi.org/10.1021/ac60111a017.

Duxbury, A. C., & Yentach, C. S., (1956). Plankton pigment monograph. *J. Mar. Res.*, *15*, 190, 191.

FAO, (2017). *Fishery and Aquaculture Statistics*. http://www.fao.org/fishery/statistics/software/fishstatj/en (accessed on 7 April 2022).

FAO/WHO, (1991). *Protein Quality Evaluation* (p. 58). Report of the joint FAO/WHO expert consultation. Bethesda, Md., USA.

Fariman, G. A., Shastan, S. J., & Zahedi, M. M., (2016). Seasonal variation of total lipid, fatty acids, fucoxanthin content, and antioxidant properties of two tropical brown algae (*Nizamuddinia zanardinii* and *Cystoseira indica*) from Iran. *J. Appl. Phycol.*, *28*, 1323–1331. doi: https://doi.org/10.1007/s10811-015-0645-y.

Fertah, M., (2017). Isolation and characterization of alginate from seaweed. In: Venkatesan, J., Anil, S., & Kim, S. K., (eds.), *Seaweed Polysaccharides: Isolation, Biological and Biomedical Applications* (pp. 11–26). Elsevier Inc, Amsterdam, Netherlands. doi: https://doi.org/10.1016/B978-0-12-809816-5.00002-5.

Fleurence, J., (1999). Seaweed proteins: Biochemical, nutritional aspects and potential uses. *Trends Food Sci. Technol.*, *10*, 25–28. doi: https://doi.org/10.1016/S0924-2244(99)00015-1.

Foley, S. A., Mulloy, B., & Tuohy, M. G., (2011). An unfractionated fucoidan from *Ascophyllum nodosum*: Extraction, characterization, and apoptotic effects *in vitro*. *J. Nat. Prod.*, *74*, 1851–1861. doi: https://doi.org/10.1021/np200124m.

Francavilla, M., Manara, P., Kamaterou, P., Monteleone, M., & Zabaniotou, A., (2014). Cascade approach of red macroalgae *Gracilaria gracilis* sustainable valorization by extraction of phycobiliproteins and pyrolysis of residue. *Bioresour. Technol.*, *184*, 305–313. doi: https://doi.org/10.1016/j.biortech.2014.10.147.

Freile-Pelegrín, Y., & Murano, E., (2005). Agars from three species of *Gracilaria* (*Rhodophyta*) from Yucatán Peninsula. *Bioresour. Technol.*, *96*, 295–302. doi: https://doi.org/10.1016/j.biortech.2004.04.010.

Freitas, A. C., Rodrigues, D., Carvalho, A. P., Pereira, L., Panteleitchouk, T., Gomes, A. M., & Duarte, A. C., (2015). Marine functional food. In: Kim, S. K., (ed.), *Handbook of Marine Biotechnology* (pp. 969–994). Springer, Berlin, Heidelberg. doi: https://doi.org/10.1007/978-3-642-53971-8.

Galland-Irmouli, A. V., Fleurence, J., Lamghari, R., Luçon, M., Rouxel, C., Barbaroux, O., Bronowicki, J. P., Villaume, C., & Guéant, J. L., (1999). Nutritional value of proteins from edible seaweed *Palmaria palmata* (Dulse). *J. Nutr. Biochem.*, *10*, 353–359. doi: https://doi.org/10.1016/S0955-2863(99)00014-5.

García, P. C. V., Ramos, F. D., Estrada, V., Villar, M. A., & Diaz, M. S., (2017). Optimization of an integrated algae-based biorefinery for the production of biodiesel, astaxanthin and PHB. *Energy*, *139*, 1159–1172. doi: https://doi.org/10.1016/j.energy.2017.08.036.

Garcia-vaquero, M., Rajauria, G., Doherty, O. J. V., & Sweeney, T., (2017). Polysaccharides from macroalgae: Recent advances, innovative technologies and challenges in extraction and purification. *Food Res. Int.*, *99*, 1011–1020. doi: https://doi.org/10.1016/j.foodres.2016.11.016.

Gilbert-López, B., Mendiola, J. A., Fontecha, J., Van, D. B. L. A. M., Sijtsma, L., Cifuentes, A., Herrero, M., & Ibáñez, E., (2015). Downstream processing of *Isochrysis galbana*: A step towards microalgal biorefinery. *Green Chem.*, *17*, 4599–4609. doi: https://doi.org/10.1039/C5GC01256B.

Glasson, C. R. K., Sims, I. M., Carnachan, S. M., De Nys, R., & Magnusson, M., (2017). A cascading biorefinery process targeting sulfated polysaccharides (ulvan) from *Ulva ohnoi*. *Algal Res.*, *27*, 383–391. doi: https://doi.org/10.1016/j.algal.2017.07.001.

Gnansounou, E., Alves, C. M., Pachón, E. R., & Vaskan, P., (2017). Comparative assessment of selected sugarcane biorefinery-centered systems in Brazil: A multi-criteria method based on sustainability indicators. *Bioresour. Technol.*, *243*, 600–610. doi: https://doi.org/10.1016/j.biortech.2017.07.004.

Goldstein, I. J., & Poretz, R. D., (1986). Isolation, physicochemical characterization, and carbohydrate-binding specificity of lectins. In: Liener, I. E., Sharon, N., & Goldstein, I. J., (eds.), *The Lectins: Properties, Functions, and Applications in Biology and Medicine* (pp. 33–247). Harcourt Brace Jovanovich, Orlando.

Gracía, J. S., Palacios, V. M., & Roldán, A., (2016). Nutritional potential of four seaweed species collected in the barbate estuary (Gulf of Cadiz, Spain). *J. Nutr. Food Sci.*, *6*, 1–7. doi: https://doi.org/10.4172/2155-9600.1000505.

Guedes, A. C., Amaro, H. M., & Malcata, F. X., (2011). Microalgae as sources of carotenoids. *Mar. Drugs*, *9*, 625–644. doi: https://doi.org/10.3390/md9040625.

Gutiérrez-Rodríguez, A. G., Juárez-Portilla, C., Olivares-Bañuelos, T., & Zepeda, R. C., (2017). Anticancer activity of seaweeds. *Drug Discov. Today.* doi: https://doi.org/10.1016/j.drudis.2017.10.019.

Hahn, T., Lang, S., Ulber, R., & Muffler, K., (2012). Novel procedures for the extraction of fucoidan from brown algae. *Process Biochem.*, *47*, 1691–1698. doi: https://doi.org/10.1016/j.procbio.2012.06.016.

Haugan, A. J., & Liaaen-Jensen, S., (1992). Naturally occurring stereoisomers of fucoxanthin. *Phytochemistry*, *31*, 1359–1361. doi: https://doi.org/10.1016/0031-9422(92)80290-U.

Hayes, M., (2013). Biological activities of proteins and marine-derived peptides from byproducts and seaweeds. In: Kim, S. K., (ed.), *Marine Proteins and Peptides: Biological Activities and Applications* (pp. 139–165). John Wiley & Sons, Ltd, Chichester.

Hehemann, J. H., Boraston, A. B., & Czjzek, M., (2014). A sweet new wave: Structures and mechanisms of enzymes that digest polysaccharides from marine algae. *Curr. Opin. Struct. Biol.*, *28*, 77–86. doi: https://doi.org/10.1016/j.sbi.2014.07.009.

Hemmingson, J. A., Falshaw, R., Furneaux, R. H., & Thompson, K., (2006). Structure and antiviral activity of the galactofucan sulfates extracted from *Undaria pinnatifida* (Phaeophyta). J. *Appl. Phycol.*, *18*, 185–193. doi: https://doi.org/10.1007/s10811-006-9096-9.

Hoffman, D. R., Boettcher, J. A., & Diersen-Schade, D. A., (2009). Toward optimizing vision and cognition in term infants by dietary docosahexaenoic and arachidonic acid supplementation: A review of randomized controlled trials. *Prostaglandins Leukot. Essent. Fat. Acids*, *81*, 151–158. doi: https://doi.org/10.1016/j.plefa.2009.05.003.

Holdt, S. L., & Kraan, S., (2011). Bioactive compounds in seaweed: Functional food applications and legislation. *J. Appl. Phycol.*, *23*, 543–597. doi: https://doi.org/10.1007/s10811-010-9632-5.

Hori, K., Miyazawa, K., & Ito, K., (1990). Some common properties of lectins from marine algae. *Hydrobiologia.*, *204*, 561–566. doi: https://doi.org/10.1007/BF00040287.

Hörtensteiner, S., & Kräutler, B., (2011). Chlorophyll βreakdown in higher plants. *Biochim. Biophys. Acta-Bioenerg.*, *1807*, 977–988. doi: https://doi.org/10.1016/j.bbabio.2010.12.007.

Huskens, D., & Schols, D., (2012). Algal lectins as potential HIV microbicide candidates. *Mar. Drugs*, *10*, 1476–1497. doi: https://doi.org/10.3390/md10071476.

Ina, A., Hayashi, K. I., Nozaki, H., & Kamei, Y., (2007). Pheophytin a, a low molecular weight compound found in the marine brown alga *Sargassum fulvellum*, promotes the differentiation of PC12 cells. *Int. J. Dev. Neurosci.*, *25*, 63–68. doi: https://doi.org/10.1016/j.ijdevneu.2006.09.323.

Isailovic, D., Li, H. W., & Yeung, E. S., (2004). Isolation and characterization of R-phycoerythrin subunits and enzymatic digests. *J. Chromatogr. A*, *1051*, 119–130. doi: https://doi.org/10.1016/j.chroma.2004.07.038.

Jaswir, I., Noviendri, D., Salleh, H. M., & Miyashita, K., (2012). Fucoxanthin extractions of brown seaweeds and analysis of their lipid fraction in methanol. *Food Sci. Technol. Res*, *18*, 251–257. doi: https://doi.org/10.3136/fstr.18.251.

Jaswir, I., Noviendri, D., Salleh, H. M., Taher, M., Miyashita, K., & Ramli, N., (2013). Analysis of fucoxanthin content and purification of all-trans-fucoxanthin from *Turbinaria turbinata* and *Sargassum plagiophyllum* by SiO_2 open column chromatography and reversed phase-HPLC. *J. Liq. Chromatogr. Relat. Technol.*, *36*, 1340–1354. doi: https://doi.org/10.1080/10826076.2012.691435.

Joubert, Y., & Fleurence, J., (2008). Simultaneous extraction of proteins and DNA by an enzymatic treatment of the cell wall of *Palmaria palmata* (Rhodophyta). *J. Appl. Phycol.*, *20*, 55–61. doi: https://doi.org/10.1007/s10811-007-9180-9.

Jung, K. A., Lim, S. R., Kim, Y., & Park, J. M., (2013). Potentials of macroalgae as feedstocks for biorefinery. *Bioresour. Technol.*, *135*, 182–190. doi: https://doi.org/10.1016/j.biortech.2012.10.025.

Kadam, S. U., Álvarez, C., Tiwari, B. K., & O'Donnell, C. P., (2017). Extraction and characterization of protein from Irish brown seaweed *Ascophyllum nodosum*. *Food Res. Int.*, *99*, 1021–1027. doi: https://doi.org/10.1016/j.foodres.2016.07.018.

Kanazawa, K., Ozaki, Y., Hashimoto, T., Das, S. K., Matsushita, S., Hirano, M., Okada, T., et al., (2008). Commercial-scale preparation of biofunctional fucoxanthin from waste parts of brown sea algae *Laminaria japonica*. *Food Sci. Technol. Res.*, *14*, 573–582. doi: https://doi.org/10.3136/fstr.14.573.

Keyrouz, R., Abasq, M. L., Bourvellec, C. L., Blanc, N., Audibert, L., Argall, E., & Hauchard, D., (2011). Total phenolic contents, radical scavenging and cyclic voltammetry of seaweeds from Brittany. *Food Chem.*, *126*, 831–836. doi: https://doi.org/10.1016/j.foodchem.2010.10.061.

Kim, S. M., Jung, Y. J., Kwon, O. N., Cha, K. H., Um, B. H., Chung, D., & Pan, C. H., (2012). A potential commercial source of fucoxanthin extracted from the microalga *Phaeodactylum tricornutum*. *Appl. Biochem. Biotechnol.*, *166*, 1843–1855. doi: https://doi.org/10.1007/s12010-012-9602-2.

Kiso, Y., (2011). Pharmacology in health foods: Effects of arachidonic acid and docosahexaenoic acid on the age-related decline in brain and cardiovascular system function. *J. Pharmacol. Sci.*, *115*, 471–475. doi: https://doi.org/10.1254/jphs.10R39FM.

Knutsen, S. H., Myslabodski, D. E., Larsen, B., & Usov, A. I., (1994). A modified system of nomenclature for red algal galactans. *Bot. Mar.*, *37*, 163–170. doi: https://doi.org/10.1515/botm.1994.37.2.163.

Konda, N. V., Singh, S., Simmons, B. A., & Klein-Marcuschamer, D., (2015). An investigation on the economic feasibility of macroalgae as a potential feedstock for biorefineries. *Bioenergy Res.*, *8*, 1046–1056. doi: https://doi.org/10.1007/s12155-015-9594-1.

Kumar, K. S., Ganesan, K., & Rao, P. V., (2008). Antioxidant potential of solvent extracts of *Kappaphycus alvarezii* (Doty) Doty - An edible seaweed. *Food Chem.*, *107*, 289–295. doi: https://doi.org/10.1016/j.foodchem.2007.08.016.

Kumar, M., Gupta, V., Kumari, P., Reddy, C. R. K., & Jha, B., (2011). Assessment of nutrient composition and antioxidant potential of *Caulerpaceae* seaweeds. *J. Food Compos. Anal.*, *24*, 270–278. doi: https://doi.org/10.1016/j.jfca.2010.07.007.

Kumar, S., Gupta, R., Kumar, G., Sahoo, D., & Kuhad, R. C., (2013). Bioethanol production from *Gracilaria verrucosa*, a red alga, in a biorefinery approach. *Bioresour. Technol.*, *135*, 150–156. doi: https://doi.org/10.1016/j.biortech.2012.10.120.

Lahaye, M., & Robic, A., (2007). Structure and function properties of ulvan, a polysaccharide from green seaweeds. *Biomacromolecules*, *8*, 1765–1774. doi: https://doi.org/10.1021/bm061185q.

Lakshmi, D. S., Trivedi, N., & Reddy, C. R. K., (2017). Synthesis and characterization of seaweed cellulose derived carboxymethyl cellulose. *Carbohydr. Polym.*, *157*, 1604–1610. doi: https://doi.org/10.1016/j.carbpol.2016.11.042.

Lanfer-Marquez, U. M., Barros, R. M. C., & Sinnecker, P., (2005). Antioxidant activity of chlorophylls and their derivatives. *Food Research International*, *38*, 885–891. doi: https://doi.org/10.1016/j.foodres.2005.02.012.

Laurienzo, P., (2010). Marine polysaccharides in pharmaceutical applications: An overview. *Mar. Drugs*, *8*, 2435–2465. doi: https://doi.org/10.3390/md8092435.

Le Guillard, C., Dumay, J., Donnay-Moreno, C., Bruzac, S., Ragon, J. Y., Fleurence, J., & Bergé, J. P., (2015). Ultrasound-assisted extraction of R-phycoerythrin from *Grateloupia turuturu* with and without enzyme addition. *Algal Res.*, *12*, 522–528. doi: https://doi.org/10.1016/j.algal.2015.11.002.

Lee, W. K., Lim, P. E., Phang, S. M., Namasivayam, P., & Ho, C. L., (2016). Agar properties of *Gracilaria* species (*Gracilariaceae*, *Rhodophyta*) collected from different natural habitats in Malaysia. *Reg. Stud. Mar. Sci.*, *7*, 123–128. doi: https://doi.org/10.1016/j.rsma.2016.06.001.

Lee, W. K., Lim, Y. Y., Leow, A. T. C., Namasivayam, P., Ong, A. J., & Ho, C. L., (2017). Biosynthesis of agar in red seaweeds: A review. *Carbohydr. Polym.*, *164*, 23–30. doi: https://doi.org/10.1016/j.carbpol.2017.01.078.

Lemke, S. L., Maki, K. C., Hughes, G., Taylor, M. L., Krul, E. S., Goldstein, D. A., Su, H., et al., (2013). Consumption of stearidonic acid-rich oil in foods increases red blood cell eicosapentaenoic acid. *J. Acad. Nutr. Diet.*, *113*, 1044–1056. doi: https://doi.org/10.1016/j.jand.2013.04.020.

Li, L., Ni, R., Shao, Y., & Mao, S., (2014). Carrageenan and its applications in drug delivery. *Carbohydr. Polym.*, *103*, 1–11. doi: https://doi.org/10.1016/j.carbpol.2013.12.008.

Lindman, B., Karlström, G., & Stigsson, L., (2010). On the mechanism of dissolution of cellulose. *J. Mol. Liq.*, *156*, 76–81. doi: https://doi.org/10.1016/j.molliq.2010.04.016.

Ma, S. Y., Park, W. S., Lee, D. S., Choi, G., Yim, M. J., Lee, J. M., Jung, W. K., et al., (2017). Fucoxanthin inhibits profibrotic protein expression *in vitro* and attenuates bleomycin-induced lung fibrosis *in vivo*. *Eur. J. Pharmacol.*, *811*, 199–207. doi: https://doi.org/10.1016/j.ejphar.2017.06.022.

Marinho-Soriano, E., Fonseca, P. C., Carneiro, M. A. A., & Moreira, W. S. C., (2006). Seasonal variation in the chemical composition of two tropical seaweeds. *Bioresour. Technol.*, *97*, 2402–2406. doi: https://doi.org/10.1016/j.biortech.2005.10.014.

Marques, I. P., Batista, A. P., Coelho, A., & Lopes Da, S. T., (2017). Co-digestion of *Rhodosporidium toruloides* biorefinery wastes for biogas production. *Process Biochem.*, *64*, 221–227. doi: https://doi.org/10.1016/j.procbio.2017.09.023.

Meland, M., & Rebours, C., (2012). Seaweed Industry in Europe. http://www.seaweed.ie/irish_seaweed_contacts/doc/Filieres_12p_UK.pdf (accessed on 7 April 2022).

Mendes, F. M., Dias, M. O. S., Ferraz, A., Milagres, A. M. F., Santos, J. C., & Bonomi, A., (2017). Techno-economic impacts of varied compositional profiles of sugarcane experimental hybrids on a biorefinery producing sugar, ethanol and electricity. *Chem. Eng. Res. Des.*, *125*, 72–78. doi: https://doi.org/10.1016/j.cherd.2017.06.023.

Mišurcová, L., Ambrožová, J., & Samek, D., (2011). Seaweed lipids as nutraceuticals. In: Kim, S. K., & Taylor, S., (eds.), *Marine Medicinal Foods: Implications and Applications, Macro and Microalgae* (pp. 339–355). Elsevier Inc, New York.

Mitra, M., Patidar, S. K., & Mishra, S., (2015). Integrated process of two stage cultivation of *Nannochloropsis* sp. for nutraceutically valuable eicosapentaenoic acid along with biodiesel. *Bioresour. Technol., 193*, 363–369. doi: https://doi.org/10.1016/j.biortech.2015.06.033.

Miyashita, K., & Hosokawa, M., (2017). Fucoxanthin in the management of obesity and its related disorders. *J. Funct. Foods, 36*, 195–202. doi: https://doi.org/10.1016/j.jff.2017.07.009.

Miyashita, K., Mikami, N., & Hosokawa, M., (2013). Chemical and nutritional characteristics of brown seaweed lipids: A review. *J. Funct. Foods, 5*, 1507–1517. doi: https://doi.org/10.1016/j.jff.2013.09.019.

Moncada, B., J., Aristizábal, M. V., & Cardona, C., (2016). Design strategies for sustainable biorefineries. *Biochem. Eng. J., 116*, 122–134. doi: https://doi.org/10.1016/j.bej.2016.06.009.

Nagarajan, S., Skillen, N. C., Irvine, J. T. S., Lawton, L. A., & Robertson, P. K. J., (2017). Cellulose II as bioethanol feedstock and its advantages over native cellulose. *Renew. Sustain. Energy Rev., 77*, 182–192. doi: https://doi.org/10.1016/j.rser.2017.03.118.

Nanaki, S., Karavas, E., Kalantzi, L., & Bikiaris, D., (2010). Miscibility study of carrageenan blends and evaluation of their effectiveness as sustained release carriers. *Carbohydr. Polym., 79*, 1157–1167. doi: https://doi.org/10.1016/j.carbpol.2009.10.067.

Nobre, B. P., Villalobos, F., Barragán, B. E., Oliveira, A. C., Batista, A. P., Marques, P. A. S. S., Mendes, R. L., et al., (2013). A biorefinery from *Nannochloropsis* sp. microalga-extraction of oils and pigments. Production of biohydrogen from the leftover biomass. *Bioresour. Technol., 135*, 128–136. doi: https://doi.org/10.1016/j.biortech.2012.11.084.

Norziah, M. H., Foo, S. L., & Karim, A. A., (2006). Rheological studies on mixtures of agar (*Gracilaria changii*) and κ-carrageenan. *Food Hydrocoll., 20*, 204–217. doi: https://doi.org/10.1016/j.foodhyd.2005.03.020.

Nunes, N., Ferraz, S., Valente, S., Barreto, M. C., & Pinheiro De, C. M. A. A., (2017). Biochemical composition, nutritional value, and antioxidant properties of seven seaweed species from Madeira archipelago. *J. Appl. Phycol., 29*, 2427–2437. doi: https://doi.org/10.1007/s10811-017-1074-x.

Nunes, N., Leça, J. M., Pereira, A. C., Pereira, V., Ferraz, S., Barreto, M. C., Marques, J. C., & Pinheiro De, C. M. A. A., (2019). Evaluation of fucoxanthin contents in seaweed biomass by vortex-assisted solid-liquid microextraction using high-performance liquid chromatography with photodiode array detection. *Algal Res., 42*, 1–6. doi: https://doi.org/10.1016/J.ALGAL.2019.101603.

Nunes, N., Valente, S., Ferraz, S., Barreto, M. C., & Pinheiro De, C. M. A. A., (2018). Nutraceutical potential of *Asparagopsis taxiformis* (Delile) Trevisan extracts and assessment of a downstream purification strategy. *Heliyon, 4*, 1–28. doi: https://doi.org/10.1016/j.heliyon.2018.e00957.

O'Sullivan, A. M., O'Callaghan, Y. C., O'Grady, M. N., Queguineur, B., Hanniffy, D., Troy, D. J., Kerry, J. P., & O'Brien, N. M., (2011). *In vitro* and cellular antioxidant activities of seaweed extracts prepared from five brown seaweeds harvested in spring from the west coast of Ireland. *Food Chem., 126*, 1064–1070. doi: https://doi.org/10.1016/j.foodchem.2010.11.127.

Okai, Y., Higashi-Okai, K., Yano, Y., & Otani, S., (1996). Suppressive effects of chlorophyllin on mutagen-induced umu C gene expression in *Salmonella typhimurium* (TA 1535/pSK

1002) and tumor promoter-dependent ornithine decarboxylase induction in BALB/c 3T3 fibroblast cells. *Mutat. Res. – Genet. Toxicol.*, *370*, 11–17. doi: https://doi.org/10.1016/S0165-1218(96)90122-6.

Özdenkçi, K., De Blasio, C., Muddassar, H. R., Melin, K., Oinas, P., Koskinen, J., Sarwar, G., & Järvinen, M., (2017). A novel biorefinery integration concept for lignocellulosic biomass. *Energy Convers. Manag.*, *149*, 974–987. doi: https://doi.org/10.1016/j.enconman.2017.04.034.

Pacheco, R., Ferreira, A. F., Pinto, T., Nobre, B. P., Loureiro, D., Moura, P., Gouveia, L., & Silva, C. M., (2015). The production of pigments and hydrogen through a *Spirogyra* sp. biorefinery. *Energy Convers. Manag.*, *89*, 789–797. doi: https://doi.org/10.1016/j.enconman.2014.10.040.

Paiva, L., Lima, E., Patarra, R. F., Neto, A. I., & Baptista, J., (2014). Edible Azorean macroalgae as source of rich nutrients with impact on human health. *Food Chem.*, *164*, 128–135. doi: https://doi.org/10.1016/j.foodchem.2014.04.119.

Pangestuti, R., & Kim, S. K., (2015). An overview of phycocolloids: The principal commercial seaweed extracts. In: Kim, S. K., & Chojnacka, K., (eds.), *Marine Algae Extracts Processes, Products, and Applications* (pp. 319–330). Wiley-VCH Verlag GmbH and Co., Weinheim, Germany.

Papageorgiou, S. K., Kouvelos, E. P., & Katsaros, F. K., (2008). Calcium alginate beads from *Laminaria digitata* for the removal of Cu^{+2} and Cd^{+2} from dilute aqueous metal solutions. *Desalination*, *224*, 293–306. doi: https://doi.org/10.1016/j.desal.2007.06.011.

Peng, J., Yuan, J. P., Wu, C. F., & Wang, J. H., (2011). Fucoxanthin, a marine carotenoid present in brown seaweeds and diatoms: Metabolism and bioactivities relevant to human health. *Mar. Drugs*, *9*, 1806–1828. doi: https://doi.org/10.3390/md9101806.

Peng, Y., Xie, E., Zheng, K., Fredimoses, M., Yang, X., Zhou, X., Wang, Y., Yang, B., Lin, X., Liu, J., & Liu, Y., (2013). Nutritional and chemical composition and antiviral activity of cultivated seaweed *Sargassum naozhouense* Tseng et Lu. *Mar. Drugs*, *11*, 20–32. doi: https://doi.org/10.3390/md11010020.

Phong, W. N., Show, P. L., Ling, T. C., Juan, J. C., Ng, E. P., & Chang, J. S., (2016). Mild cell disruption methods for bio-functional proteins recovery from microalgae - recent developments and future perspectives. *Algal Res.*, *31*, 506–516. doi: https://doi.org/10.1016/j.algal.2017.04.005.

Phull, A. R., & Kim, S. J., (2017). Fucoidan as bio-functional molecule: Insights into the anti-inflammatory potential and associated molecular mechanisms. *J. Funct. Foods*, *38*, 415–426. doi: https://doi.org/10.1016/j.jff.2017.09.051.

Rahimi, V., Karimi, K., Shafiei, M., Naghavi, R., Khoshnevisan, B., Ghanavati, H., Mohtasebi, S. S., et al., (2017). Well-to-wheel life cycle assessment of *Eruca sativa*-based biorefinery. *Renewable Energy*, *117*, 135–149. doi: https://doi.org/10.1016/j.renene.2017.10.035.

Raikova, S., Le, C. D., Beacham, T. A., Jenkins, R. W., Allen, M. J., & Chuck, C. J., (2017). Towards a marine biorefinery through the hydrothermal liquefaction of macroalgae native to the United Kingdom. *Biomass and Bioenergy*, *107*, 244–253. doi: https://doi.org/10.1016/j.biombioe.2017.10.010.

Rajauria, G., (2018). Optimization and validation of reverse phase HPLC method for qualitative and quantitative assessment of polyphenols in seaweed. *J. Pharm. Biomed. Anal.*, *148*, 230–237. doi: https://doi.org/10.1016/j.jpba.2017.10.002.

Rajauria, G., Foley, B., & Abu-Ghannam, N., (2016). Characterization of dietary fucoxanthin from *Himanthalia elongata* brown seaweed. *Food Res. Int.*, *99*, 995–1001. doi: https://doi.org/10.1016/j.foodres.2016.09.023.

Robic, A., Rondeau-Mouro, C., Sassi, J. F., Lerat, Y., & Lahaye, M., (2009b). Structure and interactions of ulvan in the cell wall of the marine green algae *Ulva rotundata* (Ulvales, *Chlorophyceae*). *Carbohydr. Polym.*, *77*, 206–216. doi: https://doi.org/10.1016/j.carbpol.2008.12.023.

Robic, A., Sassi, J. F., Dion, P., Lerat, Y., & Lahaye, M., (2009a). Seasonal variability of physicochemical and rheological properties of ulvan in two *Ulva* species (*Chlorophyta*) from the Brittany coast. *J. Phycol.*, *45*, 962–973. doi: https://doi.org/10.1111/j.1529-8817.2009.00699.x.

Roca, M., Chen, K., & Pérez-Gálvez, A., (2016). Chlorophylls. In: Carle, R., & Schweiggert, R. M., (eds.), *Handbook on Natural Pigments in Food and Beverages: Industrial Applications* (pp. 125–158). Woodhead Publishing, Cambridge.

Rodrigues, D., Freitas, A. C., Pereira, L., Rocha-Santos, T. A. P., Vasconcelos, M. W., Roriz, M., Rodríguez-Alcalá, L. M., et al., (2015). Chemical composition of red, brown and green macroalgae from Buarcos bay in Central West Coast of Portugal. *Food Chem.*, *183*, 197–207. doi: https://doi.org/10.1016/j.foodchem.2015.03.057.

Ruban, A. V., (2010). Identification of carotenoids in photosynthetic proteins: Xanthophylls of the light harvesting antenna. In: Landrum, J. T., (ed.), *Carotenoids: Physical, Chemical, and Biological Functions and Properties* (pp. 389–410). CRC Press, Boca Raton. doi: https://doi.org/doi, 10.1201/9781420052312-c12.

Sanjeewa, K. K. A., Lee, J. S., Kim, W. S., & Jeon, Y. J., (2017). The potential of brown-algae polysaccharides for the development of anticancer agents: An update on anticancer effects reported for fucoidan and laminaran. *Carbohydr. Polym.*, *177*, 451–459. doi: https://doi.org/10.1016/j.carbpol.2017.09.005.

Sankalia, M. G., Mashru, R. C., Sankalia, J. M., & Sutariya, V. B., (2006). Stability improvement of alpha-amylase entrapped in kappa-carrageenan beads: Physicochemical characterization and optimization using composite index. *Int. J. Pharm.*, *312*, 1–14. doi: https://doi.org/10.1016/j.ijpharm.2005.11.048.

Seghetta, M., Marchi, M., Thomsen, M., Bjerre, A. B., & Bastianoni, S., (2016). Modeling biogenic carbon flow in a macroalgal biorefinery system. *Algal Res.*, *18*, 144–155. doi: https://doi.org/10.1016/j.algal.2016.05.030.

Sekar, S., & Chandramohan, M., (2008). Phycobiliproteins as a commodity: Trends in applied research, patents and commercialization. *J. Appl. Phycol.*, *20*, 113–136. doi: https://doi.org/10.1007/s10811-007-9188-1.

Senthilkumar, K., Manivasagan, P., Venkatesan, J., & Kim, S. K., (2013). Brown seaweed fucoidan: Biological activity and apoptosis, growth signaling mechanism in cancer. *Int. J. Biol. Macromol.*, *60*, 366–374. doi: https://doi.org/10.1016/j.ijbiomac.2013.06.030.

Shibata, T., Fujimoto, K., Nagayama, K., Yamaguchi, K., & Nakamura, T., (2002a). Inhibitory activity of brown algal phlorotannins against hyaluronidase. *Int. J. Food Sci. Technol.*, *37*, 703–709. doi: https://doi.org/10.1046/j.1365-2621.2002.00603.x.

Shibata, T., Nagayama, K., Tanaka, R., & Yamaguchi, K., (2002b). Inhibitory effects of brown algal phlorotannins on secretory phospholipase A_2s, lipoxygenases and cyclooxygenases. *J. Appl. Phycol.*, *15*, 61–66. doi: https://doi.org/10.1023/A: 1022972221002.

Shukla, M. K., Kumar, M., Prasad, K., Reddy, C. R. K., & Jha, B., (2011). Partial characterization of sulfohydrolase from *Gracilaria dura* and evaluation of its potential application in improvement of the agar quality. *Carbohydr. Polym.*, *85*, 157–163. doi: https://doi.org/10.1016/j.carbpol.2011.02.009.

Silalertruksa, T., Pongpat, P., & Gheewala, S. H., (2017). Life cycle assessment for enhancing environmental sustainability of sugarcane biorefinery in Thailand. *J. Clean. Prod.*, *140*, 906–913. doi: https://doi.org/10.1016/j.jclepro.2016.06.010.

Singh, R. S., Walia, A. K., Khattar, J. S., Singh, D. P., & Kennedy, J. F., (2017). Cyanobacterial lectins characteristics and their role as antiviral agents. *Int. J. Biol. Macromol.*, *102*, 475–496. doi: https://doi.org/10.1016/j.ijbiomac.2017.04.041.

Singh, S., Gaikwad, K. K., Park, S. I., & Lee, Y. S., (2017). Microwave-assisted step reduced extraction of seaweed (*Gelidiella aceroso*) cellulose nanocrystals. *Int. J. Biol. Macromol.*, *99*, 506–510. doi: https://doi.org/10.1016/j.ijbiomac.2017.03.004.

Smit, A. J., (2004). Medicinal and pharmaceutical uses of seaweed natural products: A review. *J. Appl. Phycol.*, *16*, 245–262. doi: https://doi.org/10.1023/B:JAPH.0000047783.36600.ef.

Smith, A. G., & Witty, M., (2002). Laboratory methods for the study of tetrapyrroles. In: Smith, A. G., & Witty, M., (eds.), *Heme, Chlorophyll, and Bilins: Methods and Protocols* (pp. 1–12). Humana Press, New Jersey.

Sonani, R. R., Singh, N. K., Kumar, J., Thakar, D., & Madamwar, D., (2014). Concurrent purification and antioxidant activity of phycobiliproteins from *Lyngbya* sp. A09DM: An antioxidant and anti-aging potential of phycoerythrin in *Caenorhabditis elegans*. *Process Biochem.*, *49*, 1757–1766. doi: https://doi.org/10.1016/j.procbio.2014.06.022.

Stenbaek, A., & Jensen, P. E., (2010). Redox regulation of chlorophyll βiosynthesis. *Phytochemistry*, *71*, 853–859. doi: https://doi.org/10.1016/j.phytochem.2010.03.022.

Stengel, D. B., & Walker, J. M., (2015). *Natural Products from Marine Algae.* doi: https://doi.org/10.1007/978-1-4939-2684-8.

Sudha, P. N., Aisverya, S., Nithya, R., & Vijayalakshmi, K., (2014). Industrial applications of marine carbohydrates. In: Kim, S. K., (ed.), *Marine Carbohydrates: Fundamentals and Applications, Part B* (pp. 145–181). Academic Press, Cambridge.

Sudhakar, M. P., Jagatheesan, A., Perumal, K., & Arunkumar, K., (2015). Methods of phycobiliprotein extraction from *Gracilaria crassa* and its applications in food colorants. *Algal Res.*, *8*, 115–120. doi: https://doi.org/10.1016/j.algal.2015.01.011.

Takaichi, S., (2013). Distributions, biosyntheses and functions of carotenoids in algae. *Agro Food Ind. Hi. Tech.*, *24*, 55–58. doi: https://doi.org/10.3390/md9061101.

Tandeau De, M. N., Bogorad, L., Boresch, K., Bryant, D., Cohen-Bazire, G., Engelmann, T., Esenbeck, N., et al., (2003). Phycobiliproteins and phycobilisomes: The early observations. *Photosynth. Res.*, *76*, 197–205. doi: https://doi.org/10.1023/A, 1024954911473.

Trigui, M., Gasmi, L., Zouari, I., & Tounsi, S., (2013). Seasonal variation in phenolic composition, antibacterial and antioxidant activities of *Ulva rigida* (*Chlorophyta*) and assessment of antiacetylcholinesterase potential. *J. Appl. Phycol.*, *25*, 319–328. doi: https://doi.org/10.1007/s10811-012-9866-5.

Van, H. J. W., Huijgen, W. J. J., & López-Contreras, A. M., (2014). Opportunities and challenges for seaweed in the biobased economy. *Trends Biotechnol.*, *32*, 231–233. doi: https://doi.org/10.1016/j.tibtech.2014.02.007.

Vercruysse, L., Van, C. J., & Smagghe, G., (2005). ACE inhibitory peptides derived from enzymatic hydrolysates of animal muscle protein: A review. *J. Agric. Food Chem.*, *53*, 8106–8115. doi: https://doi.org/10.1021/jf0508908.

Verma, P., Kumar, M., Mishra, G., & Sahoo, D., (2017). Multivariate analysis of fatty acid and biochemical constitutes of seaweeds to characterize their potential as bioresource for biofuel and fine chemicals. *Bioresour. Technol.*, *226*, 132–144. doi: https://doi.org/10.1016/j.biortech.2016.11.044.

Vo, T. S., & Kim, S. K., (2013). Fucoidans as a natural bioactive ingredient for functional foods. *J. Funct. Foods*, *5*, 16–27. doi: https://doi.org/10.1016/j.jff.2012.08.007.

Wang, T., Jónsdóttir, R., & Ólafsdóttir, G., (2009). Total phenolic compounds, radical scavenging and metal chelation of extracts from Icelandic seaweeds. *Food Chem.*, *116*, 240–248. doi: https://doi.org/10.1016/j.foodchem.2009.02.041.

Wang, Y., Ebadian, M., Sokhansanj, S., Webb, E., & Lau, A., (2017). Impact of the biorefinery size on the logistics of corn stover supply – a scenario analysis. *Appl. Energy*, *198*, 360–376. doi: https://doi.org/10.1016/j.apenergy.2017.03.056.

Weis, W. I., & Drickamer, K., (1996). Structural basis of lectin-carbohydrate recognition. *Annu Rev. Biochem. 65*, 441–473. doi: https://doi.org/10.1146/annurev.bi.65.070196.002301.

Wiesberg, I. L., Brigagão, G. V., De Medeiros, J. L., & De Queiroz, F. A. O., (2017). Carbon dioxide utilization in a microalga-based biorefinery: Efficiency of carbon removal and economic performance under carbon taxation. *J. Environ. Manage*, *203*, 988–998. doi: https://doi.org/10.1016/j.jenvman.2017.03.005.

Wijesekara, I., Pangestuti, R., & Kim, S. K., (2011). Biological activities and potential health benefits of sulfated polysaccharides derived from marine algae. *Carbohydr. Polym.*, *84*, 14–21. doi: https://doi.org/10.1016/j.carbpol.2010.10.062.

Willstätter, R., & Page, H. J., (1914). Chlorophyll: The pigments of the brown algae. *Justus Liebigs Ann. Chem.*, *404*, 237–271.

Wu, L., Sun, J., Su, X., Yu, Q., Yu, Q., & Zhang, P., (2016). A review about the development of fucoidan in antitumor activity: Progress and challenges. *Carbohydr. Polym.*, *154*, 96–111. doi: https://doi.org/10.1016/j.carbpol.2016.08.005.

Yamaguchi, K., (1997). Recent advances in microalgal bioscience in Japan, with special reference to utilization of biomass and metabolites: A review. *J. Appl. Phycol.*, *8*, 487–502. doi: https://doi.org/10.1007/BF02186327.

Yang, J., & Li, J., (2018). Self-assembled cellulose materials for biomedicine: A review. *Carbohydr. Polym.*, *181*, 264–274. doi: https://doi.org/10.1016/j.carbpol.2017.10.067.

Young, V. R., & Pellett, P. L., (1994). Plant proteins in relation to human and amino acid nutrition. *Am J Clin Nutr*, *59*, 1203–1212. doi: https://doi.org/10.1093/ajcn/59.5.1203S.

Yuan, Y. V., & Walsh, N. A., (2006). Antioxidant and antiproliferative activities of extracts from a variety of edible seaweeds. *Food Chem. Toxicol.*, *44*, 1144–1150. doi: https://doi.org/10.1016/j.fct.2006.02.002.

Zia, K. M., Tabasum, S., Nasif, M., Sultan, N., Aslam, N., Noreen, A., & Zuber, M., (2017). A review on synthesis, properties and applications of natural polymer-based carrageenan blends and composites. *Int. J. Biol. Macromol.*, *96*, 282–301. doi: https://doi.org/10.1016/j.ijbiomac.2016.11.095.

INDEX

C

D

F

G

M

P

X

Y

Z